HIGH RISK
SAFETY TECHNOLOGY

HIGH RISK
SAFETY TECHNOLOGY

Edited by
A. E. Green
National Centre of Systems Reliability
United Kingdom Atomic Energy Authority
Culcheth, Warrington, UK

A Wiley–Interscience Publication

JOHN WILEY & SONS
Chichester · New York · Brisbane · Toronto · Singapore

363·11 HI9

295974

28 MAR 1983

Library of Congress Cataloging in Publication Data:
Main entry under title:

High risk safety technology.

  'A Wiley–Interscience publication.'
  Includes index.
  1. Industrial safety—Addresses, essays, lectures.
I. Green, A. E. (Arthur Eric)
T55.H49   363.1'1   81-19781

ISBNO 471 10153 2   AACR2

British Library Cataloguing in Publication Data:

High risk safety technology.
  1. Industrial safety—Addresses, essays,
  lectures
  I. Green, A. E.
  363.1'1   T55

ISBN 0 471 10153 2

Photo Typeset by Macmillan India Ltd., Bangalore.
Printed by Page Bros. (Norwich) Ltd.

Contributors

A. AITKEN   *Safety and Reliability Directorate, Culcheth, Warrington, Cheshire*

A. C. BARRELL   *Health and Safety Executive, 25 Chapel Street, London NW1 5DT*

A. J. BOURNE   *Safety and Reliability Directorate, Culcheth, Warrington, Cheshire*

MME A. CARNINO   *Electricité de France, 32 Rue de Monceau, Paris, France*

G. CRELLIN   *Los Alamos Technical Associates, Inc., San Jose, California, USA*

A. CROSS   *Safety and Reliability Directorate, Culcheth, Warrington, Cheshire*

B. K. DANIELS   *Safety and Reliability Directorate, Culcheth, Warrington, Cheshire*

J. A. DENNIS   *National Radiological Protection Board, Harwell, Didcot, Berkshire*

DOROTHY DICK   *British National Oil Corporation, 150 St Vincent Street, Glasgow, Scotland*

G. T. DICKSON   *Glaxo Operations Ltd, Greenford Road, Greenford, Middlesex*

F. R. FARMER   *Safety and Reliability Directorate, Culcheth, Warrington, Cheshire*

B. J. GARRICK   *Pickard, Lowe & Garrick Inc., 17840 Skypark Boulevard, Irvine, California, USA*

MME M. GRIFFON   *Commissariat a l'Energie Atomique, 31–33 Rue de la Federation, Paris*

v

G. HELSBY   *HELP Ltd, 40 Bewsey Street, Warrington, Cheshire*

G. HENSLEY   *SARA Ltd, 40 Bewsey Street, Warrington, Cheshire*

F. HEWITT   *ICI Ltd, Mond Division, Winnington, Northwich, Cheshire*

D. M. HUNNS   *Health & Safety Executive, Nuclear Installations Inspectorate, Branch 3, Silkhouse Court, Tithebarn Street, Liverpool, Merseyside*

D. HURRELL   *Department of Health and Social Security, 12–14 Russell Square, London WC1B 5EP*

G. D. KAISER   *NUS Corporation, 910 Clopper Road, Gaithersburg, Maryland 20878, USA*

S. KAPLAN   *Pickard, Lowe and Garrick Inc., 17840 Skypark Boulevard, Irvine, California, USA*

G. H. KINCHIN   *Safety and Reliability Directorate, Culcheth, Warrington, Cheshire*

T. KLETZ   *ICI Ltd, Petrochemicals Division HQ, Wilton, Middlesbrough, Cleveland TS6 8JE*

H. G. LAWLEY   *Shell UK Exploration and Production, 1 Altens Farm Road, Nigg, Aberdeen, Scotland*

E. S. LONDON   *HELP Ltd, 40 Bewsey Street, Warrington, Cheshire*

A. F. MCKINLAY   *National Radiological Protection Board, Harwell, Didcot, Berkshire*

J. G. MARSHALL   *J H Burgoyne and Partners, College Hill Chambers, College Hill, London EC4R 2RT*

D. OKRENT   *University of California, Los Angeles, California, USA*

J. A. PREECE   *Information Services, British Nuclear Fuels Ltd, Risley, Warrington, Cheshire*

J. PUGH   *HELP Ltd, 40 Bewsey Street, Warrington, Cheshire*

E. B. RANBY   *ICI Ltd, Mond Division, Winnington, Northwich, Cheshire*

J. RASMUSSEN   *Risø National Laboratories, DK-4000, Roskilde, Denmark*

P. V. RUTLEDGE   *J H Burgoyne and Partners, College Hill Chambers, College Hill, London EC4R 2RT*

G. I. SCHUËLLER — *Institut für Bauingenieurwesen III, Arcistrasse 21, Munchen 2, F R Germany*

P. J. C. SCOTT — *Health and Safety Executive, 25 Chapel Street, London NW1 5DT*

A. C. SELMAN — *Department of Health and Social Security, Euston Tower, 286 Euston road, London NW1*

A. M. SMITH — *Los Alamos Technical Associates, Inc., San Jose, California, USA*

F. W. TEATHER — *Glaxo Operations Ltd, Greenford Road, Greenford, Middlesex*

V. M. THOMAS — *National Coal Board, Ashby Road, Stanhope Bretby, Burton-on-Trent, Staffordshire*

W. VINCK — *Commission of the European Communities, Rue de la Loi 200, B-1049, Brussels, Belgium*

I. B. WALL — *Electric Power Research Institute, 3412 Hillview Avenue, Palo Alto, California, USA*

R. F. WHITE — *Safety and Reliability Directorate, Culcheth, Warrington, Cheshire*

J. A. WILLIAMS — *British Nuclear Fuels Ltd, Risley, Warrington, Cheshire*

R. WILSON — *Harvard University, Cambridge, Massachusetts 02138, USA*

K. K. WONG — *BIS Applied Systems Ltd, Quay House, Quay Street, Manchester M3 3JH*

Contents

Preface

There is a growing public awareness of the risks associated with various industries such as processing and energy. In recent years accidents such as those involving chemical plants at Flixborough in the UK, Seveso in Italy, and the incident with the Three-Mile Island nuclear reactor at Harrisburg in the USA have served to intensify this public awareness. Moreover, regulatory bodies are requiring greater control over major hazardous undertakings. For those installations with the highest hazard potential, safety submissions of a rigorously high standard will be required.

Techniques exist for the assessment of risk with the objective of improving the appropriate engineered systems to cope with the risk in question. Various industries have developed techniques to meet their own requirements but there has emerged a generic set of techniques which can be applied to high risk situations. Several decades ago these techniques were based mainly on deterministic approaches, but these have been extended to include the measurement of risk and associated reliability in a quantified manner. In the interest of the workers, the employers, and the general public in all industrialized countries, it is desirable that the technology necessary to provide the safe control of high risk installations should be available to those responsible for the design, construction, operation, and regulations of such plants. It is the purpose of this book to examine 'High Risk Safety Technology' by bringing together contributions in different fields to illustrate some of the generic ideas which been emerging. Contributions have been made by various specialists from different industries in different countries in order to present a composite scenario.

This book should be of direct interest to design engineers, regulators, users, and all those engaged in safety of plants which involve high risks in which decision making based on a disciplined technology is a prerequisite. Students who are engaged in the study of safety and the interrelated subject of reliability of systems equipment and their application will be required to have a comprehension of the concepts described if they are to develop appropriately their ideas in the field of safety technology applied to high risk situations. In this context this book forms a companion volume to that of *Reliability Technology* in the Wiley—Interscience series.

The editor wishes to thank the authors for contributing to this volume and for developing their ideas in the particular field of safety technology. In addition, the editor very much appreciates the efforts made by the authors in order to bring into being a basic source book for a number of industries which should be of direct help in assisting the reader in developing his own ideas.

A. E. GREEN

The National Centre of Systems Reliability
Safety and Reliability Directorate
United Kingdom Atomic Energy Authority
Culcheth
May 1981

PART 1

SAFETY ASSURANCE

High Risk Safety Technology
Edited by A. E. Green
© 1982 John Wiley & Sons Ltd

Chapter 1.1

The Concept of Risk

G. H. Kinchin

Definition of Risk

The word 'risk' can refer either to a hazard or to the chance of loss. We shall be concerned with both facets of the meaning of the word in considering risk assessment – both the magnitude of the hazard and the probability of its occurrence.

No human activity can be carried out without risk, and a high risk activity may be one in which the consequences of an infrequent accident for society are very large – the death of several hundred people, for example. Equally a high risk activity may be one in which the consequences of individual accidents are small, but in which the frequency of such accidents is high.

The most important consequence of an accident in considering risks, is that of death and it is this consequence that is discussed in by far the largest volume of published papers. There are, of course, endless other consequences – illness, financial loss, missing the train, etc. – and in principle the same arguments can be applied to these as to fatalities. It is with the objective of comparing one risk with another, and of ensuring that some important risks are not overlooked, that quantitative assessment of risks is attempted and acceptance criteria are sought, as described in the following sections.

Risk Statistics

Fatalities

Risks of death can be subdivided in a variety of ways. One possible subdivision separates voluntary and involuntary risks, for example. Someone who climbs mountains does so by choice and accepts whatever risk may be associated with his activity, whereas the risks arising from natural disasters, such as earthquakes or floods, cannot readily be avoided. The risk from mountain climbing is voluntary, while the latter risks are involuntary. The distinction is not, however, always clear – guiding mountain climbers may be the only occupation open to a particular individual, so that the voluntary nature of his choice is qualified, and in principle it is possible to move away from an area where the risk of flood or

3

earthquake is high. Nevertheless the point is one to be kept in mind when considering the acceptability of risk.

Before quoting statistics for different industries and activities it is desirable to consider how readily one set of figures may be compared with another and what sources of variation there may be. There is no ambiguity about early death, and there is no problem due to variable reporting standards, such as may happen in the reporting of minor accidents and ailments.

Accidental death rates vary with time, largely due to statistical fluctuations but also due to changing trends in different industries. There are also differences in the possible calculations of numbers of employees at risk – Are clerical and administrative staff included, for instance? There is no alternative to looking at the detailed descriptions of the sources of the figures such as those given in Health and Safety Statistics[1]. With something of the order of 200 deaths per year, the death rate per million employees at risk in the construction industry is shown in Table 1.1-1 the variation from year to year can be seen. Finally, it is necessary to be clear about the area to which the statistics apply. In the case of the United Kingdom it is often possible to find figures which refer only to England and Wales, or to Great Britain, but there are much bigger differences in individual accident rates in other countries, although the comparison is not an exact one (some relate to fatal accident rates per million man-years of 300 days each, some to rates per million wage-earners and some to rates per million employed). Table 1.1-2[1] shows comparative industrial fatal accident rates in a number of different countries.

Table 1.1-1   Construction Industry – fatal accidents per year per million at risk[1]

Year	1972	1973	1974	1975	1976
Accident rate	187	216	160	177	153

Table 1.1-2   Incidence rates of fatal accidents for 1972–1974 (deaths per million per year)

	Manufacturing industry	Construction industry	Mining and quarrying
Great Britain	40	190	390*
France	110	460	750
Fed. Rep. of Germany	170	360	620
Irish Republic	80	150	640
Italy	80	560	340
Netherlands	40	110	190*
Canada	170	1020	2230

* One year's figure not available

Such comparisons as these indicate that different levels of risk appear to be acceptable in different countries, and as may be seen from Table 1.1-2 there is a range of incidence rates for fatal accidents between different industries. This range is illustrated for a wider range of industries in Great Britain in Table 1.1-3.

The data shown in Tables 1.1-1 to 1.1-3 refer to industrial accident rates. The total number of accidental deaths is much larger than those defined as industrial fatalities, as may be seen from Table 1.1-4 which compares the number of fatal accidents at work reported to the Health and Safety Commission (HSC) enforcement and other authorities[1] with the total number of fatal accidents[2] for Great Britain.

Table 1.1-3 Incidence rates of fatal accidents per million at risk in 1976[1]

Quarries	326
Coal Mines	196
Coal and petroleum products	192
Railways (staff)	188
Construction	153
Agriculture	141
Ship building	133
Metal manufacture	94
Bricks, pottery, glass cement	79
Chemicals	70
Timber, furniture, etc.	35
Paper, printing and publishing	19
Clothing and footwear	7

Table 1.1-4 Numbers of fatal accidents in Great Britain for 1972–1975

Year	1972	1973	1974	1975
HSC enforcement and other authorities	812	874	786	730
Total for Great Britain	18 803	18 947	18 335	17 704

As in the case of the industrial accidental fatalities, similar care must be used in properly identifying the area to which these data apply and there are substantial differences between one country and another, such as those shown in Table 1.1-5[3]. The data refer to different years for Great Britain and the United States, but the temporal changes are small compared with the substantial differences which may be seen in the table.

The individual risk of accidental death as a function of age is shown in Figure 1.1-1 for Great Britain in the years 1971 to 1975[2]. It may be seen that apart from a relatively low risk between the ages of 5 and 15, the probability of

Table 1.1-5   Incidence rates of fatal accidents in GB and
USA (deaths per million per year)

Cause	Incidence of fatalities GB	USA
All accidents	340	630
Road accidents	140	250
Falls	110	100
Fire	18	40
Drowning	11	33
Electrocutions	2.4	6.3
Lightning	0.2	0.5

accidental death does not vary much up to the age of 65. Figure 1.1-1 also shows the corresponding risks of death due to all causes. Above the age of 45 the accidental death rate rapidly falls to less than 5% of the total death rate.

Injuries and occupational disease

Many of the accidents which can cause death are also capable of causing injury which may lead to permanent disablement or to the temporary loss of ability to work. The Health and Safety Executive[1] give figures for lost time accidents under the heading of 'Compensated industrial accidents'. For 1971 to 1975 these amounted to over 600 000 accidents per year, each giving rise on average to more than 20 days of certified incapacity. In addition to such accidents, data are also presented on occupational diseases which can lead to spells of certified incapacity. These are 'prescribed' diseases for which benefits are paid. Up to 1976, there were about 13 000 spells of incapacity notified each year resulting from prescribed diseases other than pneumoconiosis and byssionsis. All but a few hundred of these cases were due to dermatitis, inflammation of the tendons of the hand and 'beat knee'. In 1976 there were some 40 000 people receiving benefit due to pneumoconiosis and byssinosis.

Evidently the effects of such accidents and occupational diseases would be important from the point of view of industry, even if there were no human suffering associated with them, but it should be noted that in many cases the occupational diseases can lead to early death. The numbers of deaths attracting awards of industrial benefits are shown in Table 1.1-6[1].

Deaths such as these occur often long after the contraction of the industrial disease and in that sense are similar to delayed deaths which may result from excessive doses of ionizing radiations.

Risk to society

The previous sections have considered the statistics relating to the risk to individuals, and it can be argued that these in fact are the only figures that matter.

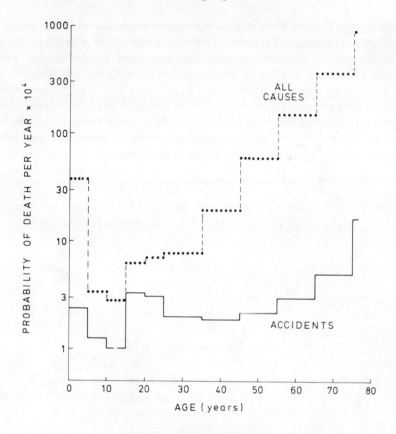

Figure 1.1-1 Probability of death

Table 1.1-6 Deaths from Pneumoconiosis and Other industrial diseases in
Great Britain

Year	1972	1973	1974	1975	1976
Number of death benefit awards	977	903	898	946	867

Nevertheless, whatever the logic behind it, there is little doubt that large
accidents, for example in the field of transport, attract far more public attention
than do the numerous small accidents which annually kill many more people. We
should therefore also examine the way in which accidents of different magnitudes
occur.

Griffiths and Fryer[3] have carried out such a study for the UK, presenting figures for the frequency of accidents resulting in more than N fatalities for fires, ship losses, railway accidents, aircraft accidents and Public Service Vehicle accidents. Of course, in treating only data for the UK, the statistics for large accidents rapidly become very poor. Fryer and Griffiths[4] have studied worldwide data on multiple fatality accidents and have used the world trends to extrapolate the United Kingdom figures in the direction of larger accidents. Their results are shown in Figure 1.1-2 for man-made events, from which it may be seen that the larger the accident, the lower the probability.

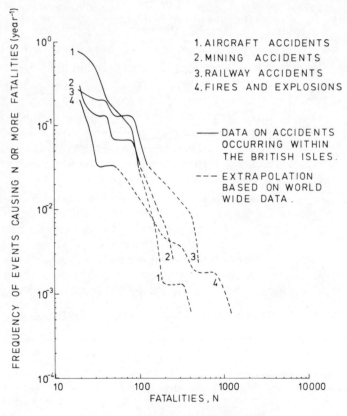

Figure 1.1-2 Multiple fatality accidents

It should be remembered that although the data are sparse for the UK, multiple fatality accidents due to natural causes such as floods, storms and landslides can be important and outweigh man-made accidents in the world scene. In making comparisons between one country and another, the relationship between risks to society and the size of the population must be kept in mind. For comparable safety standards, the annual probability of accidents killing more than 10 people would be 10 times lower in a country with a population of 5 million than in a country with a population of 50 million.

Criteria for the Acceptability of Risk

Cost–benefit analysis

In looking for criteria which might be used to determine the acceptability of given public death rates, two possibilities will be discussed. The first method concerns cost–benefit analysis, while comparison with other existing death rates will be considered in the next section.

In principle, cost–benefit analysis provides an objective way of determining how much money should be spent in order to reduce fatalities or injuries. The method works well in determining, for example, whether the cost of buying a computer is more than offset by the benefit of a reduced salary bill for the smaller number of staff needed to operate the computer, rather than to process all the information by hand. Even in such a simple case, where money can readily be used to assess both cost and benefit, there may be difficulties in assessing the monetary value of quicker turn-round, but the difficulties become more marked when it becomes necessary to place a money value on death or injury.

In a recent, interesting, consultative document[5] the National Radiological Protection Board has considered the composition of the health detriment arising from exposure to ionizing radiation with the costs of keeping the ensuing doses at a low level. In the course of this study a number of different methods of assessing the value of a human life were reviewed, and a very wide range of values was arrived at, ranging from £50 to £20 million.

By applying appropriate discount rates to the future net income of a worker, the 'present worth' of the remainder of his career can be assessed, and this can be used as a measure of the value of his life. The methodology is one commonly used in comparing the costs of projects which are spread out over different periods of time[6]. The discounting process is particularly relevant for the study of the effects of ionizing radiation since such effects are often delayed by many years, and such delays should be properly taken into account.

Other methods of arriving at the value of a human life are to study the outcome of legal cases resulting in compensation awards for severe injury or loss of life, or to study the life insurance market. The greatest range of values for a human life comes, however, from a study of government actions and expenditure which imply some value for a human life; the cost of the change of building standards following the collapse of Ronan Point high-rise flats is said to imply a value for human life of £20 million in 1972.

There are drawbacks to all of these methods of estimation, and none includes valuations by the victims, as should be the case if an estimate is to be made of what people will pay to avoid a risk, as well as what benefit will reconcile them to the risk. Thus the answer to what is apparently a straightforward question is much more difficult than it may appear, and a large measure of judgement must be used to modify values to allow for 'non-pecuniary' factors.

Given values for both human life and for injuries, then it is possible to optimize the expenditure on safety measures. In spite of the inevitable uncertainties surrounding the valuations, the exercise can nevertheless be helpful and instructive in drawing attention to particularly severe accident sequences which might otherwise be accepted in conformity with past practice.

Comparison with existing individual and societal risks

Given the statistics presented above, it is possible to put forward individual risks which may be considered acceptable, assuming that the activity giving rise to the risks is one which is beneficial to society.

Webb and McLean[7] have suggested that a risk of death of 10^{-6} per year is not taken into account by individuals in making decisions, and is therefore acceptable. The International Commission on Radiological Protection[8] indicates that a risk in the range 10^{-6} to 10^{-5} per year would be likely to be acceptable to any individual member of the public, while the level of acceptability for fatal occupational risks would be an order of magnitude higher. The Advisory Committee on Major Hazards[9] in suggesting a probability of 10^{-4} per year for a major accident to an individual plant is implying a risk to a particular individual living close to the plant of more than 10^{-5} per year. The author has proposed an individual risk of early death of 10^{-6} per year[10] and a risk for delayed death due to cancer of 3×10^{-5} per year. In the latter case death is unlikely to occur until 15 to 20 years after exposure to ionizing radiation and, by the process of discounting referred to above, a higher probability of delayed death is acceptable. It is also relevant that the chance of dying of cancer without any additional exposure to radiation is about 2.5×10^{-3} per year, very much higher than the proposed additional risk of 3×10^{-5} per year. It is evident, from the statistics in Tables 1.1-2 and 1.1-5, that acceptable risks vary from country to country as well as from industry to industry. Risks acceptable in Great Britain may be conservative for other countries.

Turning to the risk to society, in this case the comparable data for Great Britain are those shown in Figure 1.1-2. Estimates, possibly rather pessimistic, for the probability of industrial accidents of different sizes are given in the report by the Health and Safety Executive on Canvey Island[11], in the mouth of the River Thames. These estimates are shown in Figure 1.1-3 and give some measure of the possible sizes of industrial accidents, in this case related to oil refineries, to the storage of ammonia under pressure and to the storage and handling of liquid natural gas and liquid petroleum gas. It must be borne in mind that this study

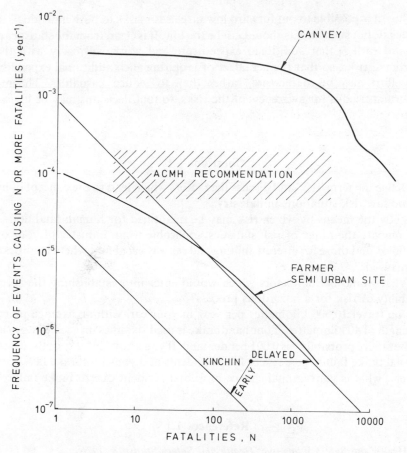

Figure 1.1-3 Criteria for multiple fatality accidents

covers only a small part of Great Britain and that the appropriate probabilities for comparison with those in Figure 1.1-2 would be higher. Figure 1.1-3 also shows the general region of probability for 'large accidents' to a single plant put forward by the Advisory Committee on Major Hazards (ACMH)[9], and suggested criteria[10] for a single nuclear reactor.

There is understandable reluctance to appear to endorse in cold blood the probability, however small, of human deaths, and criteria are sometimes expressed in the nuclear energy field in terms of probabilities of exceeding certain radiation doses or the probabilities of releasing different quantities of radioactivity. One of the best known is the Farmer criterion[12] which gives permissible probabilities for the release of various quantities of ^{131}I. These releases can be converted[13] into fatalities as shown in Figure 1.1-3.

Thus, it is possible to put forward low targets for risks to both individuals and to society, but such targets should not be *too* low. It is clear, from the statistics put forward earlier, that additional expenditure and work will carry with them additional risks, so that the net effect of requiring such additional expenditure and effort may be to increase, rather than to reduce, casualties. The most important need is for awareness of the risks, so that decisions may be logically made.

Questions 1.1

1. Define the term 'risk' and discuss the factors which lead to the concept of high and low risk situations in industry.
2. Study the means by which risk may be quantified for human fatalities. Compare the range of risk statistics applicable to an industry of your own choice, and those for aircraft, mining and railway accidents, where fatalities are involved.
3. What would be the factors which would enter into establishing the acceptability of risk for a hazardous process?
4. You travel 10,000 kilometres per year in your car, with an average journey length of 40 kilometres. Your handbrake is used six times on each journey and the failure probability is 0.001 per demand. If you know that 1 % of the cases of total brake failure give rise to fatal accidents and your footbrake fails once a year, what is your annual chance of a fatal accident due to brake failure?

References 1.1

1. Health and Safety Executive. *Health and Safety Statistics, 1976.*
2. Grist, D. R. *Individual Risk: A compilation of recent British data.* UKAEA Report, SRD R 125. 1976.
3. Griffiths, R. F. and Fryer L. S. *The Incidence of Multiple Fatality Accidents in the UK.* UKAEA Report, SRD R 110. 1978.
4. Fryer, L. S. and Griffiths, R. F. *Worldwide Data on the Incidence of Multiple Fatality Accidents.* UKAEA Report, SRD R 149. 1979.
5. National Radiological Protection Board. *The Application of Cost–Benefit Analysis to the Radiological Protection of the Public: A Consultative Document.* 1980.
6. Iliffe, C. E. 'Assessing the economics of nuclear power stations in an electricity generating system.' *IAEA Symposium on Economic Integration of Nuclear Power Stations in Electric Power Systems, Vienna, 5–9 October 1970,* Paper SM-139/32.
7. Webb, G. A. M. and McLean, A. S. *Insignificant Levels of Dose: a practical suggestion for decision making.* NRPB Report R 62. 1977.
8. The International Commission on Radiological Protection ICRP Publication 26, *Recommendations of the International Commission on Radiological Protection.* Pergamon Press. 1977.
9. Health and Safety Commission. Advisory Committee on Major Hazards. First Report. 1976.

10. Kinchin, G. H. 'Assessment of hazards in engineering work.' *Proc. Instn. Civ. Engrs.*, Part 1, **64**, 431–438, 1978.
11. Health and Safety Executive. *Canvey: An Investigation of Potential Hazards from Operations in the Canvey Island/Thurrock Area.* HMSO. 1978.
12. Farmer, F. R. 'Sitting criteria – A new approach.' *IAEA Symposium on Containment and Siting, Vienna, April 1967.* Paper SM-89/34.
13. Farmer, F. R. and Beattie J. R. 'Nuclear power reactors and the evaluation of population hazards.' *Advances in Nuclear Science and Technology*, **9**, 1, Academic Press. 1976.

High Risk Safety Technology
Edited by A. E. Green
© 1982 John Wiley & Sons Ltd

Chapter 1.2

The Regulatory Position– United Kingdom and Elsewhere

A. C. Barrell and P. J. C. Scott

Provisions in the United Kingdom Prior to the Introduction of the Health and Safety at Work Act 1974

An examination of British legislation relating to high risk technology shows that there is nothing new about industrial fire and explosion hazards, the first Explosives Act having been passed in 1875. The Petroleum (Consolidation) Act 1928 with its extensive subsidiary legislation covers petroleum storage and the Nuclear Installations Act 1965 provides for the licensing of nuclear power stations. The Factories Act 1961, the last in a long series of Factory and Workshop Acts, provides for limited control of some pressurized systems and covers certain aspects of risk from explosive or flammable materials. However this Act is concerned only with risk to employees; any protection afforded to the public is incidental. Although the 1970s have seen further, important developments there is as yet no specific or detailed legislation encompassing toxic, flammable and pressure–energy risks.

At the beginning of 1972, following discussions between government departments, informal arrangements were made to enable local planning authorities to seek the advice of the Factory Inspectorate about the safety aspects of developments involving potentially hazardous industries. The arrangements are outlined in Department of the Environment Circular 1/72, Scottish Development Department Circular 58/72, and Welsh Office Circular 3/72. The sites of the industries in question have become known as Listed Major Hazard Sites. Two particular areas in which planning authorities seek advice from the Factory Inspectorate (see Figure 1.2-1) – latterly, as we shall see, a part of the Health and Safety Executive – are in relation to new green-field sites for chemical plants or extensions to existing ones, and proposals for building schools, hospitals or houses in the vicinity of Listed Major Hazard Sites. However, as this is a voluntary agreement not all such planning applications are referred for advice.

The Study of Health and Safety at Work: The Robens Report

1972 also saw the publication of the Robens Report[1]. Two years earlier Barbara Castle, then Secretary of State for Employment and Productivity in the UK, had

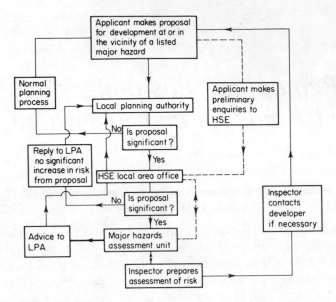

Figure 1.2-1 Consultation by local planning authorities with HSE

appointed a Committee of Inquiry under the chairmanship of Lord Robens, its terms of reference being

> To review the provision made for the safety and health of persons in the course of their employment . . . and to consider whether any changes are needed in:
> (1) the scope or nature of the relevant major enactments, or
> (2) the nature and extent of voluntary action concerned with these matters, and
> to consider whether any further steps are required to safeguard members of the public from hazards . . . arising in connection with activities in industrial and commercial premises . . . and to make recommendations.

The Committee concluded that the existing statutes and subordinate statutory instruments on safety and health at work comprised a 'haphazard mass of ill-assorted and intricate detail'. It recommended a new, comprehensive Act containing a clear statement of the general principles of responsibility for safety and health at work but otherwise mainly enabling in character. The Act should be supported by a combination of new, simpler regulations and non-statutory codes and standards. Remarking unfavourably on the existing fragmentation of administrative jurisdiction, the Committee recommended the establishment of a national authority for safety and health at work which would control a unified inspectorate.

In accordance with its terms of reference the Committee gave particular consideration to public safety and recommended that legislation on safety and health at work should apply explicitly for the protection of the general public as well as workpeople. Special attention should be given to the need to protect the public, as well as workers, from the very large-scale hazards which sometimes accompany modern industrial operations. The Committee called for the creation of a special unit within the unified inspectorate which would concern itself with all industrial situations with a potential for causing danger to the public on a large scale.

The HSW Act: The Role of the HSC and HSE

It is interesting and significant that preparation of the new comprehensive Act recommended by the Robens Committee was begun by a Conservative Government and completed by a Labour Government. The Health and Safety at Work, etc. Act 1974 (the HSW Act), which came into operation in stages beginning on 1 October 1974, applies to all work activities except work by domestic servants in private houses and places general duties on employers, the self-employed, and persons in control of certain premises to ensure so far as is reasonably practicable the health and safety of both workpeople and the general public. The Robens Committee had been particularly concerned that the existing mass of law had an unfortunate psychological effect in conditioning people to think of safety and health at work as a matter of detailed rules imposed by external agencies. The Committee emphasized that the primary responsibility for doing something about occupational accidents lies with those who create the risks and those who work with them. Thus, to encourage personal responsibility and self-generated effort the HSW Act requires every employer except the smallest to prepare and keep up to date a written statement of his general policy with respect to the health and safety of his employees and the organization and arrangements for carrying out that policy. It is implicit that an employer cannot properly prepare this statement without first making an adequate appraisal of the risks arising from the work activities for which he is responsible.

The HSW Act established two new bodies, the Health and Safety Commission (HSC) and the Health and Safety Executive (HSE). The Commission comprises a chairman and up to nine members appointed by the Secretary of State for Employment. At present there are eight members, three appointed after consultation with the CBI, three appointed after consultation with the TUC and two appointed after consultation with representatives of the local authorities. The role of the Commission is one of general direction and policy-making and it does not normally involve itself with day-to-day operational matters.

The Health and Safety Executive brought together the main inspectorates – the Factory Inspectorate, Mines and Quarries Inspectorate, Explosives Inspectorate, Alkali and Clean Air Inspectorate, Nuclear Installations Inspectorate and, latterly, the Agricultural Inspectorate – as well as their scientific and medical support staffs. Certain other bodies while not forming part of the Executive perform work for it on an agency basis; these include the Petroleum Engineering Division of the Department of Energy, the National Radiological Protection Board, and the Chief Inspecting Officer of Railways and his staff. The Executive has been described as the Health and Safety Commission's operational arm but in addition to its operational responsibilities it provides administrative and technical support for the Commission's policy-making activities.

The Health and Safety Commission is empowered under the HSW Act to appoint advisers and it used these powers to set up the Advisory Committee on Major Hazards (ACMH) which first met in January 1975. The appointment of the ACMH, the first of several such advisory committees, was stimulated particularly by the explosion at Flixborough in 1974.

The terms of reference of the ACMH are:

> To identify types of installations (excluding nuclear installations) which have the potential to present major hazards to employees or to the public or the environment and to advise on measures of control, appropriate to the nature and degree of hazard, over the establishment, siting, layout, design, operation, maintenance and development of such installations as well as over all development, both industrial and non-industrial, in the vicinity of such installations.

So far the ACMH has produced two reports. In its first report[2] the Committee suggested a scheme for the notification, survey, assessment, and appraisal of specified categories of installation for the purposes of identifying major hazards. This scheme would necessitate the making of regulations which the Secretary of State for Employment is empowered to do by the HSW Act. However the Act also requires the Health and Safety Commission to carry out certain consultations before it submits proposals to the Secretary of State for the making of regulations. Therefore in accordance with its general policy in such matters, the Commission has published a consultative document[3] containing draft regulations together with background discussion and explanation and a general invitation to submit comments and suggestions.

The second report[4] of the ACMH covers a number of matters including further thinking on notification, hazard surveys, and planning controls. The report also discusses the possibility of a scheme for licensing plant of the highest hazard but stresses that responsibility for the safety of operation should remain with industry.

The Robens Committee recommendation for a special major hazards unit has not gone unheeded. A Hazardous Installations Group (HIG) was established during 1978 as part of a reorganization within the Health and Safety Executive. (see Figure 1.2-2). The HIG has been placed under the direction of the Chief Inspector of Nuclear Installations because of the similarities between the techniques of assessment for the safety of nuclear and non-nuclear installations and because many of the engineering problems are similar. Two branches have

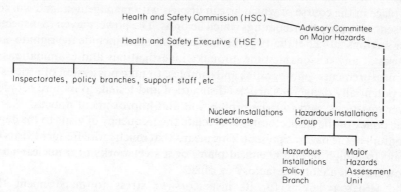

Figure 1.2-2 Part organization chart showing relationships between HSC, HSE and MHAU

been set up within the HIG, the first being the Hazardous Installations Policy Branch whose responsibilities include servicing the ACMH and development of the regulations applicable to hazardous installations. The second is the Major Hazards Assessment Unit (MHAU) whose responsibilities include the preparation of advice to local authorities about developments involving major hazard sites, the preparation of siting policies, and the collation of data relating to such installations. In due course the MHAU will also be responsible for the operation of the proposed notification scheme and the assessment of hazard surveys prepared by industry.

A further power given to the Health and Safety Commission by the HSW Act is to direct the Executive to investigate specific matters and to make a special report on them. Just such a report was that published in June 1978 on an investigation of the risk to people living in and around Canvey Island, Essex, from the existing and proposed industrial activities in the area. This report[5], which runs to some 200 pages, was the first of its kind in the UK and probably in the world.

The Enforcement of the HSW Act

The 1974 Act places very general obligations on employers and others to secure the health and safety of employees and other people who may be affected by work activities. These obligations are in addition to the more detailed requirements set out in many Acts and Regulations. To secure compliance, the HSE, like other enforcing authorities, informs, advises, persuades and, if these fail, may require action by the use of an enforcement notice. The point may finally arrive when prosecution for failure to comply with statutory requirements is the appropriate course of action.

The main instrument for securing compliance is the visit by an inspector to a workplace in the course of which he can discuss with management and workers' representatives any shortcomings which he finds. The powers given to inspectors under sections 20–22 of the 1974 Act are extensive and include the right to enter premises at any reasonable time, to make investigations and examinations, to take measurements, photographs and samples, to take possession of articles, to have potentially dangerous articles dismantled and tested, to require people to answer questions, and to issue prohibition and improvement notices.

The policy of the Inspectorates is to relate the frequency of visits to the degree and nature of the hazards present. This means that much more frequent visits will be paid to a high hazard chemical plant, or a steel works or a nuclear power station than to a clothing factory or office.

The HSE's policy is for its inspectors to stress to management their responsibility for effective control of hazards, rather than for inspectors to solve problems for management. For this reason all inspectorates are much concerned during visits to assure themselves that management has an effective policy for health and safety, a proper organization to carry it out, and arrangements to ensure safe system of work, and furthermore that it regularly monitors the results.

None the less, at any visit an inspector may need to discuss how to identify hazards and to give advice and information on methods of control. A major part of the inspector's role in securing safe and healthy workplaces is in the provision of information and advice on health and safety problems.

Improvement and Prohibition Notices

The 1974 Act gave to enforcing authorities a new power to issue an Improvement Notice requiring anybody to take specific action to remedy a contravention of health and safety legislation. It also provided for the issue of a Prohibition Notice stopping a particular activity if an inspector considered that its continuance would involve a risk of serious personal injury. In either case, the Notice may include directions as to the remedial measures to be taken. Unlike legal proceedings, Notices are intended to remove the hazard to workpeople or others rather than to punish those who have failed to comply with the law. The Notice procedure enables inspectors to act immediately on discovering a risk to health or safety. In the past their only sanction of remedial rather than punitive effect was the cumbersome and involved process of applying for a Court Order.

The enforcing authorities have made substantial use of Notice procedures. In 1979 the total numbers of notices issued were:

13 498   Improvement Notices
3 111   Immediate Prohibition Notices
538   Deferred Prohibition Notices.

Inspectors discuss the terms of a Notice both with the employer and with representatives of the workpeople who may be concerned, though of course the final responsibility for settling those terms lies with the inspector. The Act gives the right to employers or others to appeal to an Industrial Tribunal against the terms of a Notice.

An Improvement or Prohibition Notice is not necessarily issued for every case in which a contravention of health and safety legislation is found by inspectors. In most cases the first action by an inspector is to seek action by the employer or other person concerned. A Notice is likely to be issued where the employer is not prepared to give an undertaking to act, or where an inspector doubts, perhaps from previous experience, whether such assurance will in fact lead to action. There will be cases where the risk of injury is thought by the inspector to be so immediate that he cannot leave the urgency of the required action in doubt. There may also be situations in which an inspector regards it as essential to make clear precisely what are the requirements of the HSE. If a Notice is issued for any of these reasons, it is the stated intention of the HSE to prosecute if the Notice is not complied with.

Prosecutions and Penalties

Prosecutions may be considered for contraventions of the specific requirements of the Acts and Regulations which are under the umbrella of the HSW Act, or for contraventions of the general provisions of sections 2 to 9 of that Act. The latter course has enabled successful prosecutions to be taken where in pre-Act days it would have been extremely difficult to prove any specific breach of health and safety at work legislation.

The major innovation of the 1974 Act in this area has, of course, been the power to take proceedings on indictment, with the possibility of unlimited fines or in defined contraventions even imprisonment. However, most cases are brought to summary trial before magistrates' courts where the maximum fine is currently £1000.

'Reasonably Practicable'

In the 1974 Act the general duties laid on employers and others by sections 2 to 9 are qualified by the phrase 'so far as is reasonably practicable'. This phrase and similar phrases have a long legal history, occurring as far back as the Metalliferous Mines Regs. Act 1872 and the Coal Mines Act 1911. It was the subject of considerable attention in the passage of the HSW Act through Parliament and subsequently in views expressed by trades unions and employers' associations.

The meaning of the phrase and the test to be applied in considering the words were discussed in the case of *Edwards* v. *National Coal Board*[6]. Edwards had been the victim of a fall off the roof and it was common ground that, if the whole of the side of the road had been shored up, the accident could have been prevented. The headnote to the case stated:

> In considering whether or not it was 'reasonably practicable to avoid or prevent' a breach of the statute . . . the mineowner must, before the occurrence of an accident, make a computation in which the *quantum* of risk run by the worker was placed on one scale and the sacrifice of the mineowner involved in the measures necessary to avert the risk (whether in money, time or trouble) was placed in the other, and if there was a gross disproportion between them – the risk being insignificant in relation to the sacrifice – the mineowner discharged the onus which was on him.

The assessment of the quantum of risk involves consideration of both the probability of a particular event and the severity of its consequences. Thus at a major hazard site the potentially severe consequences of even a remote event will require, by the standard of 'reasonable practicability', a substantial sacrifice in measures necessary to avert the risk.

The Formulation of Siting Policies: HSE Advice to Planning Authorities

Historically, the problems posed by the proximity of high-hazard non-nuclear installations to domestic populations have in the UK been somewhat under-estimated. In 1967 the then Chief Inspector of Factories stated:

> Ever since the start of the industrial revolution, catastrophes resulting in considerable numbers of deaths have taken place from time to time in industry. In the latter part of the nineteenth century boiler explosions were common; in the early part of this century there were disastrous secondary dust explosions in such places as vegetable oil mills. Legislation, inspection and better appreciation of the hazards have greatly reduced danger in these categories, but the increased scale of modern manufacture, the vastly larger plant used, the higher speeds of much machinery, have given rise to the possibility of primary explosions through the ignition of dust clouds inside plant at least comparable in scale and violence with the earlier explosions resulting from the secondary ignition of dust clouds, formed from accumulation of dust disturbed by a primary explosion, and filling whole rooms. Again, the scale of modern manufacture has resulted in the storage and use of very large quantities, often measured in thousands of tons, of potentially hazardous materials such as acrylonitrile, liquefied petroleum gas and liquid oxygen. Even the storage of an apparently harmless substance like flour in very large silos gives rise to a major hazard, not when the silo is full but when it is nearly empty, because of the very size of the dust cloud which can form inside it and be ignited under certain adverse conditions.
>
> The taking of precautions to prevent such adverse conditions arising, the provision of explosion reliefs to minimise the effects if ignition does take place, the siting of storages and discharge points to minimise danger, all cost money, often running into many thousands of pounds. The Inspectorate, and particularly members of the Chemical Branch, are constantly called on to give advice on such matters, advice which must be realistic, having regard to the often extreme remoteness of the risk, the scale of disaster which could ensue, and the cost of remedial measures.
>
> (Crown copyright. Reproduced by permission of The Controller of Her Majesty's Stationery Office.)

The Robens Committee further considered the problems and its recommendations are referred to above. In 1972, planning authorities were advised by Department of Environment Circular 1/72 to consult the Factory Inspectorate about developments at or in the vicinity of sites at which certain hazardous materials were stored or processed ('Listed Major Hazards'). The levels of hazardous inventories which qualify a site for inclusion in the list of major hazards are set out in Table 1.2-1. For comparison, the inventories which will require notification and survey under the proposed Hazardous Installations Regulations (see also section below) are given alongside.

Table 1.2-1 Criteria for an installation to be included under the regulations†

Potentially hazardous substance or process	Storage quantity for listing as major hazards	Proposed hazardous installations regulations: Notification	Proposed hazardous installations regulations: Survey
Acrylonitrile	50 tons	20 t	200 t
Ammonia	250 tons	100 t	1000 t
Bromine	100 tons	40 t	400 t
Chlorine	25 tons	10 t	100 t
Ethylene oxide	20 tons	5 t	50 t
Hydrogen cyanide	50 tons	20 t	200 t
Liquid oxygen	135 tons	1000 t	10 000 t
Liquefied petroleum gas (LPG)	100 tons	30 t	300 t
Phosgene	5 tons	2 t	20 t
Sulphur dioxide	50 tons	20 t	200 t
Flour	200 tons*	—	—
Refined white sugar	200 tons*	—	—
Petrochemical and plastic polymer manufacture	Economic size of plant would involve such quantities of materials that the risk would invariably be present	Only where satisfying other criteria	
Aluminium and magnesium powder production			
Flammable liquids with flash point less than 21 °C	—	10 000 t	100 000 t
Flammable gasses not otherwise specified	—	15 t	150 t
Carbon disulphide		20 t	200 t
Toxic liquids or gases likely to be lethal to man in quantities of less than 1 mg		100 g	1 kg
Toxic solids likely to be lethal to man in quantities of less than 1 mg other than those which are and which will be maintained at ambient temperature and atmospheric pressure	—	100 g	1 kg
Propylene oxide	—	5 t	50 t
Organic peroxides	—	5 t	50 t
Nitrocellulose compounds		50 t	500 t
Ammonium nitrate		500 t	5000 t

Table 1.2-1 (*Continued*)

Potentially hazardous substance or process	Storage quantity for listing as major hazards	Proposed hazardous installations regulations:	
		Notification	Survey
Sodium chlorate	—	500 t	5000 t
Flammable liquids above their boiling point (at 1 bar pressure) and under pressure greater than 1.34 bar including flammable gases dissolved under pressure but not mentioned in any other category	—	20 t	200 t
Liquefied flammable gases under refrigeration which have a boiling point below 0°C at 1 bar pressure and are not otherwise specified	—	50 t	500 t
Compound fertilizers	—	500 t	5000 t
Plastic foam	—	500 t	5000 t

* There must be storage of at least 200 tons of the material in a single silo before the criterion applies. Thus a storage of 200 tons in four separate silos, each containing 50 tons of the material, does not satisfy the criterion.

† Since the time of writing there has been progress towards the implementation of the proposed Hazardous Installations Regulations. In particular, the substances and quantities of such substances which will require notification and survey have in some cases changed. The figures shown in columns 3 and 4 of Table 1.2-1 should therefore be treated as of historical record as from May, 1980

While the proposed Hazardous Installations Regulations are concerned with the design and operation of major hazard plant, they do not deal with the siting of such plant in relation to the surrounding population. They do not contain any reference to separation distances which might be thought advisable between chemical plant and houses, hospitals, schools, shops and the like.

Under the planning legislation in the UK siting decisions are the responsibility of the local planning authorities. The ACMH has strongly confirmed its view that the location of hazardous development should always be a matter for planning authorities and not the HSE to determine, since the safety implications, however important, cannot be divorced from other planning considerations[7]. However the ACMH also considered that the local authorities should be given sufficient information and advice about the risks from hazardous installations in order to make such decisions. With this in view, the ACMH has recommended that the voluntary arrangements for consultation with HSE which have been in existence since 1972 should in due course become statutory.

The mechanism of the consultation with HSE by planning authorities is illustrated in Figure 1.2-1. Originally, those significant planning applications which were sent to HSE headquarters for advice were considered by a group known as the Major Hazards Risk Appraisal Group (MHRAG), staffed by members of the Hazardous Installations Group and senior Specialist Inspectors of the Factory Inspectorate and Nuclear Installations Inspectorate. MHRAG met regularly to consider appraisals of planning applications prepared by Specialist Inspectors of the appropriate disciplines. Their task was to formulate the advice given by the Executive to local planning authorities against the background of the ACMH's thinking and with the aim of consistency.

Since the Major Hazards Assessment Unit has been set up and staffed with Inspectors from the various Inspectorates within HSE, the assessment of applications and formulation of advice to local authorities is now carried out within the Unit. The Unit receives several hundred planning applications each year (469 in 1979) on which to advise. Many of these require complex technical consideration if informed advice is to be given. This process is time-consuming and while unnecessary delays can be avoided, the period required for a proper assessment often conflicts with the time constraints within which the planning control system is legally obliged to operate.

The circulation of general guidelines on the siting of houses, schools, etc. near major hazard plants would help all those concerned with the planning process. The chemical industry would find it helpful to know what the HSE considered to be broadly acceptable separations for the siting of new chemical plant or extensions. A general policy on siting as opposed to advice on individual applications would also help local planning authorities, particularly if such a policy could be incorporated into county structure or local plans for development before individual planning applications were considered.

The formulation of such a policy requires a balance between, on the one hand, the undesirability of, e.g. a new primary school being built close to large-scale toxic storage, and on the other hand the creation of extensive buffer zones around chemical plant involving loss of land value and other amenities. To strike this balance it is necessary to start from the assumption that the hazardous installations themselves will be properly designed and operated within the requirements of the HSW Act and its related legislation. The proposed Hazardous Installations Regulations are intended to give confidence in such an assumption as the hazard survey will show whether the significant hazards and the appropriate countermeasures have been identified. There will still be a residual risk which cannot be eliminated but which can be estimated by quantitative methods: such assessment techniques are still somewhat tentative and rely on some unexplored assumptions. The probability of loss of containment of a hazardous substance can be estimated. Having assumed this loss in terms of a leakage of a toxic gas or flammable vapour at a certain rate, it is possible to estimate how far the cloud of gas will spread and how rapidly it will dilute to a safe level, or for a flammable vapour cloud, how far it will extend before ignition. From there it is possible to calculate the overpressure resulting from the explosion of a cloud of flammable vapour at various distances from the centre of the explosion and postulate certain levels of damage which may result at those distances. Other important factors which need to be taken into account in formulating the siting policies are: the number of people at risk; the number of hours per day when they are at risk; and the ease of evacuation in any emergency. In the Flixborough explosion a number of people were less seriously injured than they might have been because they were able to escape from the vicinity of the release.

There are two further principles which govern the general application of these criteria. The first is the need to consider to what extent the development proposals represent an improvement on an existing, possibly far from ideal situation. The second relates to developments in the neighbourhood of an existing hazardous installation where there may already be other land users which are closer and possibly incompatible with the hazard.

In the view of the Major Hazards Assessment Unit the existence of such intervening development should not in any way affect the advice about the possible effects of the hazardous activity on proposed developments which may appear to be less at risk than the existing ones. In other words the existing situation should never be regarded as providing grounds for failing to draw attention to the implications for development at a greater distance. This view has been fully endorsed by the ACMH in their second report[8].

Applying these principles to the immediate area around a potentially hazardous installation, it follows that nearest the site it would be preferable to allow only those developments or uses such as agricultural land, golf courses, etc. where the number of people is small, their resident time short, and evacuation at short notice is not difficult. Further away from the site, development such as factories, warehouses or motorways may be acceptable on the grounds that it is possible either to limit the number of people, or evacuate them quickly, or stop them entering the area if there is an emergency. Finally, it will be prudent to ensure greater separation for such developments as hospitals, schools or shopping precincts where large numbers of people are usually present and where there are considerable difficulties in arranging for effective evacuation.

The preparation of general siting policies relating to particular hazardous substances is a priority task for the Major Hazards Assessment Unit. It is likely that the most commonly occurring substances such as LPG and chlorine will be the first to be considered. In the long term; it may be possible to extend these general interim siting guidelines and to develop with local planning authorities an individual development control policy for the land around each hazardous installation.

Hazardous Installations (Notification and Survey) Regulations†

These Regulations, referred to above, will if made, require firms to present to HSE comprehensive information about activities at installations having significant quantities of hazardous substances. Occupiers of certain installations, where potentially hazardous activities are carried out, will have to notify HSE of their

† See footnote under Table 1.2-1.

existence, and in certain cases, perhaps one-tenth of the total they will also have to carry out a hazard survey and report the results to HSE.

Notification

The materials within the scope of the Regulations are mostly either very highly flammable, highly reactive, or very toxic. Certain reactions of materials in the gas phase which involve a high pressure energy potential are also included. Schedule 1 in the consultative document lists the various criteria which qualify an installation for inclusion under the Regulations in five main groups (see Table 1.2-1).

1. Toxic substances stored in bulk – chlorine and ammonia being two of the most commonly occurring.
2. Substances of extreme toxicity – probably only 30 or so substances would qualify in this group.
3. Highly reactive substances.
4. Liquefied flammable gases including LPG.
5. Gas phase reactions where the pressure multiplied by the volume exceeds a specified figure.

Notification will require the provision of little more information than the name and address of the company and the particular activity which meets the qualifying criterion.

Hazard Survey

Schedule 3 of the consultative document outlines the particulars which will need to be included in the report to HSE of the hazard survey. Some of the required information is factual: a description of the process sufficient to put the potential hazards in context; information about the use of land around the installation; process and site diagrams; and the management philosophy and system by which the hazards are controlled. Then, for plants which contain the significant contributions to the hazardous inventory, the process precautions which are intended to prevent those significant events which could lead to a release of hazardous material or energy. In addition, estimates are required of the maximum quantities of hazardous material which could foreseeably be released despite the successful operation of preventive or control measures, and an outline of the emergency plans which have been prepared to minimize the consequences of such a release.

There is also provision in the Regulations whereby HSE can, after considering the initial hazard survey, call for a detailed assessment of any hazard or aspects of the control system. Naturally there are likely to be informal exchanges which precede such a request, but the power to require this further assessment will underline the need for firms to ensure that the initial hazard survey is sufficiently thorough, and that the report of the survey sent to HSE adequately reflects its comprehensiveness.

Overseas Interest

There is a great deal of interest in the problems of major hazard installations overseas. In the USA, one of the more notable landmarks has been the GAO Report[9] to the US Congress on liquefied energy gases safety. There have been confrontations in the USA between pressure groups expounding the concept of 'zero risk', and industry. In Europe the accidents at Seveso, Manfredonia, and elsewhere have led to considerable concern about major hazards. A direct outcome of these has been a draft EEC directive[10] now under consideration in Brussels. The draft directive currently requires the notification of certain kinds of hazardous installations and in its final form may resemble the UK proposed Hazardous Installations Regulations described above although it is at present rather more comprehensive.

EXAMPLE 1.2-1

An explosion in a high pressure electrolyser resulted in quite severe damage to the plant and to the building in which it was housed, and the subsequent death of the plant operator. It generated considerable local concern.

The incident reinforced the anxiety which had arisen in the locality because of the presence of housing and a medium-sized school close to a factory in which a variety of chemical processes were carried out, including, until shortly before the explosion, the cyclohexane oxidation process. There was also concern that the explosion might have been even more severe, in which case greater damage could have been caused to the locality than in fact occurred. Accordingly, the HSC gave instructions for an investigation to be set up under the power contained in section 14(2)(a) of the HSW Act. The report[11] of this investigation has now been published, and can be consulted for fuller details. The incident occurred five days after the HSW Act came into force and resulted in a prosecution under section 2 of the Act.

Apart from underlining the well recognized dangers of explosion with hydrogen gas, the incident focused attention on the explosive limits of hydrogen as a contaminant of oxygen, and vice versa. These limits are appreciably different from the upper and lower limits for hydrogen in air, and special arrangements are required to ensure that dangerous concentrations cannot occur. Continuous instrumental monitoring was seen to be essential, coupled with arrangements to ensure that abnormal readings, i.e. those tending towards dangerous levels, would automatically shut down the plant. Periodical routine tests by the plant operator should not be relied on, by themselves, in such critical circumstances. They should be regarded as part of the system for checking the performance of the continuous monitors, supplementing similar tests by laboratory staff.

Another important and obvious lesson was that plant with the potential to cause a serious accident should not continue to be worked when there is clear evidence that a fault is developing which could have serious consequences. From an early stage in the investigation discussions were held with two other firms – the only other users of this design of plant in the country.

Questions 1.2

1. Give the historical build-up of the Health and Safety Commission in the UK and state what part the Robens Report played in this evolvement.
2. How are the Improvement and Prohibition Notices issued and enforced under the Health and Safety Act 1974, in the UK?
3. What is meant by the term 'Reasonably Practicable' and what is the procedure of implementing this into the approach recommended to planning authorities. Illustrate this by considering in an industrial application of your own choice the proposal for Hazardous Installations Regulations in the UK.
4. Outline the notification and survey regulations which would be expected to apply to a UK hazardous plant. What would be the extent of the survey which would be expected to apply to such a plant?
5. Develop a hypothetical explosion occurring in part of a plant in an industry known to you and assume one man is killed on the site. Using Example 1.2-1 carry out a critique of the explosion and refer to the result of a known accident condition where appropriate.

References 1.2

1. Safety and Health at Work. Report of the Committee 1970–72.
   HMSO, London. 1972.
2. Advisory Committee on Major Hazards, 1st report.
   HMSO, London. 1976.
3. Health and Safety Commission Consultative Document 'Hazardous Installations (Notification and Survey) Regulations 1978'.
4. Advisory Committee on Major Hazards, 2nd report.
   HMSO, London. 1979.
5. *Canvey: An Investigation of Potential Hazards from Operations in the Canvey Island/Thurrock area.*
   HMSO, London. 1978.
6. *Edwards* v. *National Coal Board* 1949 *I. All England Law Reports* 743.
7. Advisory Committee on Major Hazards, 2nd report.
   HMSO, London. 1979 – Paragraph 80.
8. Advisory Committee on Major Hazards, 2nd report.
   HMSO, London. 1979 – Paragraph 108.

9. Report to the Congress: Liquified Energy Gases Safety. Comptroller-General of the United States/GAO, 1978.
10. European Community: Proposal for a Council Directive on the Major Accident Hazards of Certain Industrial Activities (Council Document 8409/79 *et seq.*).
11. *The Explosion at Laporte Industries Ltd, Ilford, 5 April 1975.* A Report by HM Factory Inspectorate, HMSO, London 1976.

High Risk Safety Technology
Edited by A. E. Green
© 1982 John Wiley & Sons Ltd

Chapter 1.3

Assessment Methodology

SECTION 1.3.1 Analysis of Risk

R. F. White

Introduction

As shown in Chapter 1.1, risk is a probabilistic concept and is quantified by the product of the accident frequency and the consequence of the accident. In many cases, such as road accidents, fire, etc. (see Table 1.1.1-5) sufficient data have accumulated to enable statistical estimates to be made of individual and societal risks from these causes.

It has long been recognized that high technology installations such as industrial and nuclear reactor plants are potentially hazardous.

Many chemical plants involve the production and storage of large quantities of toxic, flammable or explosive materials. Accidents in such plants could lead to the release, deflagration or explosion of the hazardous materials which could lead to severe consequences in terms of loss of life, plant and property damage or contamination.

In nuclear reactors large amounts of radioactivity are generated by the fission process. The bulk of this activity remains in the fuel as long as it is adequately cooled. However, postulated reactor accidents which could lead to severe overheating and melting of the fuel have the potential to cause large releases of radioactivity to the environment resulting in severe consequences. The prevention of such accidents and the mitigation of their potential consequences have been the primary objectives of nuclear power plant safety design. Thus the conservative design, construction, and operation of plants are required so that accidents will be prevented or be assured to have a low probability of occurrence for normal operations. Then, to provide defence in depth, mitigation capability is required for accidents that are postulated to occur even though the design is required to include measures to prevent them. Thus, for example, the emergency core cooling system is required to be installed to mitigate postulated loss of coolant accidents even though the design of the primary reactor coolant system is required to contain substantial margin to prevent failure under normal conditions and certain abnormal conditions, e.g. anticipated transients.

Thus, in the chemical and nuclear power industries, whose plants incorporate defence in depth, the frequency of accidents is small and so it is not possible to get direct statistical measures of the risks associated with these plants.

The designs of the early nuclear power stations were judged against a safety philosophy based on the maximum credible fault. Arguments were then put forward, based on deterministic calculations, which showed that the reactor design with its defence in depth safety features was capable of bringing any such incident to a satisfactory and safe conclusion. This was a deterministic assurance of the safety of a plant, and an unquantified and therefore subjective assurance that the risk to the public was small.

The above type of analysis assumes that man-made installations behave in a deterministic way. This will only be true when all constraints and boundary conditions also behave or are applied deterministically. However, many components of engineered systems have failure modes which are stochastic, and certainly the behaviour of humans and their interactions with the system cannot be regarded as deterministic. Since it is engineered systems and human operators which apply the constraints to a process their probabilistic behaviour is superimposed on the underlying deterministic process to produce an overall probabilistic behaviour. Moreover, initiating events whether internal (failure of one or more controlling systems) or external (earthquake, aeroplane crash) are generally random occurrences. It can therefore be argued that any description of accident initiation and progression is necessarily probabilistic.

Thus, the building of a reactor implies the acceptance of some finite degree of risk. No engineering plant and no structure is entirely risk free, and there is no logical way of differentiating between 'credible' and 'incredible' accidents. The logical way of dealing with this situation is to seek to assess the whole spectrum of risks in a quantity related manner.

In 1967 Farmer[1] proposed a new approach to reactor safety based on the frequency of release of various quantities of iodine-131, the fission product that carries the greater threat to health than any of the other fission products that might be released in a reactor accident. The approach is to postulate events which would initiate accident sequences. Any initiating event, e.g. failure of pipework, loss of circulator power, sets up a whole spectrum of accident sequences depending on the various combinations of failures in the safety systems provided. The likelihood of following any one path depends on the performance of many items of plant and from a knowledge of this performance a probability can be assigned to each pathway. The full safety assessment would then comprise a spectrum of accident sequence end-points with associated frequencies and associated consequences. The risk associated with the plant is deemed to be acceptable if the cumulative frequency of accidents resulting in a given release of iodine-131 is below some arbitrary safety criterion line on a graph of accident frequency versus consequence as shown on Figure 1.3.1-1.

In 1969 Pugh[2] described how a design office used this probabilistic approach to achieve a safe and economic design of the Steam Generating Heavy Water Reactor.

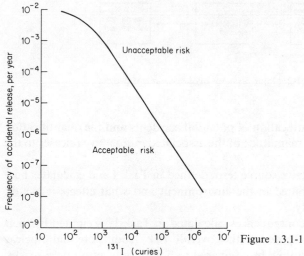

Figure 1.3.1-1 Release criterion proposed by Farmer[1]

The Farmer paper and the report by Pugh contain the elements of a methodology, i.e. event trees, which was later to be used extensively to assess the risks associated with nuclear power plants.

In the nuclear field the new technique has been applied in three major studies.

1. The Reactor Safety Study[3] which assessed the risk from US commercial operation of Light Water Reactors.
2. The Accident Initiation and Progression Analysis[4] which assessed the risk from the use in the USA of a High Temperature Gas Cooled Reactor designed by General Atomic.
3. The German Risk Study on the operation of Light Water Reactors carried out for the Federal Minister of Research and Technology by the Gasaellschaft für Reaktorsicherheit[5].

The first and third of these studies were of reactor types for which there is a substantial amount of operating experience, and the second was a reactor at its design stage.

Risk Assessment Methodology

In this section an overall picture is presented of what is involved in the organization and execution of a risk assessment project of the type carried out in the studies mentioned above.

The risk determination is divided into three major tasks shown in Figure 1.3.1-2.

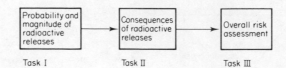

Figure 1.3.1-2 Major tasks in a risk assessment project

Task I includes the identification of potential accidents and the quantification of both the frequency and magnitude of the associated radioactive releases to the environment.

Task II uses the radioactive source term defined in Task I and calculates how the radioactivity is distributed in the environment and what effects it has on public health and property.

Task III combines the consequences calculated in Task II, weighted by their respective frequencies to produce the overall risk from potential nuclear accidents. Such results can then be compared to a variety of non-nuclear risks.

Quantification of Radioactive Releases

The objective of Task I is to generate a histogram of the form shown in Figure 1.3.1-3 which shows the frequency and magnitude of the various accidental radioactive releases. To generate a composite histogram of this type, the methodology employed must in principle be able to identify the accidents that can produce significant releases and determine their frequency.

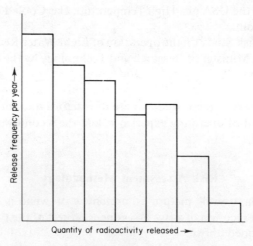

Figure 1.3.1-3 Illustrative release frequency versus
release magnitude histogram

The sub-tasks required to achieve the objective of Task I are shown on Figure 1.3.1-4. The first group of sub-tasks are shown as inputs to the block labelled 'Identification of Accident Sequences'. It should be recognized, however, that the steps indicated on Figure 1.3.1-4 will not flow as simply as implied by the block diagram. For example, it may not be immediately obvious which accident sequences are important and which make only negligible contributions to risk. Thus, many sequences may be preliminarily analysed under a set of pessimistic simplifying assumptions. Those that show up as significant contributors may then be reanalysed using more detailed, realistic methods. A number of such iterations may be necessary to determine the accident sequences that are the dominant contributors to the frequency of a given consequence.

Details of the System Design and Operation

The first requirement is a detailed knowledge of the design and operation of the plant under consideration. Such information is usually obtained from detailed drawings and documentation of the design and safety philosophy, and discussions between the safety analysts and the design office staff. Whenever possible, discussions with the operations staff and inspection of the plant provide another source of essential information.

A nuclear reactor of the type shown on Figure 1.3.1-5 will be used as an illustrative example. The reactor core, circulating pumps and heat exchangers are contained in a pressure vessel which forms the primary containment. Penetrations are provided for the control rod operating mechanisms and also for the heat exchanger connections. Numerous smaller penetrations are provided for reactor instrumentation sensors. The components of the system are all designed to codes which ensure that they will withstand the required operating temperatures, pressures, and radiation levels.

Under normal operating conditions the reactor core is maintained critical at a specified power level by means of the control rods. The heat generated in the core is removed by forced circulation of a coolant which passes to a heat exchanger where the heat is transferred to water inside the tubes. The steam produced in the steam generator passes to the steam turbine.

In the event of loss of normal coolant flow for any reason an auxiliary coolant circulator is provided together with an auxiliary heat exchanger to remove the decay heat since the reactor would be shut down under such circumstances. A non-return valve prevents the auxiliary coolant flow from bypassing the core. The coolant inside the tubes of the auxiliary heat exchanger flows to an air blast heat exchanger which rejects the heat to atmosphere.

The whole reactor is enclosed by a structure known as the secondary containment. In the event of a release of radioactivity from the primary containment the secondary containment provides a further barrier to prevent the release of radioactivity to the atmosphere.

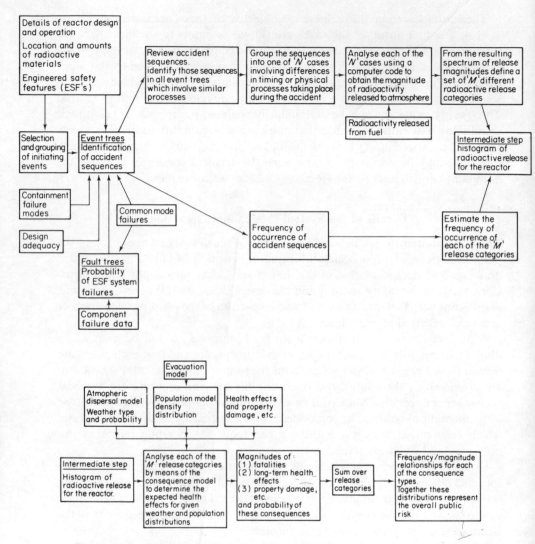

Figure 1.3.1-4 Block diagram showing sequence of tasks in a risk assessment project

The power supplies for the various running and standby components of the system are shown on Figure 1.3.1-6.

Location and Amounts of Radioactive Materials

Table 1.3.1-1 shows a typical radioactivity inventory for a 1000 MW (E) nuclear power reactor[3]. This indicates that a large proportion of the radioactive

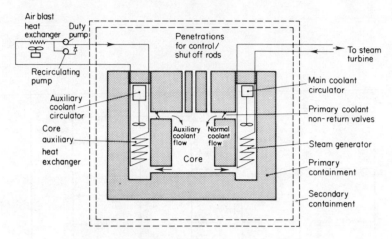

Figure 1.3.1-5 Typical features of a nuclear power reactor

materials are located in the fuel elements in the reactor core. The coolant itself possesses some radioactivity induced by gamma radiation.

However, in an overall risk assessment we must not forget the radioactivity in other locations such as that contained in the spent fuel storage pool, the shipping cask, a spent fuel sub-assembly during refuelling and in the waste gas and liquid hold-up tanks.

However, the fuel in the core has a much larger heat generation rate even when the reactor is shut down than the fuel in any of the other locations indicated in the table.

Thus, large consequence accidents would arise mainly from the release of large fractions of the core inventory.

Engineered Safety Features

Engineered safety features (ESF) may be considered to be the engineered answers to questions asked by the designer about his plant. For example, he may ask, 'What would happen if the main reactor cooling system failed?' He would conclude that there would be a gross imbalance between heat production and heat removal resulting in overheating and failure of the fuel elements accompanied by the release of radioactivity. Continued heat production would result in high system pressure culminating in failure of the containment. He would then ask, 'What would be required to mitigate such severe consequences?' He would conclude that reactor would have to be shut down rapidly and that the core would have to be cooled by some other means and that the decay heat would have to be removed from the coolant and rejected by some other means.

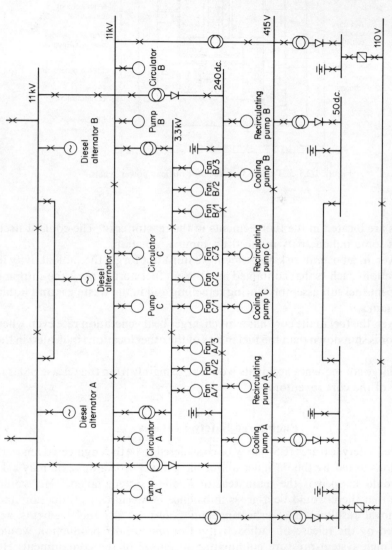

Figure 1.3.1-6 Power supplies for reactor

Table 1.3.1-1 Typical radioactivity inventory for a 1000 MW(E) nuclear power reactor

Location	Total inventory (curies)			Fraction of core inventory		
	Fuel	Gap	Total	Fuel	Gap	Total
Core[a]	8.0×10^9	1.4×10^8	8.1×10^9	9.8×10^{-1}	1.8×10^{-2}	1
Spent fuel storage pool (Max.)[b]	1.3×10^9	1.3×10^7	1.3×10^9	1.6×10^{-1}	1.6×10^{-3}	1.6×10^{-1}
Spent fuel storage pool (avg.)[c]	3.6×10^8	3.8×10^6	3.6×10^8	4.5×10^{-2}	4.8×10^{-4}	4.5×10^{-2}
Shipping cask[d]	2.2×10^7	3.1×10^5	2.2×10^7	2.7×10^{-3}	3.8×10^{-5}	2.7×10^{-3}
Refuelling[e]	2.2×10^7	2×10^5	2.2×10^7	2.7×10^{-3}	2.5×10^{-5}	2.7×10^{-3}
Waste gas storage tank	—	—	9.3×10^4	—	—	1.2×10^{-5}
Liquid waste storage tank	—	—	9.5×10^1	—	—	1.2×10^{-8}

a. Core inventory based on activity 30 minutes after shutdown.
b. Inventory of 2/3 core loading; 1/3 core with three day decay and 1/3 with 150 day decay.
c. Inventory of 1/2 core loading; 1/6 core with 150 day decay and 1/3 core with 60 day decay.
d. Inventory based on 7 PWR or 17 BER fuel assemblies with 150 day decay.
e. Inventory for one fuel assembly with three day decay.

These additional safety features could then be engineered into the design so that in the event of the postulated event the reactor could be brought to a safe shut-down condition.

Thus, an engineered safety feature is any system which has been specifically designed to mitigate the effects of an imagined initiating event.

If we restrict the present discussion to the initiating events associated with the reactor core then regardless of the design details of a particular reactor we find that there are some engineered safety features which will mitigate the effects of a large number of the initiating events which can be imagined, and in fact perform a uniform set of functions. These include:

1. Rapid reactor shut down, or reactor trip (RT)
2. Emergency electric power (EP)
3. Emergency core cooling (ECC)
4. Primary containment (PC)
5. Post-accident radioactivity removal, or clean-up plant (CP)
6. Secondary containment (SC)

Initiating Events

An initiating event is any failure of the operating system which potentially might result in the release of radioactivity. Since a large proportion of the radioactive materials are in the reactor core the initiating events of prime importance are those which might cause an imbalance between the heat generated in the fuel and the heat removed from the fuel which could ultimately lead to a melt-out situation in the core. There are three ways in which this heat imbalance could arise, assuming that one or more of the mitigating systems failed to operate.

1. The occurrence of a loss of coolant event.
2. Events which cause the heat removal capacity of the reactor cooling system to drop below the core heat generation rate.
3. Events which cause the reactor power to increase beyond the capacity of the reactor cooling system.

The many such potential failures previously identified and defined by the many years of safety analysis in the atomic energy industry and in the licensing process for commercial nuclear power plants serves as the starting point for listing the initiating events to be considered in an overall risk analysis. However, searches should be conducted for newly defined initiating events, particularly where operator actions are involved.

Care must be taken when placing initiating events into groups for the purpose of subsequent event tree development. Consider, for example the initiating event under the group heading 'Loss of Cooling Capacity'. This could be the result of loss of off-site power which would cause the main circulators to stop. This fault would be mitigated by the emergency electric supplies and the emergency cooling system. If, however, the loss of cooling capacity were the result of a flow blockage then the fault would not be mitigated by these engineered safety features. Thus, the above two cases under the general heading of loss of cooling capacity would in fact require separate event trees.

Identification of Accident Sequences

Event tree methodology is used as the principal means for identification of the significant accident sequences.

An event tree is a logic method for identifying the various possible outcomes of a given event which is called the initiating event.

The construction of an event tree is started by indicating the initiating event at the left of the diagram which in the example shown on Figure 1.3.1-7 is 'Loss of Off-site Power'. In considering the events following the initiating event one must include all the safety systems that can be utilized and define the functions that these engineered safety features are required to perform. The functions relate to the physical processes associated with the system's operation, such as the function of heat removal. The ESF functions are then indicated as event tree headings moving from left to right in roughly chronological order after the initiating event. This set of functions acts as the initial headings of the event tree. At a later stage in the analysis these ESF function headings are replaced by ESF system headings which are more specific to the associated hardware. For example, there may be more than one safety system provided to perform the same function such as a diverse reactor shutdown system. There may also be time dependencies which necessitate more than one event tree heading to describe the ESF function. For example, an emergency cooling system may require two out of three loops for

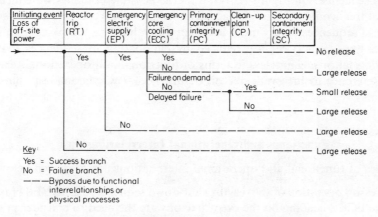

Figure 1.3.1-7 Functional level event tree for loss of off-site power

successful operation in the early stages but may require only one out of three loops for successful operation in the later stages of an accident sequence.

Several iterations between the function and system hardware event trees may be necessary to help check the adequacy of the modelling. However, the understanding and appreciation of the system gained from the function event tree provides useful groundwork for the subsequent development to the system event tree stage.

Particular attention must be paid to the ordering of the headings on an event tree. The rationale for the order of the headings on the event tree on Figure 1.3.1-7 is as follows, remembering that the initiating event is 'Loss of Off-site Power'.

1. Reactor trip (RT) is placed first because failure to shut down the fission process could result in high core temperatures which would render the emergency core cooling system ineffective event if it started.
2. Emergency electric power supply (EP) is listed next because several of the other ESFs are dependent on its successful operation, e.g. the emergency core cooling system. In the event of a release of radioactivity into the secondary containment the clean-up plant is also dependent on the emergency electric supply.

The remaining headings follow in chronological order. The necessity for correct ordering of the headings on the event tree will become more apparent when the ESF functions, e.g. EP and ECC are developed, for the same initiating event, to the ESF system event tree level (see Section 1.3.2).

The event tree is then drawn from left to right by the addition under each heading of branches corresponding to two alternatives: (i) successful performance of the ESF function (the upper YES branch); and (ii) failure of the ESF function (the lower NO branch). After the tree is drawn, paths across it can be traced by choosing a branch under each successive heading. Each path corresponds to an accident sequence. On Figure 1.3.1-7, if all the possible branches were included the event tree would indicate $(2)^6$ accident sequences. As can be seen the actual number of sequences is very much less than this value. The following discussion indicates how event trees are reduced by elimination of sequences that are either illogical or meaningless in terms of functional and operational relationships. Reference to the event tree on Figure 1.3.1-7 will be made to illustrate the points.

Functional and Operational Interrelationships

Examples of functional and operational interrelationships are given below:

1. Successful operation of the reactor shut-down sequence, e.g. all YES branches to the ECCS heading on the event tree obviate the need to consider primary containment (PC) integrity, clean-up plant (CP) operation or secondary

containment (SC) integrity and so these headings can be bypassed as indicated by the dotted line to the end of the sequence.

2 The failure of one ESF function renders one or more of the remaining ESFs ineffective even if they started, e.g. failure to trip the reactor in the event of loss of main cooling would render all other ESFs ineffective.

3. Failure of an ESF function prevents another ESF from operating, e.g. failure of the emergency electric power would prevent the start of the emergency core cooling system.

4. Failure of an ESF function causes other functional failures due to physical processes taking place, e.g. failure of the emergency core cooling system would ultimately result in high pressure which would cause the primary containment to fail. In general these physical processes introduce a time delay between the failures which can be important in assessing the consequences.

5. Time-dependent performance requirements for the physical system needed to perform the various ESF functions.

An example of this is the number of emergency cooling loops required for successful operation depending on the time after the initiating event. Another example is if the ECC system operated on demand but failed some time later. Such a time delay could affect the effectiveness of the clean-up plant. This situation can be represented on an event tree by two NO branches as shown on Figure 1.3.1-7 under the ECCS heading.

The first NO branch is labelled 'Failure on demand' in which case the clean-up plant would be ineffective.

The second NO branch is labelled 'Delayed ECCS failure' in which case the clean-up plant may be effective in reducing the release of radioactivity.

Careful consideration of these interrelationships between the engineered safety features, particularly when the event tree is developed to the ESF system level, will often reveal inter-system common mode failures which eliminate meaningless sequences on the event tree.

Descriptive Material on Accident Sequences

Each path on an event tree corresponds to an accident sequence. If all the engineered safety features operate at their design basis the accident sequence resulting from an initiating event is the design basis accident defined in the reactor licensing process and the consequences are quite small. If, however, one or more of the engineered safety features do not perform their designed function then a broad spectrum of accident sequences can occur, each with a frequency and consequence dependent on the availability of the mitigators.

In the risk assessment process, estimation of frequencies and consequences requires the determination of a number of specific parameters associated with each potential sequence.

1. The context and criterion of successful operation of systems providing accident mitigation functions. This information is particularly required in fault tree analysis.
2. System interrelationships that either preclude or make more likely various failure modes for individual sequences.
3. Parameters associated with potential radioactivity release. This information is required in the consequence model.

The information required for each accident sequence is incorporated in tabular sequence descriptions which accompany the event trees to assure that all meaningful information about each sequence is used in a quantitative assessment of the trees.

To aid in understanding the relationship between the operation and function of the ESFs on a given event tree, a chart providing a sequence by sequence description of the event tree is drawn up. This chart summarizes the results of individual sequences from a particular event tree as to whether or not the sequences result in core melting. Also the chart links the sequences with the mode of failure of the containment which is considered to be appropriate to each sequence.

To provide additional assistance in understanding the accident sequences it is useful to prepare a listing of the time phasing of certain important physical events occurring in each sequence with estimates of significant physical quantities, e.g. temperature, at these times.

Examples of this descriptive material which accompanies the event trees are given in Section 1.3.2.

Frequency of Occurrence of Accident Sequences

Referring to Figure 1.3.1-7, the first task arising from the event trees is the estimation of the accident sequence end point frequencies. These are obtained simply by combining the initiating event frequency with the probabilities of failure or success of each of the engineered safety features along each path through the event tree.

The failure probabilities of the engineered safety features are obtained by fault tree analysis which is described in Section 1.3.2.

Release Magnitudes

Referring again to Figure 1.3.1-7, the second task arising from the event trees is the estimation of the magnitude of the release associated with the end-points of each of the accident sequences.

The very large number of end-points on all the event trees rules out the analysis of each individual sequence in terms of the magnitude and type of release. At this stage in the risk assessment, therefore, it is expedient to review the accident sequences in all the event trees to find those sequences which involve similar processes. It is then possible to place the sequences into one of a number of groups involving differences in the time of the releases or the physical processes taking place during the accident. If, in each of these groups we can estimate the amount of radioactivity released from the fuel then it is possible to analyse each of the groups in turn by means of a suitable computer code to obtain the magnitude of radioactivity released to the atmosphere. From the resulting spectrum of release magnitudes it is possible to define a small number of different radioactive release categories, each representing a certain range in the magnitude of release.

We now have to return to the event trees to sum all the accident end point frequencies whose releases fall into the same release category. In this way a histogram can be built up of accident frequency against magnitude of release, typified on Figure 1.3.1-3 which was the first of the stated objectives of the risk assessment. When plotted in this form the frequency–release data of the particular reactor may be compared with some suitable criterion such as the Farmer criterion on Figure 1.3.1-1.

So far we have only traced the radioactivity to the point of release from the reactor containment. This information is useful to the designer because by considering the major contributors to the frequency of accidents in a given release category he may assess the design adequacy of the engineered safety features, or indeed some of the reactor design details.

Risk to the Public

The final part of the risk assessment project shown on Figure 1.3.1-4 is to determine the effect of each radioactive release category on the public. To do this a dispersion model of the type embodied in the TIRION computer program[6] is used.

Dispersion Model

The information about each release category is tabulated under the following headings:

1. Frequency of the release
2. Time of release after the initiating event
3. Duration of the release
4. Warning time for evacuation
5. Elevation of the release
6. Fraction of the core inventory released (for each of the biologically significant isotopes)

The computer code takes as input the above release information together with the statistics of weather and population distributions for the particular reactor site. If evacuation plans are adopted then the population distributions must be time dependent. The program has a data store on the health effects caused by contact, inhalation or ingestion of various isotopes.

The program also includes a model of atmospheric diffusion, transport, deposition, and plume rise for a general time-dependent release of a spectrum of radioactive isotopes. It is clearly not possible to foretell the nature of the weather around a nuclear power plant site on the date of an accident some time in the future. An average of the consequences for different types of weather is therefore used, each being weighted with relative probability of occurrence at the site. It is usual to adopt the Pasquill weather classification (six categories) and it is necessary to seek data on these and the probability of occurrence of various wind speed and directions for the particular site.

Since for a release of even one radioisotope there are at least three probability distributions (population, weather categories, wind directions) to be folded into the analysis it is clear that even for some specific activity release a distribution of consequences is produced. These distributions, in a comprehensive analysis, are added up for the different release categories, weighting with the frequency of each category to provide the overall risk relationship for the reactor site.

The dispersion model enables one to translate magnitude of release to dose rates for any radioisotopes, and from dose rates the consequences of each release category can be calculated in terms of the following:

1. Fatalities
2. Long-term health hazards, i.e. thyroid cancers
3. Property damage, i.e. area of agricultural land that cannot be used for a certain number of years

Together these three distributions represent the overall public risk.

The purpose of this section has been to present an overall picture of the tasks involved in a risk assessment project. More detailed accounts of some of the tasks such as the development of event trees from the ESF function to the ESF system level, fault tree analysis and quantification, and the dispersion and consequence models are given later in this chapter.

Questions 1.3.1

1. Discuss the main factors which enter into the preparation of an event tree.
2. For a process known to you, imagine a set of initiating events and construct relevant event trees. Rate the consequences as safe, low hazard, high hazard. Do the engineered safety features provide adequate protection against the potential hazards?

3. Discuss the advantages and disadvantages of event trees as a methodology for the assessment of risk.

References 1.3.1

1. Farmer, F. R. '*Siting criteria – a new approach.*' IAEA Symposium on the Containment and Siting of Nuclear Power Reactors held in Vienna, 3–7 April 1967. SM/89/34.
2. Pugh, M. C. *Probability Approach to Safety Analysis.* United Kingdom Atomic Energy Authority, Report TRG 1949 (R). 1969.
3. United States Nuclear Regulatory Commission (USNRC). *Reactor Safety Study–An assessment of accident risks in US commercial nuclear power plants.* WASH-1400 (NUREG-75/014), 1975.
4. General Atomic *HTGR Accident Initiation and Progression Analysis.* GA-A13617. 1975.
5. Gesellschaft für Reaktorsicherheit (GRS). *The German Risk Study –* Summary. 1979.
6. Kaiser, G. D. '*A description of the mathematical and physical models incorporated in TIRION – 2 –* A computer program that calculates the consequences of a release of radioactive material to the atmosphere and an example of its use.' Systems Reliability Directorate/R.63. 1976.

High Risk Safety Technology
Edited by A. E. Green
© 1982 John Wiley & Sons Ltd

SECTION 1.3.2 *Fault Trees and Event Trees*

A. Cross

Introduction

Section 1.3.1 has given an outline of the fault tree–event tree methodology. This section now focuses on some of the finer details of the method and highlights some of the problem areas. Obviously there is no way that these few pages can give a complete analysis of the sample reactor system (Figure 1.3.1-5), this would require a document the size of the Reactor Safety Study[1].

The first problem area to be encountered is one of terminology. When the foundation of the method was first proposed by Farmer[2] the aim was to establish a 'Frequency–Consequence' diagram to demonstrate the *total* risk of operating a particular plant (or group of plants). The modern trend seems to prefer a 'Probability per year–Consequence' diagram; invariably the 'per year' becomes lost, leaving a 'Probability–Consequence' diagram which has no meaning, and which will frequently cause embarrassment when the 'Probability' approaches or exceeds unity.

The second problem area is one of credibility. There is no point in trying to demonstrate that a particular end state of an event tree has a frequency of, say, 10^{-25} per year. Such a result will never be believed for the simple reason that it can never be justified. At the same time there is no hard and fast boundary between credibility and incredibility. If we are told that a particular piece of equipment has a failure probability on demand of, say, 10^{-8} to a 95% 'confidence', there *might* be good grounds for doubt (earthquakes, falling aircraft, sabotage, etc.).

The event tree is essentially a cause → effect model; i.e. given an initiating event, the possible outcomes of that event are to be modelled. As we shall see, event trees can become exceedingly complex, especially when a number of time–ordered system interactions are involved.

The fault tree is essentially an effect → cause model; i.e. given a system failure state, the model attempts to determine all of the system states that could lead to the failure state. Initially the model included only the Boolean 'AND' and 'OR' logical operators. Subsequently the Boolean 'NOT' logical operator was added, but this has caused some controversy[3][4]. Recently a number of non-Boolean

49

logical operators have been introduced, e.g. 'PRIORITY AND', 'MUTUALLY EXCLUSIVE OR', 'DELAY' and 'SUMMATION' gates. While such logical operators might be correct in relation to the model of the system, the analytical tools to quantify such fault trees are not always to hand.

Many of the mathematical assumptions behind the available analytical tools have been the subject of some discussion. A simple example of this is the 'repair rate' for simple component faults, which makes the assumption that a repair may begin as soon as the fault occurs. In reality there is often a minimum time before which no repairs can be effected. A more complex example is the discussion between Murchland and Vesely[5][6]. Frequently the difference between two mathematical models is a nicety of interest only to academics, especially when one considers the uncertainty of the physical data.

Preliminary System Analysis

The functional event tree shown in the previous section (Figure 1.3.1-7) gives the important system functions which must be considered when the 'initiating event' is assumed to be 'Loss of Site Electrical Supplies'. These functional headings must now be developed down to system and sub-system headings so that the system and sub-system dependencies can be identified. Invariably a preliminary assessment will only reveal the obvious dependencies, others will be revealed during the subsequent fault tree analysis and the analyst may have to go back and modify the event tree (perhaps several times). This iterative procedure can be frustrating and expensive, but it is necessary and can lead to substantial benefits by giving the analyst a greater understanding of the plant operations.

The functional headings of the event tree will now be examined in more detail:

(a) Reactor trip

The function of the Reactor Protective System (RPS) is to shut-down the controlled chain reaction when an undesired transient occurs. When the transient is caused by a loss of electrics, the RPS should operate automatically without needing any of the standby electric supplies. For other transients (excessive neutron flux, high temperature, etc.) an electric supply would normally be necessary for the RPS to operate.

Following a successful reactor trip, or, as it is sometimes called, reactor scram (an emotive word indicating a general state of panic!), the heat generated by the reactor decreases rapidly. This decay heat is, however, quite sufficient to cause major damage to the reactor core if it is not removed.

(b) Emergency electric power

This general function heading included two main groups of power supplies:

(i) The Guaranteed Non-Interruptable Supplies (GNI).

The GNI supplies include all of the essential electric supplies required for continuous control and protection of the reactor, and are always backed by batteries. During normal reactor operation, these battery supplies should be connected to the incoming power source so that they are kept fully charged.

In the example reactor system (Figure 1.3.1-6) the 240 V DC, 50 V DC, and 110 V AC systems are GNI systems.

(ii) The Guaranteed Interruptable Supplies (GI).

The GI supplies serve items of reactor equipment which, while they might be essential to the protection of the reactor, can tolerate a short interruption to their supply. In some cases a short delay before the operation of these equipments might be essential. The 11 kV, 3.3 kV and 415 V systems are GI systems and are supplied by three diesel-alternator sets which should energize two to five minutes after the loss of normal electric supplies.

The GI and GNI systems in our example reactor system are totally inter-dependent. The diesel-alternator sets are dependent upon the 50 V DC and 110 V AC GNI supplies for control and excitation. The diesel loading breakers are dependant upon the 240 V and 50 V DC GNI supplies. In the medium term, the GNI system is dependent upon the GI system taking over before the batteries discharge (assuming that normal power has not been restored).

(c) Emergency core cooling system (ECCS)

The ECCS is designed to remove the decay heat from the reactor core after a transient which causes (or requires) a shutdown of the normal heat removal system. The ECCS does not have the capacity to remove the full heat output from the reactor without reactor trip, hence its function can be misunderstood by the layman. Decay heat removal system might be a more accurate and less emotive title.

The example reactor system (Figure 1.3.1-5) shows one of the three ECCS loops, each having a 50% decay heat removal capacity.

The essential items of equipment on each loop consist of:

(i)   An auxiliary coolant circulator for circulating the primary coolant through the reactor core.

(ii)   An auxiliary core heat exchanger for transferring the decay heat from the primary coolant to the secondary coolant.

(iii) A secondary coolant duty pump for circulating the secondary coolant after the GI system has been energized by the diesel alternators.

(iv)   An air blast heat exchanger with three GI powered fans, for transferring the decay heat from the secondary coolant to the atmosphere.

(v)   A secondary coolant recirculating pump which operates continuously on the GNI system, for maintaining steady thermodynamic conditions in the auxiliary core heat exchanger during normal reactor operation.

(vi) An auxiliary coolant circulator cooling pump which operates continuously
      on the GNI system. This pump operates in conjunction with a fluid/air heat
      exchanger which does not require forced air convection.

(vii) Primary coolant non-return valves which control the direction of the
      primary coolant flow during normal reactor operation, and when the
      auxiliary core heat exchangers are being operated.

(d) Primary containment

The primary containment is essentially a passive structure to contain the primary
coolant, shield the operators from the radiation, and to provide a barrier to
contain radionuclides released from the fuel. The integrity of the primary
containment, if large amounts of the fuel melt, has always been a matter for some
conjecture, and no doubt will continue to be so.

(e) Clean-up plant

The clean-up plant is designed to remove radioactive nuclides and heat which
might escape from the primary containment during an accident sequence. The
clean-up plant will generally be designed to deal with small–medium releases
from the primary containment. The power supply for the clean-up plant will
probably come from the GNI/GI system used for the emergency core cooling
system.

(f) Secondary containment

The secondary containment is essentially a passive structure to contain
small–medium releases of radioactive nuclides from the primary containment.
There might be some features of the secondary containment that are active, e.g.
the ventilation system. During an accident sequence the ventilation system must
be closed; this would require power from a GNI system.

Preliminary Event Tree Analysis

Consider that partial event tree shown in Figure 1.3.2-1, which covers, say, the
first 30 minutes following the loss of normal electric supplies. The time periods
for this initial phase are dependent upon the answers to a number of questions;
for example:

How long can the GNI system provide power?
How long can the reactor fuel tolerate a complete loss of cooling before fission
products are released?
How long can the reactor core structure tolerate a complete loss of cooling
without damage that might restrict cooling at a later time?

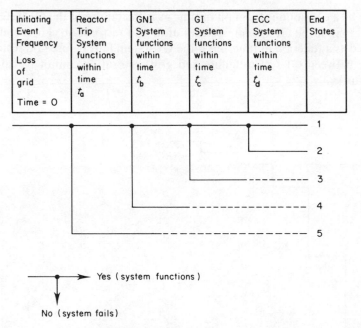

Figure 1.3.2-1 Initial event tree for loss of grid supplies

The five end states shown in Figure 1.3.2-1 are as follows:

1. All emergency systems have, so far, operated correctly. What must now be considered is whether or not the system can be restored to normal operation. For this we must consider the probability of grid restoration and the reliability of the ECC, GI, and GNI systems.
2. The RT, GNI, and GI systems have operated, but the ECC system has not functioned. The severity of damage and fission product release will depend on what can be done to retrieve the situation. Can the ECC system be operated at a lower rating? The probability of grid restoration and the reliability of the GI and GNI systems will also be relevant.
3. The RT and GNI systems have operated but the GI system has failed, the ECC system might be functional but is dependent on the GI system. The restoration of grid supplies at an early stage could prevent or limit the damage.
4. Similar to (3) but the GI and ECC systems cannot operate because the GNI system has failed.
5. Since the RT system has failed, the GNI, GI, and ECC systems would only have a limited effect.

The subject of grid restoration will arise many times throughout the analysis; the problem is how to deal with it. The probability of grid restoration is generally taken to be a continuous function of time as is the failure of the GI system. The probability that the GI system operates until grid restoration can, therefore, be calculated as a function of time, but if the GI system fails before grid restoration the time between GI system failure and grid restoration cannot be calculated deterministically.

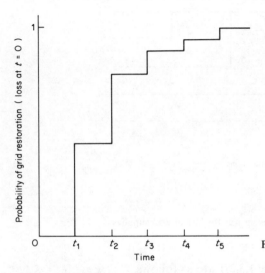

Figure 1.3.2-2 Probability of grid restoration treated as a step function

If the restoration of the grid is treated as a step function of time (Figure 1.3.2-2), then the probability of grid restoration at the end of each time interval can be modelled by introducing additional event tree headings or by subdividing the initiating event. These procedures allow the event sequences to be studied in more detail, particularly those event sequences which might be dependent upon actions carried out by the operator. The event tree shown in Figure 1.3.2-3 shows the development of the outcome for the RT, GNI, GI, and ECC systems during the time intervals 0 to t_2. Two aspects have not been considered in this event tree. The first is the performance of the GNI system following a failure of the GI system; it is pessimistic to assume that the GNI system will fail if the GI system fails after starting, but it does simplify the analysis. The second aspect not considered is the restoration of the reactor system to normal operation after the grid has been restored; this aspect will require careful treatment since the operators will play an important role in this transition, and it would be best to treat this as a separate event tree, using the relevant end states from Figure 1.3.2-3 as the initiating event.

The event tree shown in Figure 1.3.2-3 must now be developed to include the remaining time intervals and the remaining engineered safety features, i.e. the primary and secondary containments and the clean-up plant.

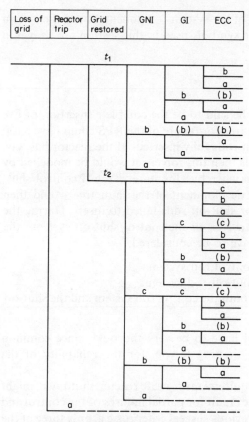

Loss of grid	Reactor trip	Grid restored	GNI	GI	ECC

a = Failure on demand
b = Start on demand, failure before t_1
c = Start on demand, failure before t_2 but after t_1

Figure 1.3.2-3 Development of event tree for loss of grid

The Reactor Safety Study[1] produced in 1975, gives many examples of event tree development. The HTGR Accident Initiation and Progression Analysis[13] produced in 1978, contains many more examples and also demonstrates how the technique has developed in a short space of time.

Preliminary Fault Tree Analysis

The basics of fault tree analysis have been described many times[7]. The only point that I would want to emphasize here, is that there is no point in starting a fault tree until one has a thorough knowledge of the system. To this end, a Failure Modes and Effects Analysis (FMEA) of the basic system components, the construction of Truth Tables to clarify some of the complex system logic, the development of Markov models and a number of other techniques, can be very useful. It would also be advisable to check that if all of the system components work, the system will meet its demand! The fault tree symbols used in this section are shown in Figures 1.3.2-7, 1.3.2-8 and 1.3.2-9.

(a) Reactor trip

If the success state of the RPS can be said to be the complete insertion of five control/shut-off rods, then the fault tree for failure of the RPS could have a top event 'Less than five control/shut-off rods fully inserted'. If the reactor has, say, six control rods and six shut-off rods then this top event would be modelled by considering all possible combinations whereby 8 (or more) of the 12 control/shut-off rods failed to insert fully. The development of the fault tree would then establish why each of the control or shut-off rods failed to insert. During the subsequent common mode failure analysis or the control/shut-off systems, the following common modes would have to be considered:

(i)   Common mode faults of the control rod system.
(ii) Common mode faults of the shut-off rod system.
(iii) Common mode faults affecting both the control rod system and the shut-off rod system.

The common mode analysis would have to be very thorough since common mode failures are the most likely limiting factors for the reliability of the system.

A more detailed analysis of the requirements for safe reactor shutdown, might show that it would be unwise to take credit for partial failures of the control and shut-off systems that satisfied the previous success criterion, e.g. only three of the six control rods insert and only two of the six shut-off rods insert. In that case the fault tree for failure of the RPS would be drawn as shown in Figure 1.3.2-4. This fault tree is more pessimistic than before, but after a thorough common mode analysis, the numerical results for failure to trip might not be significantly different.

(b) The GNI system

The preliminary assessment of the GNI system suggests that the top of the GNI system fault tree should be as shown in Figure 1.3.2-5. The assumption here, is that should any of the 240 V, 110 V or 50 V sub-systems fail to energize, then the GNI system will have failed. One major problem here is that the GNI sub-systems are interlinked, not only with each other, but also with the GI system. Some faults in one part of the system can be transferred to other parts if the protection systems fail to operate. Hence, even though the GNI system might not have failed, there might be faults on the GNI system which affect the GI system, or vice versa.

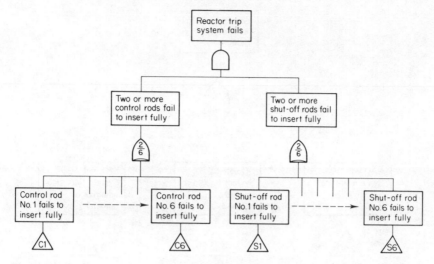

Figure 1.3.2-4 Fault tree for ile reactor trip system with two-out-of-six majority vote gates

(c) The GI system

The analysis of the GI system is in two parts:

(i) Does it start on demand?

(ii) Does it continue to function until normal grid supplies are restored?

Initially the GI system requires power from two of the three diesel-alternator sets, and two of the three ECCS loops must have power from the 11 KV, 3.3 KV, and 415 V sub-systems. At a later stage, only one diesel-alternator set will be required and only one of the ECCS loops will need power from the GI system.

The fault trees for the failures of the GI system will be complex, and the fault tree for failure of the GI system after starting will have to model the changing demands on the system; this will involve the application of Multi-Phase theory[11].

Since the fault trees for the failure of the GI system will contain failure elements of the GNI system, there is always the fact that some of the GI system failure modes are the result of the GNI system failures. If this is considered to be overpessimistic, the GI fault trees can be made conditional on the GNI system being working. Such a fault tree is non-coherent in failure[3][4].

(d) The ECC System

The analysis of the ECC system is also in two parts:

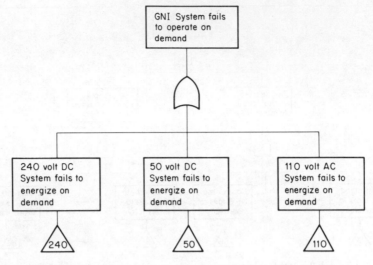

Figure 1.3.2-5 Development of the fault tree for failure of the GNI system

(i) Does it start a demand?
(ii) Does it continue to function until the main loop cooling can be restored (by restoration of the normal grid supplies). This part of the analysis will also involve an application of Multi-Phase theory.

Comments on the Preliminary Analysis

The preliminary analysis of the GNI, GI, and ECC systems might tempt one to say 'Why not look at the systems as a whole and determine whether or not they work?' The answer is as follows. If the systems work, fine; but if the systems fail, then if we are to follow the possible accident sequences through, and determine the probable consequences of each end state (on the event tree), we need to establish the plant conditions as accurately as possible. If we are to make an attempt to look at the possible operator – plant interactions, whether the operator is preventing the propagation of an accident sequence or making it worse, then we must have as much background material as possible.

Such an analysis can only be the work of a dedicated group. Initially the group will meet to look at the overall problem so that it can be broken down into units to be studied in more detail. This will be followed by an iterative process of looking for the interactions between the units being studied. Finally after publication of a draft safety analysis there will be further discussion before publication of a 'final' document, which will be the cause of further discussion.

Computer Programs for Fault Tree Analysis

The use of computer programs for fault tree analysis is becoming increasingly popular and a large number of programs are available to the fault tree analyst. However at present there is no single program, or package of interrelated programs, which can satisfy all of the requirements of the analyst; indeed it can be argued that the combined algorithms of all of the available programs might be insufficient for some problems.

There are a number of conflicting views concerning the worth of fault tree analysis programs. The truth is that such programs are tools which, if used correctly, can give the analyst useful qualitative and quantitative information. The accuracy of this information is totally dependent on the accuracy of (a) the model, (b) the data, and (c) the computer algorithm. It is, therefore, pointless for the analyst to spend a lot of time and effort in producing an accurate model with good data if he does not know the limitations of his program.

Computer programs designed for fault tree analysis generally fall into one or more of the following categories:

1. Programs for the automatic construction of fault trees. Some of these programs might be of some specialized use, but it could be another 5 or 10 years before these programs can be used generally.
2. Programs for finding minimal cut sets. There are a remarkable number of programs for finding the minimal cut sets, or the prime implicant sets, of fault trees. Many claims are made about the superior speeds of some of these programs, but they are always problem dependent and it is best to have a variety of programs to hand.
3. Programs for evaluating reliability characteristics. These programs have to be treated with some care; it is too easy to feed in failure and repair data for the basic events without thinking about the effects of combined faults. Again it is advisable to have a number of programs available, some algorithms are better suited to some types of calculations.
4. Programs for calculating basic event and minimal cut set importance. There are many ways of calculating the relative importance of the basic events and minimal cut sets. The main problem is, therefore, deciding which algorithm to use.
5. Programs for the analysis of data distributions. These programs treat the basic event reliability characteristics as variables from a know distribution, rather than the single point values used by the programs at (3) and (4) above.
6. Programs for common mode identification. These programs are designed to look for potential common mode failures of the basic events in the minimal cut sets of the fault tree.
7. Fault tree graphics programs. Such programs can reduce the tedium of drawing the fault tree, especially when the fault tree goes through a sequence of minor, or major, modifications.

The most recent review of US fault tree programs has been produced by Fussell[8] but there are many additional programs available from outside of the US. The development of a form of Boolean algebra which is not restricted to two state variables[12] seems very promising for the future.

Problem Areas

Perhaps the best known problem area is the one of common mode and common cause failures. The attack on common mode/cause is mainly one of identification and re-design[9]. Quantification of common mode/cause will remain a problem for some time, mainly because of inadequate data. The computer programs designed for common mode/cause analysis can give considerable assistance, but the onus remains on the analyst to provide the computer programs with sufficient information.

Another problem area is the supposed sanctity of numbers and the abuse of the word 'Confidence'. Heavy statistical arguments are often employed to prove certain numbers as a 'fact' to a known confidence. This might be statistically correct if all the data relevant to the problem are known, but this is seldom the case. An alternative terminology, i.e. the use of the word 'Tolerance'[10] in place of 'Confidence', would in many cases be more honest.

Phased Mission analysis has been a major problem until recently but a number of computer programs have recently been developed to assist with this problem[11].

Modelling of repair in a fault tree is still a major problem. Just because the fault tree might accurately model the failure logic of a system, it does not follow that the repair logic has been modelled accurately. The system and the partial fault tree shown in Figure 1.3.2-6 has an unavailability of 5.5×10^{-4} if the basic component failure and repair characteristics are evaluated in the usual manner. On the other hand, if the basic failure and repair data are used to calculate the frequency of the top event, the system unavailability is 0.25. While this might seem to be an artifical problem it is not untypical of some of the problems that can be encountered.

Summary

A complete hazard analysis of a complex plant using the event tree/fault tree technique will require many man-years of effort and a number of simplifying assumptions will have to be made. While the technique can be criticized, there is no doubt that it provides a systematic method for exposing the areas of potential risk. Any subsequent discussion about whether or not the plant is safe enough, can then at least be based on a sound foundation.

The event tree/fault tree technique has developed rapidly over the last 10 years, and there are many developments to come in the future. New algorithms which

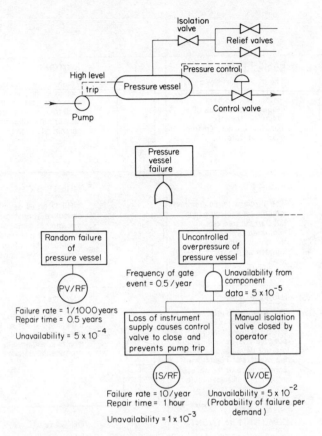

Figure 1.3.2-6 An example of non-coherence in repair

model the failure, repair, testing, and maintenance of basic component failures more realistically are required, the combination of basic component failures into system and sub-system failures also requires more research.

It is worth considering one final point. If an analysis shows that a particular category of accident can be expected, say, on average once in 1000 years, it does not follow that such an accident will not happen tomorrow. Neither does it mean that the analysis was wrong if it does happen.

Questions 1.3.2

1. Define what is meant by a fault tree.
2. Define what is meant by an event tree.
3. When would an event tree or a fault tree be used.

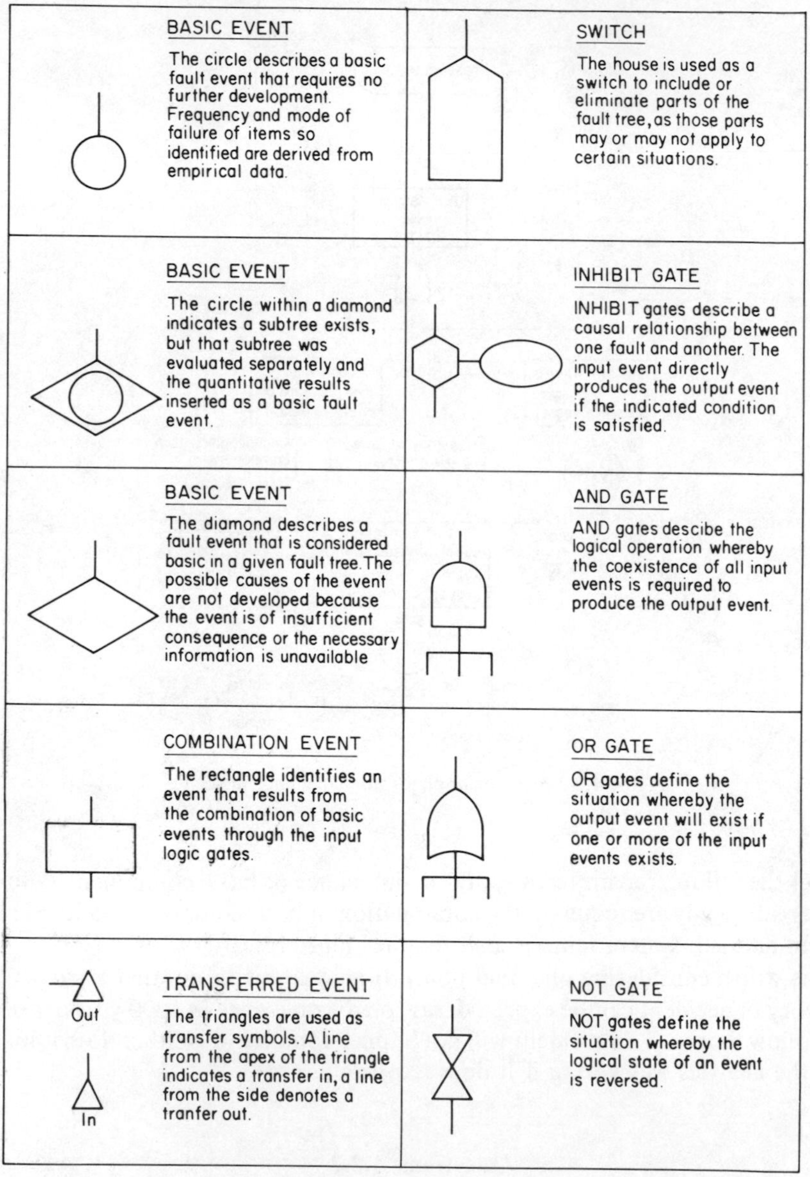

BASIC EVENT		SWITCH
The circle describes a basic fault event that requires no further development. Frequency and mode of failure of items so identified are derived from empirical data.		The house is used as a switch to include or eliminate parts of the fault tree, as those parts may or may not apply to certain situations.
BASIC EVENT The circle within a diamond indicates a subtree exists, but that subtree was evaluated separately and the quantitative results inserted as a basic fault event.		**INHIBIT GATE** INHIBIT gates describe a causal relationship between one fault and another. The input event directly produces the output event if the indicated condition is satisfied.
BASIC EVENT The diamond describes a fault event that is considered basic in a given fault tree. The possible causes of the event are not developed because the event is of insufficient consequence or the necessary information is unavailable		**AND GATE** AND gates descibe the logical operation whereby the coexistence of all input events is required to produce the output event.
COMBINATION EVENT The rectangle identifies an event that results from the combination of basic events through the input logic gates.		**OR GATE** OR gates define the situation whereby the output event will exist if one or more of the input events exists.
TRANSFERRED EVENT Out / In The triangles are used as transfer symbols. A line from the apex of the triangle indicates a transfer in, a line from the side denotes a tranfer out.		**NOT GATE** NOT gates define the situation whereby the logical state of an event is reversed.

Figure 1.3.2-7 Fault tree symbolism

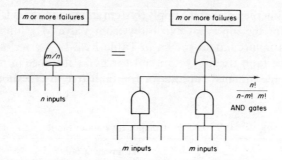

Figure 1.3.2-8 Majority gate logic

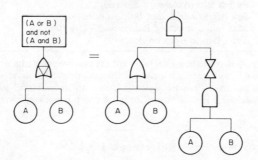

Figure 1.3.2-9 Exclusive or gate

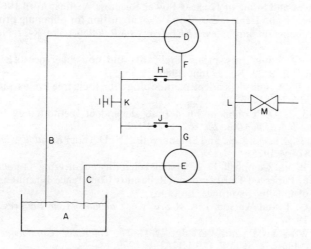

Figure 1.3.2-10 Emergency water supply system

4. Given the Emergency Water Supply System shown in Figure 1.3.2-10, draw the fault tree for the top event 'No Flow from Valve M'. Consider only those component failure modes shown in Table 1.3.2-1.
5. Examine the fault tree in Question 4 for non-coherence in repair.
6. Determine the minimal cut sets for the fault tree in Question 4.

Table 1.3.2-1   Component Failure Modes

Event*	Component fault
A	Reservoir level too low
B, C	Inlet Pipe to Pump ruptures
D, E	Pump/Motor unit fails to start on demand
F, G	Cable fault between Pump/Motor and Power Source
H, J	Switch fails to close on demand
K	Bus Bar/Power Source fails on demand
L	Manifold connecting Pumps and Valve ruptures
M	Valve fails to open on demand

* Each event is considered to be a total failure of the component. Reduced performance of equipments is not considered. It is assumed that the demand generates perfect signals where required, and that there are no earth faults on the electrical supply system or pipe/manifold blockages. Reverse flow through a pump is not considered.

References 1.3.2

1. *Reactor Safety Study – an assessment of accident risks in US commercial nuclear power plants.* WASH–1400 (NUREG–75/014) 1975.
2. Farmer, F. R. 'Siting criteria – a new approach'. IAEA Symposium on the Containment and Siting of Nuclear Power Reactors. Vienna, April 1967 SM/89/34.
3. Kumamuto, H. and Henley, E. J. 'Top-down algorithm for obtaining prime implicant sets of non-coherent fault trees.' *IEEE Trans. on Reliability* Vol. R-27, No. 4, October 1978.
4. Locks, M. O. 'Fault trees, prime implicants and non-coherence.' *IEEE Trans. on Reliability.* Vol. R-29, No. 2, June 1980.
5. Vesely, W. E. 'A time–dependent methodology for fault tree evaluation.' *Nuc. Eng. Des.,* **13,** 337–360, 1970.
6. Murchland, J. D. 'Comment on a time–dependent methodology for fault tree evaluation.' *Nuc. Eng. Des.* **22,** 167–172, 1972.
7. Barlow R. E., Fussell, J. B., and Singpurwalla, N. D. (Eds) *Reliability and Fault Tree Analysis.* SIAM 1975.
8. Fussell, J. B. and Campbell, D. J. 'System reliability engineering – a nuclear industry perspective.' Paper 6/3 Third National Reliability Conference, Birmingham UK 1981. (Institute of Quality Assurance, London).
9. Edwards, G. T. and Watson, I. A. *A Study of Common-Mode Failures.* SRD R 146 UKAEA, 1979.
10. Parry, G. W., Shaw, P., and Worledge, D. H. *The Tolerance–Confidence Relationship and Safety Analysis.* SRD R 129 UKAEA. 1979.

11. Burdick, G. R., Fussell, J. B., Rasmuson, D. M., and Wilson, J. R. 'Phased mission analysis: a review of new developments and an application.' *IEEE Trans. on Reliability*, Vol. R-26, No. 1, April 1977.
12. Caldarola, L. *Boolean Algebra with Restricted Variables*. KfK 2915, 1980.
13. *HTGR Accident Initiation and Progression Analysis Status Report*. GA-A15000 (UC-77) April 1978.

High Risk Safety Technology
Edited by A. E. Green
© 1982 John Wiley & Sons Ltd

SECTION 1.3.3 *Fault Analysis*

A. Aitken

System Response to Fault Condition

When the various faults have been identified, by use of fault and event trees or by any other means, progression of such faults requires to be studied in detail. Usually, for a dynamic system, a mathematical model is set up to obtain a measure of the behaviour of the system under normal control conditions and it is also used to simulate the effects of faults on the system. Particularly noteworthy are faults whose unchecked progression could lead to some limit being exceeded, which in turn might lead to a hazard as discussed in Section 1.3.1. Adequate protection against such faults needs to be provided. For slowly developing faults it may be sufficient to alert the operator via suitable alarms and he can then take action through appropriate controls, but for fast-acting faults automatic protection will be necessary.

Careful consideration needs to be given to the form of protection provided against each fault condition and whether its objectives are adequately met. The objectives may be defined in terms of properties of materials used in construction of, for instance, a nuclear reactor core, which may have a limit on the temperature of the fuel cladding, or the secondary containment, where pressure may be a limiting feature. These and other relevant limits define the requirement for protection and should in the ideal state represent the basis for design of the protective system. In practice, development of the mathematical model and the design of instrumentation and protective devices are concurrent and it is prudent, whereever possible, to allow a good safety margin on both.

Mathematical modelling should include, where appropriate, algorithms for performance of the protective devices so that the full feedback cycle from initiation of the fault condition to shutdown of the process is obtained. It is usually advantageous to repeat transients over a range of parameters and settings of trip devices instead of taking single value estimates. Of necessity, single value estimates for high risk systems would be pessimistic. In addition repetition over a range of values leads to better understanding of the operational sensitivity of the complete system.

A typical example of the information to be gained by analysis in this way is given in the graph of Figure 1.3.3-1 which shows a temperature transient on one type of reactor system due to a loss of coolant flow. Various delays, representing differences in settings of the trip devices, indicate that the performance required to prevent the temperature exceeding X °C has to be better than 10 s in total delay time, from initiation of fault to the start of action to terminate the excursion. A family of curves will be obtained when the mathematical model's coefficients are changed and it may happen that a setting producing no more than 5 s delay may be necessary to cater for all conditions that may arise.

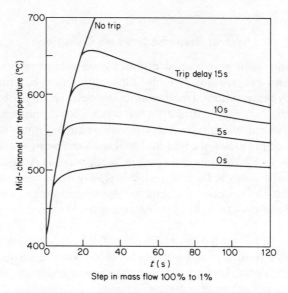

Figure 1.3.3-1 Typical transient, bottom duct fracture, gas-cooled reactor

Time Response

The total delay in the fault example given may consist of several delays lumped together, e.g.
1. Time to exceed the trip setting.
2. Allowance for trip setting accuracy.
3. Transfer function of the sensor and associated amplifier.
4. Fixed delays in operation of the trip system.

The analyst has to be careful that everything is included, particularly when the safety margin is low. In addition it may happen that the actual measurement made is not the one of primary interest but can only be correlated to the critical parameter. A good example in a nuclear reactor is that perhaps the central fuel or fuel cladding temperature limit may be important but, due to engineering difficulties, a measurement can only be made of coolant outlet temperature from the fuel channel. Assurance has to be obtained that the correlation holds for all conditions during the transient. In general such a correlation implies a longer time delay before protective action can take place.

For the example given in Figure 1.3.3-1 tripping on a measurement of coolant temperature would involve both the response time of the sensing device, e.g. thermocouple (0.1 s to 7 s typical range) and the transport time for the coolant to travel from the heated fuel element to the point of measurement (possibly seconds at the much reduced flow rate).

On the other hand, a measurement of channel flow, either directly by a flowmeter, or indirectly by measurement of differential pressure across the reactor, could give a rapid response, initiating trip action in less than 1 s.

In practice, both methods would be used to increase the probability of tripping the reactor.

Figures 1.3.3-2 and 1.3.3-3, taken from Reference 1, illustrate a good example of the use of redundant and diverse measurements to protect a potentially hazardous plant, in this case producing heavy organic chemicals. About 50

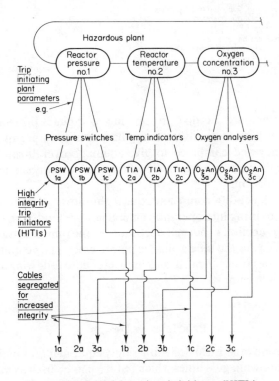

Figure 1.3.3-2 High integrity trip initiators (HITIs)

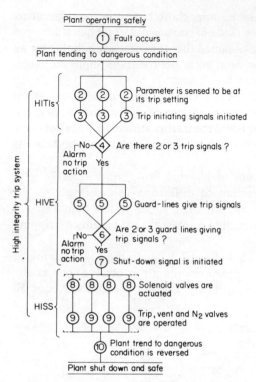

Figure 1.3.3-3 Sequence of operation leading to a trip

tripping parameters are needed to provide complete protection and the illustrations show how these diverse tripping parameters are monitored in 2-from-3 redundancy and combined to give adequate overall safety protection, while at the same time minimizing spurious tripping from any single fault element in the protective system.

Following a fault condition and consequent shutdown of a process there may be a requirement to bring into operation emergency or standby equipment which is not called on to function in normal operation. Some time constraints may need to be considered, e.g. in the time available before emergency cooling circuits are needed to remove decay heat from a nuclear reactor core following a loss of power supplies to the main pumps.

Principles in Design

Examination of good practice in designing safety equipment for high risk systems has yielded the following guidelines which in turn can be used as basic criteria in assessment[2][3].

1. Wherever possible a direct measurement should be made of the critical or limiting parameter or some closely related function.
2. Preference should be given to robust and reliable components of proven performance, particularly where they are required to operate in a hostile environment.
3. Facilities for proof testing and maintenance should be provided.
4. Equipment used for control purposes should not also be used for safety.
5. The protective system design should meet reliability requirements as defined by the safety criteria.

Normally in a high risk system redundancy (using more than one of the same kind of function) and diversity (performance of the same overall function by independent and different means) are employed throughout the safety equipment. Reliability assessment is dealt with in more detail in the next section (1.3.4) but ergonomic aspects require some attention. It is important that settings for safety control equipment are maintained at all time and generally controls, other than for continuous operation, are locked away to prevent interference. For example trip setting controls are only accessible for proof testing and calibration.

There must always be a good presentation to the operator of the state of the plant which he has to control and this also implies, in addition to normal instruments for control purposes, providing information on the status of the safety equipment. A good alarm system can alert the operator to take corrective action before the plant can get into a hazardous state. Sophisticated disturbance analysis systems are being developed to provide assistance to the operator under normal working as well as incident conditions.

However, any action through the operator is on a time scale which may be too protracted for some fault conditions and there is always a need for automatic protection, even if fault transients are on a long time scale, on the grounds that the operator may not be aware of gradual degradation or even that the information presented to him may be wrong, perhaps because of a peculiar condition in the process being controlled. ·

Response of the Operator to Alarms

Under any abnormal condition in a process the first intimation to the operator is usually from a suitable alarm in which a warning lamp (sometimes flashing) and also an audible warning are actuated. What the operator does with this information depends on the following:

1. The process he is controlling.
2. His knowledge of the state of the plant.
3. Prior knowledge of the effects of control action.
4. Human factors relating to the operator himself.

The dynamics of the process will dictate what kind of action is required, and knowledge of the plant state, gained from instrument systems providing displays, can often point the way for the operator to take the appropriate control action. Another important feature may be the training of the operator which may be on a plant simulator. For some extremes in plant conditions simulation is the only reasonable way to give operators the relevant experience. The assessor then may need to see how closely the simulator model adheres to what is reckoned to be real plant performance. Some simulators combine an actual control room with simulation of plant variables as inputs to the normal displays. These of course provide only for the pre-startup period since, during operation, the control room will be connected to the newly built plant. Extreme conditions can then only be simulated away from the control room proper and this is frequently done in a mock-up situation, e.g. for training aircraft pilots or nuclear reactor operators.

Most large installations built within the last decade use a digital computer which provides data reduction and display facilities, information being presented on a CRT. Different types of display can be called up by the operator, but on receipt of an alarm signal there is usually an automatic alarm display with alarm analysis as an additional aid to the operator. Some systems now being developed will go further in providing a cause–consequences diagram projecting forward the progression of a disturbance if unchecked.

Computerized data reduction and disturbance analysis systems, however, have a built in time delay due to the sampling interval which can be a nominal 5 s. Also such systems only deal with already predetermined sequences obtained from the dynamic analysis of the complete plant, perhaps before it is built and commissioned.

Questions 1.3.3

1. What are the basic characteristics of performance which require to be considered in the operation of an automatic protective system for shutting down a nuclear reactor under fault conditions?
2. Discuss the guidelines for the principles of design which should be considered in the assessment of the type of protective system mentioned in Question 1.
3. What is the role of operator alarms provided on a protective system?

References 1.3.3

1. Stewart, R. M. and Hensley, G. 'High integrity protective systems on Hazardous chemical plants.' Paper presented to the European Nuclear Energy Agency Committee on Reactor Safety Technology, at Munich, May 1971.
2. Green, A. E. 'The reliability assessment of automatic protective systems for the safety of nuclear reactors.' *Proceedings Annual Reliability and Maintainability Symposium, USA.* January 1981.
3. Green, A. E. *Safety Systems Reliability.* John Wiley & Sons, 1983.

High Risk Safety Technology
Edited by A. E. Green
© 1982 John Wiley and Sons Ltd

SECTION 1.3.4 *Probability Assessment*

A. J. Bourne

Introduction

The previous sections in this chapter have examined the general concepts of risk assessment, the approaches involved in setting up the appropriate event trees and fault trees and the overall aspects of associated fault analysis.

Within these approaches it has been shown that there is a need to quantify the fault behaviour of hazardous plants in terms of the overall risk. This quantification is of a probabilistic nature and leads to such concepts as the expected frequency or probability of occurrence of events.

It is the purpose of this sub-section to discuss some of the typical probabilistic models that may be used in the safety assessment of engineered systems. The probabilistic fault performance of these systems being of importance in the analytical build up of the event trees and fault trees which have been discussed in the previous sub-sections.

The initial steps in the reliability or safety analysis of a system are:
1. To define the system and its boundaries,
2. To establish the system's functional purpose and performance requirements.

The conceptual model for this purpose is illustrated in Figure 1.3.4.-1.

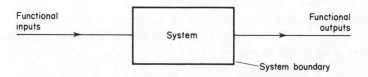

Figure 1.3.4-1 Overall system concept

The reliability of such a generalized system may then be considered as the chance or probability that the defined system functions as intended under the conditions and over the time scale of interest.

In establishing this probability of success it is necessary to examine the ways in

73

which the system may *fail* to fulfil its intended purpose. It is also necessary to examine whether the system can produce any unintended, spurious or unacceptable outputs which may also be considered as fault, or possibly accident, conditions as shown in Figure 1.3.4-2.

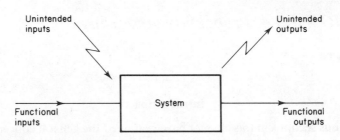

Figure 1.3.4-2 Unintended inputs and outputs

This results in two major classifications of system fault conditions:

1. Fault conditions which prevent or restrict the system from achieving its intended outputs.
2. Fault conditions which produce unwanted or spurious outputs (e.g. noise, vibration, explosion, fire, release of toxic materials, etc.).

The first set of these fault conditions is generally of concern in the overall system availability analysis or probability of success on demand while the second set of fault conditions is generally of interest in the overall safety orientated effect that the system may have on higher events in the fault tree.

With either availability or safety assessment the overall process of fault analysis is to identify all the relevant fault conditions and to evaluate their probability of occurrence. The identification leads, in its turn, to an examination of:

CAUSE and EFFECT

and the logical relationships between these aspects within the total system configuration as a result of the various possible 'states' (fault states or other states) which may exist in the elements of the system.

Measurement of Success or Failure

Obviously the precise meaning of success or failure needs to be defined for a particular situation or system before a probabilistic measure can be ascribed to this type of outcome.

The meaning of success generally arises from the detailed evaluation of the answers to two basic questions:

1. How is the system required to perform under all the relevant conditions? (Performance specification)
2. How is the system likely to perform under all the relevant conditions? (Performance achievement)

By way of simple example, suppose that a system is required to produce a power output, x, lying within a specified range such that:

$$x_1 = \text{lower limit of acceptability}$$
$$x_u = \text{upper limit of acceptability.}$$

Suppose, also, that the variations in the likely achieved values of power output can be represented by a probability distribution (probability density function), $f(x)$, then the reliability of the situation or the probability of success can be expressed as:

$$R = \int_{x_1}^{x_u} f(x)\,dx \qquad (1.3.4\text{-}1)$$

which may be illustrated as shown in Figure 1.3.4-3.

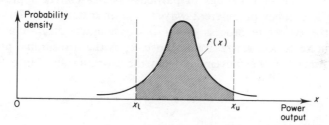

Figure 1.3.4-3 Spectrum of achieved performance against fixed required limits

The pattern of variation illustrated in Figure 1.3.4-3 can often be approximated by a simple two-state probabilistic model in those cases where only 'complete' or 'catastrophic' failure is of concern. This simple two-state model reduces the considerations shown in Figure 1.3.4-3 to those depicted in Figure 1.3.4-4.

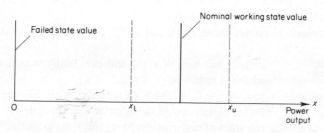

Figure 1.3.4-4 Simple two-state model of performance variation

In this case, only two complementary and discrete probabilistic measures arise, namely:

Probability of being in the failed state (F) $= p$

Probability of being in the working state (W) $= \bar{p}$

Changes of State

Overall concepts

In the two-state situation the values of p or $\bar{p}$ may be determined by direct observation of the outcomes from the events or system components of interest. Often, however, the data or information available may describe the patterns of change from one state to another or may give some indication of the mean dwell times in each state. In the latter case, the value of p will need to be deduced or calculated from the type of information available.

Changes of state or dwellings in particular states can be represented or modelled in a number of different ways[1-4]. For instance, a Markov chain model for the two-state process might be represented diagrammatically as shown in Figure 1.3.4-5 where, for instance, α_{wf} is the transitional probability at any one time for the movement from the working state into the failed state.

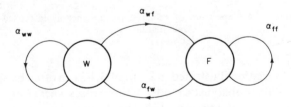

Figure 1.3.4-5 Transitions for a two-state model

For the present purpose, three principal types of changes of state will be considered;

Irreversible   – system starts in W state and eventually moves to F state where it remains,

Partly reversible   – restoration from the F state to the W state is only possible under certain conditions or at certain times.

Reversible   – the start of restoration from the F state to the W state is possible each time the F state is entered.

Irreversible changes of state

The system or element starts in the W state and after its first failure it remains in the F state. This situation applies broadly to 'one-shot' systems or for system components which are not deemed repairable.

The picture of events in the time domain may be as shown in Figure 1.3.4-6.

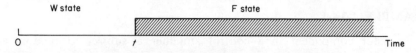

Figure 1.3.4-6 Irreversible change-of-state in the time domain

The value of p (the probability of being in the F state) is obviously dependent upon t which, in most cases, is a random variable. This random variable may be modelled by a probability density function $f_x(t)$ which is the distribution of times to first failure of the device or system being considered.

Often, in practice, the times to failure can be adequately represented by assuming that $f_x(t)$ takes the form of the simple exponential distribution, i.e.

$$f_x(t) = \theta e^{-\theta t} \qquad (1.3.4\text{-}2)$$

where

$$\theta = \text{mean constant rate of failure (failure-rate)}$$

The probability of being in the F state at time t (which, for the irreversible process, is the same as the probability of failure by time t) is given by the cumulative distribution function, $p_x(t)$, i.e.

$$p = p_x(t)$$

$$= \int_0^t f_x(t)\,dt$$

$$= \int_0^t \theta e^{-\theta t}\,dt$$

$$= 1 - e^{-\theta t} \qquad (1.3.4\text{-}3)$$

and this may often be further approximated to simply:

$$p \simeq \theta t \qquad (1.3.4\text{-}4)$$

when $\theta t \ll 1$.

Also, the probability of being in the W state at time t or the probability of survival up to time t is:

$$\bar{p} = 1 - p$$

$$= e^{-\theta t} \qquad (1.3.4\text{-}5)$$

and this is often known as the 'survival function' or the 'reliability function'.

With the simple exponential distribution, the reciprocal of θ is the mean value of the distribution and so represents the mean time to failure, λ, i.e.

$$\lambda = \frac{1}{\theta} \qquad (1.3.4\text{-}6)$$

EXAMPLE 1.3.4-1

A transistor component in a control amplifier is subject to random and catastrophic failure. Over a large number of such components it has been deduced that the times to catastrophic failure are exponentially distributed with a mean time to failure of 10^6 h. What is the probability that such a transistor will fail within periods of
(a) 200 h
(b) 5000 h
(c) 100 000 h?

The times, t, to the catastrophic failures of the transistor components are exponentially distributed. Hence, the probability density, $f(t)$, function for times to failure is given by:

$$f(t) = \frac{e^{-t/\lambda}}{\lambda}$$

where the mean time to failure, λ, is

$$\lambda = 10^6 \, \text{h}$$

The probability that a transistor will fail within a given time period, t, is given by the cumulative distribution function, $p(t)$, where:

$$p(t) = \int_0^t f(t) \, \mathrm{d}t$$

$$= 1 - e^{-t/\lambda}$$

$$\simeq \frac{t}{\lambda} \quad \text{if} \quad \frac{t}{\lambda} \ll 1$$

(a) For $t = 200$ h

$$p(t) = 0.0002$$

(b) For $t = 5000$ h

$$p(t) = 0.005$$

(c) For $t = 100\,000$ h

$$p(t) = 0.095$$

Partly reversible changes of state

The principal interest in this situation is where the restoration from the F state to the W state is only possible at certain, and perhaps regular, times. This may apply where system elements or components can only be repaired at times of scheduled shutdown or planned maintenance. More usually, however, the interest centres on protective systems or standby systems. Because these systems are normally in a dormant or passive role their current state may not be immediately apparent. Unrevealed faults may occur which can only be recognized and rectified when the system is tested on some regular basis.

The picture of events in the time domain may be as shown in Figure 1.3.4-7.

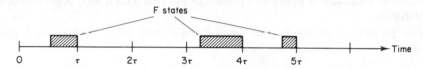

Figure 1.3.4-7 Partly reversible change-of-state in the time domain

The time at which the unrevealed faults occur after a previous test is still a random variable but now the dwell time in the F state is also influenced by the time interval, τ, between the routine tests.

The probabilistic measure of main interest with protective or standby systems tends to be the average probability of being in the failed state at any time. This, under certain conditions, will be synonymous with the mean unavailability (or mean fractional dead time) of the system.

It can be shown (References 5 and 6), given perfect testing and restoration, that this mean fractional dead time, μ_D, is given by:

$$\mu_D = \frac{1}{\tau} \int_0^\tau p_x(t)\, dt \qquad (1.3.4-7)$$

where:

$$p_x(t) = \text{the cumulative distribution function of times to failure.}$$

If, by way of illustration, the times to failure can be represented by the simple exponential distribution then:

$$p_x(t) = 1 - e^{-\theta t} \qquad (1.3.4-8)$$
$$\simeq \theta t$$

and:

$$\mu_D = \frac{1}{\tau} \int_0^\tau \theta t \, dt$$

$$\simeq \frac{\theta \tau}{2} \qquad (1.3.4\text{-}9)$$

Reversible changes of state

This situation deals with entries into the F state which are immediately recognized (revealed faults) and where some restoration process is started directly the F state entry takes place. Basically it can be used to model normal on-line processes, continuously operated systems or systems with a high degree of self-monitoring.

The picture of events in the time domain may be as shown in Figure 1.3.4-8.

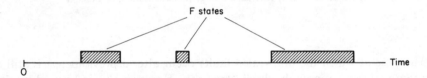

Figure 1.3.4-8 Reversible changes-of-state in the time domain

Once again the main probabilistic measure of interest tends to be the overall mean unavailability (average probability of being in the F state) or, its complement, the overall mean availability.

The situation may be modelled with a knowledge of the distribution(s) of times to failure since the last restoration, $f_f(t)$, and the distribution(s) of times to restore, $f_r(t)$.

The effective 'age' of the system following a restoration process is an important aspect of the modelling process. For instance, the restoration may leave the system 'as good as new', 'as good as old' or at some age mixture in between.

For the 'as good as new' situation (or alternating renewal process as it is sometimes described) it can be shown (Reference 6) that the mean unavailability fairly quickly reaches a steady-state constant value given by:

$$\mu_D = \frac{\mu_r}{\mu_f + \mu_r} \qquad (1.3.4\text{-}10)$$

where:

$$\mu_f = \text{mean of the distribution } f_f(t) \text{ or}$$
the mean time to failure
$$\mu_r = \text{mean of the distribution } f_r(t) \text{ or}$$
the mean time to restore

Similarly,

$$\mu_A = \frac{\mu_f}{\mu_f + \mu_r} \qquad (1.3.4\text{-}11)$$

where both μ_D and μ_A are constant values.

Where μ_f is very much greater than μ_r, the mean unavailability, or mean fractional dead time, may be approximated by:

$$\mu_D \simeq \frac{\mu_r}{\mu_f} \qquad (1.3.4\text{-}12)$$

If the failure and restoration processes follow simple exponential distributions with, say, mean failure-rate θ and mean time to restore τ_r, then:

$$\mu_D \simeq \theta \tau_r \qquad (1.3.4\text{-}13)$$

NB   In this case both the 'good as new' and 'good as old' assumptions lead to the same answer.

Failure Parameters of Interest

It can be seen that the main probabilistic measures of failure for the two-state situation and changes-of-state models that have been discussed may be summarized as follows:

p probability of failure per event or occasion or probability of being in the F state
$p_x(t)$ cumulative probability of failure by time t
θ mean failure rate
μ_D mean unavailability or mean fractional dead time
μ_A mean availability.

Some or all of these may be required as input measures into the various basic events of the fault tree or other similar logical representations of system failure.

Probabilistic Combinations Required

It has been seen that the probabilistic measures, that may be associated with individual events in a fault tree, include such parameters as mean failure rate, probability of failure per demand, probability of failure with respect to time, mean restoration time, and mean unavailability.

It has also been seen from the previous sub-sections that typical fault trees contain combinations of events. Principally, these combinations are of the logical 'AND' or 'OR' type, but other logical combinations such as the 'EXCLUSIVE OR' and 'MAJORITY-VOTE' may also be included.

The logical combinations and their manipulation may be studied in the context of Boolean algebra (see Reference 6). However, it is also necessary to consider the corresponding combinations of the appropriate probabilistic measures.

Rules of Probabilistic Combinations

It can be shown that if an event A is statistically and physically independent of an event B then the probability of occurrence of the combined event A AND B is given by:

$$p(\text{A and B}) = p(\text{A})p(\text{B}) \tag{1.3.4-14}$$

If two elements, 'a' and 'b', are now considered (see Figure 1.3.4-9) where each element can be in either the failed state or working state at any time,

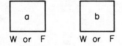

Figure 1.3.4-9 A two-element, two-state system

and if,

$$\text{probability of 'a' in F state} = p_a$$
$$\text{probability of 'b' in F state} = p_b$$

then a Truth Table (see Table 1.3.4-1) may be drawn up showing the Boolean combinational and probability combinational expressions for each combined state.

Table 1.3.4-1

Combined state no.	Element states		Boolean combination	Probability combination
	a	b		
1	W	W	$\overline{A}.\overline{B}$	$(1 - p_a)(1 - p_b)$
2	W	F	$\overline{A}.B$	$(1 - p_a)p_b$
3	F	W	$A.\overline{B}$	$p_a(1 - p_b)$
4	F	F	$A.B$	$p_a p_b$

Each of the combined states one to four in Table 1.3.4-1 is mutually exclusive of the rest and the total number of combined states (four in this case) are exhaustive. Hence, the simple summation of the probability combinations in the last column is unity as would be expected.

The probability of any set of combined states can be obtained by the summation of the probabilistic expression for each mutually exclusive state in the set.

For instance,

$$p(\text{State 2 or 3}) = (1 - p_a)p_b + p_a(1 - p_b)$$
$$= p_a + p_b - 2p_a p_b \qquad (1.3.4\text{-}15)$$

'OR' Combinations

State equations

As seen previously, the fault-tree diagram for a simple 'OR' combination of two elements is as shown in Figure 1.3.4-10.

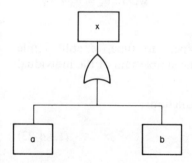

Figure 1.3.4-10 Two-element, OR-gate combination

The combination corresponds to States 2, 3 or 4 in Table 1.3.4-1 (i.e. a failure at 'x' occurs if 'a' fails or 'b' fails or both fail).

In Boolean notation, this leads the expression for the combined failure at X as follows:

$$X = \overline{A}.B + A.\overline{B} + A.B$$
$$= A + B \qquad (1.3.4\text{-}16)$$

which, in probabilistic terms, becomes:

$$p_x = (1 - p_a)p_b + p_a(1 - p_b) + p_a p_b$$
$$= p_a + p_b - p_a p_b \qquad (1.3.4\text{-}17)$$

Alternatively this may be written as:

$$p_x = 1 - (1 - p_a)(1 - p_b) \qquad (1.3.4\text{-}18)$$

and, for n elements combined in an OR function

$$p_x = 1 - \prod_{j=1}^{n} (1 - p_j)$$

$$\simeq \sum_{j=1}^{n} p_j \qquad \text{if all } p_j \ll 1 \qquad (1.3.4\text{-}19)$$

Irreversible changes of state

Let the probability of failure with respect to time for both elements follow a simple exponential distribution and assume no restoration, then:

$$p_a = 1 - e^{-\theta_a t} \qquad (1.3.4\text{-}20)$$

$$p_b = 1 - e^{-\theta_b t} \qquad (1.3.4\text{-}21)$$

whence:

$$p_x = 1 - e^{-\theta_a t} e^{-\theta_b t}$$

$$= 1 - e^{-\theta_x t} \qquad (1.3.4\text{-}22)$$

$$\text{where } \theta_x = \theta_a + \theta_b$$

Hence:

The combined failure probability, with respect to time, is still simple exponential and the combined failure rate is the simple sum of the individual element failure rates.

Generally, for n elements combined in an OR function:

$$\theta_x = \sum_{j=1}^{n} \theta_j \qquad (1.3.4\text{-}23)$$

Partly reversible changes of state

As seen previously, the mean unavailability or mean fractional dead time of element x can be evaluated from:

$$\mu_{D_x} = \frac{1}{\tau} \int_0^\tau p_x(t)\,dt \qquad (1.3.4\text{-}24)$$

where $p_x(t)$ is the cumulative distribution function for times to failure of element x and τ is the length of the minimum interval over which the test pattern is cyclic.

For two elements combined in an OR function:

$$\mu_{D_x} = \frac{1}{\tau} \int_0^\tau [1 - \overline{p_a(t)} \cdot \overline{p_b(t)}]\,dt \qquad (1.3.4\text{-}25)$$

Where times to failure follow a simple exponential distribution and where each element is tested at the same time, this expression will approximately reduce to:

$$\mu_{D_x} \simeq \frac{(\theta_a + \theta_b)\tau}{2} \qquad (1.3.4\text{-}26)$$

Under the same conditions, the mean or expected number of failures over a total time T will become:

$$\mu_{N_x} \simeq (\theta_a + \theta_b)T \qquad (1.3.4\text{-}27)$$
$$\text{if } (\theta_a + \theta_b)\tau \ll 1$$

Reversible changes of state (alternating renewal process)

For this process, and under steady-state conditions
   (probability of being in F state at any time)
$$\equiv (\text{mean unavailability})$$
i.e.

$$P_x \equiv \mu_{D_x} \qquad (1.3.4\text{-}28)$$

Therefore since,

$$P_x = P_a + P_b - P_a P_b \qquad (1.3.4\text{-}29)$$

then

$$\mu_{D_x} = \mu_{D_a} + \mu_{D_b} - \mu_{D_a}\mu_{D_b}$$
$$\simeq \mu_{D_a} + \mu_{D_b} \qquad (1.3.4\text{-}30)$$
$$\text{if } \mu_D\text{'s are small}$$

where, as seen previously,

$$\mu_{D_a} = \frac{\mu_{r_a}}{\mu_{f_a} + \mu_{r_a}} \simeq \frac{\mu_{r_a}}{\mu_{f_a}} \qquad (1.3.4\text{-}31)$$

and

$$\mu_{D_b} = \frac{\mu_{r_b}}{\mu_{f_b} + \mu_{r_b}} \simeq \frac{\mu_{r_b}}{\mu_{f_b}} \qquad (1.3.4\text{-}32)$$

Under similar approximations, the mean or expected number of failures over a total time T is:

$$\mu_{N_x} \simeq T\left(\frac{1}{\mu_{f_a}} + \frac{1}{\mu_{f_b}}\right) \qquad (1.3.4\text{-}33)$$

'AND' Combinations

State equations

As seen previously, the fault-tree diagram for a simple 'AND' combination of two elements is as shown in Figure 1.3.4-11.

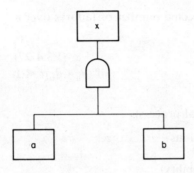

Figure 1.3.4-11 Two-element AND-gate combination

The combination corresponds to State 4 in Table 1.3.4-1 (i.e. a failure at 'x' occurs if both 'a' and 'b' fail).

In Boolean notation this yields:

$$X = A.B \qquad (1.3.4\text{-}34)$$

and probabilistically:

$$p_x = p_a p_a \qquad (1.3.4\text{-}35)$$

(Note that in this case the Boolean and probabilistic expressions are of the same form.)

For *n* elements combined in an AND function

$$p_x = \prod_{j=1}^{n} p_j \qquad (1.3.4\text{-}36)$$

Irreversible changes of state

For simple exponential times to failure where:

$$p_a = 1 - e^{-\theta_a t} \qquad (1.3.4\text{-}37)$$

$$P_b = 1 - e^{-\theta_b t} \qquad (1.3.4\text{-}38)$$

then:

$$p_x = 1 - e^{-\theta_a t} - e^{-\theta_b t} + e^{-(\theta_a + \theta_b)t}$$

$$\simeq \theta_a \theta_b t^2$$

$$\text{if } \theta t \ll 1 \quad (1.3.4\text{-}39)$$

In this case the combined failure probability with respect to time is no longer simple exponential nor is there a simple relationship for combining failure rates.

Partly reversible changes of state

As before, the combined mean unavailability or mean fractional dead time for the element x can be obtained, under the appropriate conditions, from the expression:

$$\mu_{D_x} = \frac{1}{\tau} \int_0^\tau p_x(t)\,dt \qquad (1.3.4\text{-}40)$$

which, in terms of the cumulative distribution functions for times to failure for elements 'a' and 'b', becomes:

$$\mu_{D_x} = \frac{1}{\tau} \int_0^\tau p_a(t).p_b(t).dt \qquad (1.3.4\text{-}41)$$

Where times to failure follow a simple exponential distribution and where each element is tested at the same time, this expression will approximate to:

$$\mu_{D_x} \simeq \frac{\theta_a \theta_b \tau^2}{3}$$

$$\simeq \frac{\theta^2 \tau^2}{3}$$

$$\text{where } \theta_a = \theta_b = \theta \quad (1.3.4\text{-}42)$$

and, under the same conditions, for n elements combined in an AND function:

$$\mu_{D_x} \simeq \frac{\theta^n \tau^n}{n+1} \qquad (1.3.4\text{-}43)$$

and:

$$\mu_{N_x} \simeq T\theta\tau^{n-1} \qquad (1.3.4\text{-}44)$$

Reversible changes of state (alternating renewal process)

As before, for this case:

$$\mu_{D_x} \equiv p_x \qquad (1.3.4\text{-}45)$$

Hence:

$$\mu_{D_x} = \mu_{D_a}\mu_{D_b} \qquad (1.3.4\text{-}46)$$

For a two-element AND combination, both elements have to be in the failed state at the same time to produce a combined failure at 'x'. This obviously affects the consideration of the mean number of failures that are likely to occur over some total time T.

The mean number of failures of element 'a' in a total time T is approximately:

$$\mu_{N_a} \simeq \frac{T}{\mu_{f_a}} \tag{1.3.4.-47}$$

but only a proportion of those which fall in the dead time of element 'b' constitute a combined failure. So:

$$(\mu_{N_x})_a \simeq \frac{T}{\mu_{f_a}} \times \mu_{D_b} \tag{1.3.4-48}$$

Similarly:

$$(\mu_{N_x})_b \simeq \frac{T}{\mu_{f_b}} \times \mu_{D_a} \tag{1.3.4-49}$$

Now:

$$\mu_{N_x} \simeq (\mu_{N_x})_a + (\mu_{N_x})_b \tag{1.3.4-50}$$

Therefore:

$$\mu_{N_x} \simeq T\left(\frac{\mu_{D_b}}{\mu_{f_a}} + \frac{\mu_{D_a}}{\mu_{f_b}}\right)$$

$$\simeq \frac{T}{\mu_{f_a}\mu_{f_b}}(\mu_{r_a} + \mu_{r_b}) \tag{1.3.4-51}$$

Or, for n similar elements combined in an AND function:

$$\mu_{N_x} \simeq \frac{Tn\mu_r^{n-1}}{\mu_f^n} \tag{1.3.4-52}$$

Majority-vote Combinations

State equations

Generally, for an 'n' element system, the majority-vote principle requires that 'm' or more ($m \leqslant n$) of the 'n' elements need to be successful in order to achieve system success. These are normally described simply as m-out-of-n systems.

From the failure point of view, the same systems require 'r' or more elements to fail in order to produce a system failure.

Hence a system may be described as:

$$m\text{-out-of-}n \text{ for success}$$
$$\text{or}\quad r\text{-out-of-}n \text{ for failure}$$
$$\text{where}\quad r = n - m + 1 \tag{1.3.4-53}$$

The simplest majority-vote configuration is the 2-out-of-3 system where:

$$n = 3$$
$$m = 2$$
$$r = 2$$

A fault tree for this system could be drawn as shown in Figure 1.3.4-12, but, it is often more convenient to use a special majority-vote gate as illustrated in Figure 1.3.4-13.

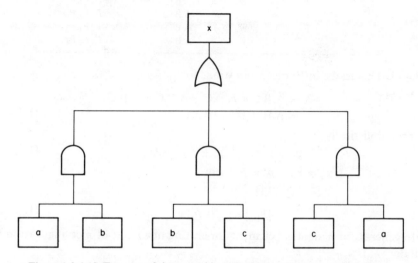

Figure 1.3.4-12 Two-out-of-three combination using AND-gates and OR-gates

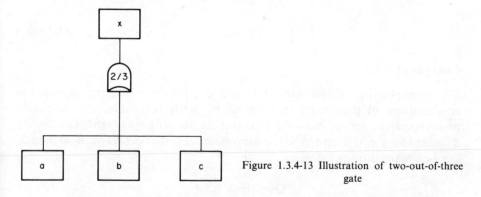

Figure 1.3.4-13 Illustration of two-out-of-three gate

The probability state equations for a 2-out-of-3 system can be derived from a truth table as shown in Table 1.3.4-2.

System, or combined state, failure corresponds to States 4, 6, 7 or 8 in Table

Table 1.3.4-2

Combined state no.	Element states			Boolean combination	Probability combination
	a	b	c		
1	W	W	W	$\overline{A}\cdot\overline{B}\cdot\overline{C}$	$\overline{P}_a\overline{P}_b\overline{P}_c$
2	W	W	F	$\overline{A}\cdot\overline{B}\cdot C$	$\overline{P}_a\overline{P}_b P_c$
3	W	F	W	$\overline{A}\cdot B\cdot\overline{C}$	$\overline{P}_a P_b\overline{P}_c$
4	W	F	F	$\overline{A}\cdot B\cdot C$	$\overline{P}_a P_b P_c$
5	F	W	W	$A\cdot\overline{B}\cdot\overline{C}$	$P_a\overline{P}_b\overline{P}_c$
6	F	W	F	$A\cdot\overline{B}\cdot C$	$P_a\overline{P}_b P_c$
7	F	F	W	$A\cdot B\cdot\overline{C}$	$P_a P_b\overline{P}_c$
8	F	F	F	$A\cdot B\cdot C$	$P_a P_b P_c$

1.3.4 -2. This leads, in Boolean notation, to:

$$X = \overline{A}.B.C + A.\overline{B}.C + A.B.\overline{C} + A.B.C$$
$$= A.B. + B.C. + C.A \qquad (1.3.4\text{-}54)$$

or probabilistically:

$$p_x = p_a p_b + p_b p_c + p_c p_a - 2p_a p_b p_c \qquad (1.3.4\text{-}55)$$

If $p_a = p_b = p_c = p$, then

$$p_x = 3p^2 - 2p^3$$
$$\simeq 3p^2 \qquad \text{if } p \ll 1 \qquad (1.3.4\text{-}56)$$

In general, for n identical elements operating in a r-out-of-n configuration for failure:

$$p_x = \sum_{j=r}^{n} \binom{n}{j} p^j (1-p)^{n-j}$$

$$\simeq \binom{n}{r} p^r \quad \text{if } p \ll 1 \qquad (1.3.4\text{-}57)$$

Changes of state

The corresponding relationships and equations for describing the appropriate combinations of parameters in irreversible, partly reversible, and reversible processes can be derived in a similar manner for majority-vote systems as already discussed for the 'OR' and 'AND' combinations. Further details of these models can be obtained from Reference 6.

Questions 1.3.4

1. A standby generator is only tested and maintained every year. In between tests it is subject to unrevealed failures. The times to the occurrence of unrevealed failures are exponentially distributed with a mean failure rate of 0.1 faults per year. Evaluate the overall:

(a) mean availability

(b) mean fractional dead time (unavailability).

2. An electrical supply system is subject to failure which causes loss of supply to a process plant. The mean time between such failures is 497 h and the mean time to repair the failures and restore the supply is 3 h. What is the average value of the availability of the supply to the process plant?

3 A control system consists of a measuring device and an actuator. The times to failure for each device are exponentially distributed. The mean time to failure for the measuring device is 5000 h and that for the actuator is 4000 h. If either unit fails the system remains in the failed state. What is

(a) the mean time to system failure?

(b) the probability of the system being in the failed state after 2000 h?

4. An electric motor contains 30 main components each of which is subject to an alternating renewal process such that when a component fails it is replaced by a new one. The mean time to failure of each component is 20 000 h and the mean time to replace it with a new one is 2 h. If the motor is out of action each time a component is being replaced, what is the mean fractional dead time?

5. A temperature measurement system consists of two nominally identical chains of measurement any one of which is adequate to provide the required information. Each chain of measurement is subject to an alternating renewal process. The means of the distributions of times to failure and times to repair for each chain of measurement are 2000 h and 20 h respectively. What is the mean unavailability of the system?

6. Three elements of a protective system each have a constant probability of 0.05 of being in the failed state at any time. What is the system probability of being in the failed state if the elements are so connected that system success is achieved when:

(a) any one or more of the three elements are successful,

(b) any two or more of the three elements are successful, and

(c) only all three elements are successful?

7. The position of an aircraft may be obtained by three different and independent navigational aids. Confidence in the accuracy of position is only obtained if any two or more of the different navigational aids produce the same result. Each navigational aid is subject to an alternating renewal process with mean times to failure of 500 h, 600 h and 900 h. The corresponding mean times to repair are 10 h, 20 h and 30 h respectively. What is the mean availability for the system in terms of it producing an acceptable accuracy of the aircraft's position?

References 1.3.4

1. Feller, W. *An Introduction to Probability Theory and its Applications.* John Wiley, New York, 1957.
2. Bazovsky, I. *Reliability Theory and Practice.* Prentice-Hall, Englewood Cliffs, NJ, 1961.

3. Cox D. R. *Renewal Theory*. John Wiley, New York, 1962.
4. Barlow, R. E. and Proschan, F. *Mathematical Theory of Reliability*. John Wiley, New York, 1965.
5. Green, A. E. and Bourne, A. J. *Safety Assessment with Reference to Automatic Protective Systems for Nuclear Reactors*. UKAEA Report No. AHSB (S) R11/, 1966.
6. Green A. E. and Bourne, A. J. *Reliability Technology*. John Wiley, London, 1972.

High Risk Safety Technology
Edited by A. E. Green
© 1982 John Wiley & Sons Ltd

SECTION 1.3.5 Consequence Assessment

G. D. Kaiser

Introduction

The previous sections of this chapter on assessment methodology have been concerned with the examination of sequences of faults in an advanced technological installation and with the estimation of the probability of occurrence of these faults. As has already been shown in Chapter 1.1, however, it is also necessary to calculate the effect of these faults on individuals, society, and the environment in order to assess the level of risk. It is therefore towards the consequences of accidents in nuclear and chemical plant that the considerations of this section are directed.

There are several ways in which the public might be harmed should there be such an accident. There may be an explosion of materials such as TNT or ammonium nitrate, in which case attention has to be directed towards calculating the effect of the resultant pressure waves on people and structures. Hazardous missiles may also be generated. A flammable vapour cloud may be released and the consequences of its deflagration or detonation must be assessed. There may be a fire following the spillage of flammable hydrocarbons or the escape of a flammable cloud and the effects of the radiant heat must be estimated. If there should be an escape of chemically toxic or radiotoxic materials to the atmosphere or to water, the dispersion of these materials in the environment must be considered and the various pathways to man must be systematically examined. A description of all these phenomena is beyond the scope of this short review, so a few important examples have been chosen for more detailed discussion.

Radiotoxic Releases

If there should be an accidental release of radionuclides into the environment, the possible pathways by which the radioactive material might reach man are many and are summarized in Figure 1.3.5-1. The various pathways may loosely be assigned to three categories:

1. Direct inhalation.

93

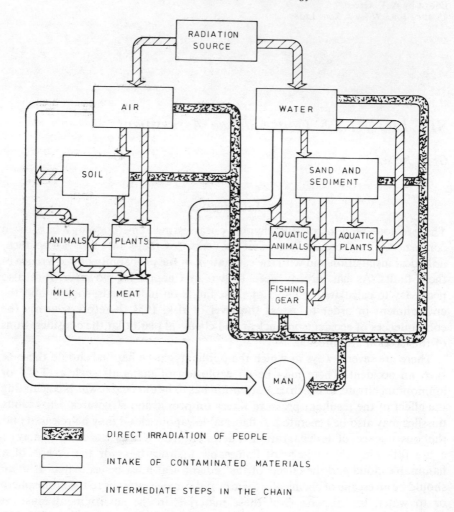

Figure 1.3.5-1 Examples of radiation pathways, from the booklet *Safety and Nuclear Power.* Published by UKAEA. Reproduced by permission of AERE, Harwell.

2. Irradiation by γ-rays from the passing cloud or from deposited fission products.
3. The ingestion of contaminated foodstuffs or water.

In general, there is time to take counter-measures to prevent the ingestion of radioactive materials and it is the inhalation and external irradiation pathways that are the most important.

If radioactive material should be released into the atmosphere, it is usually as a passive plume (unless it is accompanied by a large rate of heat release), that is, it does not appreciably modify the ambient turbulence. It is the properties of the atmospheric boundary layer, which is the layer of air adjacent to the surface, that determine how the radioactivity is subsequently dispersed and diluted. There is a vast and growing literature – see, for example, 'Meteorology and Atomic Energy'[1]-and there exist models with greatly varying levels of sophistication. Arguing simply, however, a minute particle of effluent emitted into the atmosphere will be thrown about at random under the influence of turbulent eddies of various sizes and time scales; that is, it takes a random walk. A large number of such particles will, after a large number of steps in the random walk, become distributed according to the Gaussian formula, as may be proved by applying the Central Limit Theorem. Hence, if effluent is released at a constant rate $\dot{Q}_r$ the airborne concentration χ at a point (x_d, y_c, z_h) is

$$\chi(x_d, y_c, z_h) = \frac{\dot{Q}_r \exp(-y_c^2/2\sigma_v^2(x_d))}{\pi \sigma_z(x_d)\sigma_y(x_d)\bar{u}} \left[\exp(-(z_h - h_r)^2/2\sigma_z^2(x_d)) \right.$$

$$\left. + \exp(-(z_h + h_r)^2/2\sigma_z^2(x_d)) \right] \qquad (1.3.5\text{-}1)$$

where x_d is the distance downwind, y_c is the distance across the wind, z_h is the height above the ground, h_r is the height of the release $\bar{u}$ is the mean windspeed at a height of 10 m and total reflection at the ground has been assumed. The quantities $\sigma_y(x_d)$ and $\sigma_z(x_d)$ are the lateral and vertical standard deviations and are functions of both x_d and of the weather conditions. A typical set of parametrizations of the sigmas is given in Figure 1.3.5-2. The labels on the curve correspond to different weather categories. Category A, for example, occurs on a hot, sunny day when there is plenty of convectively generated atmospheric turbulence, so that dilution occurs rapidly. Category D often occurs on a cloudy day, with a brisk wind, and is the most frequently occurring weather category in

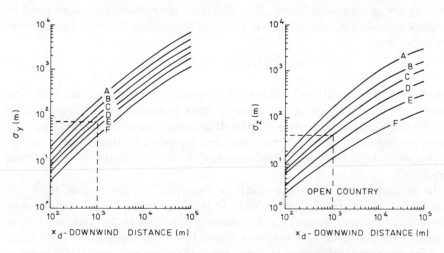

Figure 1.3.5-2 Curves of $\sigma_y(x_d)$ and $\sigma_z(x_d)$ for differing atmosphere stability categories

the UK; in this case, the dispersion is caused by mechanically generated turbulence. Category F often occurs on a cold, still night with a temperature inversion from the ground upwards, so that there is little turbulence and dilution occurs relatively slowly. The division of weather conditions into discrete categories is due to Pasquill[2] and a comprehensive discussion of how categories may be defined in terms of easily measurable quantities such as the temperature gradient in the atmosphere is given by Gifford[3]. Equation 1.3.5-1 is strictly valid only for dispersion over a flat surface in constant weather conditions. The complex effects introduced by topographical features and changing weather categories are beyond the scope of this brief review.

In studies of radiotoxicity, it is the total quantity of radioactive material χ_I inhaled that is important. If b_r is the breathing rate, which is typically $2.2 \times 10^{-4} \, m^3 \, s^{-1}$ for an adult, then

$$\chi_I = b_r \chi_T = b_r \int \chi(x_d, y_c, z_h) \, dt \qquad (1.3.5\text{-}2)$$

where χ_T is the airborne dosage and the integration is over the duration of cloud passage. As has already been mentioned, the deposited activity is important too and, in the simplest models, the deposited activity per unit area is

$$\chi_D(x_d, y_c) = V_g \chi_T(x_d, y_c, z_h = 0) \qquad (1.3.5\text{-}3)$$

where V_g is the deposition velocity and usually takes on values in the range 0.001 to $0.01 \, m \, s^{-1}$. Figure 1.3.5-3 shows some examples of the variation of χ_T with distance, assuming that Equation 1.3.5-1 has been suitably modified to take account of the depletion defined in Equation 1.3.5-3.

Once the inhaled activity $\chi_I = b_r \chi_T$ is known for each radionuclide in the release, the radiation dose received by each organ of the body can be calculated by making use of *inhalation factors*, $D_{n,k}$, which are the radiation doses received by organ k as a result of inhaling 1 Ci of radionuclide n. Thus the total dose received by organ k is

$$D_k = \sum_n D_{n,k} \, \chi_I^n + \text{external dose} \qquad (1.3.5\text{-}4)$$

where χ_I^n is the quantity of radionuclide n inhaled. There exist compilations of inhalation factors[4] which are derived from a model of the transport of radionuclides in the body. The external dose is that due to irradiation by γ-rays from passing cloud or from fission products deposited on the ground. Once χ_T and χ_D are known, there are standard methods for the calculation of this external dose[1].

In general, three important effects of radiation on the human body are considered. These are, first, early effects. In this case certain body organs such as the lung or bone marrow may quickly receive a large dose so that their working is impaired or destroyed and death or illness results. Second, after some time, perhaps 10 years or so, cancers may develop in the person affected. Third, genetic

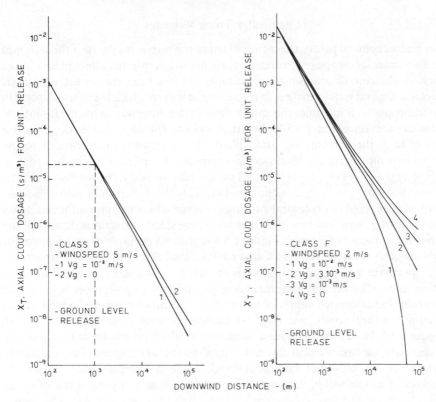

Figure 1.3.5-3 Some examples of the use of the Gaussian dispersion formula in Class D and Class F weather conditions

defects may be transmitted to succeeding generations. If the dose D_k to an organ is known, then it is usually possible to assign a probability that there will be early, late or genetic effects. These probabilities are known as dose–risk relationships and are derived from data taken from experiments with animals, from the study of people exposed in the course of medical treatment, from people exposed accidentally, or from the aftermath of atomic explosions. Many of these dose–risk relationships have been tabulated by the National Radiological Protection Board[5].

In the foregoing, then, a brief outline has been given of how, starting from a release of radioactivity, in known weather conditions, it is possible to calculate the probability that a person at a known position relative to the source of the release will suffer death or some illness such as cancer. If the population distribution is also known, the number of people in the surrounding population who are likely to suffer from early, late or genetic consequences may also be calculated–see Section 1.3.6 on overall assessment.

Chemically Toxic Releases

For many chemical plants, the main hazard to the public may arise if there should be an accidental escape of chemically toxic materials into the atmosphere – gases such as chlorine or ammonia for example. As has been shown for radiotoxic releases, the first requirement is to calculate the way in which the gases disperse in the atmosphere. Sometimes this can be done in the same way as for the radiotoxic releases – see Equation 1.3.5-1 – but it is usually the case that toxic gases are denser than the surrounding atmosphere. The necessary modifications to the dispersion modelling are discussed in Chapter 3.3 on Gas Clouds. In order to calculate the effect of a toxic material on the human body, the average airborne concentration $\chi_A = \chi_T/\tau_p$ is required where τ_p is the duration of cloud passage.

Ideally, detailed toxicological data are required which will give information on the value of χ_A as a function of τ_p that will cause death, or serious illness, or any other consequence of interest. Figure 1.3.5-4 gives an example of such a curve for chlorine; this is taken from the Canvey Island study[6]. In practice, however, the (χ_A, τ_p) curves are very poorly known for most toxic gases and much interpolation, extrapolation, and judgement is required in their application. In principle, however, the calculated values of χ_A and τ_p for a known release of a toxic gas in known weather conditions may be compared with data such as that on Figure 1.3.5-4 and used to define an area within which people are at risk of death. If necessary, the toxicological data can be modified to take account of people staying indoors, or running out of the vapour cloud. It is then a simple matter to count the number of people within this area and to estimate how many are at risk of death.

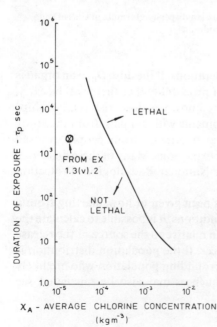

Figure 1.3.5-4 Average concentration and lethal exposure times for chlorine

Flammable Gas Clouds

The atmospheric dispersion of flammable gas clouds may be treated as for chemically toxic gas clouds. The difference with flammable clouds is that it is the instantaneous concentration χ that matters, not the time integrated or average concentrations χ_T or χ_A respectively. If χ is below a lower flammable limit (LFL), the cloud will not burn even if there is a source of ignition. For methane, for example, the LFL is about 5% by volume in air. Similarly, there is an upper flammable limit (UFL – about 15% for methane) above which the mixture is too rich to burn. At any time, it is in principle possible to calculate the contours of upper and lower flammable concentrations and estimate the quantity of material between them. The consequences of ignition – the pressure wave generated by deflagration or detonation, or the damage caused by radiant heat – may then be calculated and the number of people at risk of death or injury estimated. In practice, this is fraught with uncertainty; the problems are discussed in more detail in Chapter 3.3.

EXAMPLE 1.3.5-1

A long lived radionuclide (1 Ci) is released at ground level in Category D weather, when the mean windspeed at a height of 10 m is $5 \, \text{m s}^{-1}$. Calculate: (i) the total airborne dosage 1 km directly downwind; (ii) the total deposited activity per square metre assuming a deposition velocity of $0.003 \, \text{m s}^{-1}$; (iii) the dose to the lung if the inhalation factor is 3×10^4 rem/Ci; (iv) the probability that the release of 1000 Ci will cause lung cancer, if the dose–risk relationship is $2 \times 10^{-5} \, \text{rem}^{-1}$.

(i) From Figure 1.3.5-2, $\sigma_y = 80 \, \text{m}$ and $\sigma_z = 40 \, \text{m}$ 1 km downwind. Using Equation 1.3.5-1, or reading directly from Figure 1.3.5-3, $\chi_T = 2.0 \times 10^{-5} \, \text{Ci s m}^{-3}$.

(ii) The total deposited activity is $\chi_D = V_g \chi_T = 0.003 \chi_T = \underline{6 \times 10^{-8} \, \text{Ci m}^{-2}}$.

(iii) The dose to the lung is $D_k = \chi_T \times b_r \times \text{(inhalation factor)} = (2.0 \times 10^{-5}) \times (2.2 \times 10^{-4}) \times (3 \times 10^4) = \underline{1.32 \times 10^{-4} \, \text{rem}}$.

(iv) The probability of developing cancer for a release of 1000 Ci is $1000 \times D_k \times \text{(dose-risk relationship)} = 10^3 \times (1.32 \times 10^{-4}) \times (2.0 \times 10^{-5}) = \underline{2.64 \times 10^{-6}}$.

EXAMPLE 1.3.5-2

Given a release of $1 \, \text{kg s}^{-1}$ of chlorine for 30 minutes in the weather conditions defined in Example 1.3.5-1, is a person 1 km directly downwind likely to be exposed to a potentially fatal concentration of chlorine?

Assuming for the purposes of this calculation that density effects may be neglected, χ_T may be calculated as was done in the previous example: $\chi_T = 2.0 \times 10^{-5} \times 1800 = 3.6 \times 10^{-2} \, \text{kg s m}^{-3}$. Assuming that the duration of cloud passage τ_p equals the duration of release $= 1800 \, \text{s}$, $\chi_A = 2.0 \times 10^{-5} \, \text{kg m}^{-3}$. From Figure 1.3.5-4 it can be seen by inspection that this exposure would *not* be fatal.

Questions 1.3.5

1. A flammable gas of the same density as air is released continuously at a rate of $10\,\mathrm{kg\,s^{-1}}$. If the LFL is $0.03\,\mathrm{kg\,m^{-3}}$, estimate the maximum distance downwind at which flammable concentrations will be found in Category F weather with a windspeed of $2\,\mathrm{m\,s^{-1}}$.

2. If it is raining, at a steady rate, the total quantity of material remaining airborne at a distance x_d downwind is $Q_r \exp(-\Lambda x_d/\bar{u})$. Λ is the washout coefficient and takes on values in the range 10^{-5} to $10^{-2}\,\mathrm{s^{-1}}$ depending on the rate and type of rainfall. Assuming that the rainfall persists indefinitely, how far would a plume have to travel in a $5\,\mathrm{m\,s^{-1}}$ wind before 50% of its contents have been washed to the ground?

Notation 1.3.5

b_r	breathing rate
D_k	Radiation dose delivered to body organ k
$D_{n,k}$	Radiation dose delivered to organ k after inhaling unit quantity of nuclide n ('inhalation factor')
h_r	height of release
Q_r	quantity of material released
$\bar{U}$	mean windspeed measured at a height of 10m
V_g	deposition velocity
x_d	distance downwind
y_e	distance acrosswind
Λ	washout coefficient
τ_p	duration of vapour cloud passage
χ	airborne concentration
χ_A	Average airborne concentration
χ_D	deposited activity per unit area
χ_I	quantity of inhaled activity
χ_T	airborne dosage or time integrated concentration

References 1.3.5

1. Slade, D. H. *Meteorology and Atomic Energy.*
   United States Atomic Energy Commission. 1968.
2. Pasquill, F. *Atmospheric Diffusion.*
   Ellis Horwood Ltd, Chichester. 1974.
3. Gifford, F. A. 'A review of turbulent diffusion typing schemes.' *Nuclear Safety*, **17**, 1968.
4. Kelly, G. N. *et al. An Estimate of the Radiological Consequences of Notional Accidental Release of Radioactivity from a Fast Breeder Reactor.*
   National Radiological Protection Board (UK), NRPB R53 1977.
5. Smith, H. and Stather, J. W. *Human Exposure to Radiation Following the Release of Radioactivity from a Reaction Accident: a quantitative assessment of the biological consequences.*
   National Radiological Protection Board (UK), NRPB R52.
6. Canvey. *An Investigation of Potential Hazards from Operations in the Canvey Island/Thurrock Area.*
   HMSO (UK). 1978.

High Risk Safety Technology
Edited by A. E. Green
© 1982 John Wiley & Sons Ltd.

SECTION 1.3.6 Overall Assessment

G. D. Kaiser

Introduction

The previous section on consequence assessment is devoted to the calculation of the consequences of a single fault condition in which a known quantity of hazardous material has escaped to the environment in known weather conditions. In practice, however, a complicated advanced technological installation may fail in a number of ways. This has been illustrated in previous sections using a simplified reactor design. For a 'real' reactor, the Reactor Safety Study[1] identified about 650 accident sequences for PWRs and, even after considerable simplification, there remain nine categories of accident, each with its own characteristic release of radionuclides into the atmosphere – see Table 1.3.6-1.

The first step in the overall assessment of plant safety, then, is the identification of possible accident sequences and this has already been discussed in the sections on fault and event trees (Section 1.3.2) and fault analysis (Section 1.3.3). To each of the accident sequences $1, 2, \ldots j, \ldots n$ there is attached an estimated frequency of occurrence f_j and an estimate of the quantity of material released into the environment Q_r^j, where Q_r^j may be a single quantity, such as the mass of chlorine released after (say) the accidental failure of a storage vessel, or it may be a vector containing a number of quantities such as the activities of the various radionuclides accidentally released to the atmosphere from a nuclear installation. Table 1.3.6-1 gives an example, taken from the Rasmussen report, of the results of a comprehensive event tree analysis of a pressurized water reactor, specifying the accident sequences, the value of f_j and Q_r^j and other necessary parameters such as the height and duration of release and the sensible heat content.

Individual Risk

Consider first a single accident sequence j in weather condition i (usually specified by a combination of weather category and mean windspeed $\bar{u}$) with the wind blowing into sector k (it is usual to define 12 30° or 16 22½° sectors). Let the probability that these weather conditions and wind direction prevail be $p(i, k)$ where $p(i, k)$ can in general be constructed from readily available meteorological

Table 1.3.6-1 An example of the determination of the frequency and magnitude of releases of radionuclides into the atmosphere (from the Reactor Safety Study)

Accident category and frequency f_j	Description of accident category	Duration of release (h)	Elevation of release (m)	Energy release rate (Btu/h)	Fraction of core inventory released						
					Xe–Kr	I	Cs–Rb	Te–Sb	Ba–Sr	Ru, Rh, Co Mo, Tc	Y, La, Zr, Nb, Ce, Pr, Nd, Np, Pu, Am, Cm
PWR1a and 1b 9×10^{-7} yr^{-1}	Core melt-down and failure of containment spray and heat removal systems. 1a–steam explosion after containment failure due to overpressure: 1b–steam explosion ruptures containment.	0.5	25	20×10^6 (1a) 520×10^6 (1b)	0.9	0.7	0.4	0.4	0.05	0.4	3×10^{-3}
PWR2 8×10^{-6} yr^{-}	Core melt-down, failure of containment spray and heat removal systems. Containment fails through overpressure after commencement of core melt-down.	0.5	0	170×10^6	0.9	0.7	0.5	0.3	0.06	0.02	4×10^{-3}
PWR3 4×10^{-6} yr^{-1}	Failure of containment due to overpressure before core melt-down.	1.5	0	6×10^6	0.8	0.2	0.2	0.3	0.02	0.03	3×10^{-3}

PWR4 5×10^{-7} yr^{-1}	Core melt-down and failure of containment system properly to isolate; failure of containment spray system	3.0	0	10^6	0.6	0.09	0.04	0.03	5×10^{-3}	3×10^{-3}	4×10^{-4}
PWR5 7×10^{-7} yr^{-1}	As PWR4, but containment spray systems operate to reduce release to atmosphere	4.0	0	3×10^5	0.3	0.03	9×10^{-3}	5×10^{-3}	10^{-3}	6×10^{-4}	7×10^{-5}
PWR6 6×10^{-6} yr^{-1}	Core melt-down, failure of containment sprays. Containment maintains integrity but core melts through base.	10.0	0	0	0.3	8×10^{-4}	8×10^{-4}	9×10^{-8}	9×10^{-5}	7×10^{-5}	10^{-5}
PWR7 4×10^{-5} yr^{-1}	As PWR6, but containment sprays operate	10.0	0	0	6×10^{-3}	2×10^{-5}	10^{-5}	2×10^{-5}	10^{-6}	10^{-6}	2×10^{-7}
PWR8 4×10^{-5} yr^{-1}	Large pipe break, containment fails properly to isolate, all other engineered safeguards work, no core melt-down.	0.5	0	0	2×10^{-3}	10^{-4}	5×10^{-4}	10^{-6}	10^{-8}	0	0
PWR9 4×10^{-4} yr^{-1}	As PWR8, but all engineered safeguards function as designed. Essentially PWR design basis accident.	0.5	0	0	3×10^{-6}	10^{-7}	6×10^{-7}	10^{-9}	10^{-11}	0	0

data. The calculations outlined in the previous chapter can be carried out for an individual at a known position in sector k in order to calculate the probability $p(I|j, i, k)$ that the individual in question will suffer some consequence I, where I may be death for a chemically toxic release, or cancer for a radiotoxic release, or some other consequence of interest. $p(I|j, i, k)$ is *conditional* on the occurrence of weather condition i, wind direction k and fault sequence j and is also a function of position (x_d, y_c, z_h).

The overall risk R_I to an individual at a known position, taking into account all sequences is then simply given by

$$R_I = \sum_j f_j \sum_i p(i, k)p(I|j, i, k) \qquad (1.3.6\text{-}1)$$

and is the probability per year that, as a result of the operation of the advanced technological installation in question, an individual who is habitually to be found at the point (x_d, y_c, z_h) in sector k will suffer harm in the form of consequence I. It has been implicitly assumed that an individual in sector k is not affected if the wind does not blow into sector k. In principle, Equation 1.3.6-1 can be modified by a summation over sectors and, if need be, by allowing for the movements of the individual.

An example of this kind of individual risk analysis appears in the Canvey Island Report[2]. The Canvey Island complex contains a great variety of chemical plant and a correspondingly large number of sequences j for each of which an analysis of the type described in the foregoing must be carried out. The major problems identified by the investigating team included the possibility of a spillage of anhydrous ammonia, either from pressurized storage or from a refrigerated ship; the spillage of hydrogen fluoride; possible releases of LNG or LPG; fires occurring should there be a spillage of flammable liquid; and explosions in an ammonium nitrate store. The results of the Canvey study appear in Table 1.3.6-2.

Table 1.3.6-2   Upper limit of estimated average individual risk of death for existing installations on Canvey Island[2]

Location of individual	Existing installations	Existing installations with suggested improvements
Canvey Island	$5.3 \times 10^{-4} \text{ yr}^{-1}$	$2.7 \times 10^{-4} \text{ yr}^{-1}$
Stanford-le-Hope	$5.0 \times 10^{-4} \text{ yr}^{-1}$	$1.3 \times 10^{-4} \text{ yr}^{-1}$
South Benfleet	10^{-4} yr^{-1}	$4.0 \times 10^{-5} \text{ yr}^{-1}$

The quoted risks are upper limits because, in view of the many uncertainties encountered in consequence assessment, many conservative assumptions have been made.

Societal Risk

Once the quantity $p(I|j, i, k)$ has been calculated, it is a simple matter to calculate the expectation value of the number of people in the surrounding population who may suffer from consequence I, given the population density $N_d(x_d, y_c)$

$$N(I|j, i, k) = \iint dx_d \, dy_c \, p(I|j, i, k) N_d(x_d, y_c) \qquad (1.3.6\text{-}2)$$

$N(I|j, i, k)$ is strictly the mean of a probability distribution, but is often interpreted simply as the number of people suffering from consequence I given accident sequence j, weather condition i and wind direction k. The conditional probability associated with $N(I|j, i, k)$ is $p(i, k)$ and similar pairs of numbers can be generated for all combinations of i and k. If these pairs are then sorted into the order of increasing N and if the $p(i, k)$'s are cumulated, it is possible to generate a plot of the cumulative probability $p(N|j)$ that $N(I)$ or more consequences of type I appear in the surrounding population, conditional on the occurrence of a accident sequence j. Finally, if the following summation is made;

$$f(N) = \sum_j p(N|j) f_j \qquad (1.3.6\text{-}3)$$

then $f(N)$ is the cumulative frequency with which an accident might occur in the plant in question and cause $N(I)$ or more casualties of type I. These plots of frequency and number are often referred to as f–N lines or, in the language of the Rasmussen report, 'CCDFs' or cumulative complementary distribution functions. It is becoming increasingly common to express the risk to society as an f–N line. Figure 1.3.6-1 shows the f–N lines for early and late deaths generated by the information given in Table 1.3.6-1 together with similar f–N lines from a recent German study of PWRs[3]. Figure 1.3.6-2 gives the f–N lines from the Canvey Island study.

Comparison with Safety Requirements

In principle, the individual risk figures (see, for example, Table 1.3.6-2) and the f–N lines can be compared with target figures and lines which may be set either by law or by individual organizations. At present, it is the case that there are no absolutely binding targets against which the acceptability or otherwise of individual and societal risks can be tested – there is always a considerable element of engineering, scientific, and political judgement required.

In Chapter 1.1 it is shown that, in general, people do not tend to worry too much about individual risks of about $10^{-6} \, \text{yr}^{-1}$, while risks at the level of $10^{-4} \, \text{yr}^{-1}$ tend to attract attention, together with the demand for money to be spent on the reduction of risk. Hence it is a useful rule of thumb to say that an individual risk of $10^{-6} \, \text{yr}^{-1}$ will probably be deemed acceptable while one of

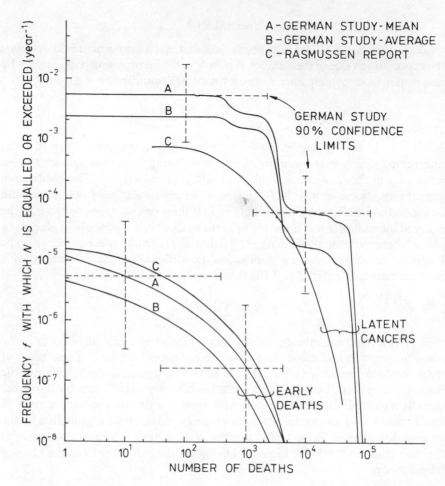

Figure 1.3.6-1 f–N lines for early deaths and latent cancers, taken from American[1] and German[3] PWR studies

$10^{-4}\,\text{yr}^{-1}$ is too high. At the $10^{-5}\,\text{yr}^{-1}$ level there is probably scope for argument, with risk and benefit finely balanced. The literature on risk appears to be moving towards a consensus here.

By contrast, there is no consensus on what constitutes acceptable societal risk. It is tempting to compare the f–N lines such as those given in Figures 1.3.6-1 and 1.3.6-2 with those generated by other human activities. Figure 1.3.6-3 shows f–N lines for certain industrial activities in the UK and worldwide[4]. It is clear that, by this criterion, both Canvey Island and nuclear reactors are acceptable. None the less, as is well known, there is considerable controversy about the acceptability of many kinds of advanced technological activities. For example, some would argue

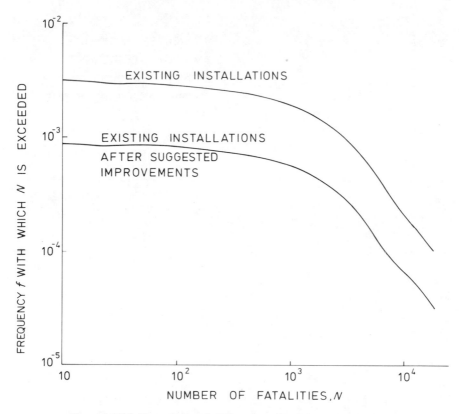

Figure 1.3.6-2 Upper bound f–N lines from the Canvey Island report

that it is much less acceptable for there to be a possibility of a large accident causing (say) tens of thousands of casualties, even at a low frequency, than for there to be a much more likely accident sequence which would cause only a few casualties – even if, taken over many years, the cumulative effect of the smaller accidents is greater than the effect of the infrequent, larger accidents.

Limitations Because of Limited Data

The implementation of a method of quantitative risk assessment such as is outlined in the foregoing requires an extremely large data base, giving information on the behaviour of complicated plant during an accident, the dispersion and dilution of a host of materials in the environment, the effect of toxic materials on the body, and so on. If the data base is sound, then there are no problems in applying the methodology. Should the data base be inadequate, however, it should be applied with caution.

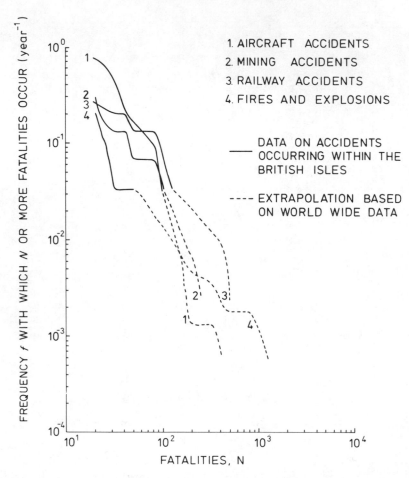

Figure 1.3.6-3 Extrapolation of British Isles Accident Data[4]

The Rasmussen report[1] on LWR safety attempts to insert 'realistic' values of probabilities into the branches on the event trees, using what are intended to be 'credible subjective probabilities' where data are lacking. For a reactor during an accident, however, components of the system may be exposed to an environment far different from any in which their performance has ever been tested so that it is questionable whether a credible subjective probability can be defined at all. As a result, the report[1] has been criticized[5] because 'we are unable to define whether the overall probability of a core melt given in the report[1] is high or low, but we are certain that the error bands are understated'. It follows that an essential element of quantitative risk assessment is the assigning of error bands to figures for individual risk or to f–N lines. In many cases, the data base is not adequate for this, in which case the results should be interpreted with due caution.

An alternative approach is that adopted by the Canvey Island team[2] who endeavoured to be conservative at all times. This can lead to the prediction of rather high values for risk – as has been seen, the individual risk in the Canvey area is predicted to be about 5×10^{-4} yr^{-1} – and the degree of conservatism is not known. It is therefore difficult to say precisely what weight should be assigned to the predictions in the Canvey report. If they are wildly conservative, expensive modifications to plant may be adopted which, later on, may be seen to be unnecessary. A review of the Canvey report[6] suggests that the degree of conservatism might be as much as a factor of 10.

Questions 1.3.6

1. Discuss the differences between individual and societal risks. State how you would decide the levels of these risks which would be acceptable.
2. From a process of your choosing where there is a potential hazard to the public, draw up a similar table to that of Table 1.3.6-1. Estimate the appropriate categories of accidents and their frequencies. Discuss in areas of difficulty the reasons which contribute to limiting quantification of the various parameters such as frequency of release.

Notation 1.3.6

f_j	frequency of occurrence of accident sequence
$f(N)$	frequency of occurrence of N or more casualties
$N_d(x_d, y_c)$	population distribution
$N(I)$	number of people suffering consequence I
$N(I\|j, i, k)$	number of people suffering consequence I given accident sequence j, weather condition i and wind direction k
$p(i, k)$	probability of weather condition i and wind direction k
$p(I\|j, i, k)$	conditional probability of an individual suffering consequence I given weather condition i, wind direction k, and accident sequence j.
$p(N\|j)$	probability that there will be N or more casualties, conditional on accident sequence j
R_I	individual risk of consequence I

References 1.3.6

1. Reactor Safety Study, *An Assessment of Accident Risks in US Commercial Nuclear Power Plants.* WASH-1400 (NUREG 75/014), United States Nuclear Regulatory Commission. 1975.
2. *Canvey: An Investigation of Potential Hazards from Operations in the Canvey Island/Thurrock Area.* HMSO (UK). 1978.
3. *The German Risk Study – Summary.* The Federal Ministry for Research and Technology. 1979.

110   *High Risk Safety Technology*

4. Fryer, L. S. and Griffiths, R. F., *Worldwide Data on the Incidence of Multiple Fatality Accidents.* SRD R149, UKAEA. 1979.
5. Lewis, H. W., *Risk Assessment Review Group Report of the US Nuclear Regulatory Commission.* NUREG/CR-0400. 1978.
6. *An Analysis of the CANVEY Report.* OYEZ Intelligence Reports, UK. 1980.

High Risk Safety Technology
Edited by A. E. Green
© 1982 John Wiley & Sons Ltd

Chapter 1.4

Operational Safety

E. S. London

Introduction

It is convenient to refer to any person who has some responsibility for control of a facility as an operator; the general manager (or director) of an establishment at which the facility is sited is simply the chief operator. None of the many responsibilities which rest on the shoulders of an operator is more onerous or more important than that relating to safety. Apart from the moral obligations and financial implications to be considered, there are many legal requirements to be satisfied. But, although the law states 'what' has to be achieved, it rarely indicates 'how'; the operator must, therefore, not only know 'what' but know, or ascertain, 'how', and then conduct his operations accordingly. However, a responsible operator is not content with aiming simply to meet legal requirements; he regards the latter as a framework to which must be welded recognized good practice and ground-rules formulated from his own experience. In this chapter some aids towards achieving operational safety will be outlined. They are neither new nor revolutionary; in fact, most have been in use in some industries for many years. An underlying principle is that the operator can make an important contribution to operational safety well before he assumes responsibility for the facility.

Safe operation, whether a facility be complex or simple, requires the provision of adequate design and operational safeguards. Design safeguards are items provided by the designer; some afford the operator with means of control, e.g. flowmeters, temperature controllers, etc., whereas others function independently of the operator, e.g. bund walls ('dikes' in Canada and the USA), explosion blast protection, pressure relief valves, etc. Operational safeguards are procedures which require decision and action by the operator; some involve use being made of installed equipment and all require organization.

A project may be considered to comprise several main stages:

Conceptual
Detailed design
Construction
Commissioning
Operation
Decommissioning

Such divisions, of course, are rarely clear-cut, and some overlapping, especially between detailed design and construction, may occur in practice. The last stage, decommissioning, is outside the scope of this chapter. Arrangements for managing the various stages vary from one project to another; an organization which is self-sufficient may handle the entire task in-house, another may sub-contract all stages up to and including commissioning and simply operate the facility, and yet another may adopt a course which is intermediate between these extremes.

It matters little how the operator acquires a facility; at the end of the day he has charge of it, and the concomitant responsibility for safety rests finally and firmly on him. From the moment the facility is handed over to him he must provide adequate operational safeguards. These he will probably formulate during the construction stage. However, the prudent operator will ensure that his contribution to safety starts at a much earlier stage; in various ways he will influence design decisions and the choice of design safeguards, and insist that operational safety is the prime objective throughout. The most important tool that is available to assist in attaining the objective is safety assessment[1].

Safety Assessment

Safety assessment is a general term used for the process of ensuring that hazards are identified and that adequate safeguards are provided at all times in respect of all persons on and off the site. (It is useful to define the term *hazard* as 'a combination of circumstances which could lead to injury or loss'. To amplify, chlorine *per se* is not a hazard; provided that it is properly contained, it is innocuous. However, if by some means (*a*) it were released in significant quantity from the containment and (*b*) the contaminated air were inhaled by an unprotected person, there would be a hazard. Similarly, water is not a hazard, but one could readily postulate combinations of circumstances in which it could lead to death by drowning, scalding, road accident, etc.)

Safety assessment has been used in certain high technology industries for more than 20 years. It is not merely a single step taken at a specified stage in a project but an on-going process throughout its entire life, i.e. from conception to eventual decommissioning – some would say from green-field back to original green-field. (Safety assessment has acquired several aliases – safety assurance, safety analysis, etc., but the name is not important.) The methodology is discussed elsewhere (Chapter 1.3) as is its application to regulatory control (Chapter 1.5), and only the operator's role needs to be outlined here.

The stages of safety assessment, in chronological order, are:

1. Preliminary Hazards Analysis (PHA)
2. Formulation of Design Safety Principles (DSPs)
3. Implementation of the DSPs in the detailed design

4. Preparation of the Safety Document (Design)
5. Independent Scrutiny of Design (on the basis of 4)
6. Preparation of the Safety Document (Operation)
7. Independent Scrutiny of Proposed Operations (on the basis of 4 and 6)
8. Periodic Reviews of Operation (Inspection; re-scrutiny of safety documentation)

Experience has shown that independent safety scrutiny (stages 5 and 7) needs a team effort, and some operators have recognized the fact and set up Safety Assessment Committees, independent of line management, to do the work. (Such committees are entirely technical and should not be confused with those called for by UK legislation[2] and which relate to joint consultation on health and safety matters.) On the satisfactory completion of stage 7 an Authority to Operate (ATO), signed by the head of the establishment or his nominee, usually results.

It is important that the operator should from the outset make a full contribution to the continuing process of safety assessment, because after handover it is he who will not only have to live with and compensate for any shortcomings in the design safeguards but be called upon to answer for any incident which may occur. The role of the operator at successive stages in the project will now be discussed.

Design – The Operator's Role

Preliminary hazards analysis

At the outset of a project, i.e. after the need for a new facility or modifications to an existing one has been identified, it is first necessary for financial approval to be obtained. This is usually sought by means of some form of Capital Expenditure Proposal (CEP). In order to ensure that the CEP includes a realistic financial provision for design safeguards, it is good practice to precede the CEP with a Preliminary Hazards Analysis (PHA). The latter is a qualitative study of the nature and degree of hazards, and it enables the requisite safeguards to be specified in terms which, although broad, are capable of being translated into budget cost figures. It is important that the operator should take the lead in carrying out this analysis and exert a strong influence on the broad specification of safeguards.

The conclusions of the PHA, together with the outline scheme, are presented in the CEP.

Design safety principles

After the project has been given financial approval, the next step, from the safety standpoint, will be the formulation of Design Safety Principles (DSPs). These are the broad concepts from which the detailed design is to be evolved, and are so worded as to state what has to be achieved and leave the designer free to use his ingenuity to decide subsequently how this may best be done. Although the designer will take the lead at this stage, the operator should get himself closely involved, and not be intimidated by the designers or others into including or excluding any principle without technical justification. For example, he should be ready to argue that rigorous application of the somewhat general principle, 'The design will make it more difficult for the operator to operate unsafely than safely', would have prevented many an incident and, similarly, that 'The requirements of safety, e.g. means of escape, will not have a lower priority than those of security, e.g. safeguarding of valuable material', would have reduced the consequences of many incidents.

Important matters to be discussed at the safety principles stage include building and/or site emergencies. For example, how should the design facilitate control of personnel and operations in the event of such emergency? Should a continuously tenable control room be provided to facilitate safe shutdown or continued operation? How should the design assist post-incident re-entry? These and many other matters relating to emergencies are vital to the operator and he has a great contribution to make in deciding them and formulating the relevant DSPs.

The operator should also influence the maintenance philosophy; non-maintainability may be essential in some areas of the facility, whereas a choice between scheduled and breakdown maintenance may be available in others. Decisions made at this early stage may have far-reaching implications on operational safety.

Any propensity on the part of the designers to postpone discussion of the treatment of arisings of wastes 'until we know more about them' should be corrected; the very likely alternative for the operator is an inheritance of storage/disposal problems with probable rapid growth.

Detailed design

During the next stage of the project, detailed design, when the DSPs are being implemented and taking shape into design safeguards on the drawing board, the operator should continue his dialogue with the designer. He should satisfy himself that every reasonably practicable design safeguard is being included, so as to lighten the burden he eventually has to carry. In short, the operator must be satisfied that the designer has achieved a reasonable balance between design and operational safeguards and, specifically, that there is not to be disproportionate dependence on operational control. One important consideration is the degree of sophistication of any automatic operation or control. In influencing the decision, the operator will attempt to weigh such factors as, say, a chronic shortage of skilled process workers in the labour catchment area of the site against the fact that sophisticated, albeit reliable, equipment may require sophisticated maintenance by engineers, who may also be or become in short supply.

Alarms, both visual and audible, have an important part to play in warning the operator of events or conditions. However, too many alarms are equally undesirable as too few; an operational control room which looks and sounds like a fairground during rush hour will cause confusion and lack of confidence, and in this matter the operator may need to moderate the designer's enthusiasm and, perhaps, his own.

Maintenance techniques must also be carefully hammered out at the design stage. The operator will ensure not only that his maintenance engineers are satisfied with the practicability of these techniques and the ease with which they can be carried out, but that the design includes the necessary provisions to enable him (the operator) to prepare the plant for maintenance, control it while such work is in progress, and bring it back fully on to line on completion.

The foregoing examples are intended to demonstrate that the operator has a significant contribution to make during the design stage. Many other topics could have been cited, e.g. work stations; access and egress; transport operations; emergency lighting needs; standby and emergency supplies of electricity, water, steam, inert gas, compressed air, etc.; the need to isolate breathing air supplies from process air; the requisite degree of diversity and redundancy of controls, etc. and the attendant need for the designer to utilize reliability assessment, etc. The list may be long for even a simple project, but throughout the design stage the operator will always be mindful that the probability of solving one problem by the introduction of another is never low.

A plant is like a baby; the designer is the midwife who brings it into the world and the operator is the guardian who has the more difficult long-term task of living with it. The author will resist the temptation to draw the analogy any further!

Safety documents

The function and content of safety documents are discussed elsewhere (Chapter 1.5). Logically, preparation of the Safety Document (Design) will form part of the design package and should be undertaken by the designer. However, sometimes the designer complains that the matter diverts him from his main task, which is to lead the design team; in such cases the operator may decide to make staff available to undertake the work of obtaining from the designer the appropriate information and marshalling it into a document. Where the designer does the job it is in the operator's interest to afford all reasonable assistance to the designer.

Preparation of the Safety Document (Operation) is clearly the responsibility of the operator and is a task for which he will need to make sufficient effort available.

Finally, it must readily be acknowledged that safety documents *per se* cannot beget safe operation. They are simply a means of persuading the designer and operator not only to think through the whole spectrum of problems and record their solutions but to present their arguments (or case) in a logical fashion for others to scrutinize.

Independent Scrutiny

This aspect of safety assessment is discussed elsewhere (Chapter 1.5) and the role of the Safety Assessment Committee (Safety Working Party) is highlighted. It is necessary here only to emphasize that the members should be appointed not only for their ability but for their belief in the importance of safety assessment. Persons who might regard it as a chore should be avoided, as they will probably become passengers. Fortunately, experience has shown that most members are enthusiastic and approach the work in a professional way. However, it must be stated that, however well-written the safety document may be, the task of assessing it is demanding on the time of individual members; some of the work will need to be done outside of committee, and main-line duties must still be performed. The chief operator must bear this in mind when complementing departments and nominating committee members.

A further point is that, in the UK, even when the independent assessors are satisfied, there is a further criterion to be met, and that is that the designer and operator have a statutory duty[2] to provide a safe plant. In practice, this means that they must provide all the safeguards that are both technically and economically feasible, and not simply those that safety assessment has shown to be necessary.

Construction – The Operator's Role

Discussion has so far centred on the operator's contribution to design. The next stage, construction, is the translation of the designer's intentions from drawings, specifications, etc., into hardware. The possible implications, if this is not accomplished accurately, are obvious. The importance of Quality Assurance is mentioned elsewhere (Chapter 1.5); if this work is undertaken by staff not under his control, the operator should liaise with them and ensure that they are fully aware of the purpose and importance of all the design safeguards and, in this respect, are equipped to work effectively.

During this period the operator himself will look out for shortcomings, perhaps ostensibly trivial, which may be vital in certain circumstances, such as the emergency cooling water valve which is accessible only to a contortionist, or the data logger which is so located in the control room that the only possible working position for the instrument engineer is that in which his weight is directed via his buttocks on to a row of vital switches.

Up to this point the operator's role has been akin to that of the 'backseat driver.' However, very early in the construction stage he prepares to move forward to take over the wheel himself; he starts to spell out the operational safeguards and will have completed the task prior to the start of commissioning. Unless it is already an operational site, he will also have to formulate a Safety Policy, and then set about the task of providing the means by which the policy will be implemented.

Safety Policy – Formulation

When formulating his safety policy the operator will take due account of the legal, moral, and financial aspects of safety. His policy will be to ensure that the design and operational safeguards not only are adequate at the outset but continue to be adequate throughout the life of the facility. To facilitate the implementation of this policy he will set up a suitable organization, the effectiveness of which will, in the long run, reflect his own degree of belief in the policy.

If the policy is to succeed, it will need to include a definite commitment to employ safety assessment and authority-to-operate procedures. Where practicable and necessary the operator will encourage the use of quantitative assessment techniques, including the art of reliability assessment. However, he will be aware that the considerable effort that has to be invested in calculating the probability of an event and the magnitude of its consequences might be better employed in seeking a means of eliminating the hazard altogether.

The operator will endeavour to demonstrate to all staff under his control that he has a positive motivation towards safety and he will strive to imbue them with it, too; he will remember that in this context actions undoubtedly speak louder than words and that it is good for morale if the staff believe that the management is sincere. The following statement hits the nail on the head. 'A practical test of the safety policy is that a manager or supervisor can recognise that he will be supported by the company if he reasonably chooses the safety of his subordinates before the demands of production[3].'

There should be a commitment to provide staff, adequate in both numbers and quality, to make the operator's input during the design stage, to ensure safety during operation and maintenance, and to carry out thoroughly the safety assessment and committee work.

In the UK the employer of five or more persons must[2] make a written statement of his general policy, organization and arrangements for health and safety at work. The statement, revised as necessary, must be brought to the notice of all employees. This statutory statement of policy relates only to employees, but the operator may consider it prudent to extend it to cover other persons on the site and the general public. (In any case, he has, of course, a statutory duty not to expose these persons to health and safety risks.)

Lastly, there should be a policy of sharing safety experience. This is a broad subject which includes *inter alia* feedback to designers; the recording of safety related information for inclusion in data banks; participation in national and international work on improvements in safety standards; publication of reports on investigations of accidents and near-misses, e.g. in the *Loss Prevention Bulletin* of the Institution of Chemical Engineers; participation in discussion of safeguards at local, national, and international level, etc.

Safety Policy – Implementation

It is the responsibility of the head of the establishment to ensure that the declared Safety Policy is implemented. The policy must be supported by each level of management and, in turn, each subordinate employee has a responsibility to work towards its effective implementation.

While the head of the establishment has overall responsibility for safe operation of the site, i.e. to provide adequate operational safeguards, each member of staff has a personal responsibility for the safety of operations under his control. This means that the duties of every person in respect of safety need to be spelt out, understood, and carried out religiously.

The head of the establishment must ensure the continuous existence and use of suitable administrative procedures and arrangements (viz. ATOs, safety assessment, etc.) to enable the design and operational safeguards to be considered adequately.

To enable him to implement his safety policy throughout the life of the project the operator will need to set up an adequate organization. This will fall under three headings: departmental structure, safety committee structure, and organization for emergencies.

Departmental structure

There will be an Operating Department, e.g. Production, which will be assisted in its task (or hindered, if the organization is inadequate) by several others: Maintenance; Safety; Technical; Services (supplies of electricity, steam, etc.); Training; Medical; Administration, etc. The corporate organization must enable the operator continuously to provide the operational safeguards and to maintain the design safeguards.

All staff in the Operating Department will be personally responsible via line management to the head of the establishment (chief operator) for the safety of all operations under their control. In the context of operational safety, the Maintenance Department, whose function is to assist in keeping the facility in an operable condition, may be regarded as an extension of the Operating Department. Except for matters of engineering theory and practice, in which they will be led by their own departmental head, they will do what they are required to do by the Operating Department. Since the Operating and Maintenance Departments work at the 'coal-face', their activities will impinge more directly on operational safety than those of other departments.

The Safety Department will provide advice to the Operating Department, which the latter will be free to accept or reject, and a range of safety services. The organization will reflect the particular needs of the establishment; there may be a separate group working full-time on safety assessment, while other services would include such matters as monitoring for flammable atmosphere; issue and maintenance of protective clothing and equipment; provision of fire prevention and fire fighting services; monitoring the environment both in and around the plant; provision of assistance to the Training Department in safety education; maintenance of safety records; notification of accidents and dangerous occurrences, etc. Safety-related problems which require in-depth study, either theoretical or practical, will be passed to the Technical Department; such problems may range from aspects of ventilation and corrosion to shortcomings in design safeguards. The Technical Department will also provide an analytical service, e.g. to check on in-plant safety-related instruments; to analyse atmospheres, say, before entry is made into confined spaces; for use as a tool in safety-related aspects of quality assurance.

The Services Department will be charged with providing to the Operating Department, on an uninterrupted basis, the requisite supplies of electricity, steam, water, compressed air, etc., on which some safeguards ultimately depend.

Personnel at all levels will need some form of programmed training in order to gain an adequate appreciation of the operations and hazards, on the one hand, and the design and operational safeguards on the other. The meaning and significance of operating instructions, operating limits, emergency instructions, etc., must be understood clearly by all concerned. Staff such as administration and medical should not be overlooked; they, too need to have an appreciation of matters so far as they may be affected or they may affect matters. There will also need to be a programme of re-training, i.e. continued training with up-dated material and refresher courses. The operator's organization will almost certainly utilize a Training Department to assist with some of these matters.

The role of the Medical Department will be to supervise the health of those who may be exposed to hazards. Among its duties will be to advise on the medical aspects of the hazards, maintain the necessary health records, provide first-aid services, give medical examinations, and assist the Training Department in safety education.

The functions of the Administration Department will include consideration of relevant political and other implications of the operator's activities, and relieving operators of mundane duties which do not require their professional qualifications, e.g. recruitment, provision of general services, etc., thereby leaving them free to concentrate on their main task.

Safety committee structure

It was mentioned earlier that some operators set up a separate safety assessment team (committee), independent of line management, to carry out in-depth safety scrutiny as a necessary precursor to the issue of an Authority to Operate (ATO). In practice the Safety Assessment Committee not only carries out this main duty but vets the output from most of the other stages of safety assessment, thereby providing an important, formal link throughout the entire safety assessment process.

There is no doubt that an active Safety Assessment Committee coupled with an ATO procedure is an important aid in the achievement of operational safety. If the operator adopts this arrangement, he should identify the Safety Assessment Committee as part of his organization by showing it, superimposed on the departmental structure, in his organization chart.

Organization for emergencies

Emergency planning is discussed in detail elsewhere (Chapter 1.7) and only one or two aspects need be mentioned here.

The operator will need to define and publicize standby arrangements and clear lines of responsibility for handling three broad types of emergencies:

1. Building emergency, i.e. when the effects of an incident are confined to the building or area of origin.
2. Site emergency, i.e. when the effects of an incident have spread beyond the building or area of origin but are confined within the site.
3. Off-site or district emergency, i.e. when the effects of an incident have spread beyond the boundaries of the site, and collaboration with external bodies is needed for control purposes.

On the basis of the assessment work done in aid of obtaining the ATO, the operator will be able to decide which incidents are so low in probability as to warrant being dismissed from further consideration; all other incidents should be identified and categorized.

The importance of periodic emergency exercises cannot be overstated. A well-prepared and fully available Site Emergency Handbook, which lays down procedures for controlling emergencies, i.e. who does what and when, may prove valuable, as may a schedule of special equipment for use in emergencies, which shows locations, responsibilities for maintenance and testing, etc.

Commissioning

Prior to commissioning, or pre-operational testing as it is sometimes called, the operator will have spelt out the operational safeguards and presented them for scrutiny in the form of the Safety Document (Operation). Most of the effort in formulating them will have been put in by the operator, although some may have been specified by the designer, e.g. certain operating limits, and others may have emerged later as a result of the independent safety scrutiny.

From the safety standpoint, the purpose of commissioning is to prove as far as is reasonably practicable that the design safeguards are adequate and that the operational safeguards are both adequate and practicable. Additionally, commissioning provides an opportunity to examine possible areas of doubt which may have been identified in the safety documents. The outcome may be that some safeguards will be judged to need reinforcement or reassessment, whereas others warrant relaxation.

The commissioning programme will define, in writing, the objectives of the exercise, spell out the steps to be taken in order to meet these objectives, and define the acceptance criteria to be used. Both normal operations and unplanned occurrences will be examined. Where necessary, simulation will be resorted to in respect of some aspects of both states. (Normal operations may be defined as operations in which the process and equipment are responding to the controls applied. An unplanned occurrence is one of the following: (*a*) failure, mechanical or electrical, of equipment or services; (*b*) mal-operation or error by an operator or service department; (*c*) fire, explosion, impact by missile, etc. An unplanned occurrence may occur either inside or outside the plant area.)

Commissioning, like operation and, indeed, design, must be a team effort, and the operator will need the full support of most of the ancillary departments referred to in the section concerned with Departmental Structure. For example, the Maintenance and Services Departments will be intimately involved. Furthermore, commissioning will not be limited to functional testing of the plant and its associated equipment, services, instrumentation, etc., but will embrace maintenance programmes and procedures, with particular reference to proof-testing methods and frequencies. (Maintenance may be defined as inspection, adjustment, repair or replacement involving departments ancillary to the Operating Department. The plant as a whole or in part may, or may not, be shut down at the time.) Commissioning, therefore, provides staff at all levels and in all departments with a familiarization and learning opportunity which is unique.

The results of the commissioning exercise will be presented in a form which facilitates comparison between the expected results, which should be set down in the programme, and those actually obtained. Also, any modifications to design or operational safeguards, which have been shown to be necessary, will be described. This document will, in effect, become an appendix to the Safety Document (Operation) and be reviewed by the safety assessment team before the Authority to Operate is raised.

It is not unusual for the period originally allocated for commissioning to tend to be compressed, as a consequence of delays in construction and allegedly immutable start-up dates. However, the prudent operator will not allow himself to be stampeded into accepting unrealistic time scales which do not permit him adequately to prove his plant, process, and procedures. He will be ever-mindful that it is he who at the end of the day will 'be carrying the can'.

Operation

All staff in the Operating Department should be instructed as to the requisite operational safeguards by means of written operating instructions and operating limits, backed up by written safety regulations and emergency instructions. Ideally, staff should be so indoctrinated that they believe that only by their strict adherence to these instructions can the safety of people on site and the general public be assured. The prime purpose of the documents is, clearly, to instruct; however, they could when presented coherently comprise the foundation of the Safety Document (Operation).

Operating instructions

The written procedures that govern every routine step that is taken in operating the component parts of a facility are known as operating instructions; of course, not all have a safety connotation. Each operator in a supervisory capacity must be fully conversant with every detail of the instructions governing his area of responsibility and be certain that all his subordinates understand them thoroughly, carry them out faithfully, and continuously have ready access to them. (A firm rule cannot be suggested regarding access, which is a matter of choosing the most suitable arrangement for the local circumstances; in one case, the issue of individual copies to each person may be justified, whereas in another the posting up of communal copies may suffice.)

Any effort that is spent by supervisors in testing operators on their knowledge of the operating instructions and, indeed, on the other operational safeguards, will yield a good return.

Operating instructions are best written in a step-by-step style. They should be totally unambiguous, require no interpretation whatsoever and, ideally, leave not the slightest margin for judgement; indeed, the acid test would be whether they would be workable by anyone with average intelligence, who can read and write and knows the identity and location of every item referred to.

Operating limits

Complementary to, and usually presented as an integral part of, the operating instructions are the operating limits. These are numerical values, or a range of values, of physical parameters appropriate to a process or plant. The values relate to safety and must reflect design limitations, such as pressure, as well as operational aspects, such as rate of reaction, inventory, and concentration of hazardous by-products.

The limits should be established at a sufficiently early stage to enable the appropriate instrumentation to be included in the design so that compliance can subsequently be effected and demonstrated. The latter aspect is considered further in Chapter 1.5. It is emphasized that the operator has a major part to play in the setting of operating limits. If the limits are to be effective, they must be workable, and the operator will have strong opinions on that subject. Furthermore, he must confer with instrument design engineers on the siting of instruments and controls and the sensitivities and displays of proposed instrumentation relative to envisaged tolerances on operating limits.

Finally, the status of operating limits must be established. If they are to be effective, then they must be sacrosanct. This can only be efficiently achieved if the limits are realistic and the requirement of compliance has been one of the foundations of a project, or modification, from the conceptual stage. The quality of staff has a bearing on this subject. The author feels strongly that any temptation to employ people who are much overqualified for their particular duties should be resisted. Such persons are prone to hold the view that operating limits are for the benefit of lesser mortals, whereas they themselves are capable of judging matters for themselves.

Clearly, the role of operating limits should also be emphasized in training programmes and in discussions between the employer, employees, and trade unions. The keynote is that the object of operating limits is to protect the employees, the plant, and ultimately the public.

Safety regulations

Safety regulations are written standing orders which cover all routine, mandatory procedures other than operating instructions. They are complementary to the latter; some relate to specific operations within the facility and others to the facility in general. They will cover matters such as the use of adequately trained staff only; the designation of authorized persons; the mandatory status of operating instructions, operating limits, and emergency instructions; the appropriate protective clothing; restrictions on eating, drinking, smoking, and the introduction of non-ferrous tools; the use of antistatic equipment; the designation of restricted areas; security; the entry of visitors; the designation of 'Proscribed Areas', as defined in certain electricity regulations; leaks and spillages; minimum acceptable manning levels; general alarm systems (other than those specified in operating instructions); the reporting of accidents; procedures which relate to potential hazards specific to the particular facility. The foregoing is not intended to be a checklist; there are many other matters. However, one important subject, which is common to most facilities, is the need to control inspection, maintenance, repair or modification. There is no doubt that this is best achieved by the use of a system of Permit to Work.

Permit to work

Reference was made earlier to the interface between operators and maintenance personnel. This is particularly important during inspections, preventive maintenance, breakdown maintenance or modifications. These, and perhaps other situations, will not have comprehensive, permanently installed safety systems for the protection of personnel and plant. In such circumstances safeguards may be exercised by a permit to work system[4].

The permit to work system is an administrative method for exercising control of personnel and plant during operations outside the steady state which have a potential for creating a hazard. Certain of these operations, for example the testing of live electrical equipment or entry into a confined space, will have legal requirements as to the procedures to be followed. There will, however, remain a large number of operations which are not covered in detail by legal requirements. The permit to work is intended to embrace both compliance with regulations and good safety practice.

The key to the system is the authorized person; it is he who will issue the permit and specify the safety procedures and controls to be exercised. This is a responsible position and requires an appreciation of the plant, the process, the implications of the proposed operation in respect of the plant status and, possibly, associated plant or the site as a whole.

The need for an operation having been agreed, the first entry on the permit is the specific and unambiguous identification of that operation and its location. This enables the authorized person to begin an assessment of the pertinent potential hazards from the plant in question, and possibly from others (by interaction). Identification of the potential hazards leads to definition of the precautions to be taken. All of this may be achieved with the aid of a checklist, which is relevant to a particular plant, process, site, and possible operations. Any specific safety actions, such as plant isolation, custody of interlock keys or removal of fuses, must be listed. In addition, requirements such as temporary protective instrumentation, safety equipment or frequency of gaseous atmosphere sampling and analysis must be listed.

For the operation, a team leader is appointed and named on the permit; in addition, the names of all other personnel directly involved should be entered. It is essential to limit the number of personnel to those necessary. Excessive numbers may inhibit evacuation or nullify a premise of the safety assessment.

The authorized person issues a permit to work for a specific operation and enters the time of commencement and duration of the permit. The team leader signs the permit to indicate that he understands the nature and location of the work, the precautions to be taken against potential hazards and the action necessary, in the event of a hazard developing.

During the period in which the work is carried out no departure from the defined status of the plant may occur without the agreement of the authorized person.

On completion of the work, the team leader returns the permit, having signed that it has been completed and that all personnel and equipment have been withdrawn. The authorized person may then liaise with plant management for its restoration to a normal state.

The permit to work is a *pro-forma* document relevant to the activities of a site; it covers precautions against potential hazards generally existent on the site, with the scope to add precautions against hazards from a particular plant, process or rare operation.

The authorized person(s) should be an individual or individuals appointed by name. If more than one authorized person exists on a site there must be demarcation of areas of responsibility and full communication to avoid any possible interactions, which, although individually safe, in coincidence could give rise to a hazard.

Emergency instructions

The mandatory procedures discussed previously must cover every routine aspect of operation which may reasonably be expected to arise during the lifetime of the facility. All other aspects will, if they have a significant safety connotation, be regarded as emergencies and, as such, will be catered for in emergency instructions.

If the design and operational safeguards have been properly assessed and been judged to be adequate, it follows that emergencies should arise only very infrequently. Emergencies may be considered to fall into two broad categories, viz. incipient emergencies, i.e. where it may still be possible to recover control and redeem the situation, and actual emergencies, i.e. where the point of no-return has been passed and effort can be directed only towards mitigating the consequences. An example of the former category would be, say, the stoppage of an important agitator, because of a power failure, followed by the standby diesel generator's subsequent refusal to start. Fire, explosion, or loss of containment of a toxic material as a consequence of a dropped load, would all be classed as actual emergencies.

With regard to incipient emergencies, the emergency instructions will, where possible, specify what further action must be taken, with the aim of preventing the development of an actual emergency.

All that needs to be added to what has been mentioned earlier concerning actual emergencies is that the instructions need to be crystal clear, and must be specific as to when and how they would be invoked, and how all concerned would be made aware without delay that they had been invoked. They must also be specific on such matters as immediate evacuation of the area and delayed evacuation, i.e. standby alert, and on the circumstances in which some attempt should be made to shut down the facility.

Safety Inspections

The object of safety inspections is to check that: (*a*) the plant and its process are what they were declared to be in the Safety Document (Design); and (*b*) the plant is being operated in accordance with the Safety Document (Operation). In other words, are the conditions under which the Authority to Operate was issued being adhered to?

The inspections should be carried out under the aegis of the Safety Assessment Committee by a small team of professional staff, independent of line management. (Line management carries out its own form of surveillance as part of its routine duties, but has no part to play in inspection, except to afford all reasonable assistance to the team.) An inspection need not follow any set pattern, but is best tailored to suit the needs of the particular facility. Beforehand, the team should be given an opportunity to study the safety documentation, the written instructions referred to on page 122 and the ATO. During the course of an inspection they should be allowed access, where practicable, to all parts of the facility for the purpose of viewing the operations, and be given the opportunity to interview supervisors and their subordinates and to examine log books, record sheets, accident reports, etc.

The team will report in writing to the Safety Assessment Committee and send a copy to the manager of the facility. Their comments should be constructive and trivial matters should be avoided; otherwise, inspection will merely be brought into disrepute.

The time expended on an inspection will probably be in the region of three days; the preparatory study of the documents, the actual visit to the facility and the preparation of the report will each require about a day. A competently operated facility should not need more than, say, two inspections per year, one of which should take place shortly before the ATO is reviewed by the Safety Assessment Committee.

The Human Element

A plant which has been well designed and constructed depends for its safety almost entirely on people. It requires operators to intervene in its functions, and it depends almost entirely on the operators as to whether it functions correctly (and safely) or, in their hands, becomes potentially hazardous. Human factors are discussed elsewhere (Chapter 1.6), but it must be emphasized here that, however adequate the staff may be in terms of quality, quantity, and training, it is nevertheless vital for the operator to maintain their morale, enthusiasm, interest, and alertness.

Questions 1.4

1. In principle, how should the operator of a hazardous plant approach the maintaining of operational safety?
2. What is the operator's role in the following:
   (a) Plant design
   (b) Plant construction

(c) Plant commissioning

(d) Formulation of safety policy?

3. Select some type of plant which is known to involve a hazardous process and discuss how the appropriate safety policy should be implemented by the operator. Describe the responsibilities of each level of management and the subordinate employees so that there is an effective implementation of the safety policy.

4. Discuss the means by which an independent safety assessment of the plant operation may be derived and indicate the various safety documents which may be taken into consideration. How would a responsible operator use the various safety documents to assist in the organization of normal plant operation, emergencies, and safety inspections? In dealing with the 'human element' what kind of instructions would be issued by the operator?

References 1.4

1. *Chemical Plants and Laboratories Safety Assessment*. UKAEA Report SRD R8.
2. Health and Safety at Work, etc. Act 1974.
3. *Effective Policies for Health and Safety*. Health and Safety Executive Report, 1980.
4. Permit to Work. Technical Note HELP 1/78.

High Risk Safety Technology
Edited by A. E. Green
© 1982 John Wiley & Sons Ltd.

Chapter 1.5

Application of Regulatory Control to High Risk Plant

G. H. Helsby, J. Pugh, and G. Hensley

Introduction

In all countries in which complex and potentially hazardous installations are operated, there has been an increasing tendency towards formalizing safety requirements by the promulgation of regulations relating to high risk plants. Risk in this context is defined as the product of the consequences of an abnormal event and the probability of that event occurring [1]. This definition and relationship will be used throughout the remainder of this chapter.

Following a summary of the current attitudes towards regulatory control of high risk plants, suggestions are made for a scheme which can be applied to all types of high risk installations. It is considered that the system is sufficiently flexible to embrace both legal requirements, where applicable, and self-imposed safety arrangements and procedures required to cover novel installations and processes to which existing legislation is either inapplicable or insufficiently comprehensive.

A number of accidents and abnormal occurrences, plus the activities of environmental pressure groups, has resulted in a greater awareness by the community of the potential hazards of advancing technology, in terms of more complex plants and larger manufacturing operations. This awareness has led to requirements that adequate consideration must be given to the safety of the design, construction, and operation of all such plants in order to ensure that the risk to both workers and the general public can be demonstrated to be acceptably low. Compliance with this requirement may be based on government legislation where relevant enactments are applicable or on voluntary self-imposed safety arrangements and procedures.

Current Situation in the United Kingdom

Prior to the coming into force of the 1974 Health and Safety at Work, etc. Act [2] regulatory control of high risk plants was divided between requirements from a number of enactments [3]-[8]. While these Acts and associated regulations and codes of practice gave requirements and guidance on standards for certain specified aspects of the design, construction, and operation of hazardous plants, none of them, except the Nuclear Installations Act[9], provided any comprehensive overall structure upon which regulatory control of high risk plant could be based. The shortcomings of legislation prior to 1974 were recognized in the Robens Committee Report[10] from which the Health and Safety at Work, etc. Act 1974 evolved.

In fairness to the majority of industrial concerns operating high risk plant it must be stated that conscientious efforts have been made, not only to conform with the legal requirements which exist, but also to attain very high standards in design and operation procedures.

Before the Health and Safety at Work, etc. Act 1974 had come fully into force a disastrous explosion occurred at a works at Flixborough on 1 June 1974. The disaster inquiry report[11] highlighted in the section 'Lessons to be learned' (pages 32–37) the need for more comprehensive systems of regulatory control to be applied to complex and hazardous installations.

Towards the end of 1974 the Advisory Committee on Major Hazards was set up by the Health and Safety Commission to consider the safety problems associated with large-scale industrial premises conducting potentially hazardous operations.

The first report of this committee[12] reviewed the situation and made various recommendations, based mainly on inventories of hazardous materials, in respect of the notification and survey of such premises. A Consultative Document[13] was published in 1978 for 'Hazardous Installations (Notification and Survey) Regulations 1978'.

A further report of the Advisory Committee on Major Hazards was published in 1979[14] which suggested future legislation in the form of regulations and possibly the introduction of a licensing scheme for selected notifiable high hazard installations.

Proposed Scheme of Regulatory Control

From the above summary of the current regulatory position in the UK it is apparent that recommendations have been made for new regulations which should give a better overall basis for control of high risk plants. Examination of Reference (14) shows that many of the recommendations for controls, actions, arrangements, and procedures are very similar to the systems which have been developed for compliance with the requirements of the Nuclear Installations Act for licensed nuclear installations. The scheme outlined in the remainder of this chapter has been broadly based on the experience gained in the control of nuclear installations plus an interpretation of what regulations are likely to arise from the work of the Advisory Committee on Major Hazards.

It is considered that even in the absence of regulations requiring specified procedures, the following suggestions would result in a demonstrably adequate regulatory control system. The scheme also has a format which can be relatively easily modified to include any mandatory requirements which may follow future legislation and it is, therefore, considered that it could be of general application to all types of high risk plants.

The implementation of a regulatory control system, whether mandatory as a result of legislation or voluntary, first requires a conscious decision by the company executive (assumed in this document to be the General Manager) to define administrative arrangements and procedures. In addition, it is necessary to formulate criteria and guidelines as a framework around which the control system will operate. It should be appreciated that the regulatory control concept does not in any way detract from the responsibilities and liabilities of an employer (or his designated representative) to (a) his employees, (b) any other person on the site, and (c) the public. The system is designed solely to assist in the achievement of these aims by detailed technical consideration of the safety aspects of high risk plants.

Administrative arrangements should include provision for suitably qualified and experienced personnel in order that the control procedures can be implemented effectively.

The basis of the proposed regulatory control consists of the preparation of sequential safety documentation and assessment of documents at each stage. If it is considered that adequate attention has been given to all the safety aspects of the design, construction, commissioning, operating, and maintenance as appropriate, a recommendation is made that permission be granted allowing procedure to the next stage.

The various aspects of the regulatory control system applicable to high risk plant will be considered as follows:

Safety Working Party

A Safety Working Party (SWP) would consist of a number of members who would have a detailed technical appreciation of the proposed plants and processes and their potential hazards to employees and the public. The SWP would receive and assess safety documentation relating to the proposed plant and advise the General Manager of the safety arguments made in relation to the adequacy of safeguards to be provided. It is emphasized that the role of a SWP is to advise the General Manager.

The SWP should have a degree of independence from the commercial and production considerations of the project. Preferably at least one suitably qualified member of the SWP should be independent of the site (e.g. from a member company, another site or a consultancy).

The SWP discussed here should not be confused with any Safety Committee set up under the Health and Safety at Work, etc. Act 1974. It is considered that this latter committee will be mainly concerned with general and everyday matters, particularly from an employer/trade union/employee viewpoint. It does, however, follow that the Safety Committee will be appraised of the proceedings of the SWP and of any recommendations that are made. The Safety Committee must also have the option of requesting the SWP to institute studies on any safety matter relevant to the particular plant or process, or to the site as a whole.

Preliminary Safety Report

This first phase of safety documentation should consider the specific plant or process and its wider implications for the site as a whole and the possible off-site impact. Sections of the report concerning the site as a whole and the off-site impact should include such considerations as siting policy, hazard assessment, and increment of risk involved. In addition the report should consider the ability of existing resources such as fire fighting and medical facilities to cope with potentially greater emergency situations. Having identified the increment of risk the new situation should be considered in the light of any existing emergency arrangements made with local authorities or adjacent industrial concerns. For a specific project, if the preferred processes were not the safest of the options available, it would be necessary to justify the selection of the process for the particular application and location.

Prior to commencement of construction of a plant it would be necessary to submit the preliminary safety report to the SWP for approval. This report would include the philosophy of design and construction of vessels, instrumentation, and mode of control. It would also indicate the approximate inventories of process materials, working temperatures and pressures and would pay particular attention to safety aspects especially if they were to involve novel solutions or hazards of a type not previously present on the site.

The objective of the preliminary safety report is to identify all significant potential hazards that could arise from the proposed plant and then to convince the assessors (SWP) that adequate safeguards will be provided and that these can be effective.

It is envisaged that the preparation of such a document may involve a Research and Development Department, a Production Department, Construction Department, and Design Offices. It is likely that the Design Office would play a major role and would coordinate the preparation of the document.

The approval of a preliminary safety report by the SWP would lead to a recommendation to the General Manager that permission to construct could be granted. A formal notification of permission to construct would then be issued by the General Manager and design work would proceed followed, when sufficient detailed design was available, by construction.

Control During Design and Construction

During design work and construction there should be Quality Assurance control to company, national or international standards and close liaison between design and construction departments, any external contractors, and suppliers of equipment or plant.

It is essential that any modifications proposed after the approval of the preliminary safety report or any particular difficulties experienced during the design or construction phases, that could have an affect on the initial safety assessment, should be referred back to the SWP for approval before implementation.

In the event of any plant item being damaged and/or repaired, careful assessment of its continuing suitability should be carried out. The reasons for reaching a conclusion that such items are still suitable for the specified application should be fully documented and should be submitted to the SWP before a final decision is taken.

Final Safety Report

On completion of the plant and before it is commissioned a Final Safety Report should be submitted to the SWP for assessment. The report should demonstrate how the aims and safety principles of the preliminary safety report have been achieved in practice.

The report should consider structural and constructional aspects, process parameters, instrumentation, and the sensitivity of the process to control from a safety viewpoint. The potential hazards identified in the preliminary safety report should be considered in relation to the installed safeguards and safety margins. In situations where reliance is to be placed on operator control the available safety margin and response times required must be clearly stated.

For particular plants it may be that a reliability assessment would be of great value in assessing the redundancy and diversity of safety-related systems.

The range of process flowsheet conditions, which have been assumed by the designer as applying to the particular plant, should be clearly stated. These will subsequently be used in the formulation of more restrictive Operating Limits. In addition, the design life of the plant and the basis for that life, e.g. corrosion allowances, should be quoted.

Finally, the assessment should consider the possibility of interaction of the plant under fault and accident conditions with other plants either on or off the site. The effect of external forces such as wind, flooding, earth tremors, missiles (generated within or outside the plant) should also be included in the assessment.

It is not intended that this report be written after the completion of the plant, rather that it should be written in parallel with the construction phase. It is, however, essential that the completed report does accurately represent the final plant and includes any modifications or significant differences in installed plant or equipment.

At the time that the Final Safety Report is submitted to the SWP, two further documents relating to the plant should also be submitted to the SWP. The first of these documents relates to construction of the plant and will certify that it has been constructed in accordance with the design and the specified quality assurance procedures and requirements. Any special inspection reports pertaining to safety should be referenced or appended to this report.

The second report, The Operational Assessment Report (OAR), to be read in conjunction with the Final Safety Report, should include such items as staffing levels in the light of the assumed operational control philosophy and identify any special training requirements for operators. The OAR should also identify items requiring routine maintenance (statutory or otherwise). It may also be convenient to include consideration of what additional emergency arrangements may be required for the site as a result of the addition of this plant.

Using the design parameters and flowsheets as a guide, formal operating limits for such parameters as temperature, pressure, concentrations of reactants, acidity, and flow rate should be drafted. Where possible compliance with such operating limits should be verifiable using installed instrumentation.

It is probable that all Works Departments would be involved in the preparation of these reports. The Design Office leading in the Final Safety Report, Construction Department for the Construction Report, and the operators for the OAR.

When the SWP has assessed the Final Safety Report, the Construction Report and the Operational Assessment Report it will, if all these are considered to be adequate, recommend to the General Manager that permission be given to commence commissioning.

Commissioning Arrangements and Commissioning Report

The general approach to commissioning and its relevance and involvement in the verification of installed safety equipment and feasibility of essential operational controls would be described in a document concerned with commissioning arrangements. This document would be submitted to the SWP who, if finding it acceptable, would recommend to the General Manager that permission be given to start commissioning.

It may be that it is appropriate, in certain instances, to commission the plant in stages and the arrangements should take cognizance of this requirement. Furthermore, it may be desirable to commission using simulates followed by trace quantities prior to production conditions.

The commissioning period should be specified in terms of time or quantity of material to be processed. Any request for an extension of these limits should be referred to the General Manager through the SWP.

At the completion of commissioning a Commissioning Report would be submitted to the SWP covering safety aspects and demonstrating that the installed plant functions correctly and can be safely operated. If the SWP is satisfied with the report it will recommend to the General Manager that a Permit to Operate be issued.

The SWP may wish to recommend a safety review of the plant after a specified period of time. Alternatively, there may be a site policy of plant review which the SWP sees no reason to vary for this plant.

Control of Operational Plants

It is suggested that a site should adopt a policy of periodic safety reviews for all plants. On sites which have operational plant before the introduction of the regulatory scheme, such operational plants should be reviewed as early as possible (see pages 135–137).

Operational and maintenance data may indicate that a revision of the originally ascribed plant life is necessary or, alternatively, that it could be extended. The safety review should specifically consider this matter.

A rigorous scheme of control over modifications should be instituted. Any proposed modification that might affect safety should be subject to the regulatory control system before it is implemented. If a modification is approved, then drawings and documentation must be updated. The implications of the modification to control and operational philosophies and existing emergency arrangements must also be considered in the approval of any such modifications.

There should be routine checking of compliance with operating limits and recording of such compliance.

A permit to work scheme incorporating a formalized and recorded system of plant hand-over should be instituted for all plant maintenance for the protection of personnel. Such a scheme could require central coordination on sites where interconnected plants exist which are supervised by different personnel.

On certain types of plant, management may wish to consider the institution of certification of personnel with specialist training as necessary. Such expertise would be recorded and updating of training carried out periodically.

Finally, fault data generated on the plant and similar plants should be collected and be capable of retrieval and interpretation. This will assist in the planning of maintenance schedules and the design of future plant. Table 1.5-1 summarizes the stages of regulatory control applied to a new plant.

Plants Operational before the Introduction of the Regulatory Scheme

In some cases a site may have high risk plants already operational before a regulatory system of control is introduced. In such circumstances it is suggested

Table 1.5-1   Sequence of regulatory control stages for a new high risk plant

Pre-design →	Design →	Construction →	Commisioning →	Operation
Management Board decision on need for new plant.	Preliminary plant design proceeds.	In most cases construction will commence before detailed design is completed.	Before commissioning is allowed to proceed a document detailing *Commissioning Arrangements* must be prepared and submitted to the SWP. If satisfied the SWP will recommend that the General Manager gives:	Compliance with Operating Limits.
Research and Development Work	Liaison between designers and research and development staff.	Liaison between designers and operators will continue during this stage.		Checks on validity of ascribed plant life.
Liaison between design and operation departments and set up of project teams.	Preparation of safety documentation now in progress leading to the production of a *Preliminary Safety Report.*	If necessary the SWP may be consulted, in particular if a novel solution to a safety principle is proposed. This would avoid unnecessary expenditure in the event of the proposed solution being inadequate.	PERMISSION TO COMMISSION	Periodic Safety Reviews.
Consultation with Site Safety Organization during feasibility studies.	Approval of this report by the Safety Working Party (SWP) is necessary before a recommendation can be made to the General Manger that construction may proceed.	Quality Assurance will monitor all aspects of the construction, supplies of plant, equipment, and materials.	When commissioning has been completed a *Commissioning Report* will be submitted to the SWP.	Monitoring of maintenance and plant hand-over procedures.
		The *Final Safety Report* will be prepared as design and construction proceed and together with the *Operational Assessment* and the *Construction Report* will be submitted to the SWP for assessment.	If the SWP is satisfied that the installed plant functions satisfactorily and can be safely operated it will recommend to the General Manager that a Permit to Operate be issued.	*Supplementary*
	PERMISSION TO CONSTRUCT			Collection of plant and equipment failure data for reliability assessment.
		If considered to be satisfactory, the SWP will recommend to the General Manager that the project may proceed to the commission stage.	PERMISSION TO OPERATE ON PRODUCTION BASIS	Feedback of above data to designers, SWP and in some cases to national reliability data banks.
				Modifications to Plant
				Agreement in principle that modification is necessary will be followed by a complete repeat of the Regulatory Control Sequence starting at the Design Stage.

that the General Manager should arrange for a complete safety review of all existing plants on his site to be carried out as soon as possible. Such reviews should consider in detail all aspects of the design and operation in terms of safeguards provided and their adequacy in respect of the potential hazards with the plant during normal operation, maintenance, and foreseeable incidents or mal-operation. Many of the criteria identified in the foregoing sections on Preliminary and Final Safety Reports will be relevant in the preparation of the safety reviews for existing plants.

The safety reviews should be presented as a Process, Plant, and Operational Safety Report for each plant which should include recommendations as to what modifications are considered to be required in respect of the safe operation of the plant. This document should be submitted to the SWP who will make recommendations to the General Manager as to its contents and the significance of suggested modifications for improved safety provisions in relation to the plant and site as a whole.

When carrying out such reviews it should be appreciated that existing plants are being assessed against safety criteria which may be more recent than the actual plant design standards. Furthermore, the methodology of the assessment may, using modern techniques, be more rigorous than the original safety study. As a result of the review it is probable that some recommendations will be made as to modifications.

It is, however, realized that in the case of existing plants it may not be reasonably practicable, or even in some cases possible, to carry out all the modifications which would be necessary to achieve safety provisions equivalent to proposed, new or recently completed plants. Some balance will have to be made between prohibitive cost and the highest standards attainable and one approach would be to divide recommendations for improvements into categories of descending order of safety priority such as the following:

1. Essential to safety – must be done as soon as possible despite the cost.
2. Desirable improvement in safety – should be done within a reasonable time, dependent on resources available and expected life of plant.
3. In line with good practice – not essential but advisable as time and money permit.

On some sites there may be several individual high risk plants. In order to allocate effectively safety assessment resources, it is suggested that a priority rating be given to such plants. This may be achieved by assessing, if necessary by approximate methods, the relative potential hazards of the plant.

Conclusions

The method of regulatory control proposed above for high risk plant has already proved to be effective in the nuclear installations field. In the event of legislation or public opinion requiring operators of other high risk plant to adopt similar standards, this system can instil adequate confidence that a logical method has been used to demonstrate that full consideration has been given to all aspects of safety and that the operator has done all that is reasonably practicable to ensure the plant does not cause harm or damage on or off the site.

The control system proposed has two essential parts. First, a group of technically qualified and experienced personnel who are capable of writing comprehensive and adequate safety reports at the required stages. Second, an independent group of assessors with relevant experience and an appreciation of the application of safety principles to this type of plant. The second group would constitute the Safety Working Party and it is essential that they have sufficient time and resources to do justice to their vital role in the system.

The regulatory control procedures can undoubtedly provide for the minimizing of the risk of loss of plant or life on site and lessen the probability of off-site injury or damage. As such this would represent actual or potential financial savings in the long term. However, such savings may be offset in the short term by the costs involved in the employment of additional, suitably experienced, staff to implement the system.

EXAMPLE 1.5-1 Regulatory Control Applied to a New Plant

On a site, located approximately 1.5 kilometres from a small town, a company already has a petrochemical plant. A proposal has been made for diversification by the installation of an additional process which requires the use of chlorine. A storage system for 10 tonnes of chlorine is envisaged.

This example concerns only one aspect of the proposal, namely the safety aspects of the chlorine storage plant.

Using the administrative regulatory procedure described in the text, and assuming the existence of safety criteria and a safety working party structure, the first safety document would be the Preliminary Safety Report.

The Preliminary Safety Report should address itself to three major aspects: siting, containment, and the consequences of loss of containment. The latter is particularly important because of the proximity of a significant population. The three aspects are interrelated in that siting may have an impact on mechanisms for breaching containment, e.g. vulnerability of supply and delivery pipes and impact from possible missile sources; while siting clearly has a relationship with dispersion of any release.

The Preliminary Safety Report could subdivide the above subjects, for example:

1. Siting should consider if there are alternative locations on the site and which is 'safest' bearing in mind mechanisms for breaching containment.
2. Possible breach mechanisms would include corrosion, overpressurization,

vehicular impact, missiles generated on site by explosion or rotating machinery.
3. The consequences, covering both on- and off-site personnel, would indicate any additional requirements of emergency arrangements and services.

In effect, for this example, the Preliminary Safety Report would represent a feasibility study with the emphasis on safety. However, on the basis of a company's safety policy and criteria, and legal requirements, the safety study would also begin to indicate the extent of safety-related engineering, i.e. part of the capital costs of the proposal. These costs may not have been apparent when the scheme was initially proposed. Indeed, it may be that such safety considerations ultimately determine the viability of the proposal for that site.

The subsequent Final Safety Report would demonstrate how the potential hazards and initiation mechanisms have been dealt with in practice. The report would also consider the consequences of different degrees of hazard, e.g. differing release rates under various weather conditions.

EXAMPLE 1.5-2 Modifications to an Existing Plant

A continuous chemical process requires modification to increase throughput while utilizing as much of the existing plant as possible.

The process requires the heating of two reactants under an inert atmosphere of nitrogen. The relative concentration of the two reactants is critical in the production of a stable product. Any residual unreacted material could lead to the evolution of flammable and toxic vapours.

A decision has been taken to achieve increased throughput by operating with additional heat input at a reduced nitrogen pressure.

The Preliminary Safety Report would indicate the broad outlines of process control which are safety related. These would include control of steam and nitrogen pressures, minotoring of concentration of reactants in the process, the product, and the waste streams. The report would indicate requirements for additional ventilation and electrical power supplies to relevant standards bearing in mind the flammable and toxic nature of the materials. The report should also consider the suitability of the present plant and ancillary equipment for operating under reduced pressure.

The Final Safety Report would consider all these aspects in detail. It would be necessary to demonstrate that the safety-related systems have the necessary reliability. In cases where common mode failure could occur, the question of diversity of control should be specifically considered. The report would discuss response times of detection systems and compare them with times to the development of a dangerous condition and response times for automatic or operator remedial action. Such matters as the reliability of nitrogen supplies and pressure regulation for inert gas and steam should be studied. The permitted degree of air in leakage, its detection and consequences should be discussed in the light of safeguards to be provided on the plant. Toxic aspects should be dealt with in the light of detection of release, ventilation provisions for normal and abnormal conditions, and the consequences to on-site personnel.

The interaction of the modified plant with others on the site should be carefully considered. The possibility of incidents on other plants initiating an incident on this plant, and vice versa, should be carefully considered particularly in view of the increased inventory of materials in the modified plant.

The Final Safety Report and the Operational Assessment Report should logically demonstrate that the potential hazards identified in the Preliminary Safety Report have been reduced to an acceptably low level by a combination of design and operational control. The documents should particularly emphasize any redundancy or diversity of safety instrumentation and controls. Evaluation of the margin of safety for automatic and operator intervention during any deviation from steady-state conditions towards danger levels should be included. The object is to produce a credible, logical, reliable, and economic solution which demonstrates the attainment of a safe plant.

Questions 1.5

1. Your company has decided to adopt a voluntary regulatory control system and you have been appointed to head a small group with the task of preparing detailed proposals for the management board. How would you proceed, bearing in mind your company's present safety organization, the availability of staff with relevant experience, and the possible need for external advice or consultations?

2. You are told that you have been appointed as Chairman of the Safety Working Party for a proposed new scheme of voluntary regulatory control. Your works contain a number of high risk plants including a large-scale high pressure hydrogenation plant and storage and handling facilities for large quantities of flammable liquids. Make suggestions as to what members you would request for your committee from within your own organization and, if possible, from sources totally independent of the site.

3. As Chairman of the Safety Working Party you are not satisfied with certain aspects of the proposed Commissioning Schedule for a new high risk plant. What actions would you take and who would you consult before finalizing your committee's recommendations for the General Manager?

4. In Examples 1.5-1 and 1.5-2, what qualifications and experience would be required by the staff responsible for the production of the necessary safety documentation and what time scale would you envisage for the preparation of the documents?

References 1.5

1. 'From systems reliability assessment to long term safety assurance.' Paper No. ZA/Z/1-5 presented at Second National Reliability Conference. Birmingham, March 1979 by G. Hensley, SARA Ltd, Bewsey Street, Warrington.
2. Health and Safety at Work etc. Act 1974.
3. Factories Act 1961, plus relevant regulations.
4. Petroleum (consolidation) Act 1928.
5. Explosives Act 1875 as amended by the Explosives Act 1923.
6. Nuclear Installations Acts 1965 and 1969.
7. The Radioactive Substances Act 1960.
8. Alkali etc. Works Regulations Act 1906.
9. Gausden, R. 'Regulatory Control of Nuclear Installations in the United Kingdom.' Paper given to Energy Engineering Convention 1975. 'Lines of Development in Energy Engineering' at Düsseldorf, May 1975.
10. *Safety and Health at Work*. Report of the Robens Committee.
11. *The Flixborough Disaster*. Report of The Court of Inquiry.
12. Advisory Committee on Major Hazards. First Report 1976.
13. Hazardous Installations (Notification and Survey) Regulations 1978 – Consultative Document.
14. Advisory Committee on Major Hazards. Second Report 1979.

High Risk Safety Technology
Edited by A. E. Green
© 1982 John Wiley & Sons Ltd

Chapter 1.6

Human Factors in High Risk Technology

SECTION 1.6.1 Human Reliability in Risk Analysis

J. Rasmussen

Introduction

Two general trends in modern man-made systems of any kind are centralization and automation in order to secure effective and economically optimal operation. This is the case for industrial production systems, for transportation systems, and for information systems used as basis for commercial or civil service decision making. Centralization leads to large systems with potential for major consequences of mal-operation, and special precautions are necessary to achieve the extremely low probability which is acceptable for accidental release of this hazard. Automation does not remove people from the systems, but merely moves them to other functions related to maintenance, repair, and to higher level supervisory control and decision making. Centralization and automation together typically imply that the effect of human decisions and acts can be very drastic. At the same time, the basis for these decisions becomes more obsure due to the increased system complexity.

Another aspect of modern industrial activities is the short time-span from product or process development to large-scale operation, which does not permit the evolution of safe systems from feedback of operational experience. The consequence is that for safe design of industrial systems, empirical design rules and equipment standards are gradually being replaced by explicit risk criteria and analytical techniques for safety assessment.

This development is based on the fact that in high risk technology, the release of the risk potential will not depend upon simple, individual equipment faults or human mistakes. In a system of balanced design, major accidents will depend on a complex chain of events including equipment faults and latent risky conditions, together with human mistakes and errors. Taken individually, such events are typically frequent enough to allow data collection or estimation. Analytical risk estimation depends on such data, together with a model of the relevant accidental chains of events in which the people in the system play a very important role.

The Anatomy of an Accident

Analytical assessment of the risk posed by an industrial installation depends upon a model of the accidental chains of events which may lead from normal operating states to the state of accident. A basic assumption behind useful models of accidents is typically that major consequences of mal-operation depend upon the loss of control of large accumulations of energy and/or toxic material in a system. Such energy models have been used for evaluation of general work accidents[1][3] and have been the basis for MORT, the comprehensive 'Management and Oversight Risk Tree analysis' developed by Johnson[2]. Johnson defines an accident in the following way: 'An accident is an unwanted transfer of energy, because of lack of barriers and/or controls producing injury to persons, property or process . . .' This energy model is useful to describe the anatomy of a typical industrial accident and to identify some of the general problems in risk assessment (see Figure 1.6.1-1). The typical accident in an installation of balanced design is not caused by only one simple equipment fault or human error; on the contrary, major events will depend on a complex chain of events including equipment faults, latent risky conditions from repair and modifications as well as human mistakes and decision errors (for analysis of actual events, see Reference (11). Analytical risk assessment depends upon the assumption that these individual events occur frequently enough to allow empirical fault rates to be collected. At present, however, the collection of human error data suffers from some methodological problems, empirical data are scarce and analysis has to be based on estimates and special assumptions.

Following Figure 1.6.1-1, the course of a typical accident has three different phases in the loss of control of the accumulated energy in the system. Transition from normal operating state into a state of abnormal or disturbed operation can be caused by an abnormal event related to equipment failure or human errors during normal work conditions such as mistakes during valving and switching, etc. Or, normal operation or changes in system state can release the effect of a latent risky condition from equipment failure or human error left over from maintenance, test or calibrations.

Industrial safety generally depends on alternative means for control when normal operation breaks down, either in the form of disturbance control algorithms performed by operators and/or the control system, or in the form of automatic transfer of the plant to a safe state by a separate protective system. For this phase, two categories of human errors are important. One is the introduction of latent risky conditions in safety systems during test, repair, and calibration. This category is more likely to be met in stand-by supply and safety systems than in active process equipment where faults more readily reveal themselves and are corrected by the operator. In practice, the overall reliability of safety systems based on redundancy is more dependent upon the probability of subsequent

ANATOMY of an ACCIDENT

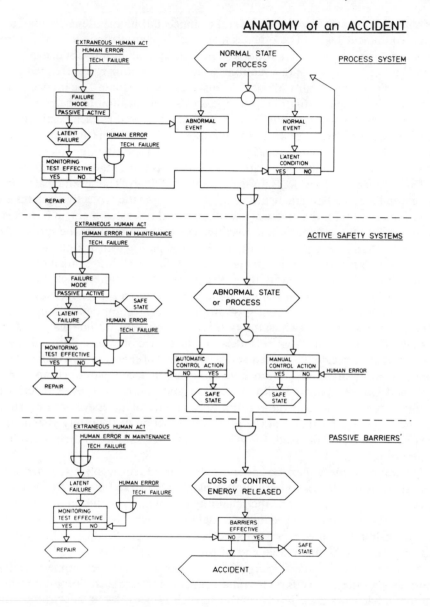

Figure 1.6.1-1 Anatomy of an accident. Accidents in modern process systems have a typical anatomy which depends on the systematic use of automatic safety systems and passive barriers. Safety depends to a large degree on the reliability of human maintenance of such protective systems

High Risk Safety Technology

introduction of latent failures in several channels during maintenance than it is upon equipment faults. The other category of human error is loss of control caused by error in protective actions. Operator intervention in abnormal plant states can be very reliable in situations which the operator meets frequently or which allow for reversibility of possible mistakes and adequate time for decisions. However, in complex situations, the operator's identification of the plant state and his decisions will typically be unreliable and unpredictable. Furthermore, there is repeated evidence (reactor operators will recall the incidents at the Dresden 2 and Three Mile Island plants) that operators' attempts to control an abnormal plant state interfere with and obstruct the automatic protection.

The dependence of accidents upon the control of flow of energy and matter in a system and the standard practice in the design of automatic protective systems are basic prerequistes for the practical use of analytical risk assessment, in particular in coping with the variability and inventiveness of the humans of the system. An automatic safety system is basically a feedback path which maintains critical variables within permitted limits, and as for all feedback loops, analysis of the performance depends primarily upon the properties of the feedback path, and only to a limited extent on the variate and partly unknown properties of the process path itself. The feasibility of automatic protection depends upon the existence of a safe state to which the system can be brought automatically and the availability of early warnings, i.e. a delay in the course of events. Such a delay is present when an accident is related to loss of control of energy or mass flow due to transport delay or to integration (pile-up) delay in accumulating components. Automatic safety systems typically monitor and control the levels of critical energy accumulations and the resulting probability of excessive accumulation depends upon the reliability of the protective function of the safety system and the total frequency of demands for safety actions – irrespective of their individual nature or cause.

Typically, the effect of exotic and unpredictable human acts (e.g. the man with the candle in the Brown's Ferry nuclear plant[4]) is masked by the frequency of trivial equipment faults, while the reliability of human activities in maintaining safety system performance becomes of vital importance. However, maintenance, test, and calibration of safety systems are well known, specified tasks which can be designed so as to be analysable and of predictable reliability.

If loss of control of major energy accumulations does in fact happen (i.e. the third phase), safety depends on barriers against the release such as containments, shields, or barriers against human entrance, such as locks, doors, safety zones, etc. Again, the protection furnished by barriers depends upon the reliability of the normal function of the barriers rather than upon the cause and nature of the accidental chain of events.

To sum up, the very low levels of probability of accidents which are necessary for large industrial installations can be attained and demonstrated analytically due to two basic assumptions: first, the three phases of control of accidental release of energy and/or matter can be based on equipment or precautions with *causally independent failure mechanisms.* This means that very low resulting probabilities for the ultimate accident depend on realistic figures for failures in each of the independent mechanisms, which can be empirically verified or based on conservative estimates. Second, the probability of accidental chains of events can be estimated using *analytical techniques based on feedback concepts.* The upper boundary for the probability can be estimated from the frequency of *opportunities* for a family or chains of events together with the *reliability* of protective systems and barriers. These two assumptions are generally the prerequisites for an acceptable analytical risk estimation including the unpredictable aspects of many human activities. Continuing this line of reasoning and considering the role of human operators as monitors and supervisors in modern automated plants, the risk taken with human operators will not be that they will cause accidents but rather that they may not succeed in preventing them.

Accepted Risk, Oversights, and Errors of Management

One major problem in analytical risk assessment is to obtain a clear and explicit formulation of the boundaries of the analysis. The final result of a risk analysis is a theoretical construct which relates empirical data describing functional and failure properties of equipment and processes to a quantitative or qualitative statement of the overall risk to be expected from the operation of the system. The analysis depends on a decision regarding the boundaries of the system to be considered: a model describing the structure of the system and its functional properties in normal and in all relevant abnormal modes of operation, including the activities of the people present in the system; together with a number of assumptions which are necessary to facilitate a systematic analysis. These assumptions, the model, and the characteristics of the sources of the empirical data used, are just as important results of the assessment study as is the resulting risk figure. Unfortunately, they are generally not explicitly formulated or considered in present practical applications of analytical techniques.

An important category of risk from the operation of a complex system may be related to major loss situations of very low probability. Typically, such situations result from complex chains of events, including the coincidence of events and the presence of *a priori* improbable failure modes. Therefore, sources of risk hidden behind an incomplete analysis become a major problem. They may reveal themselves as oversights and errors of management, as illustrated by Figure 1.6.1-2. It is important to make the assumptions of the analysis explicit and operational, especially since they will be important guides for the activities of several different categories of personnel involved in system construction and operation.

It goes without mentioning that the analysis is incorrect if the components and equipment used for construction of the real system are not of a quality equal to

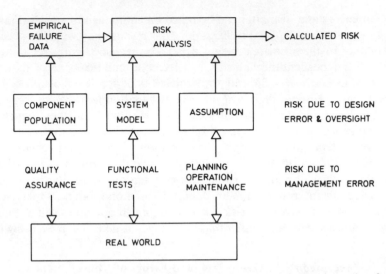

Figure 1.6.1-2 When the calculated risk has been accepted, important results of analysis are: the required component quality, the conceptual system model, and the assumptions of the analysis. This information must be the backbone of risk management serving to maintain correspondence between the analysis and the real world through quality assurance, functional tests, and planning of operational and maintenance procedures. The figure has been reproduced from[13] by permission of Plenum Publishing Company Limited

that of the population providing the empirical fault data. This means that the data basis for the risk analysis specifies the *quality assurance* to be performed, i.e. the quality categories and the acceptance tests and inspections to be required. In particular, it is possible to use the analytical risk assessment to identify selectively the items needing special care during quality assurance to counteract human errors during this task.

The quality of the risk analysis also depends on the degree of correspondence between the model used for analysis and the final, physical installation. Discrepancies between the model used for risk analysis and the actual functional conceptualization of the plant designer may be considered oversights on the part of the risk analyst, or design errors depending upon the circumstances. At the present state of the art of analytical risk assessment, great care is needed in an explicit formulation of the search strategy which has been used to identify the functional structure and failure mechanisms which are included in the analysis, i.e. the model behind the analysis. Furthermore, the value of the risk analysis depends on the quality of the *functional tests* which are performed to verify the correspondence between the plant concept which has been subject to analysis and the actual, operating system. Again, the model underlying the analysis must be explicitly described to be useful for planning effective functional tests. This is particularly important, since the correspondence should not only be guaranteed by the commissioning test at the start of plant life, but also be maintained through the entire plant life with its *maintenance and modification activities*. There exists ample evidence that oversights during modification of plants to cope with changed requirements are significant contributors to the overall risk[5]. To plan safe modifications, not only a consistent description of the physical plant and its functions is necessary, but also adequate information about the designer's intentions and reasons for selecting the existing plant configuration. Generally speaking, this information is not sufficiently available to the operating or modifying organizations, since it is normally not recorded and often fades away as soon as the designer has put the plant flow diagrams on paper. While the normal, functional descriptions and diagrams of the plant specify the functions aimed at in design, the risk analysis also specifies the forbidden functions and in a very effective way serves to delineate the designer's intentions and reasons. A systematic record of the model and assumptions of the risk analysis is therefore a prerequisite for maintaining the accepted risk level during plant life.

So far the considerations apply to the plant hardware; similar considerations also apply for the software of the system, i.e. its *organizational structure, management style, operating procedures*, and so forth. During risk analysis, several assumptions will be made on training and qualifications of personnel; on procedures and instructional systems; on test and calibration frequencies and procedures; on repair policy, and on independency of certain events, etc. These factors all depend upon staff and management decisions during plant life, not only in well planned responses to changes in demand from plant operations, to labour union requirements and to changing organizational and management styles, but also upon *ad hoc* decisions in response to acute events such as spare parts stock problems, strikes and conflicts, etc. It is therefore important to record the analytical assumptions carefully and to make the proper interpretations as a guide to the operating staff for safe operational planning.

To sum up, the main benefit to draw from an analytical risk assessment will probably not be the quantitative risk figure derived, but the possibility of using the structure and assumptions of the analysis as tools for *risk management* to secure the proper level of risk during the entire plant life.

Reliability and Risk Analysis, Methodological Differences

Analytical techniques for risk and reliability assessment depend on decomposition of the system, or its functions, into elements for which failure probabilities can be empirically obtained. The overall reliability or risk figure is then calculated by a probability model derived from models of the functional structure of the system. The problem in the present context is that the people in the system must be considered as system components and that human error data are needed.

In analytical techniques in general, different methodological problems are met in reliability and in risk assessment. The distinction is, however, not clearly maintained in most discussions of the techniques. This sometimes leads to confusion, in particular when dealing with analysis of the human role in system failures. For the reliability concept, the definition given by Green and Bourne[14] should be respected: 'Reliability is defined as that characteristic of an item expressed by the probability that it will perform its required function in the desired manner under all relevant conditions and on the occasion or during the time intervals when it is required so to perform'. In other words, *reliability analysis* is based on an analysis of failure to perform the normal, *required function* of the system. If, in the first approximation, only two states – normal and failed – are considered for the function of components and system, reliability analysis depends upon logical, combinatorial arguments; the system operates if all components in functional serial connection and one of the components in functional parallel are in a normal state. The overall system reliability can be derived by logical Boolean operations which can be adequately represented by a fault tree (or a reliability block diagram). A fault tree in fact represents a structure which is directly complementary to the normal functional structure of the system. The fault data needed for analysis are simple gross two-state probabilities for the components.

Risk analysis, on the other hand, involves the estimation of the probabilities of several categories of *accidental event sequences* related to the relevant categories of risk such as damage to people and environment as well as to loss of major equipment. This means that the overall probability related to a specific consequence must be calculated by a probability model derived from the family of relevant accidental event sequences together with data on the component failure modes involved. In this case, the combinatorial technique used in reliability assessment is inadequate in all practical cases due to the infinite number of accidental sequences formed by the possible failure modes and their combinations. The main elements of a risk analysis are therefore: the identification of relevant accident mechanisms; the modelling of the related sequences; and, finally, the combinatorial probability modelling. Only the latter will be adequately represented by fault trees. The sequence model and the criteria and structure of the search strategy need separate and explicit representation, if a risk analysis should be accessible to independent evaluation and validation. Unfortunately, this is often not respected in reported risk analysis and the analysis can only be ranked according to the professional standing of the analyst.

This distinction has important implications for the treatment of human factors in risk analysis and will be discussed in some more detail. Analytical assessment of *human reliability* depends on an analysis of the required human functions which are decomposed into elements for which error data are available. Two different categories of human tasks are typically considered by risk analysis; the first one being planned, familiar tasks such as test and maintenance, valving and manual control actions, etc., and the second being more infrequent and unfamiliar tasks appearing in response to plant malfunction.

For the first category, the structure of the normal task sequence which is the base of reliability calculation, is known or can be found by task analysis and the error rates for the relevant task elements can in principle be collected. Collection of generic error rate data for task elements is, however, more difficult for human errors than for component faults, since the significance of the errors and consequently the possibility of obtaining reports depends very much upon the potential for self-monitoring and error correction, which again depends on the structure of the entire task. Consider, for example, calibration of a safety channel. The task involves setting up test equipment, adjustment, reading and acceptance of set-point and, finally, restoration of normal operation. The situation is in a way symmetrical because the act of reading and acceptance of the set-point will detect abnormalities in the preceding task elements as well as in channel function and the operator will accordingly repeat and correct. The task is an example of the feedback effect, leading to a reliability depending mainly on the quality of the monitoring function. This feature is more or less present in most real life tasks and, consequently, there will be a task-context dependent bias for the error rates which can be obtained from real life performance.

When considering the second category, the reliability of human performance in response to infrequent demands such as plant malfunction, we face two problems. The first problem is to obtain a model of the required performance. The basis of human behaviour depends upon the frequency of a task which in turn is related to the risk involved in a situation (see Figure 1.6.1-3). Human response to an unfamiliar, rare demand is complex, it involves identification of system state, decision concerning goal and planning of action, and models acceptable for reliability analysis are at present not available. If the problems of stress in emergency situations are also considered, the problem is even worse, but the question is whether analytical assessment of this response is needed for overall risk analysis. Reliability estimates based on present knowledge are in the range of a probability of success of 0.2–0.6[6]. Therefore, if the uncertainties in the overall analysis are considered, the result will not be significantly influenced by conservatively assigning a probability of failure of 1 to human performance in unfamiliar, stressed situations. The second problem is to get empirical data. The error mechanisms for the same task element embedded in a familiar, frequent task and in an infrequent response will be basically different since the person's internal control of the task will be different (see Figure 1.6.1-3) and, therefore, error rates obtained from general error reports will not apply for infrequent responses. In conclusion, the reliability of human performance in frequent tasks such as maintenance of automatic safety equipment plays an important and realistic role in analytical risk assessment, whereas the reliability of human response to infrequent situations as, e.g. in case of emergency, is not at present accessible for analysis but can actually be omitted without having a major effect on the result.

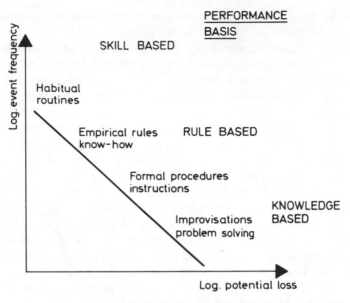

Figure 1.6.1-3 Graph illustrating the fact that the behavioural control of
human performance depends upon the familiarity and hence the frequency
of a task. In a well balanced system design there will also be an inverse
relation between the frequency of a situation and the risk involved

A major problem in *risk analysis*, mentioned above, is the identification and
modelling of the accidental sequences, the scenarios, which must be analysed and
these problems are to some extent related to the people in the system. Human acts
are typically important links in accidental chains of event leading to major
hazards. In this context it is not only the question of the reliability of required,
planned activities but also the probability of specific erroneous acts which will
initiate an accidental chain or prepare the path for chains of events which are
otherwise triggered.

Effective techniques have been developed for hazard identification based on
various heuristics related to control of energy[7]-[10]. The sequence modelling is,
however, generally based on a backward ('fault tree') or forward ('failure mode
and effect') causal event tracking through the functional structure of the system as
defined by the plant flowsheets. This linear search is effective when considering a
single or a few equipment faults and human errors but major, rare events are
probably dependent upon 5 to 10 faults and abnormal conditions[11] which
appear in a pattern which cannot be identified by a linear search but only by a
'bird's-eye' search directly for potentially risky structures.

The people of a system do not appear on the flow schematics; at best, the
required, planned activities are prescribed by written procedures. They are,

however, important elements of the accidental patterns: operators shut off safety systems, maintenance people leave stand-by equipment inoperable, painters obstruct mechanical regulators with ladders or step upon equipment, and plumbers move around with electric welding gear, etc. An accident is typically caused by a sneak path for events, created by the accidental timing of a considerable number of normal and erroneous human acts together with latent risky condition and equipment failures. A search for such patterns is similar to a design strategy. Identify the hazard potential: energy concentrations; toxic material; and the potential targets such as people or major equipment. Then design the possible and necessary chain of events which can constitute an accident and look for the constituents among the normal and accidental acts and functions in the plant area and for situations which could lead to proper timing and chaining. Very probably, it will be difficult to get quantitative, empirical data on the probability of the erroneous human decisions and acts involved in such rare accidental patterns but the analysis can serve to verify that unacceptable sneak paths through the functions of the engineered safety measures are not present or if they are, to identify necessary design modifications.

The major point to emphasize is the need for a systematic and explicit formulation of the search strategy and modelling technique which is used in risk analysis, in particular regarding the role of the personnel. A systematic solution of the completeness problem implied by this can at present only be given for systems in which the effects of erroneous human acts upon the release of an accident potential can be detected by monitoring a variable or a functional state late enough in the accidental sequence to guarantee complete coverage of possible human acts, but still at a state where the effect is reversible. As previously discussed the *risk* can then be inferred from the frequency of opportunities and the *reliability* of the protective function.

To sum up, an analytical estimation of human reliability during normal, familiar tasks is practically feasible, whereas analytical estimation of the reliability during rare, emergency situations is not at present realistic, nor is it necessary for analytical risk assessment of systems in general, since a failure probability of 1 can be used without significant problems. Risk in general can be analytically assessed – regardless of the completeness problem – if adequate protective functions allow for the use of boundary estimates based on frequency of opportunities and reliability of protection.

Categories of Human Behaviour and Human Error

Analytical techniques for reliability and risk assessment depend upon a breakdown of the particular sysem into parts or functions to a level at which a sufficient amount of data on failure properties can be obtained, from previous similar applications of the same parts or functions. For the human involvement, this decomposition results in identification of human tasks together with a set of possible errors or faults expressed in terms of their effect upon the task, such as omission of steps, commission of wrong acts or inadequate timing. Empirical data or estimates of the error rate are then sought from similar task situations, and the data are modified according to performance- or situation-modifying factors. However, if analytical techniques are to be acceptable for tasks other than stereotyped routine tasks, the different types of human error must be identified with reference to properties and functions of man himself and his input conditions should likewise be carefully analysed.

Both for analytical assessment techniques and for supporting an improved system design, there is a need for human error categories for which the information input, the internal mental function, and the human error mechanisms are determined in addition to the effect of error upon the external task. Only then will quantification of human errors have real meaning and transfer to other work situations different from those for which data are collected, will be possible. The concepts considered by an OECD/CSNI working group for analysis of event reports from nuclear reactors[12] are illustrated in Figure 1.6.1-4. The benefit from such a multi-facet definition of categories compared to a conventional hierarchical classification scheme is that good resolution in terms of the effect

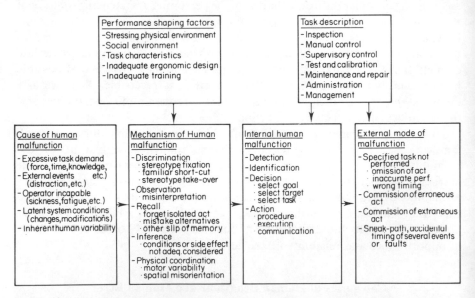

Figure 1.6.1-4 The block diagram illustrates the relations among the various categories of a multi-facet classification system for events involving human malfunction. The lower row indicates that an external cause may release a failure mechanism which in turn affects the internal human function which finally manifests itself in an external failure mode. The upper row indicates performance shaping factors determining the sensitivity of failure mechanisms for release by an inappropriate situation. The task content determines the internal and external effect of the release of a failure mechanism

upon the task can be obtained with a reasonable low number of elements in the different facets – which greatly facilitates the analysis and classing of event reports. Furthermore, the transfer of error data to new tasks is possible, since the external mode of malfunction can be inferred from the task structure together with the categories of internal mechanisms and modes of malfunction. An initial attempt at a direct practical use of this approach has been made[10] in the design of operational procedures and the interlock system for a chemical reactor. First a consequence analysis of potential errors at the external mode level was made using standard preprinted sheets (Figure 1.6.1-5). One such sheet was completed for each operational step in a critical task. For those steps having potential significant consequences, a checklist based on the classification facets of Figure 1.6.1-4 was used to support identification of relevant error mechanisms and weighting of potential causes and to set priorities for procedural changes.

For different applications, emphasis is put on different aspects of the human error event – or, more correctly, human malfunction, which is a more acceptable term as long as causes external to the man have not been explicitly excluded. In the following sections, the different facets of the classification of Figure 1.6.1-4 are discussed in more detail.

External Mode of Malfunction

The external mode of malfunction describes the immediate, observable effect of human malfunction upon task performance. It reflects the way in which the malfunction initiates the consequent chain of accidental events and is thus important for predicting the effect of human malfunction for a specific system and its actual operating state. In this respect, three main categories are indicated on Figure 1.6.1-4. 'The specified task not appropriately performed' is a category of major interest in quantification of human *reliability*; typical examples of this category have been found in reports of inappropriate test and calibration acts[13]. For example, (i) omission of the act to restore normal operation after test or (ii) inaccurate performance in set-point adjustments. Data on modes of malfunction of the category 'specific erroneous acts' are important in risk analysis for assessing accidental chains of events. In general, it is not possible to predict the wrong act which will be performed when people fail, but in some cases or tasks, the number of possible alternatives of a choice may be rather low – 2 or 3 – and the effect of error can be judged. Mistakes of alternative possibilities such as up for down, + for –, increasing for decreasing, seem to be an important category in the task of calibration. Such error modes can be identified by relating acts found by task analysis to internal human mechanisms of malfunction, and in principle data can be collected. With regard to the category 'extraneous human acts', these are more difficult to identify since they very often depend upon the detailed physical layout of the system and upon human activities of very different sorts, maintenance people dropping tools, stepping upon equipment, moving with heavy gear, etc.

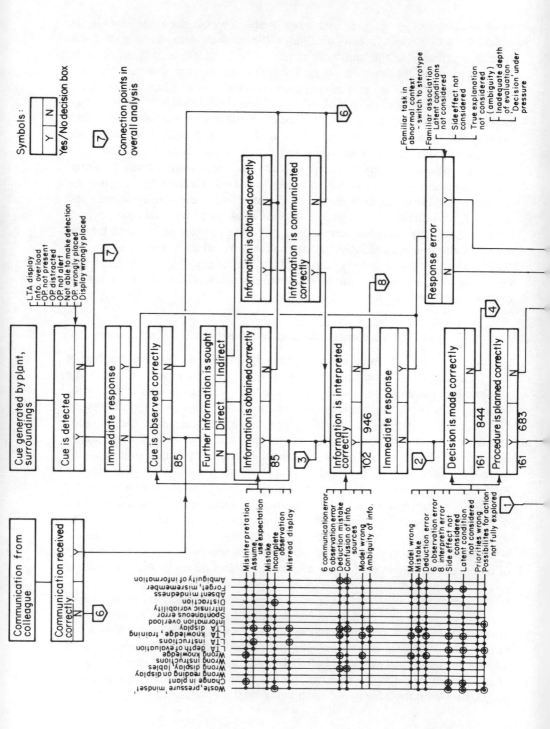

Symbols:

Y	N

Yes/No decision box

Connection points in overall analysis

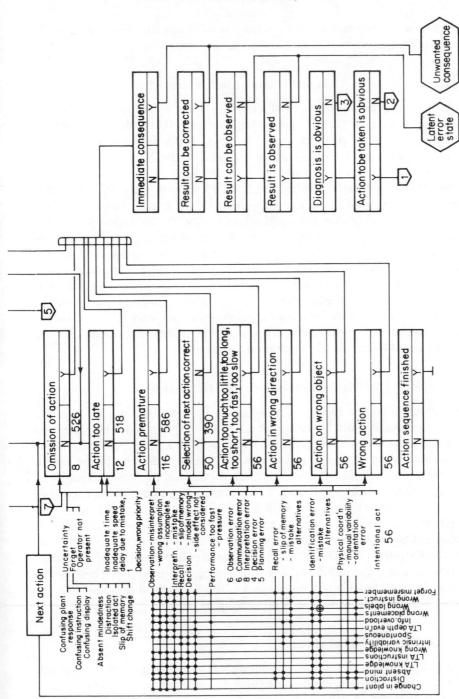

Figure 1.6.1-5 Flow diagram used to analyse critical human control task in chemical process plant based on the multi-facet description. The flow diagram includes more than thousand error classes by combinatorial use of the categories of figure 4. Reproduced from[10]

The categories of external modes of malfunction discussed here are similar to the categories of human error used by Swain in Reference (6).

Internal Human Malfunction

To predict when and why human malfunction in a task will appear, it is necessary to analyse which task the man attempts to perform internally, i.e. his mental activity needed for the performance. A part of a task or an act may be omitted because the man simply forgets a familiar act; because he does not have the necessary input information to see it was needed; or, because he does not know enough of the system to identify the demand appropriately. It is therefore necessary to characterize the internal human function that failed. In Figure 1.6.1-4 this 'internal human malfunction' is described by the main categories of detection, identification, decision, and action. In analysing an event it should be determined which of these functions were not performed as required *without* considering the cause of the malfunction, be it a change in work condition, a spontaneous slip of memory, high workload, etc. The specific causes should be considered separately. Consideration of the human function that failed is important to avoid transfer of error data between situations which have similar external task characteristics, but which imply different internal human functions and thus also different failure mechanisms and causes.

Mechanisms of Human Malfunction

The internal failure mechanisms depend upon the psychological aspects of the human performance, such as the way observed information is processed and the underlying knowledge and experience of the man. Some kind of task- and system-independent model of human performance is necessary to describe the way in which behaviour is organized and controlled. Analysis of accident and event reports strongly indicates that such a model must represent different levels of human behaviour: (1) automated, skill-based behaviour; (2) goal oriented, rule-based behaviour; (3) goal controlled, knowledge-based behaviour; and finally (4) discrimination, i.e. selection of the proper level of behavioural control. This model is illustrated in Figure 1.6.1-6. In addition to a representation of the human function of identification, decision and planning, the model emphasizes the bypassing of such functions in routine tasks. The characteristics of the different levels of behavioural control are the following:

Automated skill-based behaviour

This is a characteristic of performance during acts or activities which, following a statement of an intention, evolve without conscious control as smooth, highly integrated patterns of behaviour. Such patterns can be considered as the subroutines which are subject to higher level sequence control in more complex activities. Such subroutines cannot be decomposed into more elementary, separate acts. The highly integrated nature of skilled subroutines is reflected in studies[15] of the influence of fatigue and stress upon pilot performance. These studies indicated much less influence of fatigue and stress upon the quality of the individual subroutines than upon the higher level coordination of behavioural sequences.

The smooth course of skilled acts reflects the high level of adaptation of performance to the spatial and temporal characteristics of the physical task environment. Consequently, error data are related to a particular man/equipment configuration and data should be collected for complete, integrated subroutines with reference to interface configurations. Generic human reliability data in this domain without reference to interface configuration will only be meaningful in the form of distribution functions of the spatial and temporal precision of manual operations. Such data could in principle be related to task requirements with respect to precision for manual tasks like set-point adjustments and work on wire terminal strips, but this will probably not be practically feasible. For more complex, highly skilled subroutines such as manual steering and tracking tasks, simulated tasks in high fidelity simulators are preferable data sources[16][17].

The review of 200 cases of operational problems in nuclear reactors[18] indicates that consequences of errors in skilled subroutines, in terms of external modes of malfunction, are roughly equally frequently related to the specified task not being performed (inaccurate calibration), to the effect of erroneous act on the system upon which the man operates (e.g. short-circuits), and to extraneous effects upon other, nearby systems (e.g. dropped tools, misplaced jumpers). This means that for effective reliability and risk analysis, detailed information on the topographical layout of the related equipment is necessary. Again, these data are task and device dependent.

Goal-oriented, rule based behaviour

This is the typical level of control of stereotyped performance in familiar work situations. The task consists of skilled subroutines which are coordinated or sequenced consciously, but alternative decisions or choices are *not* considered. The rule of control can be formal prescriptions or instructions or a successful work procedure evolving from experience. This implies that an operator's functional reasoning resulting in a modification of procedures to fit changes in work situation or plant state, or in a regeneration of degenerated rules belongs to the next category of knowledge-based control.

Consequently, the present category relates only to stereotyped performance of a familiar task, such as test and calibration, manual control operations, inventory control, etc. which can be decomposed into a sequence of subroutines. Typical error mechanisms are deficiencies in coordination of subroutines (forgetting acts, especially functionally isolated acts), errors in recall of reference data (set-points), etc., and mistakes among alternative possibilities (up/down; increasing/decreasing; left/right, etc.).

In the case of omissions, the effect of the errors in terms of external modes is by definition related to the fact that the specified task is not performed; in case of mistakes, the effect of the majority of errors in the 200 cases reviewed is related to the specific wrong act. Since the number of alternatives for mistakes is typically very low (one or two), safety and reliability prediction related to rule-based tasks will be practically realistic if the task elements are cued separately by the work environment to ensure stable task content. This category of behaviour is typically the domain for application of decomposition methods for reliability prediction, such as Sandia's THERP – Technique for Human Error Rate Prediction[6]. A prerequisite for prediction will, however, be the availability of error rates. The extent to which omissions and mistakes are independent of work situations is in fact not quite clear. Distraction, preoccupation, and high workload can be important factors. Furthermore, the resulting error rate in a task is highly dependent upon the opportunity for error corrections, i.e. observability and reversibility, factors which are device and task dependent.

Knowledge-based, goal-controlled performance

This is the typical level of performance in less familiar situations, when problem solving and planning is necessary; when different goals must be considered; and decisions or choices among alternative plans should be taken.

Prediction of performance in this domain is exceptionally difficult for several reasons. The sequence of elementary activities is generated *ad hoc*, and due to human flexibility and the frequently ill-structured task conditions, it cannot be reliably predicted in advance. Only the attainment or non-attainment of the goal can be taken as normative. In addition, the goal itself may be very ambiguous in an infrequent situation of mal-operation and the decision on proper goal as well as the identification of the actual state of the system are very vulnerable to personal and situational features. Finally, the effects in terms of external mode of errors in this domain, due to error mechanisms such as inadequate consideration of side effects or latent conditions in case of demands for improvisation, are typically related to the specific erroneous acts. If a person makes an improper decision in an unusual situation, it is not practically possible to predict where, when, or how he will interfere with system operation.

Discrimination

The operator's selection of the proper mode of control of his activities is a very important phase of an activity considering the potential for errors. The important error categories include fixations, 'mental traps' and other types of errors which as their basic cause can have interference between the operator's large repertoire of stereotyped habitual – and often subconscious – responses and aspects of the actual work situation during infrequent and unique task demands. The features of such interferences are very situation and person specific, and it is not practically possible to collect statistical data for quantification of the probability of proper discrimination in an infrequent situation.

Error mechanisms

The implications of the different levels of behavioural control of prediction of human performance have been summarized on Figure 1.6.1-6. Information on the different internal mechanisms of malfunction in task- and equipment-independent terms is very important for design of error tolerant man–machine interfaces. Therefore, special emphasis has been put on this facet of human malfunction in analysis of case reports[18].

In this analysis, the categories of 'human error mechanisms' shown in Figure 1.6.1-4 were identified from a review of 200 cases of 'operational problems' in nuclear reactors, and the relatively small number of categories (10) were found to cover 90 % of the cases. This is due to the rather general nature of these categories, but used together with categories describing 'internal human malfunction' and 'external mode of malfunction', a high resolution is obtained in description of events.

The different error mechanisms are closely related to the functions illustrated on Figure 1.6.1-6. The result of the review has been discussed elsewhere, but some typical error mechanisms and their functional relations will be noted here. Two prominent mechanisms were found – equally frequent, covering in all 10 % of the cases – during the execution of the actions at the skill-based level, being variability or lack of precision in manual acts (such as too little or too much force in manual valving or inaccuracy in adjustments) and inadequate spatial orientation (knowing where things are but inadvertently mistaking positions).

At the rule-based level, a typical error found during execution of a procedure was mistakes among alternatives, such as up/down, increasing/decreasing, $+/-$, A/B, etc. This category appeared in about 5 % of the 200 cases. Typically, it will be the effect of error in running memory. Another 'slip of memory' is found in the function of controlling a procedural sequence. Then, acts or steps which are functionally isolated from the overall structure of the task are likely to be omitted. Omissions of such acts, e.g. not restoring valves after test, not recording 'jumpers' during repair, etc. account for more than 30 % of the reviewed 200 cases.

In the knowledge-based domain, a typical error mechanism is inadequate consideration of side effects of the procedure chosen, e.g. when improvising in response to changes in the work situation or in attempts to improve a procedure, or inadequate consideration of possibly latent conditions in the system. Such errors were found in 15–20 % of the cases, mostly during planning of procedures

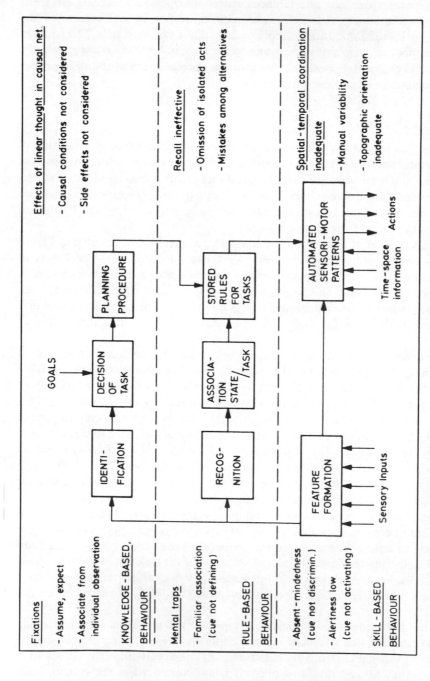

Figure 1.6.1-6 Schematic illustration of the different levels of internal control of human activities. The typical mechanisms of human malfunction are indicated from a study of 200 US Licencee Event Reports. Reproduced from[18] by permission of John Wiley & Sons Ltd

but also during identification of abnormal system states. The error mechanism can reasonably be attributed to difficulties of unsupported, linear natural language reasoning in keeping track of the propagation of effects in a causal network.

The implications of these error mechanisms for reliability and risk analysis have been discussed above. The error mechanisms related to the *discrimination* among the appropriate levels of behavioural control will be discussed in more detail, since they typically play an important role in major accidental situations in control rooms. The inappropriate responses of operators to unfamiliar plant states are related to the different use of the observed information in the three domains. In the *skill-based domain*, when humans in a way perform as multivariable control systems manipulating physical objects (which may be graphical figures of a display) or navigating through the environment, the sensory information acts as *signals*; i.e. analogue representations of spatial variables. When humans are not operating on physical objects directly but by means of displays and keys on a control console, the implication will be that in the skill-based domain the control console itself is operated on instead of the underlying process. In the *rule-based domain*, performance is based on recognition of states of the system with association to known rules or plans. When operation is performed directly on physical objects, states are perceived very reliably. If, however, preselected information is available only as individual variables displayed on a control console, reliable state identification will demand conscious inference on the part of the operator, and this cannot be expected from operators during all routine situations. Instead, characteristic indications of the available information, including convenient sounds and noises, will act as *signs* representing system states. Signs may be labelled as states, events, tasks or perhaps other names related to the physical states by experience or convention, just as traffic lights are, for example. Finally, for *knowledge-based* performance observed information from a data display acts as *symbols*, i.e. representations which can be treated directly by symbolic data processing for problem solving as, for instance, when operators read temperatures and pressures prior to consulting a steam table. It is clear from this that the role of presented data varies fundamentally with the task situation and the operator's background in training and know-how. The degree to which an operator may develop automated patterns of skill and the size of his repertoire of signs for rule-based response depends very much upon the detailed visual and auditory appearance of the system and consequently, due to this variability, the operator's selection of the proper level of response in case of abnormal system state cannot be predicted during system design.

The internal mechanisms related to inappropriate discrimination are difficult to identify from case reports, but three broad categories indicated on Figure 1.6.1-5 appear to be typical. 'Stereotype fixation' refers to the situation when heavily trained routines or habits are used without realizing at all that special circumstances require attention; the cues are not alerting the operator. 'Familiar association short-cut' takes place when it is consciously recognized that an abnormal situation is present, but cues are incorrectly interpreted as signs indicating familiar states as when, for instance, instrument indication of abnormal plant state is referred to as instrument failure. 'Stereotype take-over' takes place when special circumstances have been realized and proper decisions taken, but relapse to stereotype action links occurs during performance. If actions following a conscious decision take some time, e.g. if the operator has to move to another place to perform the action, the mind may return to other matters and the performance will be vulnerable to interference from stereotype activities which have steps overlapping the intended task.

To sum up, the level of control of behaviour and the related mechanisms of malfunction have important implications for analytical assessment of human performance. In the skill-based situations, prediction is feasible if interface-specific data are available; in the rule-based domain, decomposition techniques are applicable, but due to the problems of discrimination, only for well trained and planned task situations. For problem solving in the knowledge-based domain, analytical prediction is only possible in some cases in the form of asymptotic boundary considerations based on feedback design concepts, i.e. the frequency of error opportunities together with the reliability of error detection and correction (see Figure 1.6.1-7).

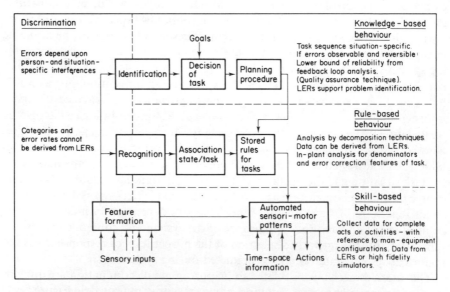

Figure 1.6.1-7 The feasibility of quantification and prediction of human performance depend on the internal control of the performance in question. The methodological characteristics indicated in the figure have been concluded from analysis of legally required event reports from US nuclear power plants, Licencee Event Reports (LERs)

Causes of Human Malfunction; Performance Shaping Factors

The categories of human malfunction discussed so far can be seen as various dimensions of a mismatch in a man–machine resource/demand system. These aspects are important for the design of improved man–machine interfaces, since improvements can be obtained by increasing the margin to misfit in the dimensions which have proved sensitive from operational experience, independent of the actual causes to human mal-operation. However, to be able to predict the frequency or probability of human malfunction in a task by means of empirical data, the identification of causes of human errors is important. In Figure 1.6.1-4 a distinction is drawn between causes like abnormal task demand, incapable operator, external causes, and intrinsic human variability.

Unfortunately, it is generally not possible to identify the causes of human malfunction from case reports. Special follow-up studies are needed as well as careful analysis of the general work situation as discussed in section 1.6.2. In general, it is advantageous to distinguish clearly between *causes*, which are changes or events followed by a chain of events, and more general *factors* which influence the flow of events by modifying human behaviour or probabilities of responses. Such performance shaping factors[6] are important to characterize the quality of a work situation; a few typical categories are indicated on Figure 1.6.1-4, but in general they cannot be identified in any satisfactory detail from case reports (see Section 1.6.2).

Design for Error Tolerance

The human operator is an extremely flexible and adaptive component in an industrial system and, in general, he develops a skill and pattern of performance which will fit the peculiarities of the system in a very effective way. This is the very reason for having human operators in industrial plants but, paradoxically enough, it is the same features which make him the imp of the system when considering incidents and accidents. The immediate impression one gets when reviewing case reports is that such reports do not represent the errors operators commit but by and large the errors which are not corrected, either because they had an effect which was irreversible – and therefore had to be reported, because of damage to equipment or of a major automatic safety action, or because the errors were not immediately apparent to the person himself, but were found later by inspection or test. It is in a way trivial that operators do correct the errors which they themselves can see. On the other hand, a special feature of a work situation is the potential for feedback correction of errors which result in heavily biased incident reports, and this aspect is generally not adequately considered in empirical data collection.

Human errors are in a way the inevitable side effects of the human adaptability. Basically adaption depends upon variability and selection; unsuccessful trials manifest themselves as errors only when selection is ineffective due to irreversible or latent effects of the trials. Whenever a system designer meets a technical component having unacceptable intrinsic variability, his normal reaction will be to compensate for this by the systematic and careful application of the feedback principle. To design human error tolerant systems, the same precaution should be applied to compensate for human variability. If the possibility of operators to monitor their own performance were considered explicitly during task design in an ordinary engineering way of thinking, a large fraction of reported cases would not reach the printed page. Furthermore, the empirical basis existing in event reports and case stories should be used in a more systematic way than is presently the case. The man–machine interaction can be seen as a complex, multidimensional resource/demand fit. If we do not consider emotional or legal aspects of guilt, we can relax with respect to the question as to whether system malfunction is caused by human error or by man–machine misfit due to inappropriate design. To evaluate means for improvement, it is more important to find the nature or dimensions of the misfits than to identify their causes. Fortunately, to identify the modes of human malfunction – what went wrong – is much more realistic from case reports than to trace the causes as to why it happened. If the analysis is based on a simultaneous consideration of facets related to task elements and to internal human functions and error mechanisms, more effective guides for design of error-tolerant systems can be derived. The taxonomy illustrated by Figure 1.6.1-4 is a step in this direction[12].

Design for Analysability

It appears from the previous discussion that analytical prediction of human performance is only feasible under certain restrictive assumptions and conditions. At the same time, the potential for major losses, especially in terms of damage to persons and environment in modern centralized installations, gives rise to design targets for safety which cannot be directly empirically verified due to the extremely low acceptable probability. Therefore, the situation can be foreseen when designers will have to accept these assumptions and conditions as 'criteria for analysability' during plant conception. The discussion of the previous sections can be summarized in the following criteria.

The *reliability* of a human task can be analysed by decomposition techniques if it is a planned task performed in a well trained, rule-based way and the separate steps of the task are cued individually so as to counteract systematic changes or 'optimizations' of performance. When applying error rates for the task elements, the potential for self-monitoring and error correction must be taken into account.

When a task is performed in the knowledge-based domain, only the goal is normative, i.e. serves as reference when judging the quality of performance, and task sequence will be very flexible and will depend upon features which are very situation and person specific. In this case, identification of task elements for error rate attribution is generally irrelevant since they cannot be separated from their context in a meaningful way. Analysis of goal-controlled, knowledge-based performance is only feasible if a feedback concept is adopted. The task must be designed so as to make the effect of errors immediately observable and reversible; then the *reliability* of the task can be evaluated by boundary considerations related to the frequency of error opportunities – i.e. of task demand – and the reliability of the error correction act alone.

The reliability of human responses to infrequent, unfamiliar situations even when emergency procedures are issued and trained, depends upon the reliability of the discrimination among possible levels of performance – when to 'follow the book' and when to 'think for yourself'. Since this reliability may be low, design for analysability in practice means that emergency responses should not be allocated operators unless proper support and ample time allow them to depend upon feedback error correction so that overall reliability of the task can be judged from the reliability of their error correction.

Finally, since data on specific erroneous human acts will not be available in practice, design for analysability means that such human acts must not be significant contributors to the accidental chains of events which are relevant to the analytical *risk* assessment. If they are not masked by equipment faults, they must be blocked by barriers or monitored by interlocks or by monitoring and correction functions which are designed so as to be accessible for analytical reliability assessment.

Basically, design for risk analysability provides guides for task allocation between operator and the automatic control system. Protection of persons and plant during emergency situations is automated and operators are transferred to a more planned and less stressed task of testing, calibrating, and maintaining the automatic equipment, a task which is feasible for reliability analysis. Such automatic protective equipment will by nature be based on conservative design concepts to cover all relevant circumstances in a predictable way. A role of operators may then be to protect operation from unnecessary automatic safety action by proper compensatory actions in due time. Such a design concept, where safety depends upon automatic control and optimal operation on human control, is only feasible if responsibilities of operators and designers can be separated and formulated, and criteria for risk analysability can be clearly stated.

Questions 1.6.1

1. For a plant known to you – or for a situation during high speed highway driving – try to identify hazards by means of the energy model. Which parts of the anatomy of an accident are affected by the safety precautions taken? For your car driving situations consider car and road design features as well as driving rules and legal requirements.

   Are all the barriers to energy release and safety precautions independent?

Describe some forms of interdependence of control functions and barriers.

2. Consider the various oversights and errors which may occur in the management of a plant or process. Consider, in particular, maintenance and repair policies and planning of modifications. Sketch the structure of an information feedback system which will be required for adequate management monitoring of the fulfilment of the plant's safety requirements – including the effects of management's own decisions.

3. Compare the methodological problems in reliability and risk analysis of human performance. Take, e.g. the situation that a control room operator calls a roving operator by the plant communication system to order him to shut valve no. xxx in the basement.
   What information will be needed to estimate the reliability of the operation, including the potential for recovery?
   What information will be needed to estimate the integrated risk from the operator mistakenly shutting a wrong valve? (Integrated risk = Σ Probability × Consequence, taken over possible mistakes.)

4. Try to relate the different categories of the human error classification of Figure 1.6.1-4 by considering a specific plant operator task familiar to you – or take one of your own personal tasks. Postulate various mechanisms of human malfunction of Figure 1.6.1-4 and for each find causes of release of these mechanisms which are likely to be found in the work situation. Likewise, determine the internal mental functions related to the task which are sensitive to the assumed error mechanisms, and find the external effect upon the task performance. What would be the most appropriate way to improve the reliability – To remove the cause? To change task procedure to make it more insensitive? To add a feedback path to improve probability of error correction?

5. Consider some case stories of human errors from a plant known to you, or from your own activities. What changes would be necessary to improve the possibility for detection and error correction before unacceptable effects occur? Change the work procedure or equipment to the effect that the errors would be detected by making a subsequent step difficult or impossible? Why has it not been done previously? Is it economically or technically unfeasible?

6. Consider calibration during plant operation of an automatic shutdown system of a chemical plant. An analysis of the actual performance results in a breakdown of the task into 100 separate acts related to equipment handling, switching, reading instruments, etc:

Steps 1–30:	Collect calibration equipment, arrange on trolley, getting access to calibration site.
Step 31:	Bypass output terminals of safety channel.
Step 32:	Switch safety channel input to calibration terminals.
Steps 33–75:	Connect calibration equipment, adjust ramp generators, switch digital voltmeter to correct range, find reference values in the calibration instruction, etc.

Step 76:   Read trip value of channel.
Step 77:   Compare to reference value.
Step 78:   If deviation, adjust.
Steps 79–85:   Disconnect calibration equipment from channel.
Step 86:   Switch channel to operation.
Step 87:   Remove bypass from output.
Steps 88–100:   Collect equipment and tools and depart from site.

For illustrative purposes we assume the same probability of failure for each of the elementary steps, i.e. 10^{-2}.

(a) Estimate the average frequency of errors per calibration.
(b) Estimate the probability that a spurious shutdown of the plant is caused by human error during calibration.
(c) Estimate the probability that a miscalibrated channel is not detected by the calibration task.
(d) Estimate the probability that the channel is left in an inoperable state after calibration.
(e) If control room operators are asked to verify independently the safety channel switching state after calibration by monitoring of signal lights, how would the results be changed?
(f) Compare the results from Question 6(c) and (d); which will be the most significant contributions to safety channel unavailability?

References 1.6.1

1. Gibson, J. J. 'Contribution of experimental psychology to formulation of problem of safety.' *Behavioural Approaches to Accident Research*. Assn. for the Aid of Crippled Children, 1961.
2. Johnson, W. G. *The Management Oversight and Risk Tree*, MORT. USAEC, SAN 821–2, UC-41. 1973. Also available from Marcel Dekker, 1980.
3. Haddon, W. Jr. 'The prevention of accidents.' *Preventive Medicine*. Little, Brown. 1966.
4. Scott, R. L. 'Brown's Ferry nuclear power plant fire on 22 March 1975.' *Nuclear Safety*, **17**(5), 592–611. 1976.
5. Kletz, T. A. 'A three-pronged approach to plant modification.' *CEP Loss Prevention*, **10**, 91–100. 1976.
6. Swain, A. *Sandia Human Factors Program for Weapon Development*. SAND 76–0326 Sandia Laboratories, USA.
7. Lawley, H. G. 'Operability Studies and Hazard Analysis.' *CEP Loss Prevention*, **1**, 105–116. 1974.
8. Powers, G. J. and Tompkins, F. C. 'Computer aided synthesis of fault trees for complex processing systems.' In E. J. Henley and J. W. Lynn (Eds). *Generic Techniques in Systems Reliability Assessment*. Noordhoff, Leyden. 1976.
9. Nielsen, D. S. 'Use of cause-consequence charts in practical systems analysis.' In *Reliability and Fault Tree Analysis, Theoretical and Applied Aspects of Systems Reliability and Safety Assessment*. Papers of the Conference on Reliability and Fault

Tree Analysis, Berkley, 3–7 September 1974. Society for Industrial and Applied Mathematics, Philadelphia 1975, pp. 849–880. 1974.

10. Taylor, J. R. 'Background to risk analysis.'
Working Paper to be published. Risø, 1980.

11. Taylor, J. R. 'A study of failure causes based on US power reactor abnormal occurrence reports.' In *Reliability of Nuclear Power Plants*. International Atomic Energy Agency, Vienna, 1975.

12. Rasmussen, J. and Pedersen O. M; Carnino, A. and Griffon, M.; Gagnolet, A. *Classification System for Reporting Events involving Human Malfunction*. Risø-M-2240. 1981.

13. Rasmussen, J. 'Notes on Human Error Analysis and Prediction.' In G. Apostolakis and G. Volta (Eds). *Synthesis and Analysis Methods for Safety and Reliability Studies*. Plenum Press, London. 1979.

14. Green, A. E. and Bourne, A. J. *Reliability Technology*. Wiley-Interscience, 1972.

15. Bartlett, F. C. 'Fatigue Following Highly Skilled work.' *Proc. of the Royal Society of London*, **131**, 247–257. 1943.

16. Regulinski, T. L. 'Human performance reliability modelling in time continuous domain.' In Henley and Lynn (Eds). Generic Techniques in System Reliability Assessment. Noordhoff. 1973.

17. General Physics, Anonymous, *Performance Measurement System for Training Simulators*. First Progress Report, General Physics Corporation, GP-R-610. 1976.

18. Rasmussen, J. 'What can be learned from human error reports?' In K. D. Duncan, M. M. Gruneberg, and D. Wallis (Eds). *Changes in Working Life*. John Wiley & Sons, 1980.

High Risk Safety Technology
Edited by A. E. Green

SECTION 1.6.2 Causes of Human Error

A. Carnino and M. Griffon

Introduction

When studying how to prevent accidents at work and how to improve the safety of industrial installations, the basic source of information is the incidents or even 'near misses' that take place in any industrial installation. Human interaction with the industrial processes which lead to abnormal operation, degraded functioning or faulty equipment can also be deduced from it.

In most industrial installations, all this information exists and is retrieved to learn experience from it. It is collected and stored manually or computerized in the installation. Usually, when dealing with the safety or the availability of the plant, to the list of incidents is added their severity or impact on the operation of the plant or on the workers in charge of maintaining and operating the process. Incidents which had a direct effect as well as incidents which could have potentially affected the behaviour of the installation are taken into account. In this last case, complementary information is added: which action or system or component helped in restoring a safe operation, such as human intervention, automatic protective devices, safeguard systems, etc.

This source of information is absolutely fundamental in acquiring experience and preventing new outages or incidents on the installation. It has also been found most fruitful to exchange these data from one industry to another as many similarities are revealed and learning in general benefits by the extended catchment of experience.

Analysis of Reported Events in Incident Collection

As we are dealing in this chapter with the human influence on industrial processes, it is important to analyse in this context the events reported.

Usually human influence is only reported when it has a negative effect on the safety or availability of the installation under consideration. When a human action gives a positive effect, such as restoration of normal operation or the avoidance of degraded operation, it is not reported in such data collection, being considered as 'normal' or 'natural' operational performance. This fact explains why it is so difficult to use such events on a statistical basis: the data is biased and incomplete.

The reported incidents are those due to technical failures and human interventions. It is most important to realize the complexity of the events involved in an incident. Any simple dichotomy between man and machine seems to be elusive in practice. The concept is only usable.

The field of interest for finding information on human interactions in incidents must necessarily be narrowed to that of direct negative human influence. When analysing the primary events reported one can classify these human 'errors' (as they must be considered) into the following categories:

1. Operating errors
2. Maintenance errors
3. Test errors
4. Design errors.

Experience reveals that for the large and complex installations which are typical of the nuclear industry about 70% of the total reported incidents can be interpreted into these four categories. This is illustrated in Table 1.6.2-1 using the events reported in Reference 1. Also, this table shows the difficulties encountered in trying to classify these incidents; some errors could be put in a different category or could belong to several categories.

The need is clearly demonstrated for each reported incident to be examined in much more detail so as to improve knowledge of the situation in which it occurred and additionally to shed light on the underlying causal factors.

The analysis of such an event reporting scheme highlights how little one can rely on the texts when the descriptive detail given is insufficient to yield an understanding of what really occurred.

Causal Analysis of Incidents Reportedly Attributed to Human Error

The analysis of human behaviour in an industrial process is a very complex undertaking, particularly in relation to understanding how an incident can occur. The following explanations address this task of causal analysis. (See too Figure 1.6.2-1.) *or components*

All system or facilities are designed in order to fulfil a specific mission. When this mission ceases to be accomplished, we call this state 'an incident'. This word 'incident' means that the installation is in a perturbated operating state. The perturbated operating state is the result of inappropriate actions by either human beings or the hardware of the installation. These actions are called 'failures'. They are then respectively called human failures and component failures.

Human failures may be classified by the same scheme used to classify the

Table 1.6.2-1   Classification of Errors

Nature of the error	Sub-categories of errors	Number of errors	Percentage of the total human errors reported
Erroneous operations	Valving errors – erroneous positioning	47	48%
	Reversal of cables or wiring or instrumentation connections	7	
	Omission of inhibition or reciprocal	9	
	Addition of extraneous elements or omission of elements	7	
	Programming errors	2	
	False labelling	1	
		73	
Operating errors	Mal-operation	13	
	Errors in procedures or misinterpretation	14	
	External events (electric shocks, cutting of cables, etc.)	3	20%
		30	
Maintenance errors and test errors	Pure maintenance (parts, positioning-omission, tests, etc.)	27	27%
	Fabrication defects	1	
	Calibration	13	
		41	
Design errors		7	5%

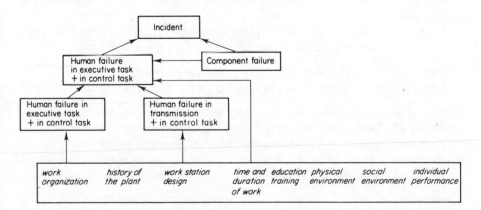

Figure 1.6.2-1 Different causes leading to an incident

various tasks allocated to human beings in a system. When the human actions directly influence the behaviour of the plant these are called executive operations and their failures are therefore called executive human failures. Other actions may influence indirectly plant behaviour such as commands or information gathering or information communication, etc. The corresponding operation would induce an incident if and only if it is followed by an executive task. Then any human executive failure can be completely explained by a human error in giving the commands or gathering the information or transmitting it. Some other actions may have a combined influence on the behaviour of the installation, such as human failures in control tasks.

In the analysis, one must now consider the characteristics of the work situation which have caused each of the human failures encountered. These characteristics can be grouped into the following eight classes:

1. Work organization
2. Design of the work situation
3. Time and duration of the work
4. Personnel education and training
5. Physical environment
6. Social environment
7. History of the plant
8. Individual performances of the personnel (psychological, physiological)

These classes are just general headings which are explained in Table 1.6.2-2 in more detail. In no sense are these classes claimed to be autonomous.

When carrying out detailed incident analysis, it is not uncommon to find many of these causal aspects present together but to fully reveal these factors can be very difficult. Special interviews with operators, knowledge of the work situation and technical background are absolutely necessary in order to conduct such an analysis.

In order (i) to reconstruct the evolution of an incident and (ii) to find the major causes of the human behaviour, two main steps are necessary:

—Identification of all types of failures (technical or human)
—Determination of their causes in relation to the human elements

These steps can be conducted in many ways, but usually the most fruitful methods are fault trees and event trees. A study of such incidents in the nuclear field showed that in decreasing order of importance the most frequent causes were:

—Procedures
—Work organization
—Lack of efficient controls

Since the first cause, namely 'procedures', seems to be a major concern for both

Table 1.6.2-2   Classification of Work Situation (see Reference 2)

1. *Work Organization*
   Content of the usual task.
   Content of the official task: oral or written procedures.
   Content of the real task to be performed.
   Quantity of information to be retrieved.
   Vigilance level.
   Level of repetition or monotony.
   Importance of the task: main or auxiliary task.
   Immediacy of the task in relation to the operation of the plant.
   Effect of the task on the next operations.
   Conflict between production and safety.
   Beginning of the task.
   End of the task.
   Interference between different or similar tasks.
   Responsibility.
   Level of initiative related to the task.

2. *Design of the work situation*
   Height and distance of the working panels.
   Arrangement of indications with respect to the visual field.
   Design of sound alarms.
   Compliance with existing stereotypes and colour standards.
   Design and readability of instrumentation.
   Readability of visual display screens.
   Readability of procedures.
   Interference between operational zones.
   Restrictions imposed by protective clothing, machine guards, etc.

3. *Time and duration of the work*
   Timetable (duration, variability, etc.).
   Shift work or not (frequency).
   Night work (permanent, occasional, etc.).
   Duration of breaks (frequency).
   Overtime (frequency, duration, transportation time).
   Work to be done in a given time (hours/days/weeks).
   Time of the incident in the daytime, in the week, in the year.
   Working after an unscheduled outage.

4. *Education and training*
   Duration of the training at the work station.
   Schooling.
   Professional training.
   Training method to perform the task.
   Number of years in the work situation.
   Number of years in the firm.

5. *Physical environment*
   Noise.
   Vibrations.
   Light – colours.
   Heat.
   Radiation.
   Smokes, dust, toxic products.
   Ground cleanliness.
   View on external environment.
   Items or people moving around the work situation.

Table 1.6.2-2 (*Continued*)

---

6. *Social environment*
   Activities outside the firm (hobbies, sports, cultural activities, etc.).
   Family life (married, divorced, single).
   Stability of employment.
   Performance stimulants and incentives within the firm (salaries, promotion, etc.).
   Company turnover.
   Conflicts in the firm.
   Participation in decisions.
   Participation in unions.
   Information sharing within the organization.
   Attitude towards hierarchy.
   Attitude towards other workers.
   Extent to which person must work on his own, physically and/or intellectually.

7. *History of the plant*
   Plant running for a long time without problems.
   Plant running with a lot of incidents and outages.

8. *Individual Characteristics*
   Age, nationality, sex.
   Tiredness.
   Degree of dependability.
   Degree of self-assurance.
   How does operative perform under stress?
   Extent of initiative.
   Ability to concentrate.
   Quality of vision – sight, colour blindness.
   Quality of hearing.

---

operators and engineers, the next section is devoted to discussing this in some detail.

A Major Contribution: Procedures

The problem of procedures is a large and important one: procedures are the rules which apply at the interface where man and system meet. Three major areas have to be dealt with:
1. The technical content
2. The presentation
3. The potential errors and their consequences.

These areas must be addressed whether the procedures are oral or written.

When determining the *technical content* the following principles should be respected:

The actions required should be presented in logical order. Also, there should be appropriate reference to any parallel actions together with clear indications of the timings of such actions and the means by which their completions are communicated.

The measurements required should be clearly defined together with their acceptable parametric limits.

As many check points as possible should be integrated which will help in detecting potentially hazardous error and which would prevent continuation of the procedure until such an error is corrected.

Consideration should be given to ensuring that the main actions or measurements are able to be performed with maximum ease.

The division of task responsibilities between team members carrying out the procedures should be clearly laid down in terms of the individuals' respective roles at each action (execution of the task, control, surveillance, etc.).

On-plant exercises should be implemented to prove the coherence and adequacy of the specified actions and measurements and to test also the divisions of responsibilities.

Particularly where actions cannot be reversed once performed, specific means should be proved to confirm that the chosen procedure is the correct one for the particular situation at that time.

Positive guidance should be included against the occurrence of failure or above limit measurement during the course of the procedure.

This last point is certainly a very important one: usually the designers of procedures suppose that everything is going to work the way it should. But the analyses of incidents or accidents frequently show this expectation to be false. Furthermore, it is a basic principle of ergonomics that operators on a shift must be positively instructed and trained what to do in a perturbated situation, and the more especially under stress conditions.

The *presentation of procedures* is also very important. There again one can list a series of recommendations, which is certainly not exhaustive but which will assist in procedure presentation. One should:

1. adapt the presentation to the level of education and training required and to that which is consistent with the professional background of the recipients;
2. take into account, using ergonomic principles, the wishes of the operators concerned; for example, with respect to levels of detail requested on checklists, provision of task descriptions, use of colours, vertical or horizontal action sequence formats etc.;
3. by using different lettering or other means discriminate clearly between principal and auxilary actions;
4. show, by means of suitable coding, the different parallel actions taking place at the same time, the permissive or non-permissive steps, and the task partitioning between team members;
5. Allocate special columns or places for the measurements, the acceptable limits, and the control actions taken;
6. nominate distinctive symbols or colours, or use columnar segregation, indicate all noted directions from the norm;

7. clearly define which actions should be taken in case of failure or unavailability of components;
8. arrange the administrative control of the results of the procedures, especially for those associated with periodic testing, so that any needed repair or calibration is undertaken, that any abnormal condition is detected and that the associated system or process is left in the correct condition;
9. train the operators to apply procedures correctly (special courses, demonstrations, simulators, etc.) and teach them the importance of the work they are asked to perform in relation to the safety or the availability of the plant.

In order to prevent the situation where errors in procedures lead to unacceptable consequences for systems or processses, a *cause – consequence study* should be performed at the design stage. This may be done in various ways but task decomposition seems to be the most fruitful one. It shows the possible recovery factors and indicates the actions essential for the good functioning of the system: preventive actions can then be introduced into the procedure so as to minimize the effects of identified errors. These preventive actions can be of many kinds: double checks, control, non-permissive actions, detection of errors by measurements, etc.

The list of recommendations given here for the improvement of procedures at the man–machine interface is certainly not exhaustive and must be individually applied to the separate cases of normal operation procedures, test and maintenance procedures, incident and accident procedures. It also has to be an iterative process between the different phases: technical content, presentation and cause/consequence studies. The optimization cannot be achieved always at the first trial and experience in the field as well as operator's experience must be integrated into this iterative process. Inevitably, operational experience will modify certain procedures during the life of a plant, but any modifications must be fully assessed in the system context before being finally confirmed.

Questions 1.6.2

1. Classify the main types of human error which can occur in association with a land-based plant and give an example in each category.
2. In incident analysis what has been found to be the most frequent underlying cause of human error? What principles can be identified to mitigate this cause?
3. Choose a work situation which is known to you and list in order of importance the work organizational factors which influence human error. Give a relevant example from your own experience to illustrate one or more of these factors.
4. To what extent is social environment an important influence on human performance? Discuss how this environment may have positive and negative motivations on the reliability of human performance.

References 1.6.2

1. Carnino, A. and Raggenbass, A. 'Composantes humaines dans les phénoménes initiateurs d'accidents nucléaires.'
   IAEA SM 215/35. 7 March 1977.
2. Griffon, M. 'Facteurs humaines de la Surete.' (Private communication) CEPN No. 18, January 1979.

High Risk Safety Technology
Edited by A. E. Green
© 1982 John Wiley & Sons Ltd

SECTION 1.6.3 Discussions around a Human Factors Data-Base. An Interim Solution: The Method of Paired Comparisons

D. M. Hunns

Introduction

Again and again in the field of systems reliability prediction the same question arises, 'How do we include the influence of the human being?' On the hardware side during the last two decades there has been a progressive development and proving of quantitative as well as qualitative assessment techniques. But on the human factors side, despite the fact that its importance always has been well recognized, this same period has seen a virtual drought in the comparable development of analytical techniques. As the technology of predictive reliability assessment continues to gain acceptance and establish its importance in the world the deficiencies on the human factors side are becoming more and more recognized as an area of major concern. This imbalance in progress is probably a reflection of the complexity of the human factors problem rather than simply an indication of lack of effort. Three principle areas of difficulty (see Figure 1.6.3-1) may be defined:

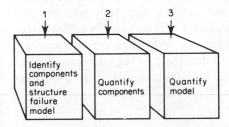

Figure 1.6.3-1 Areas of difficulty

1. System modelling – the extremely large number of possible transfer functions embodied in a human being, and the relative freedom he has in terms of establishing his connections to a system, mean that the analyst must conceive and handle a large set of possible system configurations. In comparison, taking

181

the case of a purely hardware system, the analyst normally has only one configuration (or possibly a limited set of configurations) to consider; furthermore the configuration(s) generally will be pre-defined.

2. Quantification of the human factors' components identified in the system model.

3. Synthesis of total system performance using the components' data in the model.

It is the writer's opinion that the system modelling continues to be outstandingly the area of greatest difficulty. Analytically to predict the full set of significant event chains associated with a given human/hardware system is not yet within our capabilities. Much could be written on the subject but it is not within the brief of this present paper. Of the remaining two areas of difficulty, namely the components' quantification and system performance synthesis, the latter would appear to represent the least of the problem areas. Difficulties with synthesis can arise where the components of the model lack independence; elements of interaction undermine the applicability of the traditional combinational arithmetic which is, of course, based on the assumption of independence between components. However, by taking care at the system modelling stage (a) to model wherever possible at a component level where interaction between components is at a minimum and (b) where interaction cannot be avoided, to express the interaction components as clearly identified conditional elements and quantify accordingly, it is possible virtually to eliminate the potential difficulties of the system synthesis operation.

There remains the middle area of difficulty, that of quantifying the human factors' components identified in the system model. It is with this subject that this paper is concerned. We start with the assumption that the system modelling has been carried out satisfactorily and that the stage has been reached where the analyst is faced with the task of components' quantification. From where does he obtain his data? It is virtually certain that however hard he may try he will fail to find a data source seriously offering to satisfy his needs. Odd elements of general or ill-defined information he will find but nothing in a fully relevant and comprehensive form. One may well ask why this should be so when we have large and usable sources of data on the hardware side. Why then are we so poorly provided for on the human side? Answers to this question begin to emerge when we examine the unique difficulties associated with the task of compiling a human factors data-base.

The Human Factors Data Problem

Human failure or error inevitably carries connotations of blame and personal deficiency. We do not like to be observed making errors and therefore, as potential subjects, we tend to oppose instinctively the idea of personal performance monitoring. In many workplaces the data collector will fail to overcome such reluctance. Supposing, however, that he does, the collector then must arrange his surveillance so that for all intents and purposes his subjects quickly forget its presence. If there is a continuingly conscious awareness of the monitoring, performance is likely to be abnormal. Only a minority of situations offer ready possibilities for 'transparent' performance monitoring. Another unique factor which compounds the difficulties of the data collector is that of error recovery. The majority of human errors are corrected by the person concerned before becoming evident to a third party. An effective data collection activity should be sensitive to initial errors whether recovered or not – otherwise much valuable data stands to be lost. The implementation of a surveillance scheme with this type of sensitivity but which does not impose its presence on the subject can pose formidable problems. A fundamentally important requirement of any data collection activity is that a complete data record should include not only the quantification of a given event but also a comprehensive and objective account of the determining factors surrounding the event. Many of these factors, and often the most critical ones, will be bound up in the mind and psychology of the individual concerned at the time of the event. In many instances it is highly likely, even with the most willing and intelligent cooperation, that he will find himself totally incapable of post-analysing these factors. Furthermore, repeated demands made upon him to reveal this type of information will have the undesirable effect of maintaining in him a continuing awareness of the monitoring activity.

The issues raised in the preceding paragraph are concerned with the problems of persuading the workplace to yield the information which is required. These problems are all associated with the requirements for transparency, sensitivity and comprehensiveness of monitoring (summarized in Figure 1.6.3-2) and taken together they represent formidable opposition to data collection. However, all this presupposes that the data analyst does have a clear specification of the information sets which, through his data collection activities, he is seeking. It presupposes that in each situation of surveillance he is able to establish and measure the set of performance influencing factors which do indeed constitute the basic, autonomous building bricks of that situation. But to do this successfully requires that the analyst should have a proficient understanding of the fundamental causal factors of human failure. At the present time we do not possess this necessary level of knowledge. For this same reason we have failed to propose a satisfactory taxonomy for classifying events and performance-influencing factors into a data store structure; and it is to be expected that this situation will continue until we have cleared a better understanding of these basic mechanisms of human failure. So not only is the human factors data collection activity frustrated by the unyielding nature of the observation environment but at an even earlier stage the process is confounded because no fundamental basis exists to define the type of information which is really required.

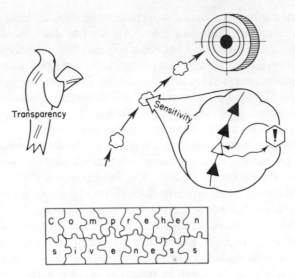

Figure 1.6.3-2 Requirements for human factors data
collection

The working out of a basic data structure is clearly the starting point on the long road towards establishing an operable human factors data-base. The source material for deducing the necessary understanding is the psychological literature and the documentation of past accidents. An immediate product of such studies is qualitative data. As information is accumulated, statistically significant quantitative information might be gleaned in limited areas; but the ultimate aim is to define the fundamental information structure for a workable data-base. Only when some progress has been made here will it be sensible to address the problems of how to obtain actual data from the real operational environments. The search for a fundamental human factors data-base structure has already started but the task is a long-term one. The provision of a substantial data-set within the deduced structure will be very much longer term.

Data Bank Theory

It is unrealistic to regard a data bank as some continuously expandable repository whose usefulness is a function of information quantity. This is close to the idea that the usefulness of a piece of string should be judged by its length. A data bank exists primarily as an assembly of reference points, rather like a multidimensional string net where the knots are defined data points and the converging strings individually represent the associated situation factors which give each data node its identity (see Figure 1.6.3-3). A system component to be quantified firstly must be translated into the factors-set which precisely represents the component in a

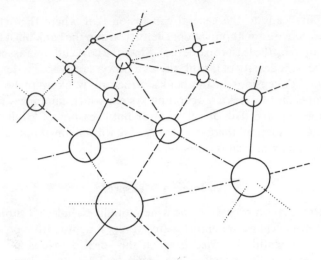

Figure 1.6.3-3 Data bank concept

form recognizable by the data bank. This factors-set then is offered to the structure of the bank and a 'best fit' is achieved. An exact fit might occur but it is very much more likely that there will be some misalignment. In which case an extrapolation process must be undertaken and an estimate obtained.

This, broadly, is how a data bank operates. The accuracy of the estimate (the data retrieved) is a function of: (i) the exactness with which the interrogating component can be translated into the language of the data bank; (ii) the effectiveness of the extrapolation process; (iii) the closeness of the reference points (the data density) in the particular interrogation area; and (iv) the accuracies of the relevant reference data points themselves. If an effective extrapolation process is operational it follows that the data density within the bank can afford to be relatively low. The effectiveness of the extrapolation process in relation to the data accuracy required is the major factor determining the necessary density, and to a fair degree the size, of the stored data-base.

Extrapolation processes vary in their degrees of sophistication. Most commonly the process is carried out informally 'by eye', the applicant to the data bank using his own estimating judgement. However, while no doubt this type of informal process is effective and acceptable from the point of view of the individual applicant, the data produced cannot be directly assessed for applicability or accuracy by those one stage removed from its derivation. The extrapolation process lacks the objectivity necessary for this 'third party' evaluation to be made. Furthermore the estimating judgement facility varies from person to person. This means that if two individuals are to retrieve data with a certain minimum accuracy from the bank, the data density within the bank must be at the level compatible with the applicant who has the inferior extrapolation capability. This leads to the general conclusion that where the extrapolation process is subjective estimation, the data density within the bank must be raised to the level required by the least capable user. Since the cost and time associated with increasing the data density of a bank rise as a power law whose index, related to the number of dimensions in the bank's structure, is likely to be large, it is obviously important to seek an operating system which minimizes rather than maximizes the required data density. So for both reasons of 'capital cost' and 'retrieved-data evaluation' there is a strong case for maximizing the power and objectivity of the extrapolation process.

An Interim Measure?

The ideal extrapolation process is one which totally excludes all subjectivity. In the human factors field we are unfortunately a very long way from possessing the precise kind of scientific knowledge which this ideal demands and so, at the present time, drastic compromises are inevitable. Also it is logical that any extrapolation process developed should be tailored to the structure of the data-base with which it is to operate. Again, at this time, the concepts of such a structure are unresolved; it would seem that until these concepts are formed we lack the necessary basis on which to build an extrapolation process. Nevertheless we do have a certain limited amount of human factors data available to us. There are the generalized data of the type listed below whose origins are a consensus of subjective belief expressed by a group of people who have been closely concerned with the subject of human reliability.

Classification of error type	Typical Probability of error
Processes involving creative thinking, unfamiliar operations where time is short; high stress situations	10^{-0}–10^{-1}
Errors of omission where dependence is placed on situation cues and memory	10^{-2}
Errors of commission such as operating wrong button, reading wrong dial, etc.	10^{-3}
Errors in regularly performed, common-place tasks	10^{-4}
Extraordinary errors – of the type difficult to conceive how they could occur; stress-free, powerful cues militating for success.	$< 10^{-5}$

We also have data, for example, on the error rates of secretarial typists and operators producing punched cards for computers – we have the results of a survey which sets out to discover the frequencies with which train drivers failed to stop at danger signals – we have data on the error rates of printed-circuit-board assembly line workers – and other sources of information could be listed. Each individual then may supplement this list with his own subjective data, bringing to bear, as best he can, his own lifetime's accumulation of human factors experience. Certainly we have data but it seems always with insufficient qualification for us to assess and apply them with confidence; and yet on the other hand, we cannot ignore this kind of knowledge even though it may appear very insubstantial. Is it therefore possible to aggregate what we have into some crude form of data bank and operate a workable retrieval extrapolation process? If the answer is yes, we have a viable interim data resource which at least can go some way towards meeting our needs until longer term researches begin to bear fruit. Of course, such a data bank has been in existence notionally for many years. However, since its data density has been very low, its structure nebulous and the extrapolation process based entirely on subjective estimation it is not surprising that its use has been restricted to a very limited set of expert estimators and the results always subject to debate and uncertainty. So can we find an extrapolation process which, despite the low data density and virtually non-existent data structure, is nevertheless a significant step forward in power and objectivity? In seeking a process within this broad specification we obviously must continue to capitalize on the subjective data-bases which we all possess. In the field of psychology, techniques long since have been in existence which operate the judgement faculty in an organized manner. One such technique is the method of paired comparisons first presented by Thurstone[8] in 1927. With certain modifications this method offers interesting possibilities as a workable extrapolation process.

The Method of Paired Comparisons

General description

The method of paired comparisons is a psychological scaling technique. It makes exclusive use of the human judgement dimension to obtain the ranking and scaling of a given set of items (stimuli) with respect to a specified attribute. The stimuli (these could be human factors' components from a system failure model) are presented in pairs to the assessor and for each pair he must decide which is the greater, or lesser, with respect to the given performance attribute (the assigned attribute in our case would be a reliability metric like failure probability). The stimuli are presented in all possible combinations of two and the complete set of 'greater than' or 'less than' decisions is recorded. For n stimuli the information set would comprise $n(n-1)/2$ decisions. The entire process is then repeated many times in order to produce a statistically usable population of information sets. In theory it would be possible to use the same assessor for the repeated runs but normally the aim would be to use a different assessor on each occasion. This total information can then be processed to produce a scaled ranking of the stimuli (see

Figure 1.6.3-4). The stimulus assessed as the lowest with respect to the defined attribute will be allocated a scale value of zero which serves as the base of a nominal scale range along which the other attributes are positioned. The numerical magnitude and resolution of this scale potentially vary with the size of the population of information sets which produce it. The feature of importance is not the scale magnitude but the relative positionings of the stimuli along it; the number of units within the scale merely reflects the resolution offered.

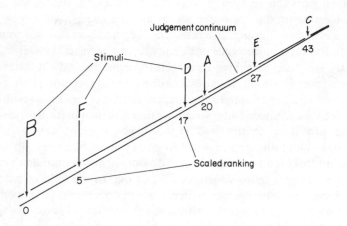

Figure 1.6.3-4 Scaled ranking of stimuli

Now the scaled ranking is not, of course, absolute data; it is simply a nominal scale of relative values; it represents relative spacing along some notional continuum; if the theory is correct it is in fact a faithful reproduction of the spacings of the stimuli in the 'communal mind' which produced the population of information sets. To convert the scaled ranking into absolute quantification (a scale of failure probabilities, say) requires that it should be accordingly calibrated at one or more points. Also required is a knowledge of the relationship between psychological scales and probability scales. The work of Pontecorvo[5] indicates that this relationship is logarithmic and indeed it is possible to make plausible hypothetical propositions from which his findings may be exactly reproduced. Applying the logarithmic concept it is possible to convert the scaled ranking into an absolute probability scale using a minimum of two calibration points (see Figure 1.6.3-5). This requires that absolute values must be given to two of the stimuli involved in the scaled ranking process. Error sensitivity is minimized if the two stimuli used are at the extreme opposite ends of the scale. In this way the method of paired comparisons forms the basis of an organized subjective extrapolation process.

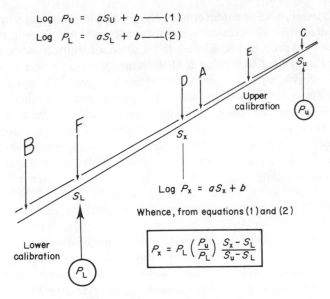

$$\text{Log } P_U = aS_U + b \quad\text{---(1)}$$
$$\text{Log } P_L = aS_L + b \quad\text{---(2)}$$

$$\text{Log } P_x = aS_x + b$$

Whence, from equations (1) and (2)

$$P_x = P_L \left(\frac{P_u}{P_L}\right) \frac{S_x - S_L}{S_u - S_L}$$

Figure 1.6.3-5 Probability calibration

As described so far it would appear that use of the method always demands that the paired presentations must be run many times in order to obtain a statistically analysable consensus of information sets. Where the data density supporting the process is low (for example, eight stimuli for which only two have known data) then this is undoubtedly necessary. However, suppose that we are in the situation where we have eight stimuli, say, seven of which we know in terms of data. In this case it is possible to run the paired presentations once, to a single assessor, and by this means a ranking will be produced. It will not be scaled and it will be the instantaneous subjective extraction of one person. Nevertheless, providing that the unknown stimulus is not ranked at either extreme end, its data range will be specified by the stimuli appearing on either side of it. If the unknown stimulus is end-ranked then a 'greater than' or 'less than' evaluation is yielded. Although not single-valued this type of boundary data nevertheless can be very useful in analytical work and is an attractively small price to pay for the advantage of obtaining results without operating the machinery of consensus. Thus here is described an alternative conceptual approach to using the method of paired comparisons as an extrapolation process.

A desirable feature of any technique which attempts to make organized use of subjective judgement is that there should be inbuilt an indication of whether or not the contributing individual really is exercising a disciplined use of his judgement machinery. Such an indication should be able to shed light on the consistency of the thought processes underlying the decision making. When a contributor exhibits no apparent consistency he may be treated as a 'rogue' input and eliminated from the consensus data-set. Methods for yielding this type of insight have been proposed which seek to judge a contribution on the basis of the extent of its agreement with other contributions – this is often referred to by the expression, 'between-judge agreement'. However, this does appear to be a doubtful criterion of assessment since a judgement situation where stimuli are naturally closely spaced along the given attribute scale would be expected to produce a variety of radically opposed opinions even though each individual judge was exercising his own judgement faculty with complete consistency. A far better basis for deciding the bona fide nature of the subjective information obtained is to somehow monitor the thought-process consistency behind each individual's contribution. It then may be possible to combine these individual evaluations into some corporate assessment of the total consensus data-set which constitutes the subsequent scaled ranking.

The method of paired comparisons, by virtue of the sequential paired presentations process, offers an interesting means of achieving this. The technique, fully described by Kendall[4], looks for decision inconsistencies in the rankings of the stimuli pairs. For example, take three stimuli, A, B, and C, producing three pairs to be ranked, AB, BC, and AC. If the judge ranked A above B and B above C in order to be consistent he should rank A above C. If he reversed this latter ranking he would have produced what Kendall[4] describes as a circular triad (see Figure 1.6.3-6). The number of circular triads occurring in a given set of decisions is a tell-tale of the structural rigidity of the thought processes used by the associated judge.

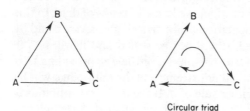

Circular triad

Figure 1.6.3-6 Decision consistency monitoring

For a given single paired comparisons run there is a theoretical maximum number of circular triads which a judge might produce. This number is a function of the number of stimuli in the run. However, if he were ranking the pairs in an absolutely random manner the mean number of circular triads registered over a large number of runs would not be the theoretical maximum; it would be some lesser number. Kendall[4] presents the frequency distributions which describe this effect. Therefore if a particular judge achieved the maximum possible number of circular triads in his contribution we would not mark him as exhibiting zero consistency but rather we would have strong grounds to suspect him of organized subversion. In other words it is possible to compare the actual number of triads produced by a judge with the frequency distribution of triads produced by random decision taking and thereby deduce the likelihood of a structured thought process. To obtain the equivalent probability for the total population of, say, r information sets, taken as a whole, the single-run frequency distribution must be combined with itself r times in order to obtain the composite distribution; this is then used in a similar fashion as before but now comparing with the total number of triads accumulated from the r runs. Since the r-fold combination of the single-run frequency distribution is analytically very complex it is expedient to produce these composite distributions by Monte Carlo simulation.

Thus, the method of paired comparisons offers itself as an organized subjective technique for using a consensus of experience to scale-rank a given set of stimuli with respect to a specified performance attribute. If operated in the non-consensus, single-run mode it will produce a ranking of the stimuli but without the scaling. In either case it is possible to analyse the information given to indicate the likelihood of the judgements having been guided by structured and disciplined thought processes. The rationale of a logarithmic relationship between relative psychological and probability scales enables the scaled ranking to be converted into measures of absolute probability by calibrating with a minimum of two known values. The method, therefore, is potentially usable as an extrapolation technique for use with a low data-density human factors data bank.

The Law of Comparative Judgement

Consider a performance attribute like 'frequency of occurrence'. Now consider, with reference to this attribute, the following stimulus: 'making a telephone call and finding the line engaged'. You will at once be able to place this stimulus into the generalized category of, say, very often or, perhaps, just occasionally, etc. Now consider another stimulus, for example, 'winning a raffle'. Almost straightaway you will place this also into a generalized category with respect to frequency of occurrence – perhaps in this case it will be in the category of 'very rarely' – and you will in the same process distinguish a clear mental separation between the two stimuli. Add to these two stimuli a third stimulus, 'catching the traffic lights at red', and you will now be attempting in the mind's eye to position it in relation to 'frequency of finding a telephone line engaged' and 'frequency of winning a raffle'. You will in fact be visualizing your judgments quite involuntarily along some mental continuum which is labelled 'frequency of occurrence'. In fact, from the very moment when the performance attribute, 'frequency of occurrence', is first defined you will already have set up the embryo of this continuum; it is the fundamental process of comprehension; unless you could successfully find stimuli recorded in the mind which relate readily to the given attribute you would fail to register an understanding. So if indeed you have personal experience which relates to a suggested performance attribute then from the moment when you are called to focus upon it you will establish the basis of a mental continuum whose length and position in space are determined by largely subconscious agglomerates of experience. The various scenarios of these associations will not be specifically reproduced in the conscious experience section of the mind until the external stimuli to be rated draw the continuum into sharper focus. The whole process is one of comparisons which involuntarily positions stimuli along a unidimensional psychological continuum defined by the judgement attribute of interest. This is, basically, the first of the postulates of Thurstone's[8] law of comparative judgment. In more precise terms this states, 'Each stimulus when presented to an observer gives rise to a discriminal process which has some value on the psychological continuum of interest' – see Figure 1.6.3-7.

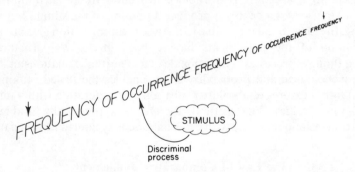

Figure 1.6.3-7 The psychological continuum

Notionally we might define some nominal scale unit and zero point for the psychological continuum. Then it could be said that a discriminal process (resolution of unidimensional spatial position by comparisons) will produce a scale value for the particular stimulus. The second postulate, fundamental to the law of comparative judgement, asserts that 'because of monetary fluctuations in the organism this notional scale value will vary'. It is further proposed that if the same stimulus is presented to an observer a large number of times a frequency distribution of scale values will be produced which will be Gaussian in form (see Figure 1.6.3-8). At first sight the idea that the frequency distribution should be Gaussian may seem somewhat tenuous, although most of us would accept the idea that there will be some variational pattern. However, the condensed evidence from experiments relating to physical and mental responsiveness consistently reveals the one common characteristic, viz., the greater the deviation from the mean the less likely is its occurrence. When the mechanisms of variation have equal freedom on either side of the mean, the frequency distributions will be symmetrical. Symmetrical, bell-shaped, distributions can be modelled by the Gaussian probability density function. Argued in these terms the postulate of Gaussian variation along the psychological continuum does appear a reasonable basis on which to proceed. The mean and standard deviation of such a distribution are termed respectively the mean scale value and the discriminal dispersion.

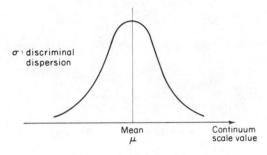

σ : discriminal dispersion

Mean
μ

Continuum
scale value

Figure 1.6.3-8 Gaussian variation of discriminal process

Consider the theoretical distributions of discriminal processes for any two stimuli, j and k, as shown in Figure 1.6.3-9. The means of these distributions, μ_j and μ_k, are the respective mean scale values along the psychological continuum of interest. If the two stimuli were presented together to the observer, each would excite a discriminal process with notional scale values, d_j and d_k. Subtracting these two values, one from the other, will produce what may be termed the instantaneous discriminal difference. By repeating this process a large number of times it will be found that the discriminal differences also follow a Gaussian form of frequency distribution. This automatically arises from the fact that the difference of two Gaussian distributions is a third Gaussian distribution. Furthermore, the mean of this difference distribution will be given simply by subtracting the means of the contributing distributions. Thus, the mean scale separation μ_{k-j} between the two stimuli, j and k, is given by:

$$\mu_{k-j} = \mu_k - \mu_j$$

Now, since the distributions for the discriminal processes of j and k will always, to some degree, overlap it is to be expected accordingly that a proportion of the discriminal differences will be negative. That is to say, on a proportion of occasions the observer will reverse his judgement and will give, say, a higher scale value to j instead of k. The theoretical distribution of discriminal differences therefore will take the form shown in Figure 1.6.3-9, where the straddling of the zero difference represents this effect.

If the observer, being presented with the two stimuli on *r* separate occasions

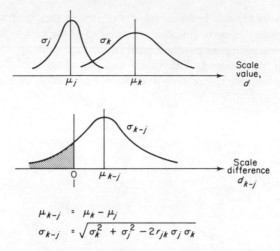

$$\mu_{k-j} = \mu_k - \mu_j$$

$$\sigma_{k-j} = \sqrt{\sigma_k^2 + \sigma_j^2 - 2r_{jk}\,\sigma_j\,\sigma_k}$$

Figure 1.6.3-9 Distribution of differences

(arranged so that he is unlikely to remember his previous decisions), simply notes each time whether it is j or k which he rates higher on the continuum, then if he chooses k r_k times, the ratio r_k/r is an estimate of the area under the positive (unshaded) section of the distribution of discriminal differences. The ratio is in fact an estimate of the probability that k will be rated greater than j. It will be an estimate of $P(d_{k-j})$ where this is given by the cumulative Gaussian expression,

$$P(d_{k-j}) = \int_0^\infty \frac{1}{\sqrt{2\pi}\,\sigma_{k-j}} e^{-\frac{(\mu_{k-j} - d_{k-j})^2}{2\sigma_{k-j}^2}} \cdot d(d_{k-j}) \qquad (1.6.3\text{-}1)$$

k − j is the standard deviation and in this case $d_{k-j} = 0$, hence,

$$P(0) = \int_0^\infty \frac{1}{\sqrt{2\pi}\,\sigma_{k-j}} e^{-\frac{(\mu_{k-j})^2}{2\sigma_{k-j}^2}} d(d_{k-j}) \qquad (1.6.3\text{-}2)$$

This expression may be transformed into the unit normal form and given by the expression,

$$P(0) = \int_0^\infty \frac{1}{\sqrt{2\pi}} e^{-\frac{x_{kj}^2}{2}} dx_{kj}$$

where $x_{kj} = \mu_{k-j}/\sigma_{k-j}$ is the unit normal deviate corresponding to the value of $P(0)$ in the cumulative normal distribution.

Thus the mean scale separation (the mean discriminal difference), μ_{k-j}, for the two stimuli, j and k, is given as,

$$\mu_{k-j} = x_{k-j} \cdot \sigma_{k-j} \qquad (1.6.3\text{-}3)$$

It can be shown that the standard deviation, σ_{k-j}, of the difference distribution relates to the standard deviations, σ_j and σ_k, of the associated absolute distribution in the following way,

$$\sigma_{k-j} = \sigma_k^2 + \sigma_j^2 - 2r_{jk} \cdot \sigma_k \cdot \sigma_j \qquad (1.6.3\text{-}4)$$

Whence, substituting 1.6.3-4 in 1.6.3-3

$$\mu_{k-j} = x_{kj} \sqrt{\sigma_k^2 + \sigma_j^2 - 2r_{jk}\sigma_k\sigma_j} \qquad (1.6.3\text{-}5)$$

which is the complete form of the law of comparative judgement.

μ_{k-j} : the mean separation along the psychological continuum for stimuli j and k.

σ_j, σ_k : the discriminal dispersions for the stimuli j and k.

x_{kj} : the unit normal deviate corresponding to the probability (derived from a theoretically infinite population of observations) that k will be ranked above j.

r_{jk} : the correlation between pairs of discriminal processes, d_j and d_k.

The correlation coefficient, r_{jk}, makes allowance for variational mechanisms in the observer which would similarly affect both scale values registered for stimuli j and k. For example, at a particular time he might give j a lower than usual position along the continuum and the same influence may cause him also to depress the positioning of k. This will have the effect of reducing the spread of the difference distribution which is accordingly reflected in the way r_{jk} acts in the expression, Equation 1.6.3-4, for the difference standard deviation. Of course, in theory, r_{jk}, might be a negative coefficient.

Consider that we have six stimuli to be scale-ranked. Six stimuli produce $n(n-1)/2 = 15$ separate pairs. The pairs are presented over and over again to an observer who diligently registers his paired ranking preferences. From this total data, estimates can be deduced for the paired ranking probabilities. Fifteen probabilities will be produced and each may be converted into its equivalent unit normal deviate, x_{kj}. Thus 15 versions of Equation 1.6.3-5, can be written. The unknowns involved amount to 15 correlation coefficients, 6 discriminal dispersions and 5 (5 spaces separate 6 points) differential scale quantities. It is the latter, of course, which must be deduced to give the required scaled ranking. So we have, overall, 26 unknowns and only 15 equations. Indeed, regardless of the number n of stimuli there always will be more unknowns than equations to solve them. In its complete form the law of comparative judgement is not solvable. Simplifying hypotheses are thus necessary in order to make the law usable.

An alternative to repeating the trials many times with a single observer, of course, is to use a different observer for each repeated run. It is reasonable to assume that the same types of variation will operate and therefore that the basic rationale of the law will continue to apply. By so doing, certain advantages are apparent. First, the problem of making successive runs independent of prior memory is overcome. Second, a wider catchment of experience is tapped. Third, because the population of jk decisions for a given pair of stimuli derive from individually composed psychological continua it would not be surprising to find the correlation coefficient, r_{jk}, virtually equal to zero. This is a most important construct because if it is valid, and the r_{jk} term in the law can be legitimately neglected, the equations then exceed the unknowns and the law becomes solvable. A further simplification can be made if it is additionally assumed that the discriminal dispersions are all sufficiently of similar magnitude to be effectively regarded as equal. The validity of such an assumption ultimately can be proven only by successive tests of the law, with and without the assumption being applied. However, there is a strong intuitive case for supposing that the assumption is more likely to apply when the trials are replicated with a population of assessors rather than with a single assessor. That is to say, it is more likely to be valid for discriminal distributions made up from the single contributions of a population of assessors rather than for those made up from a population of contributions of a single assessor.

By applying the two simplifications, namely (i) the correlation coefficients may be neglected and (ii) the discriminal dispersions for all the stimuli may be regarded as equal, $\sigma_a = \sigma_b = \ldots = \sigma_k = \sigma$, the law reduces to a very simple form as given in Equation 1.6.3-6

$$\mu_{k-j} = x_{kj} \cdot \sqrt{2}\sigma \qquad (1.6.3\text{-}6)$$

Since $\sqrt{2}\sigma$ is a constant term for all pairs of stimuli it acts simply as a common multiplication factor for the derived scale differences. Since this factor will not affect the relative divisions of the scale then for all intents and purposes it can be neglected. Whence Equation 1.6.3-6 becomes effectively,

$$\mu_{k-j} \equiv x_{kj} \qquad (1.6.3\text{-}7)$$

This is, in fact, Thurstone's[8] Case V of the law of comparative judgement. Since, for example, 6 stimuli will produce 15 values for x_{kj} (the unit normal deviate) and only 5 differential scale values are required it is clear that the solution is overdetermined. From a practical standpoint this excess of solutions is a very desirable feature. It enables an averaging process to take place which to some extent compensates for the fact that the paired ranking probabilities, not coming from an infinite population of observers, will never be more than sample estimates and therefore subject to error. Furthermore some data will be lost when the observers either make inconsistent judgements within their own paired rankings or alternatively when they return 100% agreement about one or more paired rankings. 100% agreement produces a probability estimate of unity and the indeterminate value of infinity for the unit normal deviate; such data are, of course, unusable for scaling.

Obtaining the Absolute Probability Scale

The proposed continuum of interest is given as failure probability. We may use the method of paired comparisons to scale-rank a set of defined components (stimuli) along this continuum. What we will obtain is a replication of the ranking and relative linear positioning of the components as carried in the 'communal' mind which produced the paired comparisons data. The scale separations will reflect the communal mind's vision of the relative probabilities of failure for the components. The scale units therefore in some way must be relatable to units of relative probability. The question is, what is the relationship?

The experiment by Pontecorvo[5] suggests that this relationship is logarithmic. He asked a group of assessors to scale-rank various pieces of engineering equipment from the point of view of 'time to repair'. A fundamental stipulation was that all assessors had a working familiarity with all the equipments. Each assessor was given a percentage 'scale' along which to express his judgements. Thus Pontecorvo[5] obtained a substantial and ranging population of subjectively generated data. By plotting the averaged scale values (s) against a set of 'actual' mean repair times (R), these having been calculated from records kept for the equipments, a scatter of points was obtained whose best fit took the form,

$$\log R = As + B$$

R: actual mean repair times
   s: mean scale-rank values judged by assessors
A, B: constants

Now, it hardly can be claimed that this single result is conclusive. The details of how Pontecorvo[5] controlled his experiment, the amount of information supplied to his assessors and the true credentials of his actual data (R) are not discussed in his paper[5]. However the following hypothetical arguments do give independent support to his claims. The foundation of subjective estimation is experience, direct or indirect stored in the mind of the assessor. Any event witnessed or enacted is registered in the memory; similarly, subsequent repetitions of that event are registered. In addition, events reported as 'hear-say', together with the degree of belief attributed to them, also are stored. In these ways a population of evidence relating to the occurrence of a given event accumulates in the mind. The result is to produce a 'feel' for the frequency or probability of occurrence of that event. It is as though the mind's eye views the evidence as an amorphous mass, rather like a pile of bricks, clearly feeling the bulk but not discriminating, in any numerate sense, the individual contributions from which it is formed. So, given an event like 'punctured car tyre' the mind at once feels a level of magnitude for its likelihood. However, if asked to quantify this experience and therefrom derive a figure for the number of punctures that might be expected during the next 20 years of car ownership it will be found that this 'feel' is not directly helpful. The mind then attempts to recall the individual events which formed this amorphous conviction – a mental search process which is demanding and generally not rewarding unless the experienced events-population is low. But if asked simply to make a comparison, for example between '(A) the probability of a punctured car tyre' and '(B) the probability of finding a man-eating tiger in your garden in England', the 'feel' mechanism again operates easily and directly to produce a ranking. This leads to a first, basic, hypothetical statement.

As distinguished in the mind, the amorphous masses of subjective experience are readily compared even though they are not directly quantifiable.

Let us add a third event, for example '(C) the probability of locking yourself out of the house'. It is likely that again the reader will be able to rank these events without too much difficulty. Although it may not be immediately obvious there is little doubt that the ranking will be established by the assessor mentally progressing through a series of tests (in this case, three) which compares each event with each other event. A little self-searching should clearly reveal this process; furthermore the same self-searching should reveal also that different paired comparisons generate different strengths of feeling. In other words, not only can we compare our responses to stimuli but also we can distinguish different strengths of comparative feeling. Therefore, if, for the three stimuli quoted above, '(B) the probability of finding a man-eating tiger in your garden in England' and '(A) the probability of a punctured car tyre' were judged respectively to be the least and the most probable and if, in addition, it was felt that '(C) the probability of locking yourself out of the house' produced a (B) − (C) conviction which was substantially stronger than the (C) − (A) conviction, this could be expressed undimensionally as shown in Figure 1.6.3-10.

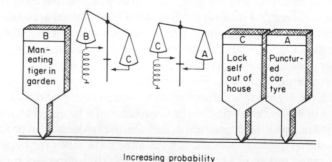

Increasing probability

Figure 1.6.3-10 Strengths of comparative feeling

The probability continuum exists in the mind by virtue of the 'mental weigh-scales' mechanism which enables it to be constructed and proportioned. Again, however, although we can distinguish clearly different strengths of comparative feeling, specific quantification of these strengths is frustratingly elusive. We can feel confident when the strengths appear to be similar, leading to equal spacings along the continuum, but when confronted with different strengths of conviction, although the mental vision of linear spacing may be geometrically clear, it is not in a form which permits easy representation on a numerically calibrated scale. Nevertheless, Pontecorvo[5] asked his assessors to attempt to do this, to undertake the double task involving both ranking and scaling; it is likely, if asked, that these assessors would have expressed far more confidence in the former activity than in the latter. Our 'mental weigh-scales' are not calibrated in a way which is consciously measurable. This leads to a second, basic, hypothetical statement.·

> In ranking pairs of stimuli with respect to some common continuum, individual paired-comparisons generate different strengths of conviction in the mind of the assessor; although these strengths are felt clearly on a continuous scale of magnitude they are not directly quantifiable.

It is here that Thurstone's[8] law of comparative judgement offers to assist. Requiring that the assessor or assessors take ranking decisions only, the rationale of the law produces automatically a statistical estimate of the scale values proportionally separating the given stimuli along the defined continuum. In effect it calibrates the 'mental weigh-scales' of the assessor(s).

Let us suppose that the 'weight' of an agglomerate of experience held in the mind is proportional to the number of building bricks, the number of repeated experiences, which formed the agglomerate. As in this case the agglomerate relates to probability or 'chance of occurrence', this implies that the experience must include events and non-events together in order to establish a feel for the required fraction. This being so, it is reasonable to further suppose that when a paired-comparison is made in the mind, that is when two agglomerates of experience are isolated and 'weighed', the strength of ranking conviction, the mental weigh-scale value, is related to the ratio of the subjectively held probabilities.

Accepting that Thurstone's[8] law of comparative judgement faithfully extracts these 'mental weigh-scale values', consider now the assessor (or, communal mind of assessors) who generates weigh-scale units as shown in Figure 1.6.3-11

Equal units between A and B and B and C imply that the assessor felt that B was more probable than A with the same conviction that he felt C was more probable than B. In other words he felt that

$$\frac{P_B}{P_A} = \frac{P_C}{P_B} = r$$

where r symbolizes the probability ratio equivalent to one scale unit.

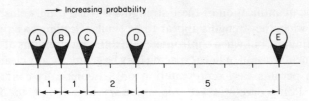

Figure 1.6.3-11 Mental weigh-scaling of ranking

Between C and D the assessor expressed two units of conviction, which, on the same scale, implies,

$$\frac{P_D}{P_C} = \left(\frac{P_B}{P_A}\right)^2 = \left(\frac{P_C}{P_B}\right)^2 = r^2$$

Similarly

$$\frac{P_E}{P_D} = r^5 \quad \text{and} \quad \frac{P_E}{P_A} = r^9$$

These concepts may be expressed in the third, basic, hypothetical statement, as follows:

Given that a scaled-ranking effectively exists for a set of stimuli in the (communal) mind of the assessor(s), it is postulated that where one unit of scale value represents a probability ratio of r, s units of scale value represent a probability ratio of r^s.

In Figure 1.6.3-12 five scale-ranked stimuli are shown with probabilities P_a, P_l, P_b, P_x, P_u, of which P_l and P_u are known.

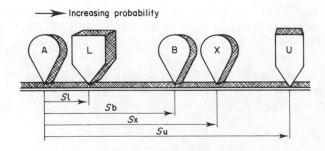

Figure 1.6.3-12 Generalized scaled-ranking

From the foregoing, and defining r as previously, the following may be written.

$$\frac{P_u}{P_1} = r^{(S_u - S_1)}$$

i.e.
$$r = \left(\frac{P_u}{P_1}\right)^{\frac{1}{S_u - S_1}}$$
(1.6.3-8)

Similary

$$\frac{P_x}{P_1} = r^{(S_x - S_1)}$$
(1.6.3-9)

Substitution Equation 1.6.3-8 into Equation 1.6.3-9

$$P_x = P_1\left(\frac{P_u}{P_1}\right)^{\frac{S_x - S_1}{S_u - S_1}}$$
(1.6.3-10)

which is identical to the equation quoted in Figure 1.6.3-5

From Equation 1.6.3-9, $\log_{10}\dfrac{P_x}{P_1} = (S_x - S_1)\log_{10} r$

whence,

$$\log_{10} P_x = A\,S_x + B$$
(1.6.3-11)

A and B are constants where $A = \log_{10} r$
$$B = \log_{10}(P_1 . r^{-S_1})$$

Equation 1.6.3-11 is of identical form to that derived from Pontecorvo's[5] experiment. Both the experiment and the foregoing hypothetical development are individually, and for very separate reasons, open to doubt. However, it could be fairly claimed that two fundamentally different approaches which, although both on tenuous grounds, do nevertheless produce the identical conclusion, go far to testify to the validity of this conclusion. On the basis of the foregoing evidence, therefore, the logarithmic relationship is proposed as a plausible basis on which to calibrate paired-comparisons scale values into a range of equivalent, absolute probability estimates. A minimum of two probability values (P_1 and P_u) are required to be known for the calibration of the scaled-ranking. Equation 1.6.3-10 may be used to calculate the probabilities (P_x) of all other rank-scaled stimuli.

It is to be expected that, of the two, the lower probability calibration point (P_1) will be the more difficult to find accurately while, at the same time, being within the repertoire of experience of the assessors. A stimulus like 'probability of driving through traffic lights while at red' (evidence collected indicates this to be of the order of $\leqslant 10^{-4}$ per opportunity) is a reasonable choice. For P_u a stimulus like 'probability of throwing a six with a single throw of a dice' is exactly known, of course, adequately high for most purposes and within the repertoire of experience of virtually any assessor. In principle it would be reasonable to include in a trial a mixture of events associated with both human and hardware failures, using the latter as calibration points. The underlying proviso, of course, is that all stimuli should be sufficiently within the experiences of the assessors to produce 'knowing' response.

Computer Implementation

An experimental version of the total approach has been compiled into a series of computer programs which operate in an interactive mode with the users. Following loading of the required set of stimuli these are paired and presented randomly to each subject, soliciting from him a sequence of decisions. These decisions are automatically analysed to produce a ranking and the subject's consistency significance probability, as described earlier. The decisions are also entered into the totalizer which accumulates the contributions from all the subjects. Using the terminology of Torgerson[10] this totalizer is referred to as the raw frequency F matrix. When all the subjects nominated have registered their contributions a group consistency significance is calculated. The F matrix is normalized into a probability P matrix and thence, using the values of the cumulative normal distribution, each element is converted into its equivalent unit normal deviate. This conversion produces what is termed the basic transformation X matrix. Each element of the X matrix corresponds to a pair of stimuli and comprises an estimate (not the actual value since the population of judges will be far from infinite) of the scale separation between the two stimuli; in effect each element stands as an estimate of one equation of the law (Thurstone's[8] Case V in this case). Now, in order to obtain a scaled ranking of n stimuli, $n-1$ data components are required, this representing the number of spaces along the scale. The X matrix actually provides up to $n(n-1)/2$ estimated scale values, giving up to $n/2$ estimates to be averaged for each linear spacing along the scaled ranking. These averaged estimates of the ranking scale values are obtained using Torgerson's[10] traditional procedure for incomplete matrices. (An incomplete X matrix occurs where perfect judges' agreement is recorded for a particular paired ranking. This produces 'all and nothing' votes in the F matrix giving probabilities of one and zero in the P matrix which, in turn, correspond to infinite unit normal deviates. Such values represent indeterminate estimates of scale separations and in this context must be considered as 'zero' data. Therefore they are registered as blanks in the X matrix).

At this point the computer presents a numeric and pictorial output of the calculated scaled ranking (the complete procedure up to this stage is summarized in Figure 1.6.3-13) and asks the user to quantify and two of the ranked stimuli. The logarithmic transition is applied but before finally listing the stimuli calibrated in

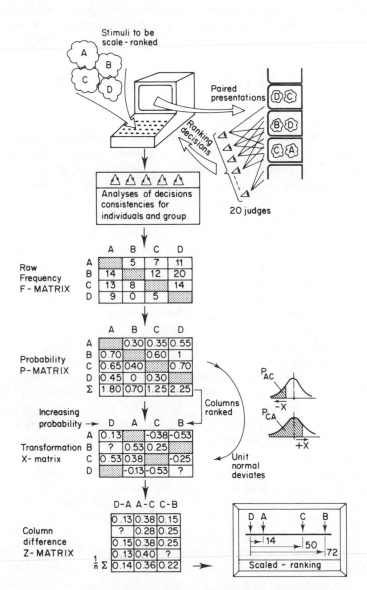

Figure 1.6.3-13 Computer calculation sequence for scaled-ranking

absolute probabilities an automatic check is performed to establish that the calibrated scale does not extrapolate probabilities of greater than unity or less than 10^{-6}. Finally the probability values are listed and against each quantity two accuracy sensitivity factors are appended, giving the percentage changes respectively for a 1% change in the upper and lower probability calibration points.

Report of a Trial Paired Comparisons Exercises

An opportunity was created at a technical meeting to obtain the cooperation of the audience in a full-scale trial run of the paired comparisons data extrapolation technique. The subject matter chosen was a well documented railway accident which occurred on the 25 August 1861 in Clayton tunnel on the London to Brighton main line. It involved a rear-end collision with appalling consequences despite the presence of a signalling system specifically designed to prevent such an occurrence. This particular example was selected because the scenario and the perversely unrelenting sequence of classical human errors could be rapidly and interestingly communicated to the audience. It was selected also because a reasonable estimate could be made of the actual event probability; to some extent this offered a means of checking the success of the exercise. Researches, admittedly crude, indicated that over the 20 years during which the particular signalling system was in operation there probably had been between 500 000 and 800 000 opportunities for the accident; history tells us that this occurred once, after which the system was changed. Taking 650 000 as the mean estimate for the number of accident opportunities, the corresponding 'ball-park' figure of the accident probability is 1.5×10^{-6}. This figure was deduced before the paired comparisons trial took place. Although such prior knowledge would have made little difference it was not revealed to the audience of judges until after the completion of the exercise.

The 62-strong audience, all male and mainly comprising experienced professional engineers and technologists, received first a description of the signalling system, its purpose, equipment and method of operation. Then the actual accident sequence was described and analysed into four specific error components which were to be quantified by the paired comparisons extrapolation technique. To these were added two 'calibration' components whose quoted occurrence probabilities, derived from specific data gathering exercises, could be accepted with reasonable degrees of confidence. Thus, with 6 stimuli in total the 15 possible pairs were presented to the audience in random sequence. The paired presentations succeeded each other at approximately 15 s intervals; for each presentation each of the 62 judges wrote down his paired-ranking decision on a card, working to a system which made it difficult to look back at prior judgements in order to assist a current judgement. Finally the cards were collected for analysis by computer. Up to this point the complete exercise had occupied approximately 30 minutes. It should be recorded that before the paired comparisons run started the audience knew that their individual judgements would be tested for consistency using the 'circular triad' technique described by Kendall[(4)] (Figure 1.6.3-6). Although the statement was clearly made that anonymous contributions were quite as acceptable as named ones it is interesting to note that not one unnamed card was submitted.

The system and accident sequence were described as follows.

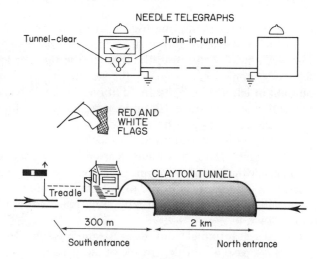

Figure 1.6.3-14 Signalling arrangements for rear-end collision prevention

The essential components of the system are illustrated in Figure 1.6.3-14. Basically it was an information system designed to prevent the possibility of a rear-end collision between two trains in a tunnel. The idea simply was to permit only one train at a time to occupy the tunnel in any one direction. Trains travelling in opposite directions would be permitted to pass in the tunnel. A signalman, in his cabin, was situated at each end of the tunnel. A train entering either end would be observed by the signalman there who would immediately notify the signalman at the other end. This second signalman would watch for the emerging train and at once notify this event to the first signalman. Meantime, it was the first signalman's responsiblity to prevent a second train from entering the tunnel until he received the 'tunnel clear' information. Each signalman communicated with an approaching train by means of a semaphore signal situated some 300 m from his cabin and the tunnel entrance. Normally these signals displayed the 'danger' indication (semaphore arm horizontal). When the tunnel was clear and he saw a train approaching the signalman would clear his signal (semaphore arm raised) thereby signalling the right of way to the driver. On passing the signal the train operated a treadle on the rail which automatically reset the signal to the 'danger' position. Should the tunnel not be clear on the approach of a train the signal would be left at danger and the driver, observing this, would stop at the signal cabin until permitted to proceed. Each signalman was supplied with a white (clear to proceed) and a red (danger, stop) flag for direct communication with the drivers. In addition; should the treadle fail to reset the semaphore signal to 'danger' on the passing of a train a warning bell would ring in the signalman's cabin.

The two signalmen communicated with each other by means of a pair of electrically connected 'needle telegraph' instruments. The normal position for lever and needle was central. Operating the lever to the right or left gave the 'train-in-tunnel' and 'tunnel-clear' signals, the needles in both instruments inclining to the right or left accordingly. In addition, the single-beat bell at the receiving instrument would respond. By use of bell codes a limited set of additional communications between the signalmen was possible. Finally, each signalman possessed a set of time-tables and a clock.

On Sunday, 25 August 1861, a violent rear-end collision occurred approximately 200 m inside the southern entrance of Clayton tunnel: 23 people died and 175 were seriously injured. It is ironic to note that without the signalling system this particular accident would not have occurred. Briefly, the accident sequence was as follows:

Train 1 approached a clear tunnel from the south and was signalled to proceed by signalman 1. Train 1 passed over the treadle of the semaphore signal but the signal failed to reset to 'danger'. Immediately the warning bell rang in signalman 1's cabin. Meanwhile train 1 had entered the tunnel and so before acting on the alarm signalman 1 sent the 'train-in-tunnel' signal to signalman 2, who acknowledged. Then signalman 1 turned his attention to the semaphore signal only to find that train 2 had appeared, much closer behind train 1 than expected, had passed the still 'clear' semaphore signal and was about to pass his cabin. In great haste he displayed his red flag as the train entered the tunnel. He then had no means of knowing whether or not the driver of Train 2 had seen his danger signal. He again sent the 'train-in-tunnel' signal to signalman 2, who again acknowledged. Signalman 1 hoped that signalman 2 had registered 'two-trains-in-tunnel'. Signalman 2 apparently concluded that his first acknowledgement had not yet been received and so the signal had been repeated. In due course train 1 emerged from the northern entrance of the tunnel and so signalman 2 sent the 'tunnel clear' signal. Signalman 1 concluded that both trains had passed through and permitted the entry of train 3 which had now approached. The driver of train 2, however, had seen the red flag, had stopped in the tunnel and was in the process of reversing back out . . .

The basic components which combined to produce the accident were defined and modelled as shown in Figure 16.3-15. A hardware analysis of the treadle signal system led to an estimated probability of 0.005 for failure on demand. The

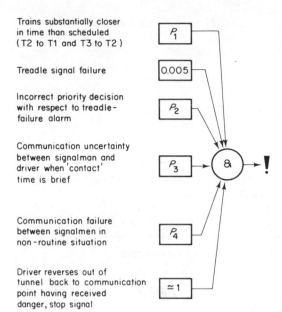

Trains substantially closer in time than scheduled (T2 to T1 and T3 to T2)

Treadle signal failure

Incorrect priority decision with respect to treadle-failure alarm

Communication uncertainty between signalman and driver when 'contact' time is brief

Communication failure between signalmen in non-routine situation

Driver reverses out of tunnel back to communication point having received danger, stop signal

Figure 1.6.3-15 Accident model

probability that driver 2 would reverse back out of the tunnel, having seen an urgent waving of the red flag, was considered to be sufficiently certain to be treated as unity. Thus, four components remained to be quantified. Two calibration components were added to these giving the total set of six stimuli; these are summarized pictorially in Figure 1.6.3-16. It was these pictorial summaries which were presented in paired combinations to the audience for ranking.

Immediately before commencing the comparisons run it was stressed to the audience that there were no right answers to be sought in making the judgements – the aim was for each person to honestly express his subjective feel on each occasion without recourse to calculation or precedent. The results obtained are summarized in Figure 1.6.3-17. With six stimuli it can be shown that the maximum possible number of inconsistent decisions (circular triads) is eight. The top of Figure 1.6.3.-17 shows that nearly 40% of the judges returned wholly consistent sets of ranking decisions and 55% were either totally consistent or produced only one inconsistent decision. Statistically this gives a 95%–98% confidence to the proposition that the individual results came not by chance but from judgements disciplined by structures of positive experience. On a group basis this confidence was found to be substantially greater that 99%. There seems little doubt that the exercise succeeded in extracting a workable bulk of genuine subjective knowledge from the 62 judges.

The scaled ranking was deduced for the complete set of contributions from all

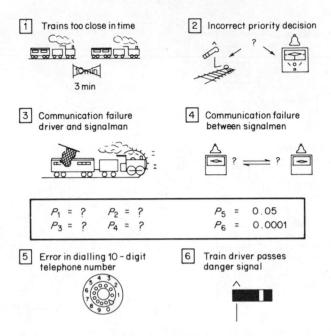

Figure 1.6.3-16 Pictorial summaries of paired comparisons
stimuli

62 judges and two further versions were produced using only totally consistent sets of data and then adding to these the sets of data with one inconsistency. In each of the three cases, shown in the three columns in Figure 1.6.3-17, the calibration components (see Figure 1.6.3-16) were used to determine the extrapolated probability values, P_1-P_4, and these, in turn, were combined as in the accident model of Figure 1.6.3-15 to deduce the accident probability. The subjectively derived estimate of 1.4×10^{-6} is remarkably close to the measured estimate of 1.5×10^{-6}.

Summary and Conclusions

When the results of this first trial paired comparisons exercise (the first by the writer; Blanchard[1], earlier, had used his own version of the method to obtain subjective probability quantification) are viewed in the context of (i) the somewhat uncertain precision of the field data used to calculate the 'true' probability of the accident and (ii) the relatively simplistic failure model given to represent the accident sequence the closeness of the correlation obtained between the 'actual' and 'predicted' values must be regarded as fortuitous, not in itself of significance. Nevertheless, having given due regard to these factors, the result

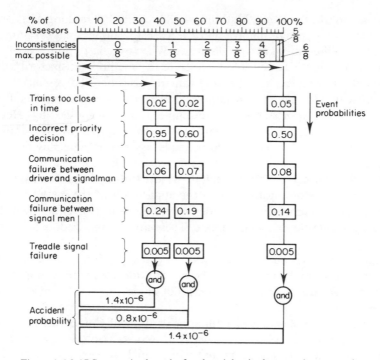

Figure 1.6.3-17 Summarized results for the trial paired comparisons exercise

remains encouraging; of course many more trials are needed before any confident assessment of the method can be made.

One feature of interest which did emerge convincingly from the trial was the high overall consistency of judgements produced by the individual members of the audience. The results showed ranging differences of opinion between assessors but at the same time indicated that the majority of individuals were making their judgements in structured manners. This is exactly the form of results anticipated from the method. The summary in Figure 1.6.3-17 shows that almost 40% of the audience made judgements without any inconsistencies, this proportion rising to over 70% if those with one or two inconsistencies (from a possible maximum of eight) are included. These results strongly indicate that the exercise was successful in extracting a consensus of structured knowledge from the assessors.

In viewing this conclusion it is important to recall that the entire operation, including introducing the stimuli in their situational contexts, occupied just 30 minutes.

Fairly obviously there is a practical limit to the number of stimuli which could be scale-ranked in one session. The trial described involved six stimuli and therefore $\frac{1}{2}n(n-1) = \frac{1}{2}6(5) = 15$ paired ranking decisions. Ten stimuli would require 45 ranking decisions, 20 stimuli would require 190 ranking decisions, and so on. Assessor fatigue probably sets 10 as a sensible upper limit for numbers of stimuli to be scale-ranked at a single sitting. When over-large numbers of stimuli cannot be avoided it is possible to operate the method satisfactorily using a substantially incomplete set of paired ranking decisions. Torgerson's[10] traditional procedure for incomplete matrices provides a viable means of analysis but there are other approaches, for example those proposed by Hevner[3], Thurstone[9], Uhrbrock and Richardson[11]. In addition to the strategy of reducing the number of decisions to be made, assessor fatigue may be eased by giving attention to the ways in which the paired stimuli are presented. The matter is discussed by Torgerson[10] and dealt with more fully by Ross[7] and Wherry[12].

A number of assumptions underly the application of the method of paired comparisons, as described and extended here, all of which are either hypothetical or, in an absolute sense, indeterminate. These are summarized as follows.

1. A *Gaussian* variation of responses along the psychological continuum is assumed, this applying whether the distribution comprises a population of responses from a single assessor or a set of single responses from a population of assessors.

2. In the Case V application of Thurstone's[8] Law of Comparative Judgement the correlation coefficient for each paired ranking is assumed to be negligible – that is to say, a temporary bias which may exist in the mind of the assessor for one stimulus will not be similarly transferred to any other stimulus.

3. Again in the Case V application of Thurstone's[8] Law, the assumption is made that the discriminal dispersions for all stimuli may be taken as equal – in other words the distribution of responses for each stimulus along the psychological continuum (of the single or 'communal' mind, as the case may be) has approximately the same Gaussian profile. Should this not be an acceptable assumption the more difficult Case III (discriminal dispersions unequal) or Case IV (discriminal dispersion unequal but differences small and similar) versions of the law could be tried.

   It is argued intuitively that assumptions (2) and (3) are more likely to be valid as acceptable working approximations when the responses come from a population of assessors rather than when they derive from the mind of a single assessor.

4. A logarithmic relationship is postulated to exist between the absolute probability value (P) for a stimulus and the corresponding nominal scale value (s) sublimated on the psychological continuum – i.e. where A and B are constants, $\log P = As + B$.

5. It is assumed that the population of paired comparisons' data-sets (for 6 stimuli, 15 decisions from one assessor is a single data-set) is sufficiently large to provide scale-value estimates virtually equal to the 'true' mean values. The

'true' mean values are those deriving from a notionally infinite population of these data-sets.

In the implementation described for the paired comparisons method no attention has been given to the relationship between numbers of data-sets and the commensurate confidence attributable to the scale value estimates. This would be a useful extension to the work. However, in the absence of such knowledge it is suggested that 20 data-sets could be taken as a practical working level with a nominal working minimum of 10.

The first four assumptions all lack scientific qualification and yet, using available supporting evidence to supplement what self-experience suggests to be reasonable, each one can be persuasively argued. This does not remove the need for future attempts to gain experimental verification but it does inspire sufficient confidence to give the method serious consideration and to proceed with a programme of trials. One important area of study would be to determine the extent to which each assumption influences the results produced.

However, one fundamental factor transcends all arguments relating to the structural theory of the method. Common to any subjective consensus technique it concerns the extent to which each stimulus presented produces the same perception in the mind of each assessor. The results of a consensus are entirely dependent on the degree to which 'common perception' is achieved. Ideally this should be tested as part of the application of the method but so far the writer has found no technique for doing this. It is clear that a monitor of judgement consistency is not helpful here. The matter remains an important area for future work if the method is to become a dependable and versatile data bank extrapolation tool. However, by choosing the assessors from the environment which contains the identified stimuli (for example, if the stimuli derive from the operations of a particular class of process plant then the assessors would be operators from that same type of plant) the chances are greatly heightened that 'common perception' will be achieved. Even without a formal test of this factor the method could be applied with fair confidence and usefulness in this way. Uncooperative or incompetent assessors in most instances should be detectable by the 'individual' judgements consistency monitor and their contributions selectively ignored. Nevertheless, even in such a favourable application there should be no relaxing of the strict need to express each stimulus as comprehensively and unambiguously as can be practically achieved. A basic guide here is to incline towards forming the stimuli at simple task levels. For example, the stimulus 'getting lost on a journey' is far more liable to 'common perception' failure than the sub-component stimulus 'not spotting a turning on an unfamiliar but planned route while driving the car'.

The specific advantages of the method may be listed as follows.

1. The method extracts subjective knowledge in the comparisons domain where the human judgement capability is naturally strong.

2. Operated as a consensus method it systematically combines any collection of experiences and expresses the communal minds' judgments in the form of absolute quantifications.

3. A powerful means is offered to check the structural quality of the subjective data contributed by each assessor. Judgement inconsistencies are directly recognized and this knowledge used to reveal assessors who, in the specific instance, are poorly qualified to contribute data.

4. Absolute probability quantification can be produced for a large number and variety of stimuli using a minimum of two known calibration points. Desirable features of the calibration points chosen are that they should be accurately known, reasonably separated as 'high' and 'low' and well within the experiences of the assessors. The extrapolated probabilities may be toleranced by sensitivity calculations using the accuracy tolerances attributed to the calibration stimuli. In theory, event probabilities in the human factors domain could be extrapolated from calibration points of hardware failure probabilities.

5. Although the method has more obvious application as a consensus technique it could be used by a single assessor who simply wished to obtain boundary values for the probability of a given event. In an interactive mode with a computer he would select perhaps four or five stimuli of 'known' and ranging probabilities and carry out a single paired comparisons run between these stimuli and the unknown. The resulting ranking would position the unknown between defined probability bounds and also enable the assessor to rate his own performance by checking that his ranking of the 'knowns' is consistent with their values.

6. Stimuli may be defined as required. It may never be practical to collect field data for a human operation which is tightly bound within an error recovery cycle – for example, the number of times that a printed circuit board assembly worker omits a component but recovers the error before the board finally leaves him. It is however possible to specify such a stimulus in a paired comparisons consensus exercise. Equally well, stimuli may be defined as fully conditional events so that for the purposes of synthesis in a system analysis their probabilities may be combined as independent variables.

7. Even without the final probability calibration stage the scaled-ranking provides a useful 'reliability spectrum', revealing and putting into numerical perspective the 'weak links' within the set of stimuli assessed.

This section has set out to review the problems associated with setting up a quantitative human factors data-base and, in the sense of building a comprehensive body of data measured from the 'field', has identified this as a long-term undertaking. As an interim measure it is proposed that we should capitalize upon subjective knowledge and the method of paired comparisons is offered as one means by which this might be done in an organized and structured manner. The theory has been comprehensively described together with an account of one full-scale trial. Further proving trials are essential and some extra theoretical developments may be desirable. However, as a potentially powerful data extrapolation technique, the method appears to be a promising canditate to fulfil a currently much needed role.

Questions 1.6.3

1. Discuss the extent to which, in your opinion, human error probability data are important to the task of safety assessment. Identify the principal difficulties which confront the prospective data collector in the field. Suggest and appraise possible data sources and collection methods which might be used – in particular discuss the potential usefulness and accessibility of subjectively-held data. Draw conclusions as to the policies and strategies which might be adopted in order to cope with the problem.

2. Give a general outline of the paired comparisons method, describing the concepts of the scaled-ranking process and explaining the basis on which the scales may be used to produce absolute quantification. Discuss the strengths and weaknesses both of the theory and of the practical application of the method. Extract and debate those assumptions which, in your view, are most critical to the viability of the method. Draw conclusions as to the usefulness of this approach as a data extrapolation technique. Suggest, if you can, one area outside the reliability field where the general technique (or part of it) could be advantageously applied.

3. Thirty-two personnel working on, or associated with, a certain process plant took part in a paired comparisons exercise. They compared the likelihoods of five postulated operational events. The events were:
   A. Following the testing of an alarm system the test engineer fails to restore the alarm 'inhibit/normal' switch to the 'normal' position;
   B. The sudden and catastrophic failure of a plant section produces a violent release of a white pungent vapour which rapidly envelops the whole plant; the operators respond by shutting down the plant totally, failing to recognize that isolation of the failed section was all that was necessary.
   C. Without apparent reason an experienced plant operator goes to the area of a fully revealed flammable gas release and deliberately establishes a naked flame.
   D. From a row of six level gauges an operator notes down in his log the reading of gauge G3, describing this as the reading of gauge G4.
   E. In a routine very regularly performed operation involving liquor transfer from tank A to tank B the operator instead vents tank A to drain.

The votes registered in the paired comparisons exercise are summarized in the

raw frequency matrix given below. An entry in a cell, thus A $\boxed{n}$

means that *n* judges voted A to be more likely to occur than B.

	A	B	C	D	E
A		12	29	19	23
B	20		32	21	25
C	3	0		4	9
D	13	11	28		22
	9	7	23	10	

Tabulated Values for the Cumulative Normal Distribution Function

Probability p(t)	Unit Normal Deviate t	Probability p(t)	Unit Normal Deviate t
0	0	0.30	0.839
0.01	0.026	0.31	0.877
0.02	0.051	0.32	0.913
0.03	0.078	0.33	0.952
0.04	0.103	0.34	0.995
0.05	0.126	0.35	1.037
0.06	0.151	0.36	1.079
0.07	0.174	0.37	1.124
0.08	0.200	0.38	1.173
0.09	0.225	0.39	1.225
0.10	0.250	0.40	1.280
0.11	0.276	0.41	1.339
0.12	0.302	0.42	1.403
0.13	0.329	0.43	1.474
0.14	0.355	0.44	1.552
0.15	0.382	0.45	1.644
0.16	0.409	0.46	1.748
0.17	0.440	0.47	1.878
0.18	0.470	0.48	2.050
0.19	0.498	0.49	2.327
0.20	0.525		
0.21	0.552		
0.22	0.581		
0.23	0.611		
0.24	0.642		
0.25	0.673		
0.26	0.703		
0.27	0.736		
0.28	0.770		
0.29	0.803		

(a) Using Thurstone's Case V of the law of comparative judgement deduce the scaled-ranking for the five events (a tabulation of one-half of the symmetrical cumulative normal distribution function is reproduced on the previous page).

(b) Given that the probabilities of events A and E are known to be, respectively, 0.02 and 0.0005, calculate the probability values for the other events.

(c) Attempt to identify the defined events A to E in relation to a classification of error type. Do the probability values calculated accord generally with typical values which might be expected?

References 1.6.3

1. Blanchard, R. E. Mitchell, M. B. and Smith, R. L. *Likelihood-of-Accomplishment Scale for a Sample of Man–Machine Activities.* Dunlop and Associates, Inc., Western Division, 1454 Cloverfield Boulevard, Santa Monica, California, 1966.
2. Embrey, D. E. *Human Reliability in Complex Systems: an overview.* UKAEA Report, NCSR RIO. July, 1976.
3. Hevner, K. 'An empirical study of three psychophysical methods.' *J. Gen. Psychol.*, **4**, 191–212. 1930.
4. Kendall, M. G. *Rank Correlation Methods*, Griffin. 1948.
5. Pontecorvo, A. B. 'A method of predicting human reliability.' *Annals of Reliability and Maintainability.* 1965.
6. Rolt, L. T. C. *Red for Danger.* David and Charles. 1966.
7. Ross, R. T. 'Optimum orders for the presentation of pairs in the method of paired comparisons.' *J. Educ. Psychol.*, **25**, 375–382. 1934.
8. Thurstone, L. L. 'A law of comparative judgement.' *Psychol. Rev.*, **34**, 273–286. 1927.
9. Thurstone, L. L. 'Rank order as a psychophysical method.' *J Exp. Psychol.*, **14**, 187–201. 1931.
10. Torgerson, W. S. *Theory and Method of Scaling.* John Wiley. 1967.
11. Uhrbrock, R. S. and Richardson, M. W. 'Item analysis.' *Person. J.*, **12**, 141–154. 1933.
12. Wherry, R. J. 'Orders for the presentation of pairs in the method of paired comparisons.' *J. Exp. Psychol.*, **23**, 651–660. 1938.

High Risk Safety Technology
Edited by A. E. Green
© 1982 John Wiley & Sons Ltd

Chapter 1.7

Emergency Planning

E. B. Ranby and F. Hewitt

Introduction

This chapter can only serve as an introduction to the many aspects of Emergency Planning. It will not supply a procedure tailor-made for your works, but by discussing some general principles and indicating where more information is available, it aims to stimulate the kind of thinking which could generate your own specific plan.

The material is aimed at land-based, fixed industrial installations. Thus special requirements for marine and transportation emergencies have not been covered – although many of the principles will still be applicable.

The reference literature quoted is primarily taken from 1970 onwards, with a few exceptions. For example, Reference 1 is a detailed book devoted to 'Emergency and Disaster Planning' in its most general context. Here are covered principles and numerous checklists in the context of peace and war, industrial emergencies and natural disasters (winds, earthquakes, and floods), bomb threats, and civil disturbances. This is followed by a glimpse into the psychological reaction of people. In the 1970s all of these topics have received individual attention from government, emergency services, academic and industrial writers.

In the UK a succession of statutory Instruments and Central Government Departmental circulars have resulted in local authorities appointing Emergency Planning Officers who have set up a framework of county plans to deal with a variety of major emergency situations. These events by definition 'create problems far beyond what is reasonable to expect the three emergency services – of Police, Fire and Ambulance – to clear up unaided[2]'.

In the main, the policy adopted has been for County Councils to produce a flexible master plan setting out the responsibilities of every other authority, service and organization involved, and for these organizations in turn to define their own response within these guide lines. The master plan may include invaluable lists of specialized equipment or resources which are available at short notice within the country.

The immediate problem of finance in an emergency situation was resolved under section 138 of the Local Government Act 1972, whereby local authorities may incur expenditure for dealing with such situations, subject to *post facto* report to the Secretary of State.

The Fire Service also has a framework document on Major Disaster Planning[53].

The recently revised BSI Code of Practice for Fire Precautions in chemical plant (BS 5908) in its section on emergency procedure, advises that plans should be made for four categories of incident: internally controlled; internal incident but external assistance needed; effects extend beyond works; and occurrence outside the works boundary. Throughout this chapter, the terms works, site, plant, etc. will be used in the manner defined by ACMH[13].

All police forces have routines to be followed in a variety of emergency situations, and their responsibilities and principles are clearly listed[3].

In the UK, National Health Service hospitals have been required to make arrangements for dealing with casualties from major accidents since 1954. In the USA, the accreditation of hospitals has long demanded a written disaster plan and two drills each year. Reference 4 gives a well referenced account of disaster planning in hospitals, which brings together the contributions of Savage and Rutherford and many others in the medical field.

Emergency or Disaster Planning has attracted increasing academic study in the context of natural disasters such as earthquakes, floods, tornadoes, etc. Sociologists in particular have been most prolific, and much information is covered in Reference 5. The Disaster Research Centre at the Ohio State University has for many years studied how people behave and groups react to natural disasters. But more recently this centre has started work to examine the socio-behavioural aspects of disasters resulting from chemical agents[6].

Finally, industry itself has generated numerous publications to aid emergency planning. One of the single, most useful documents for the chemical industry is *Recommended Procedures for Handling Major Emergencies*.[35] This is now in its second edition, and provides clear advice on setting up emergency procedures. This publication covers all the basic aspects in the body of the report, and in the appendices, describes the responsibilities of outside services and gives typical checklists for key personnel from the works. A similar clear listing of the allocation of control duties between the police, fire, ambulance and a particular works is contained in Diggle[7].

For the nuclear field, there are several international papers on aspects of emergency planning[8], including contributions from the UK, USA, Switzerland, Sweden, and Italy. An American state of the art review is given by Moeller and Selby[9] and an extensive nuclear checklist is given by Reference 10.

Further classified references on emergency planning are contained in Lees[11].

The need for emergency planning

Within the high risk technology industries, the need for satisfactory emergency plans should be self-evident. No matter how well a process is controlled and safeguarded by instruments and procedures, it is inevitable that there is a residual risk which is capable of causing a variety of emergencies. These emergencies are usually contained within the works, but the scale of many industrial activities is such that there is often a potential for affecting people outside the works boundary. In the Nuclear Industry a site licence is necessary before an installation can be operated. One of the conditions of licence is to provide and demonstrate adequate emergency exercises. Demonstration emergency exercises have to be held yearly to the satisfaction of the Nuclear Installations Inspectorate of the Health and Safety Executive[12]. More recently the Department of Energy has asked that descriptions of the District Emergency Schemes for licensed nuclear sites be made available for interested members of the public who might be affected. In certain areas this has been achieved by putting reference copies in appropriate libraries.

As far as the Chemical Industry is concerned, the statutory need has been recognized in the Appendix 1 of ACMH[13] and certain requirements have been included in proposals for major hazards legislation in the UK[14] and the EEC[57].

At present these drafts do not give guidance for the preparation of emergency plans. But a recent HSE publication[15] contains a list of six important questions about Emergency Procedures, which does indicate some of the official thinking on this subject.

The basics of emergency planning

Table 1.7-1 presents our view of the basic essentials of Emergency Planning, which follows a broad cross-section of publications and practical experience. These basics attempt to cover the total planning concept; actual works plans deal exclusively with the central issue of coping with the emergency (part B in Table 1.7-1). This is absolutely necessary; but within the overall context of planning, it is

Table 1.7-1 Basics of emergency planning

A. Assess the risk

— Define and locate hazards
— Estimate consequences
— Establish resources

B. Outline plan objectives and actions

— Rescue and safeguard people
— Control and minimize incident
— Secure safe rehabilitation of area

C. Maintain plan

— Train
— Exercise
— Review

also imperative that all realistic risks of emergencies have been assessed and that the plan remains a living article with appropriate maintenance once it has been written.

Assess the Risk

This first phase is not a detailed quantitative assessment of the probability of an incident but rather a balanced judgement of likely credible events or potential emergencies for which plans must be prepared.

Define and locate hazards

Those organizations which practise Hazard Studies will have identified the significant hazards of the process during the development of the design [16]. The most serious of these will probably involve malfunctions which result in explosions and fires, or leaks of toxic gas or radioactivity. This scheme of Hazard Studies is ideal for the design of a new plant, but on existing installations it is perhaps more appropriate to carry out a Safety Audit along the lines of CIA[17].

It is worth noting that proposals for the survey stage of UK Major Hazard Legislation require that those routes to the loss of containment of specified quantities of chemicals be formally identified. Once the hazards have been defined and their location identified, then we have the starting point for the generation of realistic plans. The information should also be recorded on a map of the site and be available for internal and external emergency services.

Estimate consequences

After the hazards have been identified it is necessary to consider their likely consequences. Industrial hazards requiring emergency procedures are generally either fire, explosion, toxic gas, or radioactive release. Occasionally a combination of these circumstances is possible. Realistic scenarios should be drawn up for just a few of the large credible events so that the different consequences can be investigated. As Diggle[7] notes, 'better plans are likely to emerge from a study of a small number of potentially large incidents of very low probability rather than the more confused picture of a large number of potentially small incidents'.

Invariably, the possible consequences of an emergency will have been considered at the plant design stage. Often it is possible to limit the inventory of hazardous material which might be released, either by suitable process flowsheet development, or by dividing the plant up into isolatable sections. Kletz[18] discusses some situations in which emergency isolation valves should be used. Lees[19] comments on the human error by an operator and his response to emergency conditions that there is a small amount of data on emergency behaviour in general, but overall conclusion is that the probability of effective action by the operator in an emergency is not high enough to rely on for safety, though it is high enough to be very useful. The available data are based on Ronan's[20] study of real and simulated aircraft in flight emergencies, which gave a 16% error rate on critical conditions. However Berkun[21], showed that experienced soldiers performed better under high stress than under low stress, but the opposite was obtained with new soldiers. Somewhat encouraging generalizations were also drawn from Drabek's[22] study of a simulation of a police communication system under stress.

So in predicting the consequences of an incident, suitable allowance should be made for the intrinsic safety design features and provision for the system to be shut down quickly.

In assessing possible fire situations, several questions have been posed in reference[23] to show how reasoned judgements can be sought by considering the type of fire, danger of spreading, vulnerability of equipment and structural effects. In a similar planning context, Duff and Husband[24] have discussed their approach to the assessment of toxic concentrations of fume or gas at ground level.

Part 3 of this book presents in detail, the techniques required for more precise estimates of the consequences of specific hazards.

The consequence of an incident must always be judged within its local environment. Population densities, adjacent industry, prevailing winds are individually capable of transforming a dangerous but controllable situation into a major emergency.

Complete assessment of the consequences of an incident which affects areas outside the works, can only be made after discussions with the external authorities. Only they can estimate the outcome on institutions and essential services. They will certainly need to know the magnitude of credible incidents, but also a feature which is too often forgotten – the possible time scale for emergencies.

The speed with which an incident can escalate needs to be carefully thought out, also the total time that a works or a community could be under an emergency situation should be realistically estimated. Little did the townspeople of Mississauga realize that they would be away from home for up to six days. In fact, some of the problems which arose were associated with people wanting to re-enter the evacuated areas and arguing about the validity of evacuation orders[25].

Establish resources

The resources to tackle a minor emergency will presumably be internal to the works. Organization of that type of response is entirely within the works control.

These resources may include a fire fighting team, a rescue team, and a first aid team. The fire fighters can be either part-time (i.e. called from their 'normal' occupation to form the team) or full-time, perhaps supplied with a small fire engine or suitably equipped patrol vehicle.

The rescue team can be built up from the works engineering resources, e.g. scaffolders, crane drivers (with their cranes), boilermakers, etc. The works first aiders can be combined as a separate team which can then act in concert with the fire fighters and rescuers.

The gatemen/security force, have a part to play in keeping open access and communications to the scene of an emergency.

These internal human resources should be backed up by stocks of protective clothing, breathing sets, specialized foams and water supplies compatible with the Fire Brigade equipment.

In the UK, the main purpose of a Works Brigade is to provide a very fast initial response to a fire, so that a small incident does not grow into a large one. But it is imperative that the County Fire Brigade is also called as soon as any fire incident is detected. This initiates a further chain of assistance, and if events do escalate, then full County emergency resources, volunteer organizations and other Industrial Mutual Aid schemes can be activated. In these circumstances, agreed and coordinated plans become absolutely vital. Therefore County officials should always be consulted while establishing resources and during the preparation of works emergency plans.

Blanchard[26], motivated by the emergency at Three Mile Island, compares the relative self-sufficiency of some German chemical firms with those in the UK, where individual companies carry much smaller internal services, but rely and cooperate more fully with local authorities. This type of cooperation is described further by Hill *et al.*[27] who discuss the mutual interdependence of the local authoriy and industry in the emergency context. Appropriately enough, the latter paper is written by senior officials from the County Constabulary, the County Fire Department, and one of the largest chemical sites in Europe. In return for local authority and community service help for on-site emergencies, the industry offers a wealth of expertise and equipment to be used for major emergencies outside the site.

Another example of mutual aid is referred to by Underwood *et al.*[28] who describes how the Northern Ohio River Industrial Mutual Aid Council (NORIMAC) exists to help one another and the surrounding community.

Outline Plan Objectives and Actions

Almost without exception, papers about emergency planning stress the importance of simplicity and flexibility. Planning must be in depth not detail, but also Diggle[29] stresses that planning for a major emergency can be infinite. 'Excellence not perfection should be the goal' says Underwood *et al.*[28]. One way of achieving these almost paradoxical requirements is to restrict the plan to the basic requirements of WHO does WHAT. There may also be a need to allocate basic WHERE places. But the HOW and WHEN must be left to be resolved in the heat of the moment. The HOWS and WHENS will however come up for discussion in exercises, and obviously this may result in a feedback to modify the basic plan. But the temptation to pack too much detail into the body of an emergency plan must be resisted. If necessary, certain lists can be included as appendices (see also suggestions on page 229), but even here, it is essential that information be presented concisely in a way that regular updating is not a burden, and such that it can be easily extracted under emotional circumstances.

A further general principle which has arisen from sociological studies of natural disasters (flood, drought, etc.) is to plan according to how people will react, not plan and hope that people will conform to the plan[30]. A similar warning is given by Perry[31] in a well referenced article about evacuation in natural disasters. This specialist in human behaviour observes that too often administratively devised plans turn out to be based on misconceptions of how people react. Hopefully, as sociologists turn their attention to the effects of industrial emergencies, we shall be able to systematically improve our conceptions. Some lessons can already be learnt from Blansham's[32] study of a hospital organizational response to a disaster. During an emergency in which the occupants of one hospital were evacuated into another one, it was observed that bureaucracy diminished, division of labour was simplified, decision making decentralized, and communication patterns became more informal and less hierarchical.

The simplest recommendation is that plans should follow normal procedures as far as possible, because it has been shown that few people can act or think clearly in the confusion and tragedy of a disaster situation[33].

The key Personnel and Actions outlined by the CIA[35] are very sound and applicable to a wide variety of industries. This scheme, and its checklist of responsibilities, has also been endorsed and presented in Lees[11], so its appendices will not be repeated again here. But a similar general framework of WHERE, WHO, and WHAT is summarized in Table 1.7-2. The initial key steps to initiate the Emergency Procedure, are also more simply conveyed by a tabular presentation and a possible approach is illustrated in Figure 1.7-1. The reader should study both Table 1.7-2 and Figure 1.7-1 before passing on to the next section, which discusses the short- and long-term objectives of the plan. Note in particular the general principle at the bottom of Figure 1.7-1, of always initiating the Forward Control Point and embryo Emergency Headquarters at the start of all incidents.

Rescue and safeguard people

Ironically enough, although the first objective in emergency situations is to rescue and safeguard people, the first action must be to raise the alarm. This is often contrary to natural instincts, and conscious training is given to the police force, who are so often first on the scene in domestic emergencies, so that they quickly appraise the situation and report to HQ[34]. In the industrial emergency, clear

Table 1.7-2 General framework

	FORWARD CONTROL POINTS	EMERGENCY CONTROL CENTRE	SEPARATE ASSEMBLY POINTS
WHERE	Adjacent to incident	Upwind and away from incident	Remote from incident
	INCIDENT CONTROLLERS	MAIN CONTROLLERS	KEY SUPPORT PEOPLE
WHO	Operational management: — Fire — Works — Police — Ambulance	Senior officials from: — Police — Works — Fire — Ambulance — HSE	Administration of: — Technical liaison — Evacuated employees — Casualty collection — Personnel information — Public relations — Service reinforcements
WHAT	ACTION AT SCENE	OVERALL CONTROL/LIAISON	OTHER ESSENTIAL ACTIVITIES

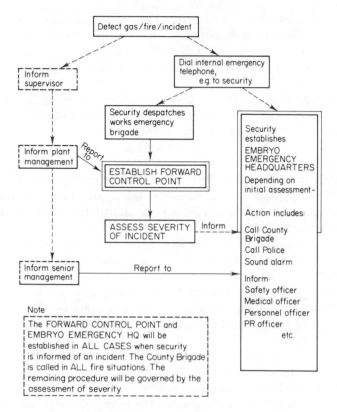

Figure 1.7-1 Initial key steps in emergency procedure

guidance is given by the CIA[35] for raising the alarm, declaring the emergency, and making the emergency known inside and outside the works, to key personnel outside working hours and to neighbouring firms.

In the UK the initial 999 call can in one move alert Police, Fire, and Ambulance services.

However, if a toxic release occurs, then initially only a minimum attendance from the three emergency services should respond to the works[23]. Augmented resources should report direct to the Separate Assembly Point, which is remote from the incident. After initial intelligence and liaison, rapid assistance can then be summoned without the danger of crews approaching from the wrong direction, directly into a hazardous area. To distinguish this type of emergency from a fire or explosion it is convenient to have a prearranged codeword, for example 'Operation Cloudburst'[26].

Each of the emergency services deals with different aspects in the preservation of life at emergencies. The Police coordinate activities and provide an administrative framework within which the Fire service can rescue people, and the Ambulance service transport victims to the hospital. In the initial stages of emergency there is often a certain flexibility about who does what, and these transitional responsibilities have been discussed from the police viewpoint by Fisher[54].

The Fire Brigade invariably takes the dominant role in rescue operations. Even in non-fire emergencies, their expertise and equipment make them the natural leaders for search, rescue, and recovery of trapped people. Some county disaster plans designate the Fire Brigade with general rescue responsibilities in order to overcome the lack of precision in the law[53].

The ambulance service will alert the casualty receiving hospitals. The term 'designated' is appplied to the nearest hospital, and two others are called 'supporting'. A mobile medical team may be transported from the designated hospital by the ambulance service.

The first ambulance team to arrive at a disaster must not necessarily drive off with the first batch of injured they find[36]. The sorting and assessing of injured, or 'Triage' must first be carried out. The Disaster Triage Label is colour coded and taped to a limb; white is for dead victims, then red, yellow, green respectively for patients needing immediate, urgent, or non-urgent treatment[36][37]. This triage label is also used to record what drugs have been given, and other immediate relevant information.

The police will organize the documentation and identification of survivors, casualties, and victims of the incident. As part of this activity, they will also set up a casualty enquiry bureau and notify next of kin.

The GPO makes extra lines available to satisfy the enquiries from relatives and friends, and an arrangement exists whereby all UK Police inform the London Telecommunications Region whenever they establish a casualty enquiry bureau. The telephone number of this bureau is then passed to all post office operators in the UK[38].

Control and minimize incident

Control and minimize the incident does not only mean, for example, put out the fire and stop it spreading. Control also implies coping with the inevitable side effects such as traffic disruptions resulting from sealing off the area; clearing routes for ambulance and fire vehicles; sightseers; effects on nearby industries and essential services. Minimizing the incident may require evacuation of adjacent areas, or defusing panic and concern by providing authoritative PR information.

At the Forward Control Point see Table 1.7-2 the senior Fire Brigade Officer present will be in control of all operations for the extinction of the fire. (Fire Services Act 1974 section 30(3)). The works people assist and advise the Fire Officer in this job.

The Works Incident Controller will usually organize the shutting down and evacuation of appropriate adjacent plants, prior to the arrival of external services. Whenever possible, sources of leaks should be identified and the nearest isolations effected.

The Fire Service use their normal radio communication system: Ambulance controls normally have a special emergency frequency so that the designated hospital, Forward Control and Main Control Officers share one channel. The Police have the responsibility of coordinating these services with those from the works.

Meanwhile, at the Emergency Control Centre (see Figure 1.7-2), assuming that administrative aspects such as establishing a chronological record of the emergency and external communications have been made, there will be a continuous review of possible developments and actions which can minimize the consequences.

For example, the dispersion (i.e. dilution by mixing with air) of leaks of flammable or toxic gas can play a large part in minimizing an emergency. If the gas concentration can be reduced to below the lower explosive limit (for flammables) or approach the threshold limit value (for toxics) then the area affected is greatly reduced.

Two main methods of gas dispersion are in use. Steam curtains can be generated from fixed multiperforated pipes around the potential leak sources. Water curtains can be provided either from a similar fixed installation for use with high pressure water or can be set up by the Fire Brigade using 'fog nozzles'. Steam curtains have the advantage of providing buoyancy, but with the marked disadvantage of needing very large quantities of steam, these are suitable only for the largest of installations.

The emergency control of gas emission from a collected pool of volatile liquid will depend on the factors causing the gas to be evolved. Covering the free pool surface with plastic sheet or carefully applied fire fighting foam can sometimes reduce the emission rate[55].

Reference to advice on coping with LPG fires is given in Reference 39.

Particularly in the case of toxic gas releases, the question of possible evacuation of adjacent sections of the community has to be considered. The works plan must provide for advising the Police on this topic. The Police will arrange any necessary evacuation outside the works. Alternatively, mobile loudspeakers are used to warn the public to stay indoors after a minor release.

Although discussing incentives for evacuation in the face of impending natural disaster rather than man made emergencies, Perry[31] does contain many observations about the problems of ensuring citizen cooperation with authorities. Similar practical problems have been described by Police Chief Burrows[25], in his account of the evacuation of 216 000 people, from a railway accident involving both flammables and toxic chlorine gas at Mississauga, Canada.

Public relations aspects are usually handled by a company spokesman experienced in these matters. The CIA has given concise guidance on dealing with the media in Reference (56). It is important to provide for these matters seriously, because otherwise, as recounted by Van Cleve[40], hysterical reports cause unnecessary worry to families of employees working on the plant, and reporters will have to resort to going to hospitals or talking to families of the injured, in order to piece together their version of events. Some of the difficulties encountered in communicating information in various earthquake situations in Japan are described by Abe[41]. He found that people tended to be as motivated by other people as by radio information, even though the latter was defined as providing quicker and more reliable information. At Mississauga more than 250 members of the media were present and 12 police officers were assigned to the task of public relations. Burrows[25] explains how the capability of the media to assist in an emergency by their communications must not be underestimated. He does, however, also warn that command decisions must get to the field force before going to the media and government representatives, otherwise this can lead to inevitable problems (e.g. congestion and confusion at road blocks).

A further practical point is made by Maas[42]; good public relations are an on-going activity. The Works Manager must continually try to view his plants from the outside, and must also foster good relations internally so that in an emergency the public relations people can adequately translate plant matters to the press.

Secure safe rehabilitation

This aspect is handled by the Works Main Controller, but it cannot simply be left to the end as a mopping up operation. A subsequent accident investigation will be necessary and evidence should not be disturbed until the Factory Inspector has been consulted. A photographic record taken during the incident can be useful in this context. Webb[43] recommends providing for interviewing the staff after event – preferably by someone they know. Process data should be retrieved if at all possible. The police and Fire Brigade will have maintained comprehensive logs of the actual emergency.

Technical aspects signalling the end of the emergency will be sorted out between the Fire Brigade and the Works. Particular care is required to establish that no dangerous traces of toxic, or flammable materials remain, nor that toxic compounds have been introduced as products of combustion. In the nuclear industry decontamination and re-entry requires even more stringent precautions.

Key personnel at a plant may be seriously injured in a major emergency. Additional shift staff can invariably be called upon to help, but it is also desirable to keep records of the current location of day staff who have previously worked on a plant, but have now transferred elsewhere within the company.

After a serious emergency both public and employees will need reassurance that the incident was controlled in a competent manner, and that appropriate action has been taken to prevent a recurrence. This will only be successful if confidence and mutual trust have been established over a period of years.

Maintain Plan

Section 2(2a) of the Health and Safety at Work Act imposes a duty on employers to provide and maintain safe plant and safe systems of work. Plant maintenance is generally acknowledged to be essential in industry, with a mixture of breakdown and preventive measures depending upon the circumstances. But maintenance of safe systems of work is too often just limited to actions after the event. This is both legally and pragmatically inappropriate; hence the following three sections on 'preventive maintenance' for the plan.

People must receive initial training in emergency procedures, then at suitable intervals this knowledge must be exercised and the basic plan reviewed and brought up to date.

Train

Training is required to develop knowledge of the total plan. Also it is essential to establish the necessary confidence and expertise so that individuals can carry out their allocated duties. Permanent site employees, adjacent plant workers or resident contractors and visitors (including those from within the same organization) all need different levels of information, ranging from mere recognition to precise action. The master plan can provide for some of these differences by having a section of basic information, which covers separately what everyone on the works should know, and what all supervisors should know and do. The general information will be brief, and restricted to a description of the emergency signals and recommended action. This information should then be made available in a suitable form to the different people.

The emergency plan itself will be simple and limited to providing an overall framework of control, communications, and allocation of who does what. All participants must be trained in this total view. But they should also be given a thorough practical familiarity with their own emergency equipment, as well as simulated experience in dealing with the 'How' – for example thinking through the implications of a particular situation. One way of satisfying this training requirement, as well as keeping the main plan simple, is to keep the more extensive information for key individuals as separate entries in one section of the document. With a suitable layout, these entries then provide emergency drills for individual or collective use. The emergency drill should define and identify the person involved, his equipment, main duties, and actions to be taken. It may also provide an *aide mémoire* for the activity, and list the individual responsibility for training, and the knowledge of other emergency drills required. A simple example of an emergency drill is given in Table 1.7-3.

A description of practical fire training for emergency squads is given by Underwood *et al.*[28], who describes seven types of realistic practice models. Very detailed and practical action checklists for training supervisors, in the control of

Table 1.7-3 Example of an emergency drill

GAS DETECTION OFFICER

1. PERSONNEL INVOLVED

   Shift Analysts

2. RESPONSIBILITY FOR TRAINING

   Analytical Laboratory Manager

3. MAIN DUTIES

   To carry out gas analysis to assist in assessing the severity of the emergency. The analyst will operate under the direction of the Works Main Controller.

4. MEANS OF CONTACT

   By telephone, Ext. 1234.

5. ACTION TO BE TAKEN

   (a) On hearing alarm or on receipt of telephone message report to Works Emergency Headquarters, Room A1 or B2.
   (b) Note location of source on site map, the wind direction and the area likely to be affected.
   (c) Go downwind of emission.
   (d) Determine gas concentrations at works boundary and beyond in the sector affected to determine limits of the sector.
   (e) Radio gas concentrations back to Works Emergency Headquarters.
   (f) Continue to monitor as directed by Factory Controller until given 'stand down message' or until relieved.

6. EQUIPMENT AVAILABLE

   (a) Transport equipped with gas detection equipment.
   (b) Respirators.
   (c) Radio.
   (d) Gas detection equipment.
   (e) Local street plan.

fires in refineries and petrochemical plants, are contained in Searson[44]. Fisher[45] offers the perspectives of significant involvement in practical training for contingency situations in the Police Force.

Exercise

Emergencies are regular happenings for the Fire, Police and Ambulance services, but hopefully they are rare events for industry. Therefore the latter must periodically test and exercise procedures, otherwise the familiarity to cope with the extraordinary event will disappear. Different aspects of the emergency plan can be split up into self-contained parts so that manageable practices can be held at appropriate frequencies. Blanchard[26] contains some information about the frequency of exercises in chemical firms in Germany and the UK, and Duff and Husband[24] comment upon exercises in the area of toxic gas emissions. Suggested frequencies vary between local internal exercises at two-monthly intervals to annual exercises involving external authorities. The range of necessary exercises is also described from the police viewpoint by Fisher[45], who describes how a full-scale exercise can involve several people in many months of preparation. 'Tactical Exercises Without Troops' (TEWTS) or 'Table Top Exercises' are especially recommended for testing the broad principles of the plan with minimum resources under varying scenarios. The 'Pollokshaws Disaster' is a composite, plane/train/petrol tanker scenario together with a discussion of the responses from the Police, Fire, Ambulance by Hamilton *et al.*[46]. A simpler ficticious chemical spillage is described by Feates[47]. Scenarios for industrial site emergency exercises can be similarly prepared.

It is also very important to bring the external emergency personnel on site in an informal way; not only do they need to learn their way about the plant, but also personal relationships should be established to provide a basis for cooperation in an actual emergency. This type of familiarization visit can often be arranged within the regular test run schedules which local Fire Brigades carry out. Individuals, or groups of people can exercise their own part of the plan, by running through the requirements of their Emergency Drills.

Review

Plants change, staffing levels alter or external resources regroup, people's ideas progress; and so periodically plans should be reviewed. If another similar works experiences an emergency, it is appropriate to question whether your plans would have coped. Especially if you have an emergency yourselves, or a detailed table top exercise, then it may become necessary to modify the plans or merely add a couple of items to a checklist. Much of the smoothness of the evacuation at Mississauga[25] must have resulted from the development of police plans through six significant incidents in the previous decade.

Lateral thinking about emergency plans can be encouraged by an examination of different types of emergencies. Accounts of the Moorgate tube accident from the Police, Fire and Medical viewpoint are contained in Howard[48]. In a different manner, three major disasters—Aberfan landslide, Hixon railway accident, and the Summerland fire—are described by Turner[49] and classified, to study the conditions under which large-scale intelligence failures develop. The emergency planning aspects of Flixborough have recently been recounted from the Police and the Fire viewpoint[50][51]. The Ambulance experience at Flixborough was described by Fozard[52].

Even more lateral thinking can be encouraged by attempting some of the questions given at the end of this chapter.

Acknowledgements

Grateful acknowledgements are due to the ICI Mond Division Technical and Personnel Director for permission to publish this work, and to many colleagues for their contributions and comments.

Questions 1.7

1. Create a brief scenario for a major industrial emergency such as a petroleum tank farm explosion and fire, a toxic gas or a nuclear radiation release.
2. Using one of the above scenarios, draft the response from the viewpoint of one of the following groups:
   Police, Fire, Ambulance and hospitals, Works Management, a County Executive plus the Coordinator of Voluntary Organizations.

Assume an explosion has caused a fire and toxic gas release, for the following three questions.

3. Draw up a checklist to be used for training the thought processes, required to carry out the functions of one of the Controllers listed in Table 1.7-2.
4. Produce separate checklists of information likely to be required for each of the assembly points listed in Table 1.7-2.
5. Working in a group starting from a common scenario, allocate functional responses as in Question 2, and carry out a discussion of the emergency.

References 1.7

1. Healy, R. J. *Emergency and Disaster Planning.* Wiley, New York. 1969.
2. Buttery, J. F. D. 'Major accidents and natural disasters.' Home Office Circular No. ES 7/1975, Home Office, London. 1975.
3. Gibson, W. H. 'Disaster planning.' *J. Soc. Occup. Med.*, **26**, 136–138. 1976.
4. Williams, D. J. 'Major disasters. Disaster planning in hospitals.' *B. Journal Hospital Medicine*, **22** (4), 308–22. 1979.
5. Quarrantelli, E. L. *Disasters, Theory and Research.* Sage Publications Inc., Beverly Hills, California, USA. 1978.
6. Quarrantelli, E. L. *et al.* 'Initial findings from a study of socio-behavioral preparations and planning for acute chemical hazard disasters.' *J. Haz. Mat.*, **3**, 77–90. 1979.
7. Diggle, W. M. 'Major emergencies in a petrochemical complex: planning action by emergency services.' *Process Industry Hazards*, I. Ch. E., Rugby. 1976.
8. *Handling of radiation accidents. Proceedings of a Symposium held in Vienna.* International Atomic Energy Agency, Vienna. 1977.
9. Moeller, D. W. and Selby, J. M. 'Planning for nuclear emergencies.' *Nuclear Safety*, **17**(1), 1–14. 1976.
10. *Guide and checklist for development and evaluation of state and local government radiological emergency response plans in support of fixed nuclear facilities.* NUREG–75/111, US Department of Commerce, Springfield. 1977.
11. Lees, F. P. *Loss Prevention in the Process Industries.* Butterworth, London. 1980.
12. Orchard, H. C. and Walker, C. W. Experience in the Provision and Exercising of Emergency Arrangements at CEGB Nuclear Power Stations. *Handling of Radiation Accidents 1977.* Int. Atomic Energy Agency, Vienna. 1977.

13. Advisory Committee on Major Hazards, *Second Report*. Health and Safety Commission, London. 1979.
14. 'Hazardous Installations (Notification and Survey) Regulations 1978.' Health and Safety Commission, London. Consultative Document. 1978.
15. 'Effective Health and Safety Policies.' Accident Prevention Advisory Unit of HM Factory Inspector. Health and Safety Executive, London. 1980.
16. Robinson, B. W. 'Risk assessment in the chemical industry.' In *Advanced Seminar on Risk and Safety Assessment in Industrial Activities*. Commission of the European Communities, Joint Research Centre, Ispra, Italy. 1978.
17. *Safety Audits, A Guide for the Chemical Industry*. Chemical Industries Association, London. 1973.
18. Kletz, T. A. 'Emergency isolation valves for chemical plant.' *Chemical Engineering progress*, **71**(9) 63–71. 1975.
19. Lees, F. P. 'Contribution of the control system to loss prevention.' *Inst. Chem. Eng. Symp. Ser*. No. 49, 13–20. 1977.
20. Ronan, W. W. *Training for emergency procedures in multi engine aircraft*. AIR-153-53-FR-44, American Institute for Research, Pittsburgh, Penna. 1953.
21. Berkun, M. M. 'Performance decrement under psychological stress.' *Human Factors*, **6,** 21–30. 1964.
22. Drabek, T. E. *Laboratory Simulation of a Police Communications System Under Stress*. Ohio State University, Ohio. 1969.
23. Brannon, A. R. 'Preplanning for fire emergencies in the chemical industry.' *Fire Engineers Journal*, June 1974, 18–25.
24. Duff, G. M. S. and Husband, P. 'Emergency Planning.' In *Loss Prevention and Safety Prevention in the Process Industries* (First Int. Loss Prevention Symposium) C. H. Buschmann (Ed.). Elsevier, Amsterdam. 1974.
25. Burrows, D. K. 'An Overview of the Derailment of Train 54, and the Evacuation of 216,000 Residents of the City of Mississauga, Nov. 10 1979.' Paper at Transchem 80, held Teeside Polytechnic, April 1980.
26. Blanchard, R. 'Coping with a catastrophe.' *Int. Mngt*, **35**(4), 12–16. 1980.
27. Hill, H. R., Bruce, D. J. and Diggle, W. M. 'The interaction between local authority services and industry to deal with major emergency.' In *Loss Prevention and Safety Promotion in the Process Industry* (Second Int. Loss Prevention Symposium), Dechema, Frankfurt. 1977.
28. Underwood, H. C., Souwine, R. E., and Johnson, C. D. 'Organise for plant emergencies.' *Chem. Eng. (NY)*, **83**(21), 11 Oct., 118–130. 1976.
29. Diggle, W. M. 'Inplant emergency plans.' Paper at Conference on Contingency Planning for Disaster Control, Oyez IBC, London. 1980.
30. Dynes, R., Quarrantelli, E. and Kreps, G. A. *A Perspective on Disaster Planning*. Disaster Research Centre Report, Series 11, Ohio University, Ohio, USA. 1972.
31. Perry, R. W. 'Incentives for evacuation in natural disasters.' *J. Am. Planning Assoc*. **45**(4), 440–447. 1979.
32. Blansham, S. A. 'A time model: hospital organisational response to disaster.' In E. L. Quarrantelli (Ed.). *Disasters Theory and Research*. Sage Publications Inc., Beverly Hills, California USA. 1978.
33. Bennett, G. 'Human reaction to disaster.' Paper presented at a Meeting of the Disaster Aid Working Party of the London Technical Group. 1972.
34. 'Major disaster: a planned response.' Film. AVE Department, University Hospital, Nottingham. 1975.
35. *Recommended Procedures for Handling Major Emergencies*. Chemical Industries Association, London. 1976.
36. Savage, P. E. A. 'Disaster planning: a review.' *Injury*, **3**(1), 49–55. 1971.

37. Savage, P. E. A. 'Future disaster planning in the UK.' *Disaster*, **3**(1), 75–77. 1979.
38. *Casualty Enquiry Service. Provision of Lines.* Post Office London Telecommunications Region Civil Emergency Group, London. 1980.
39. Clark, R. 'Emergency services planning.' *FPA Conference, LPG: The Hazards and Precautions* FPA, London. 1980.
40. Van Cleve, G. 'Disaster plan for a high pressure process plant.' *Safety in High Pressure Polyethylene Plants.* CEP, A. I. Ch. E, New York. 1973.
41. Abe, K. 'Levels of trust and reactions to various sources of information in catastrophic situations.' In *Disasters, Theory and Research.*
    E. L. Quarrantelli (Ed.) Sage Publications Inc., Beverly Hills, California. 1978.
42. Maas, W. 'Emergency planning.' In *Loss Prevention and Safety Promotion in the Process Industries* (First Int. Loss Prevention Symposium) C. H. Buschmann (Ed.). Elsevier, Amsterdam. 1974.
43. Webb, H. E. 'What to do when disaster strikes.' *Chemical Engineering*, **84**(16), 46–58. 1977.
44. Searson, A. H. 'Control of major fire emergencies in refineries and petrochemical plants: action checklist for emergency supervisors.' In *Loss Prevention and Safety Promotion in the Process Industries.* (Second Inst. Loss Prevention Symposium), Dechema, Frankfurt. 1977.
45. Fisher, B. E. 'Training on personnel, exercises and studies of contingency planning: practical experiences of a British emergency planning officer.' *Mass Emergencies*, **3**, 83–86. 1978.
46. Hamilton, J. *et al.* 'Pollokshaws disaster.' In *A Guide to Disaster Management.* W. Sillar (Ed). Action for Disaster. Glasgow. 1973.
47. Feates, F. S. 'Emergency procedures.' In *Major Chemical Hazards, Proceedings of a Seminar at the Lorch Foundation.* Harwell, Oxfordshire. 1978.
48. Howard, J. 'Developments in disaster.' *Proceedings of an International Conference, Action for Disaster.* Glasgow. 1976.
49. Turner, B. A. 'The organisational and interorganisational Development of Disasters.' *Administrative Science Quarterly*, **21**(3), 378–397. 1976.
50. Waddington, G. M. 'Flixborough explosion – 1st June 1974.' Paper at Conference on Contingency Planning for Disaster Control, Oyez IBC, London. 1980.
51. Carey, R. F. 'Flixborough and after – The lessons for the fire service.' Paper at Conference on Contingency Planning for Disaster Control, Oyez IBC, London. 1980.
52. Fozard, M. 'The ambulance experience at Flixborough.' In *Developments in Disaster Management. Proceedings of an International Conference.* J. Howard (Ed.) Action for Disaster. Glasgow. 1976.
53. 'Major disaster planning.' Central Fire Brigades Advisory Council, Home Office Fire Department, London. 1978.
54. Fisher, B. E. 'Mass emergency problems and planning in the UK from the perspective of the police.' *Mass Emergencies*, **3**, 41–48. 1978.
55. Harris, N. C. 'The Control of vapour emission from liquified gas spillages,' Third International Symposium on Loss Prevention and Safety Promotion in the Process Industries, Basle 1980.
56. *Emergency – dealing with Press, Radio and Television*, Chemical Industries Association, London. 1980.
57. 'Proposal for a Council Directive on the Major Accident Hazards of certain Industrial Activities,' COM(79)384 Final Commission of the European Communities, Brussels. 1979.

High Risk Safety Technology
Edited by A. E. Green
© 1982 John Wiley & Sons Ltd

Chapter 1.8

Reassuring the Public

SECTION 1.8.1 *Public Relations in the UK*

J. A. Preece

The Reality and the Perception of Risk

Man is the rational animal but, being an animal, he is irrational too. Thus our attitude to risk reflects both sides of our nature. Our rationality ensures that the way we respond to a particular hazard is usually based on some kind of assessment even though we may not be aware of having given the matter conscious thought.

We fly, recognizing that aeroplanes sometimes crash, since experience tells us that the chances of our aeroplane crashing are small and that flying is relatively safe because of the expertise that goes into the design and operation of aircraft today.

We eat, we walk, we drive, we go on holiday, we work, we play, even though we know that people die of food poisoning, fall and break their legs, are killed in road accidents, are trapped in burning hotels, are maimed in factories, drown in the sea, are struck down by lightning on the golf course and by rogue balls on the cricket field. We rationalize that the chances of our becoming the victims of such disasters are small though we have not submitted each risk to probabilistic assessment.

Most of us, however, are subject also to some kind or other of irrational, unreasoning fear. These instinctive terrors are not susceptible to the processes of reason. My wife is terrified of spiders though she knows perfectly well that no spider she is likely to encounter can do her the slightest harm. She is not interested in the facts and her mind is closed to any process of reassurance on the subject.

Those of us who have to explain and justify high technology industry to the public and seek to allay its apprehensions about the associated risks need to be mindful of both the rational and the irrational that exists in most people. In the jargon we are concerned not just with risk but with the perception of risk, which can be based on either or both of these inherent characteristics. Nuclear power is probably the industry where risks and perceived risks are subject to the most continuous and intensive debate. Ths shape of the arguments, if not the actual terms of the debate, will be familiar to other industries, especially oil, chemicals, and pharmaceuticals.

In the public's attitudes to nuclear power both rational and irrational concepts are important. In a democratic society a nuclear contribution to a nation's energy supplies must depend on positive answers to three key questions: Do we need it? Is it economical? Is it safe? The first two, vital though they are in the nuclear debate as a whole, are outside the scope of this chapter. However firm and positive the answers to them may be in a particular country at a particular time, unless a confident and positive answer can be given to the third, and accepted by a majority of the electorate, there will be no nuclear programme. Of course the question requires qualification. Since absolute safety can never be guaranteed, the issue must be whether nuclear power can be acceptably safe, and the answer inevitably involves value judgement. In seeking to reassure the public that such acceptable safety standards are achievable we have to identify the target audiences and address each one appropriately, working towards a value judgement that is consistent with those applied to other risks arising from other activities.

The decision makers in a democracy are politicians who, whether at national or local level, are laymen and non-expert about technology and risk assessment. True, they have at their call the best available expertise and should give due weight to informed advice. Nevertheless, as politicians, they also have to be responsive to public opinion. The processes by which politicians seek to lead, shape and guide public opinion are complex and often more effective than is generally reckoned, but in the last analysis politicians will only take decisions which they believe the public will accept. So ultimately it is public opinion, or what is believed to be public opinion, that determines policy.

There is only a limited number of subjects on which one individual can care passionately, and indeed many may not care passionately about anything at all, except perhaps their own convenience and prosperity. Therefore on any particular issue, such as the safety of nuclear power, what we mean when we talk of the climate of public opinion is whether, among those who are really concerned, the ones who are satisfied outnumber those who remain unconvinced.

Part of the strategy of the opponents of nuclear power is to appeal to the latent passions of the unconcerned and uncommitted, most frequently over safety issues. Safety, in fact, is the crucial battlefield over which the war of words about nuclear power is waged.

Safety and Critics

A favourite ploy of anti-nuclear organizations is to postulate extreme nuclear accidents under the most adverse circumstances and to dwell in detail on their effects in terms of death and injury, while brushing aside the question of just how improbable such extreme accidents may be in view of the many safeguards incorporated in the design and operation of nuclear plant. An example of this technique was provided in Britain by the Oxford-based Political Ecology Research Group (PERG) in an investigation into the hazards associated with the transport at sea of irradiated nuclear fuel[1]. The investigation was commissioned by the avowedly anti-nuclear organization Greenpeace, which was engaged in a campaign to arouse public opinion against the import of spent nuclear fuel through Barrow-in-Furness for reprocessing at Windscale. The PERG report postulated a severe accident to a ship carrying nuclear fuel, one involving a fire and release of radioactivity of an order that could, it was claimed, cause several tens of thousands of late cancers in an area up to 100 km from the point of release. Yet it took no account of the extreme improbability or even (as British Nuclear Fuels Limited would claim) the impossibility of their assumptions. Indeed it specifically stated: 'We leave it to the reader to ascribe probabilities and therefore to assess that part of the risk equation of probability and consequence.'

This technique was used by PERG and other objecting organizations in their arguments at the UK Windscale Inquiry in 1977 into the application by British Nuclear Fuels Limited to build a new reprocessing plant for thermal oxide nuclear fuel. An aerial view of the Windscale and Calder Works at Sellafield is shown in Figure 1.8.1-1. In his report the Inspector, Mr Justice Parker, roundly rejected this mode of argument (para. 11.3)[2]:

Figure 1.8.1-1 British Nuclear Fuels Limited, Windscale and Calder Works, Sellafield

. . . whilst I accept that even the remotest of possibilities can happen, I reject completely the suggestion, which was made at the Inquiry, that if something is possible it is also inevitable. It is a very remote possibility that a particular accident will occur if its occurrence requires, for example, the failure not only of a primary source of supply such as electricity or water, but also the failure of one or more alternative sources of supply and one or more automatic control systems together with more than one independent act of negligence. Improbable as it may be, however, it can happen. It is not however, inevitable that it will happen. If it were, everyone who for example drives his car tomorrow would be dead by tomorrow evening because it may truly be said of each of them that it is possible that he will be involved in a fatal accident during the day.

British Nuclear Fuels Limited in fact, in its response to the PERG report, concluded that the catastrophic release of radioactivity which PERG considered might arise from a prolonged shipboard fire was based on the occurrence of a totally unrealistic sequence of events. BNFL concluded that the report was grossly inaccurate and misleading and aimed to create alarm by postulating dire effects without attempting to show how they might come about.

The PERG report was well publicized by Greenpeace and was widely reported in the local Press and on radio and television in Cumbria. Some concern was thereby aroused about the safety of a transport activity which had proceeded in complete safety for 14 years and which is subject to the most stringent international regulations. BNFL's response to the report was also reported by the media[3], although not so prominently, and it is doubtful whether the considered, thoughtful, and factual reply ever caught up with the emotive presentation of the PERG report by Greenpeace. BNFL's critique of the report did, however, lead to second thoughts on the subject by PERG itself which in a later report commented[4]:

Two major criticisms have been made of our report:
(i) that the accident therein considered (an extended fire), is impossible on a BNFL ship because of the high standard of fire resistance and lack of flammable cargo, . . .
(ii) PERG did not adequately present the probabilities of the rare combination of events that would have to lead to such a fire and the release from spent fuel flasks.

We accept these as valid criticisms.

The Role of the Media

In the public debate on nuclear safety it is easier to arouse concern than to allay it, easier to devise disaster scenarios than to place them in proper perspective. The natural disposition of the Press and other media in a free society is of great assistance to pressure groups. A free Press is one of the strengths of the democratic system, its key role being to challenge, to examine, to invigilate and to provide a platform from which interest groups may address the public. Newspapers, together with news and current affairs programmes on radio and television, are run by journalists who apply the traditional journalistic criteria to what constitutes news. In this context catastrophe and disaster are obviously news. To a lesser extent so are warnings of potential catastrophe and disaster. They make 'good copy', 'compulsive viewing', to use the terms of the trade. Of course good journalists aim to give both sides of the story. They take care to obtain reactions from the industry to all their 'stories' based on forebodings from the anti-nuclear camp. But in journalistic terms it is the allegation, not the rebuttal of the allegation, that makes the 'story'. The reply may form just a footnote, to be tagged on at the end. Or it may provide sufficient journalistic justification for a story slanted on the line of 'A new row broke out today about nuclear safety . . .'.

'Giving both sides of the question' on matters concerning the nuclear debate often means, to the media, giving equal weight to, on the one hand an individual crank, a man who may be of no academic standing and repute or whose ideas may have been totally demolished by his peers, and, on the other, the considered view of governments, of international competent authorities, and the mass of expert scientific opinion. In fact the balance may become grotesquely distorted with all the weight and emphasis allotted to the spurious opinions of a self-styled 'expert' of no standing whatsoever, with the forces of responsible, informed, and recognized authority forced into what may appear as a weak defensive position against a zealous crusader who alone has the public interest at heart. It is feasible, of course, that once in a while, the political and scientific establishment will together get it totally wrong. Occasionally a lone voice in the wilderness may have the true message. But not every time, as the techniques of the media, based on journalistic assumptions of what makes news, would tend to suggest.

Those who work in the nuclear industry in the free world have come to recognize that they are exposed to far more critical standards of judgement by the media than those who work in other industries with equally hazardous technologies and less impressive safety records. This is presumably one of the effects of instinctive, rather than rational reactions to the special nature of radioactivity. It also reflects the origins of civil nuclear power in the technology that gave the world the unique horrors of the atom bomb. There are very few working in the nuclear field who resent being required to perform to high standards of safety. There are many who resent the fact that they are not given credit for the industry's remarkable safety record and resent also the lack of comparable media pursuit of industries which consistently fail to match this record. It is difficult to avoid the conclusion that the media apply double standards to the nuclear industry.

Sir John Hill, when Chairman of the United Kingdom Atomic Energy Authority, remarked upon this strange tendency. 'It is a common pattern of anti-nuclear articles to detail a long list of all the difficulties, delays and mishaps that have occurred in the world nuclear industry over the past 20 years, with the added question 'But what if. . . . ?' in relation to every one. This approach, if applied to other industries would persuade us that the world is too dangerous a place to live in and that it is a pity that life ever started on this planet[5].'

The nuclear industry as a whole has been reticent about criticizing media reporting of its affairs – with good reason. The public is rightly not over-sympathetic to those who habitually complain about being misunderstood or misrepresented by the Press. It is not particularly impressive company in which to be included. On a more practical level there is a feeling that journalists if provoked by criticism from the nuclear industry, would probably give the industry a worse Press than it already receives.

Nevertheless one politician who has a better grasp than most of the way nuclear affairs get portrayed in the media has spoken out about the blinkered way in which some journalists approach the subject. Norman Lamont, the former British Parliamentary Under-Secretary of State for Energy, has spoken of sections of the Press which, in dealing with nuclear matters 'sacrifice balance to sensationalism, accuracy to prejudice and substance to image'. Calling for a more responsible Press attitude towards the nuclear industry[6] Lamont said the Government did not seek an uncritical or unenquiring Press.

Cracks in nuclear reactors, the disposal of nuclear waste are just a few of the many safety issues which surround nuclear power and which are rightly the concern of the Press. The public has every right to know what is going on in the nuclear industry, especially when things go wrong. The duty of the Press is to inform. My plea is simply for more balanced and more accurate reporting. It is a plea to the Press to exercise their enormous power more responsibly. Never, in my opinion, has more information been made available to the Press on nuclear matters than is available today. If the media fail to take account of this information then the first casualty will be public understanding of the real issues.

The Windscale Inquiry: An Information Breakthrough

In Britain two of the most common complaints against the nuclear industry by its critics are of excessive secrecy and a glossing over of the hazards of nuclear power. In the early days there was some substance in both charges, but they have ceased to be true. The industry's reticence of its early days had roots in the military origins of applied nuclear fission and the necessary security in which defence matters have to be wrapped. When scientists and engineers nurtured in this security conscious environment moved into the civil nuclear field they found it hard to accept that different standards applied and that the public had a right to be kept informed. It was also perhaps understandable that, when they did communicate, the industry's experts felt that the complexities of their subject could only be communicated to laymen in terms which did not provide for the subtle gradations of what we now call risk assessment.

The most significant milestone on the road to massive disclosure of information by the industry and frank and detailed discussions of the hazards associated with nuclear power in the UK was the Windscale Inquiry of 1977[7]. In 100 days of hearings something like 4 million words of evidence were given by 120 witnesses, much of it of a detailed factual nature, and much too that was placed on the public record for the first time. The Windscale Inquiry created a pattern for a full and comprehensive examination of major nuclear projects in the United Kingdom, with a rigorous challenge of the industry's case by the full array of environmental and anti-nuclear interests. Yet, with hindsight, the fact that the Inquiry was held at all seems almost fortuitous.

British Nuclear Fuels Limited's proposals for a new thermal oxide fuel reprocessing plant (THORP) as part of an overall development plan for its

Windscale and Calder site in Cumbria were first made public in 1974 and at that time aroused little interest among Members of Parliament or in the Press. Irradiated fuel had been reprocessed at Windscale for more than 20 years and it was accepted government policy that reprocessing should develop to meet the needs of the UK electricity generating industry which was expanding its nuclear capacity, and also to cope with overseas reprocessing business for which BNFL was encouraged to compete.

Initiating the normal procedure for obtaining planning consent for a major industrial development, BNFL was encouraged by a favourable response from government departments and the local planning authorities. There was every reason to believe that the Company would obtain government approval for the construction of its new facilities without any form of inquiry. Further public statements by the Company about its investment programme were reported by only two national newspapers, both serious journals. The coverage, restrained and factual, concentrated on the possibility of BNFL obtaining valuable reprocessing business from Japan.

The situation changed dramatically nearly 12 months later when the mass-circulation national newspaper, the *Daily Mirror*, ran a front page lead story written and displayed in its avowedly sensational style which claimed that by accepting foreign nuclear fuel for reprocessing BNFL would turn the UK into 'the world's nuclear dustbin'. That report alone changed the situation overnight. It was followed up by every national newspaper and on radio and television, and quickly picked up as an issue in Parliament, and BNFL's planned Windscale development become almost overnight a matter of acute controversy. The introduction to the *Daily Mirror* self-styled 'shock report' made two claims: that the plans to import 'huge quantities of nuclear waste into Britain' were secret, and that a storm of protest was growing over the proposal. Setting aside the paradox of a storm of protest growing over a project that was claimed to be secret, the statement was in one sense self-fulfilling in that a storm of protest certainly did grow as the result of the sensational treatment and the decision of the anti-nuclear movement to promote the development into an issue of confidence. As for secrecy, David Fishlock, Science Editor of the Financial Times, dismissed this as a false charge, writing after the event: 'The negotiations had been widely reported in the business Press but had attracted no interest elsewhere.[8]'

Despite a major public information campaign launched by BNEL to allay concern about the THORP proposal, which the Company saw as a logical extension of well-established activities, the Government decided that the plan had to be submitted to the test of a full-scale public inquiry.

This inquiry bore slight resemblance to the normal investigation into an industrial development project under land-use planning procedures. It raised issues of fundamental importance to the future of the nuclear power programme in the United Kingdom. The key questions posed at the outset by Judge Parker, to which the subsequent proceedings were largely directed, were:

1. Should oxide fuel from UK reactors be reprocessed in Britain at all, whether at
   Windscale or elsewhere?
2. If yes, should such reprocessing be carried on at Windscale?
3. If yes, should the reprocessing plant be about double the estimated size
   required to handle UK oxide fuels and be used, as to the spare capacity, for
   reprocessing foreign fuels?

BNFL naturally considered the answer should be 'yes' to all three questions.
Objectors ranged from those with an outright 'no' to all three through various
permutations of 'yesses and noes" to the 'not yets' and the 'yesses, but subject to
preconditions', these preconditions being enormously varied.

As the Parker Inquiry weighed the evidence and listened to the pros and cons of
the argument public attendance at the Inquiry rapidly fell away after the opening
day and so did Press coverage. Only one national newspaper carried daily reports;
others carried periodic news agency reports of particular evidence, or from time
to time sent non-specialist reporters to write what journalists call 'colour pieces'.
Thus although an unprecedented amount of factual material about the nuclear
industry was put on the public record for the first time at the Windscale Inquiry, it
made no wide public impression at that time, though it has proved a tempting
information mine since then for participants on both sides of the nuclear debate.
(See Table 1.8-1)

Table 1.8-1 Windscale inquiry 1977

	BNFL	Supporters	Govt. depts.	Objectors
Witnesses	17	19	21	84
Reference documents	300	200	75	1100
Distribution of time*	30%	10%	10%	40%

* Opening and closing: 10%

Judge Parker's report on the Windscale Inquiry was a forthright recommendation in favour of the THORP project and an emphatic rejection of all the
arguments of the objectors. Parker's report was accepted and the Windscale
expansion endorsed by two separate votes in the House of Commons.

Objectors at the Windscale Inquiry who had been impressed by Parker's
courtesy, patience, and fairness during the course of the proceedings, reacted
angrily to his report. Naturally they disliked his conclusions, but they also
expressed resentment at the peremptory way in which they considered their
arguments were brushed aside. A common theme among them was that the
Inspector had failed to understand that in the nuclear debate value judgements
about the morality of nuclear power have an equally important place as do
technicalities and economics.

The debate continues about the issues raised at the Windscale Inquiry. Indeed
the Inquiry has ensured that no subsequent major nuclear development in Britain

will escape the attentions of a searching public examination. The mould has been set and it is likely that future inquiries over, say, the proposed building of the first British pressurized water reactor and a demonstration fast reactor will be longer, even more comprehensive, and provide even greater opportunities for the international anti-nuclear lobby to air its views and require the nuclear industry to justify itself in revelatory detail.

Face to Face with the Public

Meanwhile the day-to-day process of disseminating information about the nuclear industry, which received such an impetus at the Inquiry, continues on an increasing scale. Implicit in the industry's decision to allot considerable resources to this process in the conviction that reassurance on safety and the other fundamental issues can only come from understanding based on knowledge.

A number of organizations within the nuclear industry have professional teams of communicators, such as the author heads for BNFL, which work at this task full-time, seeking to disseminate information by various channels – to the media, to schools and colleges, by advertisements, by exhibitions, by the distribution of literature and films, and by talks to many hundreds of interest groups, from Women's Institutes and Rotary Clubs to learned societies and legislators.

Yet the task of explanation and persuasion cannot be left to the professional communicators and publicists alone. Those who work within the industry as managers, engineers, scientists, administrators, and operatives have a role to play as individuals in helping to satisfy the public that nuclear energy is a safe industry by any reasonable standards. The industry's employees are in fact in the best possible position to do this, by example as well as by precept, since they have chosen to spend their lives working in a nuclear environment, many of them subject to far more radiation exposure than any member of the public is likely to receive from the industry. They can reasonably claim that they are hardly likely to have chosen this work if they thought it created unacceptable risks. The truth is that those who know most about the industry and its safety standards are least worried about its hazards. Thus communities which have become used to living near nuclear installations are among the industry's strongest supporters, and not just for the immediate economic benefit a large industrial plant can bring to a locality.

In Britain these local good relations derive in part from the activities of the local liaison committees which have existed since the earliest days of the industry as channels of communication between nuclear installations and the neighbouring communities. Initially these were conceived as practical means for the exchange of information between the management of a nuclear site and the local public services – fire, police, medical, emergency planning and so on – who would be involved in implementing evacuation and other measures designed to protect the population in the event of an emergency developing at the plant with consequences that might extend beyond the site boundary. The maintenance of these important working relationships still forms an important part of the functions of these committees, but over the years their role has gradually widened.

Nowhere has this process gone further than at Windscale itself (or, more properly now, at Sellafield, the name conferred on the Windscale and Calder Site in 1981). Judge Parker was critical of the performance of the Windscale Local Liaison Committee, saying in his report: 'As a vehicle for keeping the public informed or reassuring the public the committee has plainly failed to carry out its terms of reference.' BNFL and the local authorities in Cumbria reacted swiftly in response to these strictures and a number of changes were made in the committee's constitution, responsibilities, and procedures to make it a more effective body. In particular its membership was widened to include elected members of local authorities, sub-committees were set up to cover the key issues of emergency procedures and environmental matters, meetings were held more frequently and, most important of all, the meetings of the full committee were opened to the public and the Press. The effect of these changes has been to transform the work of the committee in a very short period of time. The operations and development plans of the works are discussed in considerable detail. Incidents at the site, which are reported to the Press as well as to the regulatory authorities, are discussed. The elected representatives raise issues of public concern on behalf of their constituents. All these discussions are fully reported by the local media and often reach the national Press since Fleet Street's interest in Windscale can easily be aroused. While the effect of this at face value is to reinforce Windscale's reputation as a controversial site, in the long run the pattern of open management which is displayed is a valuable corrective to charges of secrecy which the anti-nuclear lobby still sometimes tries to level at the industry.

There are local liaison committees associated with all the nuclear power stations which the Central Electricity Generating Board (CEGB) has in operation or under construction, with broadly similar representation and functions. The proceedings of these committees are not, however, open to the public or to the Press, although usually a Press briefing is held after each meeting. This in no way indicates an attitude of secrecy by the CEGB towards the running of its nuclear sites. The managerial staff maintain close and helpful liaison with local reporters; as at Windscale, emergency plans for the station are on deposit for all to see in local libraries and council offices, and the nuclear power stations themselves receive a constant stream of visitors in organized parties. Some of the stations in holiday areas have become tourist attractions in their own right, and holidaymakers can virtually visit them on demand if sufficient places are available on the regular tours conducted by smartly uniformed and knowledgeable station guides, usually the wives of station staff.

The CEGB has gone even further in its efforts to reassure the public that its power stations are fully prepared and trained to tackle any conveivable emergency, to bring it under control and to minimize its effects. TV cameras and newspaper reporters have been allowed to observe the realistic emergency exercises held regularly under the auspices of the Nuclear Installations Inspectorate (NII). In these exercises the NII present station staff with an emergency scenario and monitor the effectiveness of the response and the state of training and operational efficiency that is revealed. On each occasion the media have observed one of these exercises their conclusion has been that the public may have confidence both in the operation of nuclear plant and the effectiveness of the regulatory procedures and institutions which exist to maintain an expert and independent check on the industry.[9].

In the early days of civil nuclear power in Britain, after the Queen in 1956 had formally opened Calder Hall power station, the first connected to a national electricity grid, the industry was untroubled by concern about public opinion. On the one hand was the atom bomb, which all civilized people regarded as evil (though most concluded a necessary evil for the West to have in its armoury, given the military and political state of the world). On the other hand was civil nuclear power, the peaceful atom, which all concerned could surely agree was a Good Thing. Nowadays that may seem a simplistic approach. Uniquely among the major sources of energy, nuclear power is the subject of controversy in most societies where public opinion is permitted to express itself.

The nuclear electric industry recognize that concern about the safety of the nuclear process is legitimate, and if reasonably expressed deserves reasonable answer. That answer lies in readily available information, presented and interpreted in ways which different sections of the public, with widely ranging standards of technical appreciation, may understand. These information processes cost money and ironically attract criticism from the very anti-nuclear organizations which criticize the nuclear industry for excessive secrecy. Friends of the Earth complained in 1980 that the nuclear industry in the UK was spending £5 million a year on information work – an estimate probably not very wide of the mark – but far from applauding this outlay of resources as a responsible demonstration of accountability to the public, criticized it as 'a whitewashing operation', and contrasted that level of expenditure with the more limited financial resources of anti-nuclear groups.

In the battle of words over safety in the nuclear industry, it is truth itself that is at greatest peril – a one in one risk *per diem* of mortality, it sometimes seems. Those who speak on behalf of nuclear power will do so most effectively by a straightforward and honest approach, not ducking the facts when they happen to be unpalatable, and leaving the tactics of exaggeration, evasion, half-truth, and misinterpretation for the other side in the great nuclear debate to use if it so pleases.

Questions 1.8.1

1. Describe the general motivations which cause human beings to accept or reject risk.
2. What is meant by the 'perception of risk'? From an industry which may be

known to you build up the concept of the perception of risk on a particular aspect.

3. In using mass media to communicate with the public describe the advantages and disadvantages when trying to convey information on high risk and hazardous situations.
4. What methods can be best employed to give the public confidence that the appropriate measures have been taken to minimize the risk of such situations to acceptable levels?
5. Using an industry with which you may be familiar as an example draw up:
   (a) an action plan for an environmental organization seeking to arouse concern about risks;
   (b) a plan for the industry to refute the claims of its critics.

References 1.8.1

1. 'An investigation into the hazards associated with the maritime transport of spent nuclear reactor fuel to the British Isles.' RR-3 Political Ecology Research Group, Oxford, 1980.
2. *The Windscale Inquiry Report*, Vol. 1. Mr Justice Parker, HMSO 1978.
2. Statement and Critique of PERG report RR-3 by British Nuclear Fuels Limited, 1980.
4. *Barrow Hazards Survey* RR-5. Political Ecology Research Group 1980.
5. Hill, Sir John. 'The reality of nuclear power.' *Blackwoods Magazine*, February 1980.
6. Lamont, Norman MP. Speech to Institution of Nuclear Engineers, London, 28 November 1980 (Department of Energy).
7. Bolter, H. E. Company Secretary, British Nuclear Fuels Limited: European Nuclear Conference, Hamburg, May 1979.
8. Fishlock, David. 'Exploring Europe – nuclear energy.' (European Research Centre, University of Sussex) 1979.
9. 'Red alert at Dungeness'. *Sunday Times*, 18 May 1980.

High Risk Safety Technology
Edited by A. E. Green
© 1982 John Wiley & Sons Ltd

SECTION 1.8.2 Government Sponsored Assessment

I. B. Wall

Background

Assurance of the safety of the public and the workers has been an integral part of the design and licensing procedures for nuclear reactors within the United States since their inception. The need for detailed safety studies was recognized from the beginning. First, dispersion of the core inventory of radioactive material to the environment was clearly a potential hazard to the public and the operators. Second, the general public was fearful of nuclear 'things'. After all, the public's introduction to nuclear power was the bomb which terminated the Second World War.

The technical approach to safety assurance has evolved as the technology has matured. Early hazards reports were organized around maximum hypothetical accidents. The WASH-740 and TID-14884 reports are the most significant examples of such analyses. The former report[1], published in 1957, calculated the consequences of the limiting accident scenarios for a 200 MW(E) nuclear power plant located 30 miles from a large city. The widely publicized results – 3400 deaths; 43 000 injuries; and $7000 million property damage – were stated by the authors to be an upper bound. The authors were unable to quantify (in a calculated sense) the probability of such an event. Technical experts were unanimous that the probability was small; opinions ranged from 1 chance in 100 000 to 1 in 1000 million per reactor-year. The WASH-740 report provided technical input to US Congress in their establishment of the liability limit in the Price–Anderson Indemnity Act. For the renewal of Price–Anderson in 1967, an update of WASH-740 was commissioned but never completed. TID-14884[2] presented a hypothetical accident scenario whose calculation was the beginning point for reactor site evaluation. With several modifications, this scenario and calculation are still the bases for US Nuclear Regulatory Commission's (USNRC) siting policy. In the late 1960s, the concept of probability was introduced qualitatively to reactor safety studies by introducing the concept of a 'maximum credible accident'. Designers were required to accommodate credible accidents (e.g. pipe severances) but not incredible ones (e.g. reactor pressure vessel ruptures). With a change of words to 'design basis accidents', this technical

247

approach still underlies the regulatory requirements of USNRC today.

The arbitrary nature of these safety criteria, the potential inconsistencies in the judgements on relative probabilities, and the lack of definition for 'safety' became increasingly evident during the 1960s. Many important witnesses such as Commissioners of the US Atomic Energy Commission (USAEC) and members of the Advisory Committee on Reactor Safeguards (ACRS) were manifestly uncomfortable in responding to the question, 'Are reactors safe?' from the Congressional Joint Committee on Atomic Energy. In 1971, the Congressional Joint Committee on Atomic Energy expressed to the USAEC[3] the need for a quantitative statement on the safety of nuclear power reactors. At the urging of Ralph Lapp[3] the USAEC initiated the Reactor Safety Study in the late summer of 1972.

The study was to be performed independently of the USAEC's operating and regulatory organizations, and Professor Norman C. Rasmussen of Massachusetts Institute of Technology was appointed as director and later Mr Saul Levine of the USAEC was appointed as project staff director. Although funds and other assistance were provided by the USAEC, the study operated under a general charter provided by the Commission, but received no other direction from it. (This independence was preserved by the US Nuclear Regulatory Commission when it assumed sponsorship of the study in January 1975.) The statement of objectives given to the Reactor Safety Study (RSS) by the commission was (in part) as follows:

The principal objective of the study is to try to reach some meaningful conclusions about the risks of nuclear accidents using current technology. It is recognized, however, that the present state of knowledge probably will not permit a complete analysis of low-probability accidents in nuclear plants with the precision that would be desirable. Where this is the case, the study will consider the uncertainty in present knowledge and the consequent range in the predictions, as well as delineating outstanding problems. In this way, any uncertainties in the results of this study can be placed in perspective. Thus, although the results of this study of necessity will be imprecise in some aspects, the study nevertheless will provide an important first step in the development of quantitative risk analysis methods.

As confidence within the study group grew in the ability to achieve a meaningful risk assessment, the Reactor Safety Study added the following more specific objectives under its original, broadly stated charter:

1. Perform a quantitative assessment of the risk to the public from reactor accidents. This requires analyses directed toward determining both the probabilities and the consequences of such accidents.
2. Perform a more realistic assessment as opposed to the 'conservatively-oriented' safety approach taken in previous studies of this type and the licensing process for nuclear power plants.
3. Develop the methodological approaches needed to perform these assessments and gain an understanding of their limitations.

4. Identify areas in which future safety research might be fruitfully directed.
5. Provide an independent check of the effectiveness of the reactor safety practices of industry and the government.

Organization of Reactor Safety Study

Under the direction of Rasmussen and Levine, the study work was performed by about 10 USAEC (USNRC) employees augmented by about 40 personnel from contractors and national laboratories all working at USAEC Headquarters in Germantown, Maryland. This group did the bulk of the work on the systems analysis and the off-site consequence modelling. Other selected tasks were performed by contractors and national laboratories throughout United States working under the direction of the headquarters group. Guidance on specialized topics was obtained from many consultants. The total effort was about 70 professional man-years at a cost of about $4 million over a period of three years.

The final report[4] consisted of a Main Report plus 11 technical appendices, totalling about 3000 pages. Additionally, a 12-page Executive Summary was prepared. The report has been widely distributed, both within United States and overseas; about 20 000 copies of the Executive Summary, about 10 000 copies of Main Report and several thousand of each technical appendix.

The study was organized into seven basic tasks as shown in Figure 1.8.2-1. While there were more detailed tasks than those shown, this figure gives a general appreciation of the principal efforts involved. The first task (Box 1) was to identify all the ways in which a reactor's fission product inventory could be released from the reactor into the environment. The term 'accident sequence' was coined to describe each one of the many different ways in which this event might happen. These accident sequences were largely defined by preparing event trees which are described in Section 1.3.2. It was also necessary to assign probabilities to the occurrence of the various events involved in each accident sequence. For the most part, failure rates based upon experience data for components, human error, and contributions from testing and maintenance, were combined appropriately by means of fault trees, described in Section 1.3.2, to determine the unavailability and unreliability of engineered systems; this procedure is identified by Box 2. The output of these first two boxes defined a set of accident sequences and their probabilities of occurrence. In Box 3, supplementary analyses were performed to define the magnitude of the fission products released from the fuel and how they would be transported through the containment barriers for final release into the environment. At the output of Box 3, a probability histogram was defined for the magnitude of the radioactive release to the environment.

The fission products would be released either to geosphere due to the molten fuel melting through the containment base-mat or to the atmosphere due to a breach in the containment. In the former case, the radionuclides could potentially contaminate ground-water, and be ingested by man. This pathway was found to cause relatively small health effects so the RSS only represented it by a simple

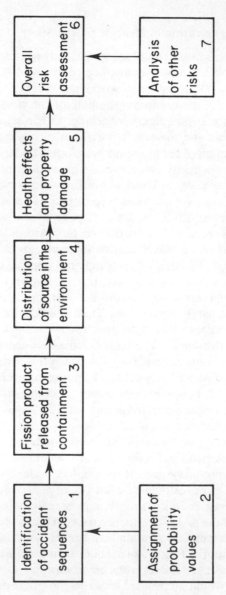

Figure 1.8.2-1 Seven basic tasks in reactor safety study

model. An atmospheric release can potentially cause large consequences and a sophisticated model was developed as indicated in Box 4. This consequence model calculated the airborne and ground concentrations of fission products versus the time after the accident and distance from the reactor. Probability distributions for weather conditions and population densities that typify commercial reactor sites were used. Finally, in Box 5, the health effects and property damage resulting from such radioactive distribution were estimated. The final output was a set of complementary cumulative distribution functions for the various health effects (e.g. Figure 1.8.2-2) and property damage which comprehensively stated the overall risk. Finally, other accident risks (e.g. dam failures) within society were examined to place the reactor risks into perspective. This comparison was deemed desirable since, although many risks are recorded in society today, e.g. automobile fatalities, they are not truly comparable to the reactor risk, which has low probability but potentially high consequences.

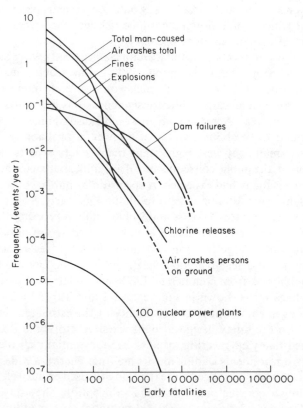

Figure 1.8.2-2 Frequency of man-caused events involving
early fatalities

The study group quickly found that, in order to perform a realistic probabilistic risk assessment, it was necessary to focus upon specific nuclear power plant designs. Detailed designs including operating procedures had an important influence upon the reliability performance of safety systems. Accordingly, the RSS analysed two reactors, the Surry I pressurized water reactor (788 MW(E)) and the Peach Bottom II boiling water reactor (1065 MW(E)). The basis for their selection was that they were the largest plants of each type that were operating. The study extrapolated the results for these two plants to the first 100 LWRs. For several reasons, it was judged that such extrapolation would probably be conservative. Although the Three Mile Island accident and later risk studies may indicate some questions on this judgement.

The Findings of the Reactor Safety Study

At the inception of the study, the general perceptions within the nuclear industry were that: (i) the probabilities of serious reactor accidents were very low; (ii) the public consequences of such accidents would be catastrophic; and (iii) the initiating event which would dominate public risk was the severance of a large pipe (large loss of coolant accident (LOCA)).

In direct response to its charter, the Reactor Safety Study was able to reach a meaningful conclusion about the public risks from reactor accidents. The probability of a severe reactor accident which would severely damage the reactor but would not cause significant public consequences was estimated to be about 5×10^{-5} per reactor-year. Accidents which would cause significant public health effects were found to be about a factor of 5 lower in probability. While these probabilities are small, they were higher than many safety experts had expected. On the other hand, the public consequences of a significant reactor accident were not catastrophic as many had expected. A spectrum of public consequences was estimated ranging from essentially zero to about 3300 early deaths, 1500 latent cancer fatalities per year for 30 years and $14000 million property damage. The probability of the latter consequences was estimated to be about 10^{-9} per reactor-year. In fact, a large fraction (e.g. about 90%) of severe accidents was estimated to cause no detectable public health effects. The 'bottom line' finding was that the public risk from commercial LWRs was relatively low compared to other societal hazards as shown in Figure 1.8.2-1 and Table 1.8.2-1. The study rendered no judgement about the acceptability of the estimated reactor risks.

With respect to the safety design of light water reactors, the Reactor Safety Study contained many engineering insights. First, essentially all the risk to the public derives from accidents leading to core melting. There is a widespread belief that this conclusion was prejudged since the study focused exclusively on accident sequences leading to core melting. This is a misconception. The study started with an open mind by considering over 100000 potential core meltdown and non-meltdown accident sequences in a systematic way. By eliminating the physically

Table 1.8.2-1   Consequences of reactor accidents for various probabilities for 100 reactors

	Consequences		
Chance per year	Latent cancer[b] fatalities (per year)	Thyroid nodules[b] (per year)	Genetic effects[c] (per year)
One in 200[a]	< 1.0	< 1.0	< 1.0
One in 10 000	170	1400	25
One in 100 000	460	3500	60
One in 1 000 000	860	6000	110
One in 10 000 000	1500	8000	170
Spontaneous incidence	17 000	8000	8000

a. This is the predicted chance per year of core melt for 100 reactors.
b. This rate would occur approximately in the 10 to 40 year period after a potential accident.
c. This rate would apply to the first generation born after the accident. Subsequent generations would experience effects at decreasing rates.

meaningless sequences and by discriminating against the less probable and less consequential ones, it was found that public risk was determined by less than 10 accident sequences for each of the two reactor types analysed. All these dominant accident sequences involved fuel melting. The underlying physical reason is that, if fuel is not melted, the bulk of the radioactive material remains entrained in it and only a small fraction of radionuclides generated in the fuel can be released.

The second insight is that these dominant accident sequences are initiated by small loss-of-coolant-accidents (LOCA), transients, and systems interactions, and not by large LOCAs whose study had been the centrepiece of the reactor safety analysis and licensing for the past decade or so.

Third, the unavailability of most engineered safety systems (ESF) was found to be relatively high, (e.g. median unavailabilities in the range of 10^{-4} to 10^{-1} per demand) and to be dominated by human error and test/maintenance outages often in a common cause failure mode. Due to the high degree of redundancy, which had been required, system unavailability due to hardware failures was generally found to be less likely.

The dominance of core meltdown accidents leads to a fourth important insight. Core meltdown accidents would impose different and larger thermal and mechanical loads on containment than are required to be considered by current containment design criteria. The study found that current containment designs would *ultimately* fail in meltdown accidents. Although there would be a melt-through of the concrete base-mat, the *initial* mode (e.g. overpressurization) of containment failure in core meltdown would determine the magnitude of public consequences. (Recent analyses, based upon new experimental data, suggest that, for most US reactors, the molten core might not penetrate the base-mat.)

The broad spectrum of public consequences depended upon the initial containment failure mode.

Since publication of the Reactor Safety Study in 1975, there has been substantial work related to the probabilistic risk assessment of light water reactors. In general, this work does not alter either the conclusion that reactor risks are relatively low or the above four engineering insights. First, early results from risk assessments currently being performed within the US are generally estimating a probability of severe core damage in the same range of that estimated by RSS. Second, risk assessments of other LWR designs have shown that the dominant accident sequences are dependent upon the detailed design of the reactor and balance of plant, but the dominant initiating events have continued to be transients and small LOCAs, and human error has continued to be a significant contributor. Finally, there has been an increasing awareness that the RSS calculation of the release of radioactive material to the atmosphere has several simplifying assumptions and almost certainly overstates their magnitude[5]. The experimental evidence for this statement is somewhat circumstantial at this time, and realistic predictions of the smaller release magnitude require more refined models and probably some further experiments. Nevertheless, it is important to note that a modest reduction (e.g. factor of 10) would have a large effect upon the prediction of early fatalities (see Figure 1.8.2-2) and other health effects having threshold doses. The effect upon the predictions of latent cancer fatalities, genetic effects, etc. would be smaller. Such changes, if substantiated, would radically change our perception of reactor risks and should have manifold ramifications upon reactor siting, emergency response planning, and containment design.

Peer Review

For scientific work, the peer review process plays an essential role in quality control of technical output. This process was especially important for such a monumental and innovative work as the Reactor Safety Study on such a highly politicized subject as public risk and reactor safety. Recognizing that the public credibility of the RSS would thereby be enhanced, the RSS staff sought peer review. For several reasons, the peer review was only marginally successful.

The draft report was widely circulated, and comments were actively sought from numerous government agencies, industrial organizations, and groups critical of nuclear power. About 90 organizations and individuals responded with comments totalling about 1800 pages. The comments were generally constructive and of considerable assistance in preparing some revisions to the draft report. The most significant change was the development of a new model for the prediction of off-site consequences. The major comments and the RSS staff response were summarized in a new appendix to the final report.

After publication of the final Reactor Safety Study report, there continued to be substantial criticism from segments of the scientific community, interveners, and some political leaders. Following a June 1976 hearing[6] before the Subcommittee on Energy and Environment of the US House of Representatives, chaired by Congressman Morris Udall, a risk assessment review group was established by the USNRC to examine key aspects of the RSS. Seven scientists, under the chairmanship of Professor Harold W. Lewis (University of California, Santa Barbara), formed the group which held a series of public hearings to receive testimony from all viewpoints on most aspects of the study. In their report[7], the review group sharply criticized certain aspects of the Reactor Safety Study (e.g. the Executive Summary, the peer review process, certain statistical analyses, lack of scrutability), but found that the methodology was sound and strongly recommended that it be more widely used by USNRC to re-examine the fabric of its regulations to make them more rational. They also noted that the engineering insights on the importance of transients, small LOCAs and human error had been inadequately reflected in USNRC's policies and research. The US Nuclear Regulatory Commission accepted these findings, withdrew support for the Executive Summary, and directed its staff to make use of the RSS 'as appropriate, that is, where the data base is adequate and analytical techniques permit'.

Accomplishments

In spite of intensive reviews, no major technical deficiency has been identified in the Reactor Safety Study since it was published seven years ago. The need to improve several aspects (e.g. seismic model, statistical models) has been noted in the Risk Assessment Review Group report and elsewhere. Many of the later probabilistic risk assessments have introduced refinements to the methodology. However, none of these deficiencies or refinements would substantively change either the bottom line risk or the engineering insights. The fact that the study became the centrepiece of a fierce political controversy over the use of nuclear power in the US has obscured the general acceptance of its methodology and findings within the knowledgeable scientific community. It is clear that the study achieved its major objectives and provided many new insights into reactor safety.

In the near term, although it was not an objective, the Reactor Safety Study had only limited success in reassuring the public about reactor safety. The reasons are subtle and somewhat intangible. First, as documented[7] by the Risk Assessment Review Group, the peer review process was inadequate due to a lack of scrutability and a co-mingling of technical review and social commentary. Second, it is questionable whether any report on such a complex subject and politicized issue, however carefully written, could communicate adequately subtle probabilistic arguments to the public. Even within the technical community, probability and statistics are widely misunderstood.

Long term, the Reactor Safety Study should have an overwhelming success in achieving safer reactors. The Risk Assessment Review Group found[7] that the RSS was largely successful in making the study of reactor safety more rational and in establishing the topology of many accident sequences. The Three Mile Island 2 accident in March 1979 was basically predicted in the Reactor Safety Study. Since the TMI-2 accident, the US Nuclear Regulatory Commission has placed increased emphasis upon the use of probabilistic risk assessment to guide reactor safety design and siting. Upon the request of USNRC or their own initiative, many electrical utilities are currently performing risk assessments of their nuclear power plants. Through its Interim Reliability Evaluation Program (IREP) or the follow-on National Reliability Evaluation Program (NREP), the USNRC envisages that, within a few years, such assessments will be done for all operating plants. The logical follow-on would be a requirement that a probabilistic risk assessment be included in all Final Safety Analysis Reports.

A major function of such assessments is a focusing of design and regulatory attention upon significant issues. In this respect, it is instructive to compare in Table 1.8.2-2 the four major engineering insights from the Reactor Safety Study to the past focus of USNRC reactor licensing requirements. There is clearly a poor congruence for each insight which is currently stimulating many modifications to USNRC requirements. First, the USNRC has promulgated[8] interim requirements related to hydrogen control and certain degraded core considerations and has initiated[8] possible rule-making hearings on degraded or melted cores. Second, as part of its post-TMI action plan, the nuclear industry was required[9] to perform many new analyses of small-LOCA accidents and new small-LOCA experiments have been performed at LOFT[10]. Third, the man–machine interface is now receiving enormous attention[11]–[13]. Fourth, alternative containment designs (e.g. filtered vented containment systems) are being

Table 1.8.2-2 Comparison of RSS engineering insights to historical focus of USNRC licensing

Reactor Safety Study Engineering Insights	USNRC Licensing Focus
Public risk dominated by core melting accidents	Non-core melting accidents
Dominant accidents initiated by: —Small LOCAs —Transients —System Interactions	Large LOCAs
Dominant failure modes are caused by human error and test/maintenance	Single hardware failures
Containment will ultimately fail under mechanical/thermal loads from core melting accidents	Containment not evaluated for such loads

evaluated by using the accident sequences developed in the Reactor Safety Study[14]–[16].

Over a period of time, one can expect (*a*) a shifting of licensing attention away from a set of hypothetical Design Basis Accidents towards the realistic assessment of a broad spectrum of potential accident sequences and (*b*) subtle design changes to reflect the engineering insights of the RSS. These changes, largely the legacy of the Reactor Safety Study, should lead, over the long term, to safer reactors and to greater public confidence.

Questions 1.8.2

1. Give the factors that involved an extension of the 'maximum credible accident' concept to the quantification of risk in the USA.
2. Describe the basic tasks which may be undertaken in a safety study for a nuclear reactor. Furthermore, give some indication of the type of organization which is involved in bringing such a study to a successful conclusion.
3. Discuss the findings of the WASH-1400 Safety Study and the implications for the future in the case of light water moderated reactors.

References 1.8.2

1. *Theoretical Possibilities and Consequences of Major Accidents in Large Nuclear Power Plants.* WASH-740, US Atomic Energy Commission, March 1957.
2. DiNunno, J. J. *et al. Calculation of Distance Factors for Power and Test Reactor Sites.* TID-14844, US Atomic Energy Commission, March 1962.
3. Pastore, Senator John O., Chairman, Joint Committee on Atomic Energy. Letter dated 7 October 1971 to Dr. James Schlesinger, Chairman USAEC. Lapp, R. E. Letters dated 26 February 1972 and 8 April 1972, to Congressman Chet Holifield, Joint Committee on Atomic Energy. Letters dated 7 April 1972 and 15 April 1972 to Dr. James Schlesinger, Chairman USAEC. Dr. James Schlesinger letter dated 17 August 1972 to Senator John O. Pastore, Chairman, Joint Committee on Atomic Energy.
4. '*Reactor Safety Study-An Assessment of Accident Risks in US Commercial Nuclear Power Plants.* WASH-1400 (NUREG-75/014), October 1975.
5. Starr, C., Levenson, M., and Wall, I. B. 'Realistic estimates of the consequences of nuclear accidents, Briefing for USNRC Commission, 18 November 1980. Transcript available from USNRC.
6. *Observations on the Reactor Safety Study.* Report prepared by Subcommittee on Energy and Environment of the Committee on Interior and Insular Affairs of US House of Representatives, January 1977.
7. Lewis, H. W. *et al.* 'Risk Assessment Review Group Report to the US Nuclear Regulatory Commission.' NUREG/CR-0400, September 1978.
8. *Federal Register*, Volume 45, No. 193, pages 65466-65477, 2 October, 1980.
9. *TMI-2 Lessons Learned Task Force Status Report and Short-term Recommendations.* NUREG-0578, July 1979.
10. Dao, L. T. L. and Carpenter, J. M. *Experimental Data Report for LOFT Nuclear Small Break Experiment L3-5/L3-5A.* NUREG/CR-1695, November 1980.
11. *Functional Criteria for Emergency Response Facilities.* NUREG-0696, February 1981.

12. Long, A. B. *et al. Summary and Evaluation of Scoping and Feasibility Studies for Disturbance Analysis and Surveillance Systems (DASS)*. Electric Power Research Institute Report NP-1684, December 1980.
13. *Guidelines for Control Room Design Reviews*. NUREG-0700, 1981.
14. Carlson, D. D. and Hickman, J. W. *A Value-Impact Assessment of Containment Concepts*. NUREG/CR-0165, June 1978.
15. Benjamin, A. S. *Program Plan for the Investigation of Vent-Filtered Containment Conceptual Designs for Light Water Reactors*. NUREG/CR-1029, October 1979.
16. Levy, S. *et al. Review of Proposed Improvements, Including Filter/Vent of BWR Pressure Suppression and PWR Ice Containments*. Electric Power Research Institute Report NP-1747, April 1981.

High Risk Safety Technology
Edited by A. E. Green
© 1982 John Wiley & Sons Ltd

Chapter 1.9

Learning from Experience

SECTION 1.9.1 Data Banks for Events, Incidents, and Reliability

B. K. Daniels

Introduction

Much of this book is concerned with the usage of various data in risk assessment, hazard analysis, and the continuing debate on the benefits and penalties which accompany modern high technology industries. The data are both qualitative and quantified, and range in quality from expert opinion to scientifically verified and statistically acceptable levels. In its most general sense a data bank should contain the full range of information potentially required by the user. New sources of data will need to be identified and regular contributions made to the stock. Old, outdated or misleading items must be identified and so labelled. The means of accessing the data bank by the user community should reflect the frequency and style of usage and the spectrum of data types and qualities requested.

This chapter considers the types of data which are needed by the user community of a data bank. The many sources of data are identified, and methods of collecting and classification considered. The data must be stored and accessible to the users and the limitations of alternative systems are identified. A large number of data banks exist and examples of their development, information content, and accessibility are given.

What Data are Needed?

The function of the Reliability Technologist, the Equipment Designer, the Emergency Services, and a Planning or Siting Authority are very different. Their needs for data in carrying out their functions related to safety in industry are also very diverse. The data range from generic failure rates through component performance parameters, codes of practice and hazard potential lists to statutory instruments. Thus, in determining the data stock for a particular bank, the function of the end users is a major factor. Not only is the discipline of the user of interest, the safety implications of the analysis to be performed also influence the depth, range, and accuracy requirements placed on the data stock.

Thus, if the data bank is only intended to serve the Reliability Technologist, the more selective can be the inputting of data to the bank and the more specific the responses made to an enquiry. Where there is a broad base of users to be served, then this inevitably leads to lower selectivity on data input, perhaps affecting quality, and less specificity in providing the response to a user request for data.

Whatever the discipline of the data user, there are two broad categories of data which may be requested. Traditionally the qualitative data have been provided by the standard library service and the training and experience of the users. Some qualitative data have also been supplied in book form, particularly generic failure rate data, physical and chemical constants, and engineering design information such as steam tables. Each practitioner has a preferred set of reference material which covers the day-to-day requirements of the job. However, the acquisition, maintenance, and retrieval of less frequently used material can be expensive, time consuming, and subject to error if carried out on an individual basis. This is the main function of a data bank to identify, acquire, store, and make available the collective expertise of all its users. 'User' is here taken in its most general sense to include provider, manager, expert reviewer, and end user of the data stock.

Qualitative Data

There are many sources of qualitative data which have application in the control of risk in industry. The collective experience in a particular field is frequently expressed as one or more of:

A code of Practice
Design Standard
Safety Principle
Statutory Instrument
Accident Report

Codes of Practice are prepared by a committee of experts in that field who produce the composite document based on their combined judgement. The aim is to provide a user with a set of rules. Following the set of rules is expected to lead to a safe design. Careful application of codes in the right environment is a useful, perhaps essential, first step in achieving an acceptable standard of safety. While conforming to a code will not guarantee the degree of hazard control achieved, the deliberate or accidental misapplication of a code by others can alert an assessor to those areas of a system which should be more closely studied.

As with Codes of Practice, Design Standards are based on the judgement of a group of experts. Standards are often used to achieve a safe equipment design and may cover such aspects as new material inspection, assembly standards, production line quality assurance and installation, operational, maintenance, and documentation procedures. The basic limitation of this approach is when a designer originates a new idea. New designs may lead to new risks. How is the new design to be compared to the experience gained with an older technology?

Already the judgement of experts has been mentioned in the compilation of Codes and Standards. It is frequently the case that these documents must be interpreted by the user for a particular application. Eventually such additional expert judgement, and infrequently the reasoning behind the decision, may appear as a revised or new standard. This is particularly true of Statutory Instruments, where a particular interpretation may be tested in a court of law. The outcome of that test may then become part of the Statutory Instrument either by incorporation in case law, where the legal system so provides, or by amendment to the instrument itself. Where a licensing authority exists for a particular type of plant, for example the Nuclear Installations Inspectorate in the UK, then there may be a reflected influence of the legal system on the need for plant operators to provide all relevant codes, standards, interpretations, adherences, and non-compliance as documentary evidence forming an integral part of the licensing procedure. Thus a major purpose of a plant oriented data bank would be to provide the current and historical status of the licence terms and satisfactory compliance.

In rapidly evolving technologies, such as electronics, there is an ever present need to update staff with the latest developments. Professional and trade journals aim to fulfil this role and may provide a useful intermediate level of qualitative data both in timing and in technical acceptability. Current awareness at the technical and practical level is an important consideration in the development of a data bank service.

In older industries, Safety Principles have been established over many years. For example, in the Nuclear Industry considerable use is made of automatic protective systems which are designed to shut down a nuclear reactor in the event of a plant failure which could have unsafe consequences. Three principles have been established:

1. No single equipment fault may invalidate the protection provided.
2. Each hazardous plant condition should be monitored by at least two parameters.
3. Power failure should move the protective system to a safer condition.

These principles have also been adopted in other industries[1].

That Codes, Standards, Principles, and Statutory Instruments offer insufficient coverage for the design and operation of modern industries is clearly evident in the incidents which occur around the world. Rasmussen[2], identifies the human role in the achievement of plant safety. He categorizes the overall risk to be expected from operation of the system as:

— Accepted risk identified by analysis.
— Risk due to incomplete knowledge, assumptions, and data.

— Risk due to management errors which allow the plant to diverge from analysed and assumed conditions.

Thus, the overall aim of a data bank should be to maximize the proportion of risk which can be analysed through better models, data, and total experience and minimize the contribution due to incomplete knowledge and management errors.

Rasmussen identifies the most important result of a risk study as the models, data, and assumptions provided for the plant management.

Thus qualitative data are an important starting point in any reliability, risk, or hazard analysis. Regular updating is essential. However, too frequent or copious updating of data can hide essential principles in a mass of ill-informed, undigested, and conflicting detail. To quote one example of data explosion, Kenton (1979)[3] reported over 1000 new documents on the Three Mile Island reactor 2 incident received in the five months following the incident by the Nuclear Safety Analysis Centre operated by the Electric Power Research Institute in California.

Quantitative Data

The major advantage which a Quantified Assessment can claim over one which is totally Qualitative is that the error prone process of subjective judgement is replaced by a more logical mathematical treatment of the relationships between safety, risk, and reliability. The analyst identifies potential hazards, ranks the safety implications and can compare the quantified result with the known recorded performance of similar plant around the world.

What does the analyst need to be able to carry out a quantified assessment?

1. An understanding of the plant, its processes, and equipment.
2. Knowledge of the environments in which the plant must operate.
3. Analysis methodology.
4. Suitable models for the plant, its components, management policies, and human factors.
5. A categorization of component stress levels.
6. Adequate data to plug into the models.
7. Plant synthesis methodology.
8. Targets for the plant safety.
9. Experience data for similar plants or equipment.

The first stage of any plant assessment should be a Capability Study. This serves two main purposes, it enables the analyst to become familiar with the particular plant, and, second, to compare it with the experience of similar plant. It can lead to the discovery of omissions, inconsistencies or overspecification in the system design at relatively low cost. The bulk of this work is carried out at a Qualitative level, but may include some Quantitative assessment to assist in sizing, or ranging based on acceptable engineering standards. A third, and often cost-saving attribute of the capability study, is in identifying those areas of the plant which could benefit from an analysis in depth. This identification may make a comparison of the estimated fail to danger rates of major system components in the particular plant configuration with the set targets for plant safety and with the achieved levels of safety and reliability performance for similar plants.

The detail of the analysis methodology is given elsewhere[4]. The choice of model to a large extent depends on the availability of data for the particular component. For example, Figure 1.9.1-1 shows the 'bath tub' curve which represents the change in failure rate with time for a typical plant component. It is usual for analysis methods to assume that all components are in their useful life phase, with a constant failure rate, and this leads to the negative exponential distribution model for reliability, see for example reference (4).

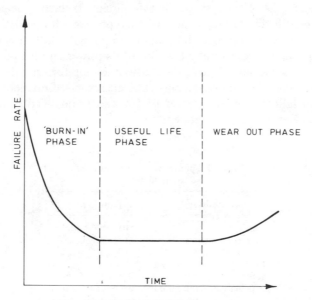

Figure 1.9.1-1 Typical failure rate curve

$$\text{Thus} \quad P_f = 1 - e^{-\theta t}$$

where θ is the constant failure rate in faults per unit time.

The analyst will therefore expect a reliability data bank to provide the appropriate values of θ for each system, or system component of interest. If the data bank provides a failure rate for the exact system, then at a coarse level no further analysis is necessary. Where, however, the data bank cannot provide the system level data, then it is necessary to decompose the system into sub-systems or components to a level where data are available, and then to synthesize a system failure rate. Thus, it may be necessary for the analyst to interact with a data bank to be able to construct the system model with the least analysis effort and the highest generic level of data.

Where safety is of paramount importance, the overall system analysis is unlikely to be completed by a generic study. It may be necessary to take into account the failure modes of components in a Failure Modes and Effects Analysis. The environment and stress levels in which the component is expected to function may need to be taken into account. The analyst will again turn to the data bank for these further levels of details. Take, for example, the electronic circuit component – a resistor. Table 1.9.1-1 shows how increasing level of detail may be brought into the analysis, and the effect that this can have on the component failure rate. Here it is presumed that the equipment is operating in a safety-related function, and the analyst has determined that a short-circuit failure in a particular resistor has systemwide safety-related effects. The generic failure rate can be modified by taking account of the type of resistor used, the purchasing quality which takes into account burn in of components, reduction in failure rate due to operating the resistor at a fraction of its power and temperature ratings within the particular circuit design and also allowing for only the short-circuit failure mode. This closer look at a particular electronic component shows how a reduction in failure rate by a factor of 500 can be justified in the particular environment for that one component.

Table 1.9.1-1 Failure rate examples; effect of some factors

Resistor	Failure rate
Generic	$0.05/10^6$ hours
Film construction	$0.017/10^6$ hours
Improved quality	$0.005/10^6$ hours
Low electrical stress and ambient temperature	$0.0006/10^6$ hours
Short-circuit failure mode	$0.0001/10^6$ hours

However, to be able to model the resistor failure rate dependencies on these various factors, a considerable data collection exercise has been carried out. The electronics field has particularly well-developed stress models and this is largely due to the standardization of component types across many manufacturers. Quality control in manufacture, and the rigorous development and application of procedures can lead to improvements in failure rates in the installed electronic component. The well-developed handbook produced by the US Department of Defense as MIL-HDBK-217C[5][6] covers most electronic component types and some light mechanical components. A continuing programme of updating the models, their parameters and the changing types of components and methods of manufacture ensures the component failure rate data keep pace with equipment performance with a minimum delay.

Where the assumption of constant failure rate cannot be made then models representing the reliability growth or wear-out phases of life are necessary. These failure distributions generally have more than one parameter and the data bank would be expected to provide the relevant model and parameter data for those components and equipments which must be represented in the system model. Unfortunately, the move away from a simple negative exponential model considerably complicates the problem of analysis and synthesis. At this time the most satisfactory technique available is simulation, which readily incorporates the majority of failure and repair distributions which are available. The more complex the distribution type, the more data generally are required to fix the parameters with any degree of confidence.

For use in Capability and Stress Analyses, the design functional envelope for the system and its components should be available to the analyst. It is a fundamental assumption for reliability data, that failure rates are only valid when components are operated within their design envelope. The effect of stress on failure rate may not be linear. With modern technologies there are substantial economies to be made by operating closer to hazard limits, and this is frequently coupled with larger inventories of hazardous materials. Using again the example of electronic components, quoting a Military or Commercial Standard, such as BS 9000, can be sufficient to establish the envelope. However, the analyst should also have access to these standards through the data bank. Where no such convenient standard exists, then the data bank must allow the envelope data to reside alongside the reliability and experience data. Not only are these envelope data essential to Capability and Stress Analyses, they serve a useful selection function in identifying existing data stocks for similar equipments and systems which can be used in analyses at a coarser level.

A particular example of the effects of functional parameters on the safety performance of a system is given in Figure 1.9.1-2. An explosion will occur in a chemical plant if a gas concentration rises above a certain level. The hazard is known, recognized within the plant, and an automatic protective system has been designed to detect the onset of the hazard and achieve a safe shutdown of the plant. A sensor is placed to detect the gas concentration. The sensor has an appreciable response time and works on a sampling basis. Also, there is a known problem in the difference between the set trip point due to instrumentation calibration errors. The shutdown action on the plant also takes a finite time. Figure 1.9.1-2 illustrates that for an achievable gas concentration rate of increase in a fault situation, the automatic protective system will fail to safely shut down the plant due to the cumulative effects of operating and trip level differences, errors, and delays. Thus, a full analysis of this example plant calls for data on rate

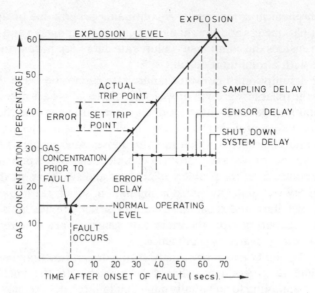

Figure 1.9.1-2 Response of a plant and its protective system

of rise of concentration (which may be calculated from physical data), equipment errors (which may be estimated from detail designs and component specifications), and system component response times. Where such data are available, they should be retained alongside the reliability parameters.

In establishing acceptable risk levels for a particular system the risk levels which have previously been accepted for comparable systems may be needed. There are two major components to establishing risk:

1. The events for comparable systems over some time period.
2. The population of comparable systems over the same period.

Since no two major plants are identical, there must be a subjective, or expert, judgement applied to which systems are comparable, the relevant time period, and which events are included (e.g. the commissioning period or the whole construction period may be excluded). The analyst should be aware that uncritical usage of any data bank in a quantified assessment is at least as error prone as the best qualitative assessment, and may be considerably worse. However 'good' the data bank, or 'clean' the source, the eventual responsibility for selection and application must reside with the end user.

Where do the Data Originate?

There is only one source of risk, hazard, safety, and reliability data and that is from experience of operating actual plant and equipment. If data are required, which are not currently in the data bank, then the only way to acquire them is to locate the appropriate equipment, obtain agreement to research the data, identify the records, retrieve the raw data, carry out some processing, preferably statistically based, and finally come up with the numbers first desired. Since all the data are derived from experience they must lag behind current technology. The better the quality required of the data the more systems must be researched. The longer the active or useful life of components for which the authenticated data apply the greater the lag will tend to be. This is the fundamental difficulty with safety and reliability data. The more stringent the requirements placed on equipment for low failure probability the more difficult it becomes to collect viable data. While there are large numbers of interactive constraints on the acquisition of good data, and not wishing to overemphasize the problems, a large number of data banks have foundered over the years because they have failed to acquire data on a timely basis and also maintain an acceptable degree of quality control.

There are basically two ways to determine what data are to be collected:

1. End user
2. Original source

Where the full requirements of the end user can be set down, then on-going data collection schemes can be acquiring data on a phased or continuous basis. The user, being aware of what data are currently available, and when new data types will become available, can take account of the lags. But what about the ownership of the data and particularly confidentiality? Where the supplier of raw data is in commercial competition with the user of the data, then the supplier will insist on some protection being applied to the data and may require legal and financial guarantees to be applied. Where a licensing authority is involved, it may be a condition of the licence that data are supplied of events having safety implications. The licensing authority may choose to circulate this information to all the licensees. This is particularly true of the Nuclear and Aerospace Industries. However, there is still no guarantee that statutory data collection schemes operate better than voluntary schemes. End user driven schemes have a major advantage in that it is known that all data collected will be used at least once. Historically, the majority of quantified data have been acquired on the end user principle, but the difficulty of establishing the requirements of end users should not be underestimated.

The opposite extreme in many respects to end user driven data collection is original source driven schemes. The data bank adopts a very broadly specified policy of the type of data to be stored. Whenever a new source of data becomes available, a decision is made on whether it falls within the policy areas, and suitable data are acquired and placed in the bank. The major advantage of this approach is that it can better deal with the unexpected request for new data.

However, there are problems with this approach. The total volume of data which may be acquired causes consequent difficulty in locating data for a user, and the overall quality control of the data stock. Historically, qualitative data have been acquired on this basis and this is particularly true of library systems.

Examples of data banks which fall into both categories are given later in this chapter, and an approach which combines the benefits of both systems is considered. The form in which the records exist will now be examined, and methods of retrieving the data highlighted.

Where a plant has potentially hazardous states then there is usually a record keeping system which is part of the operators' function. In many low hazard plants there is a maintenance function to be performed, and the engineering staff will keep records of work carried out, most probably to fulfil financial control, job planning, or wages calculation. These on-the-job and at-the-time records are the most authenticated sources of data available. Acquisition and processing of these records are discussed later.

An additional and often untapped source is the plant operators and maintenance staff themselves. This is a very fruitful source of qualitative data, and techniques are available for converting expert opinion into quantified data. Examples are briefly described in the next two sections.

Expert Opinion, Quantification by Paired Comparisons

While experts may not be able to quantify their experience directly, for example a single person is unlikely to give a reasonable failure rate for a particular system, their combined judgements can be effectively used to rank a given set of systems in failure rate order. It has been found that human judgement performs well in deciding that system 'A' will be more reliable (less hazardous, more available, of lower risk) than system 'B', based on the experience of similar systems, whereas human estimation of quantified values is generally poor. Given a number of systems, a single person can fairly readily produce a list ordered in sequence determined by some given criterion. In the method of Paired Comparison[7], the expert is presented with all possible combinations of pairs of systems and is asked to judge which system best meets the comparison requirements. Thus for 8 systems the expert would be presented with 28 pairs in random order.

The method provides for a number of assessors to be presented with the same pairs and the individual judgements are recorded. Normally at least two of the systems would have well-established data values associated with them and are used as calibration points in the final group ranking. The method takes account of differences in judgement by experts and can detect when the judgements are no better than a random selection, and where there are large differences in ranking. The sensitivity of the final result to the inclusion and exclusion of extremes in judgement can be tested both for an unexpected degree of agreement or excessive differences in judgement.

A number of field trials of the technique have been performed and are reported in reference (7). The method has also been used in safety and reliability assessment of industrial plant, and has proved particularly effective in establishing human factors related data.

Where there is need to establish data values, and access to other sources cannot be arranged, or there is time pressure, then the method of Paired Comparisons provides a useful technique. It combines existing data and the knowledge of experts in an ordered and cost effective manner. While the method does not rely on feedback of the results to the assessing experts, the individuals concerned have shown great interest in how their judgement compared with the overall result.

The Paired Comparisons method is a psychological technique. The method has worked extremely well in practice when the presentation of the pairs has been clear to the assessors, and when the systems are within the scope of experience of the majority of assessors. There is a tendency to divergence in individual judgements when material is presented to an international group, even though the level of expertise in the audience may be higher than an equivalent sized group of largely national experts. This effect may be due entirely to language and translation problems, or may be due to wider social differences.

Expert Opinion, Quantification by the Delphi Technique

The Delphi method[8], utilizes feedback of information to a group of experts to achieve a consensus opinion.

Typically a group of expert assessors are presented with a series of questionnaires that propose a structure of data and ask for estimates of data values. The assessors' response is anonymized in the combined result. The result of the first round of questionnaires is returned to each assessor, and the opportunity is given to revise any of the original responses. This revision, combination cycle goes through typically up to five interactions to finalize the group judgement.

Thus the Delphi technique provides an organized method of soliciting, grouping, correcting, and improving the judgements of individual experts. The method is well tried and several hundred exercises have been published.

This technique has been used by the Reliability Sub-Committee of the Nuclear Power Engineering Committee of the Institute of Electrical and Electronic Engineers in the USA to produce a guide to reliability data for nuclear power stations[9].

The main problem in the nuclear industry is the general high reliability of equipment and the correspondingly low number of safety-related faults which occur. To construct a data bank purely based on nuclear plant data would take a large number of years. However, the data are required now for assessment and licensing. The Delphi technique offered the means of using the data available from all sources and a group of over 200 experts to derive initial failure rates and failure modes for components used in nuclear power plants, and particularly those used in the automatic protective system. In the first round of the exercise, a questionnaire was prepared and issued to the assessors asking for estimates of failure rates, modes and factors, the expertise of the assessor, and the sources of any recorded data used in or supplied with the judgement. The assessors were encouraged to add comments on the estimates provided, and also on the format of the questionnaire.

All the responses were analysed by a coordinating group and an overall summary produced. The individual responses were confidential, and assessors were encouraged to provide estimates for only those plant components where they judged they had significant expertise.

The second round consisted of a revised questionnaire which took account of the comments received in round one and in response to the overall summary. The assessors were given the opportunity to confirm or change earlier estimates. A total of four iterations were made and the final results are presented in book form.

The complete exercise produced some 1500 independent failure rates. Table 1.9.1-2 summarizes the data classifications and stocks. For each failure rate is given a recommended figure bounded by lower and higher confidence limits and an overall maximum to be used under all applications. Each major system component is presented in a single chapter, and is hierarchially classified into sections, each having a number of sub-sections and up to two further sub-levels. Thus, component classification structure is by a five level hierarchic code, each component type having one–five level code. See, for example Figure 1.9.1-3.

Unfortunately this coding is not unique since annunciators can also appear under code 9.22. This is a common problem in coded systems.

For each component type a number of failure modes are identified. The classification of failure modes is again hierarchic, with three major sub-headings each potentially having a number of independent sub-failure modes. An example is given in Figure 1.9.1-4 for 'Blowers.' At the lowest level, the failure modes are assumed independent and derived failure rates are obtained for Degraded Mode by summing the rates for low, high, and erratic airflow failure modes. The total failure rates provided by this work exceed those given in Table 1.9.1-2 which excludes all the derived failure rates.

In a similar manner, failure rates are derived for higher level component codes by summing the failure rates for all lower level independent failure modes at lower component code hierarchic levels.

The failure rate may be specified in terms of elapsed time units of failures per million hours, and/or in terms of failures per million cycles/demands.

The Delphi technique is well proved, and the massive programme initiated by the IEEE Std. 500–1977[9] demonstrates its organizational effectiveness in generating large sets of data in short time periods and at reasonable cost (Figure 1.9.1-3). However, the IEEE have also initiated a long-term programme of data collection which will initially back up the expert opinion and eventually displace

Table 1.9.1-2 Summary of data in IEEE Std 500–1977

Chapter title	Number of sections	No. of sub-sections	Failure modes		Environ-mental factors	Failure rates	
			Time	Cycle		Time	Cycle
Annunciator modules	7	57	6	0	3	34	0
Batteries and chargers	3	15	12	1	4	64	1
Blowers	4	9	8	0	3	41	0
Circuit breakers, interrupters and relays	7	19	13	7	3	98	42
Motors and generators	4	18	18	1	3	93	3
Heaters	3	16	3	3	3	31	1
Transformers	7	16	11	0	3	656	0
Valve operators and actuators	3	8	7	9	4	61	88
Instruments, controls, and sensors	27	107	31	12	6	211	7
Cables, joints, terminations, and penetrations	4	8	14	0	3	81	0
					Totals	1370	142

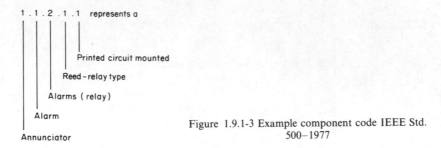

Figure 1.9.1-3 Example component code IEEE Std. 500–1977

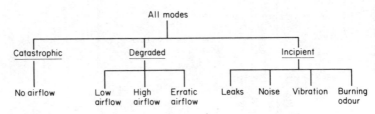

Figure 1.9.1-4 Example failure modes from IEEE Std. 550 – 1977 for blowers

it. Used with appropriate caution, this set of generic data is a valuable addition to the total stocks of reliability data.

Extracting Data from Records

Many records are produced in the normal operation of a modern high technology plant. There have been numerous attempts to use these records as the basis of a Reliability Data Collection scheme. A small proportion of these schemes are highly successful, but the majority fail for a variety of reasons.

Even in a well organized record system some effort is needed to extract the relevant information and to carry out processing to produce reliability data. Effort costs money, so a sponsor for the work must be found. If the data bank is to sponsor the work then it must generate income, and the only product it has to sell is the data bank service. Therefore, in some way the end user of data must pay for data collection, in addition to the storage and retrieval overheads of operating the data service. It may be that the organization which has the records is willing to sponsor the data collection directly because the resulting Reliability Data are of immediate use and no other suitable source is available at the same or lower cost. If the end user of the data is not the original source, then problems of confidentiality can occur. Where cost sharing of a data bank operation is among a number of potential commercial rivals, then the problems of data exchange will need to be addressed. The exchange of data is discussed in more detail in the next section.

Assuming that data collection funding can be found, there remains the problem of data quality. Frequently in data collection schemes which span a number of years of plant records there is a significant variation in quality. For example, the introduction of a bonus incentive scheme may mean that previous good descriptions of the equipment failure, repair action, parts replacement and testing can no longer be obtained because the maintenance team earns more bonus by doing work other than filling in forms. The reverse may also occur, that as part of an introduced bonus scheme allowance is made for record keeping and standards improve. Both problems have occurred in the work of the U.K. National Centre of Systems Reliability. How can such variations be handled? Sometimes it is simply not possible. Siting the data collection activity at the actual plant and having access to the staff completing the records can improve the standard of the data. However this proximity can in itself create new problems, such as the accusation of 'spying' on the activities of the people concerned. Most successful schemes are cooperative ventures between data bank staff and the personnel owning and operating the plant sourcing data.

Thus, it is not sufficient to have funding. Wholehearted support for the collection activity must be provided by the full workforce of the target plant.

How is this level of support achieved? The personal interest of a member of staff of the plant is a good starting point. The ability to demonstrate the direct benefits which have accrued to the same, or perhaps a competing organization from similar data collection schemes is an additional aid – within the limits of commercial confidentiality. The friend in the plant is a local focus to stimulate and maintain interest in the work and for anticipating problems at an early stage in their development. Thus, good management at the start, and throughout the running of a scheme is essential. How is this interest maintained over a period of weeks/months/years? Experience shows that intermediate results should be provided to both the sponsoring and data sourcing organizations on a regular and timely basis. Effective means of achieving this are the preparation of summary reports which can be supplied to the management, and equally important, the regular contact between data bank staff, the record producers, and end users.

Closely related to data collection costs and good scheme management is the choice of data bank staff for data collection work. It has been found essential to have at least one member of the team who has a detailed knowledge of the particular plant equipment, processes and record systems. In a small scheme, this will most probably be the local liaison person. In large schemes the diversity of equipment types and processes may require the involvement of a number of specialists who can advise when requested. In this case, the specialists may come from both the data sourcing organization and from the data bank. A further requirement is that a member of the data bank staff is assigned to advise on procedures related to inputting, reporting, and searching the data bank. On a long-term scheme a member of the data bank would be located at the source site for the duration of the exercise. Thus, each data scheme will normally involve a minimum of two persons, one from the sourcing organization and one from the data bank.

A number of approaches have been adopted by the National Centre of Systems Reliability in the UK to allocate staff to data collection. Based on the general guidelines indicated, data bank staff have considerable engineering, chemical, and allied experience. Therefore, the dialogue between site and data bank can be at a suitable technical level. The person assigned to a particular scheme may be an experienced member of the data bank permanent staff, or a less experienced person who is supervised closely by the source liaison staff and data bank staff. Very successful data schemes have been operated for a number of years using engineering students during their industrial experience period which forms part of their professional qualification requirements[10].

Having touched briefly on the problems of running a scheme, the types of data which can be found are now considered. Taking as an example Reliability Data, there are three basic sets of data which need to be obtained. First, the total population of equipment on the site for which data are to be collected needs to be established. This Inventory can of itself be a difficult task where records span a number of years, equipment is replaced, goes out of service, or extra equipment is brought into service. Second, the operating experience of each Inventory item will need to be established. This experience may be measured in terms of calendar time of in-service, out-of-service, and repair periods, or perhaps in terms of demands or cycles of operation. Third, the failure modes, if any, of the equipment and the working environments will need to be listed and related to the experience data. From these three base data can be calculated failure rate, failure distribution, environmental and working stress dependencies which constitute the elements of reliability data.

The general principles apply equally to extracting any data from records. The success of such a scheme, measured in the quality of data obtained, depends heavily on the record keeping standards.

The best reliability data have been obtained from data collection schemes where the opportunity has arisen to advise on the type of records to be kept. In this case, the standard plant documentation is revamped to provide the additional prompts which are essential to establishing experience, failure mode, and environment data. These specific reliability data collection exercises call for a very high level of support from the plant workforce, and the data bank also must take great care to feed back the results of the work on a regular basis.

Data Exchanges and Access

Data collection is a costly exercise, therefore there is a great incentive to make the maximum use of any fully processed data. However, since the data collected are an asset, there is understandable reluctance to free publication. Where the funding of data bank operation is on a cost sharing basis, then the benefit to each organization is access to wider ranges of data than each individual organization could afford, from its own resources.

Perhaps there should only be one central data bank to which all data is supplied and from which it is possible to purchase access rights. This approach has worked well in the library field where there are also requirements to deposit copies of new material to establish copyright. An increasing number of exchanges are taking place on indexes, abstracts, and bibliographic type data. For example, Table 1.9.1-3 lists a selection of the data-bases available to users on a multinational basis. To date, it has not been technically possible to hold all these data-bases in one central location. Nor has there been the political will to start the development of the computer system, or ease the flow of data across national boundaries. The market for such a service is global, therefore it has made technical and economic sense to provide smaller centres, and improve the communications between the user location and the many data centres to be accessed.

Table 1.9.1-3 Partial list of on-line data-bases

AB/INFORM	Business management and administration
AFEE	Water pollution and related subjects
AGREP	Permanent Inventory of Agricultural Research Projects in the European Community
AGRICOLA	Worldwide journal and monographic literature in agriculture and related subject fields
AGRIS	Agricultural science and technology
AIM/ARM	Educational and vocational instructional material
AIR POLLUTION	Air pollution effects, control and prevention
ALBO	Register of Italian barristers and solicitors
ALICE	Catalogue of books in the Italian language – all subjects
ALUMINIUM	World Aluminium Abstracts
AMERICA	Recent USA and Canadian history and current affairs
APILIT	Worldwide petroleum industry literature
APIPAT	Citations to refining patents, mainly US
AQUALINE	All aspects of water research and engineering
AQUATIC SCIENCES & FISHERIES ABSTRACTS	Marine biology and limnology
ARIANE	Building and construction engineering data bank
ART	Art bibliography
ARZ-DB	Drugs and active ingredients data bank
ASI	US Government Statistic Index
BAUFO	Building research projects
BEI	British Education Index – British educational journls
BIBLIO-DATA	Books and serials published in the Federal Republic of Germany
BID	Bibliography on data processing and law
BIIPAM	Engineering and related subjects
BIOSIS	All aspects of bio-sciences in particular Biology
BIPA	Banque d'Information Politique et d'Actualite: French political chronicle, speeches, press-cuttings
BUL-L	Documentation on linguistics (Germany)

Table 1.9.1-3 (*Continued*)

CAB	Commonwealth Agricultural Bureaux: agricultural sciences and related subjects
CAB ABSTRACTS/ ANIMALS	Animal and veterinary sciences and nutrition
CANCERLINE	Consists of Cancer Literature abstracts, International documentation and selected current projects in progress
CA REGISTRY NAME FILE	Comprehensive file of chemical compounds
CASEARCH	Pure and applied chemistry including thesaurus
CAS NAMES	Chemical compound names
CBAC	Chemical Biological Activities
CDI	Comprehensive Dissertation Index for USA academic doctoral theses
CEC	Education of gifted or deprived children
CEE	Case-law of the uropean Communities' Court of Justice
CETIM	Centre Technique des Industries Mécaniques; mechanical engineering
CHEMCON	Chemical Abstracts
CHEMLINE	Dictionary of chemical substances with CAS Registry numbers.
CIS INDEX	USA Congressional publications
CIS-BIT	Centre International d'Infornations de Sécurité et d'Hygiene du Travail; work safety and health
CISI-AFO	Financial and stock exchange data for companies quoted in France.
CISI-BIR	Information on environmental research
CISI-BRUIT	Acoustic characteristics of building material
CISI-ELECNUC	Characteristics of nuclear power stations worldwide
CISI-MEDIAM	Readership data on French medical media
CISI-MEDIAP	Readership data on French publicity media
CISI-OFCE	French companies and external trade
CISI-PI	OECD main economic indicators
CISI-SCE	OECD external trade statistics
CISI-TRANSINOVE	Transferable technology
CIVILE	Case-law of the civil section of the Rome Supreme Court of Appeal
CLAIMS/CHEM	USA Chemical Patents
CLAIMS/GEM	USA Electrical, Mechanical Patents
CLINPROT	Clinical Protocols; investigations of anti-cancer agents
COMPENDEX	Computerized Engineering Index; all branches of engineering
CONFERENCE PAPERS INDEX	Scientific and technical papers presented at conferences, etc.
CONFERENCE PROCEEDINGS INDEX	Conference proceedings received by British Library
CONSTA	Case-law of Italan State Council
CORTEC	Case-law of Italian Audit Office
COSTIT	Case-law of Italian Constitutional Court
CRECORD	Current activities USA Congress
CRIS	Agriculture related research in USA
DAFSA	Company ownership
DKF	Documentation on transportation vehicles and automobiles
DKI	Plastics, rubber, fibres
DOMA	Mechanical and production engineering
DOTTR	Summaries from the legal documentation institute of the CNR, Florence
DRE	Information on electro-technology

Table 1.9.1-3 (*Continued*)

DZF	Optical, photographic, biomedical, precision engineering
EABS	Euroabstracts; study reports and publication of results of research financed by the CEC, the European Coal and Steel Community, and Euratom
ECDIN	Data bank on pollution-causing chemical substances
EDB	Energy Information Data-Base
EDF-DOC	Electricité de France; electrical engineering
EIS	Industrial Plant Statistics in USA
ELECOMPS	Electronic Components – factual data
ELSPECS	Electronic component specification and approvals
ENDOC	Environmental centres in Community
ENERGYLINE	Energy and energy-related subjects
ENEX	Register of experts on environment.
ENREP	Current environment research projects in Community
ENSDF	Evaluated Nuclear Structures Data File, including decay data for all isotopes
ENVIROBIB	Environment Periodicals bibliography
ENVIROLINE	Environment-related issues
EPIC	Programs for calculation of physical properties of chemical compounds
ERGODATA	Biometric data bank
ERIC	Education Related report and periodical literature index
EUROCOPI	Data bank on data processing programs in physics, chemistry, engineering, etc.
EURODICAUTOM	Multilingual terminology data bank
EUROFILE	Inventory of data-bases and banks available in Europe
EXCERPTA MEDICA/ EM BASE	Bio-medical sciences literature
F&S INDEXES	Business and Economics
FEDERAL INDEX	US Government activity and documents
FRANCE ACTUALITE	Articles from the French Press
FRANCIS	Current information in social and human sciences
FSTA	Food Science and Technology Abstracts
GEODE	International bibliography on earth sciences
GEODIM	Geophysical bibliography on the Alps
GEOREF	Geological References
GRANTS	USA Federal, State, Local Governments, commercial and private grant aided programes
HEALTH	Health planning and administration
HISTORICAL ABSTRACTS	USA History to 1945
IALINE	Industrial processing of agricultural products
IBIS	Data on production and distribution companies in 130 countries
IDIS FILES/SOCIAL MEDICINE	Social and industrial medicine and public health
INFORM	Business management periodical literature
INIS	International Nuclear Information System
INKA-ASTRO	Astronomy and astrophysics
INKA-CONF	Conference announcements in energy, nuclear science, aeronautics, astronautics, space research, physics, mathematics, and astronomy
INKA-CORP	Corporations and affiliations in same fields as above
INKA-DATACOMP	Data compilations in energy and physics
INKA-HEP	High-energy physics data-base
INKA-MATH	Mathematics and related subjects data-base
INKA-MATHDI	Mathematical didactics data-base

Table 1.9.1-3 (*Continued*)

INKA-NUCLEAR	Nuclear science data-base
INKA-PHYS	Physics and related fields data-base
INKA-PLASMA	Plasma physics and technology data-base
INKA-SPACE	Conference papers on aeronautics, astronautics, and space research
INKA-SURVAC	Surface and vacuum physics data-base
INPADOC-IFS	Patents: INPADOC Family File data-base
INPADOC-IPG	Patents: INPADOC Patent Gazette
INPI	Patents: register of INPI (F)
INSPEC	Physics, electronics, computing, mathematics
INTERNATIONAL STATISTICS	Non-USA economics, finance, production
IPA	International Pharmaceuticals Abstracts
ISDS	International Serials Data Service
ISMEC	Information Service in Mechanical Engineering
ITIS	Business data for 90 countries
KOMPASS	Commercial information about 55 000 French companies
LANGUAGE & LANGUAGE BEHAVIOUR ABSTRACTS	
LC MARC	Books and serials catalogued by US Library of Congress
LEXR	Italian legislation; regional
LEXS	Italian legislation; national
LIBCON/E	USA Library of Congress English Language material listing
LIBCON/F	USA Library of Congress non-English material listing
LIT-KRAN	Hospital management, organization, and economics
	Shipping information. Lloyd's register, incidents, movements.
MEDLARS/MEDLINE	All fields of medical literature
MERLIN-GERIN I	Business and management
MERLIN-GERIN II	Electronics and electricity
MeSH	Medical Subject Headings; controlled vocabulary for MEDLARS/MEDLINE
MERITO	Case-law of Italian tribunals
METADEX	Metallurgy and related areas of science and technology
METEOROLOGICAL ABSTRACTS	Meteorology and geoastrophysics
NAR	Nutrition Abstracts and Review
NASA	All aspects of aerospace and related fields
NEW YORK TIMES	News and related information
NORIANE	Documentation on standards
NPL-SGTE	Thermodynamics
NTIS	US Government-sponsored research, development, and engineering reports
NUCLEAR SCIENCE ABSTRACTS	Nuclear Science and related areas
OCEANIC	All aspects of ocean studies
ORLIS	Documentation or urban and regional planning
PA	Psychological Abstracts; behavioural issues concerning humans and animals
PAPERCHEM	Paper industries periodicals references
PASCAL	General coverage of science and technology
PATENT REGISTER	European patents and applications
P/E NEWS	Petroleum and Energy daily and weekly sources

Table 1.9.1-3 (*Continued*)

PENALE	Case-law of criminal section of the Rome Supreme Court of Appeal
PETROLEUM ABSTRACTS	Oil and gas exploration
PHARMACEUTICAL NEWS INDEX	Important new developments in pharmaceutical, cosmetic, and medical fields
PLURIDATA	Chemical data bank
POLLUTION	Pollution and related subjects and issues
POPINFORM	Population index and contraceptive technology
PREDICASTS	Weekly business and USA financial current awareness
PSYCHOLOGICAL ABSTRACTS	Psychology
RAPRA	Technical, commercial, and research aspects of rubber and plastics
RBUPC	Register of Research in British Universities, Polytechnics, and Colleges; projects in physical, biological, and social sciences
REBI	Bibliographic file on Italy of the Rome Supreme Court of Appeal
REBIS	Bibliographic file on other countries of the Rome Supreme Court of Appeal
RESEDA	Agricultural economics
RINGDOC	Chemical, medical and pharmaceutical information
RIV	Abstracts from law periodicals by the Rome Supreme Court of Appeal
RSWB	Townplanning
RTECS	Registry of Toxic Effects of Chemical Substances
SANI	Register of Italian industrial, commercial, and other companies
SAE ABSTRACTS	Transport industries
SANP	National defaulters file in Italy
SAOE	Information on Italian export/import companies
SATELDATA	Satellite technology, i.e. Performance and launch data
SCISEARCH	All natural sciences and techniques
SDILINE	Selective Dissemination of Information on MEDLINE
SDIM	Documentation on metallurgy and metals
SDOI	Italian foreign supply and demand file
SDON	Italian supply and demand file
SGB	Finance and economics
SIBB	Official acts on joint-stock companies in Italy
SIBV	Italian financial market and stock exchange
SOCIAL SCISEARCH	Social and behavioural sciences
SPACECOMPS	Electronic components for spacecraft
SPIN	Physics Information Notices
SSIE	Smithsonian Science Information Exchange, life, physical, and social sciences
SUSIS	Sports and Sport Sciences
TELEDOC	Information on telecommunications, electronics, etc.
THERMODATA	Thermodynamic values of elements, components, and alloys in minerals
TITLEX	Titles of Italian decrees in force from 1860 to today
TIT1	Other Italian decrees, 1860–1939
TIT2	Other Italian decrees, 1939 to today
TITUS	Documentation on textiles
TOXBACK	Back files of TOXLINE
TOXLINE	Documentation on toxicology and related subjects
TRIBUT	Case-law of the Italian Central Commission on taxes
TULSA	Oil and gas exploration

Table 1.9.1-3 (*Continued*)

UK MARC	Monograph literature Legally Deposited in the UK
URBAMET	Townplanning
WORLD PATENT LATEST	Current information from WPI
WPI	World Patents Index: Patents in all fields
ZDE	Electro-technical engineering, data processing, control engineering

The political emphasis within the European Economic Community is to provide a communications facility which charges for service on the throughput of data, rather than the traditionally established distance tariffs for telephone, telegraph, postal, and goods services. The Euronet international data transmission network implements this communication link between the user and data-base. Similar services are well established elsewhere, and the national telecommunications authorities are providing an ever increasing range of data transmission services which interlink to form international network.

Additional communications channels via satellite links are now becoming practical, and the first business communications satellite has been placed in orbit by a group which includes a computer manufacturer, communications channel provider, and end users. Similar facilities are proposed for the direct broadcasting from satellites into the home and which will provide many more television channels and the potential for international viewing. The trend is away from physical transport of documents over long distances to the electronic transmission of the information and the local conversion via the television set to the well accepted visual formats or as a last resort to printed paper.

With the advent of the new data transmission networks, there is also considerable effort being applied to the storage and retrieval of data via computer systems and the establishment of a display technology suitable for mass production and usage. Table 1.9.1-4 shows the cost per character of storage for a number of existing storage technologies. New systems are being introduced which make use of unused bandwidth in broadcast television channels. For example, in the UK the television broadcasting authorities are providing access to hundreds of pages of data on each television channel using the lines of the picture which currently carry no visual information except to separate successive picture frames. The full data-base is cycled every few minutes and each suitably equipped TV set can select a page for display. The data pages can also be used to sub-title foreign language programmes, provide special services for the handicapped, and newsflashes while viewing the normal TV programme. Using very similarly equipped TV sets, the UK telecommunications authority, British Telecom, has devised an information system which displays data as graphics or text on the domestic TV and uses the home telephone line as communications medium to the computer holding the data. Selection of the data is via a small keypad equivalent to the push-button style telephone, and pages are hierarchially accessible via

Table 1.9.1-4 Costs of data storage on various media (order of magnitude)

Storage technology	Cost range (per character)	Size range (million characters)	Random access time (seconds)
Register	£10	0.001	0.000 000 01
Cache memory	£1	0.01	0.000 000 01
Main memory	£0.1	1–10	0.000 000 5
Bulk memory	£0.01	10–100	0.000 001
Magnetic disc	£0.001–£0.0001	10–10 000	0.1
Mass archive	£0.0001–£0.000 01	100 000–1 000 000	10
Video disc	£0.000 000 1	10 000–1 000 000	0.1–10
Paper	£0.000 001	1–100 000 000 (400 km of shelves)	1000–100 000

multi-digit absolute or relative codes. Display may be of an index page or a destination data page and users are charged for page accesses. This service is now used by travel agencies to provide the latest information on availability of holidays, price changes, travel services and delays, and it is possible to order and pay for goods listed by the system.

With wider public access to data on-line, there is concern over the diversity of computer storage systems which are employed. The user currently may need to become familiar with several command languages. There is an approach to standardization now taking place. In Europe, the group of data-base suppliers on Euronet, known as DIANE, have devised a common command language which is appropriate to bibliographic style data-bases. A number of the data-base providers have now implemented this language and the Europen Community is promoting the extension of the standard language[11]. There are proposals to extend the coverage of this language to quantified data-bases.

Professional institutions such as the Institute of Electrical Engineers (IEE) in London, have been active in providing current awareness information services for their members. Typically this began with printed abstract documents intended for libraries and information sections, but has now progressed to operating on-line computerized data-bases. The INSPEC service listed in Table 1.9.1-3 is originated by the IEE and installed in many data centres worldwide. Part of this service includes the establishing of links with other professional bodies and standardization authorities. However, in such multidisciplinary fields as Reliability Technology, Safety, Hazard Control, and Risk Assessment the coverage provided by each institution can be insufficient, and the combined coverage by all existing institutions ineffective. To promote the newer disciplines, centres of competence are necessary and the UK National Centre of System Reliability was set up to include such a role. Its activities include the running of the data bank SYREL described later.

In the field of reliability data, an active group of workers have formed the European Reliability Data Association which was formally constituted in 1979. This work is now sponsored by the European Economic Community Joint Research Centre at Ispra, Italy[12]. Other work at that Research Centre includes the setting up of a reliability data system, ERDS, for the European Nuclear programme of pressurized water reactors[13]. The ERDS system comprises four main sections:

1. Component Event Data Bank
2. Abnormal Occurrences Reporting System
3. Generic Reliability Parameter Data Bank
4. Operating Unit Status Reports

The source of data for this system will be the national reliability data systems operating in France, Italy, the UK, Germany, Sweden, and it is also the aim to include data from the USA. Currently the feasibility of translating the various national coding systems into a common ERDS code is being researched and sample data sets processed.

Historical Review of a Data Bank Development

The Data Bank operated by the National Centre of Systems Reliability in the United Kingdom evolved originally from a fault reporting system set up by the United Kingdom Atomic Energy Authority (UKAEA). The original scheme was qualitative and although some comparisons and trend analyses were possible, its main purpose was to provide a communication link between sites and design offices on safety-related incidents.

By 1967 the need was recognized for the development of a system to collect plant data on a more detailed basis for reliability analysis work. In order to make the maximum use of existing expertise and requirements, a Reliability Data Working Party was set up with representation from the UKAEA, the Central Electricity Generating Board, the South of Scotland Electricity Board, and the Inspectorate of Nuclear Installations. The working party in conjunction with their respective organizations, set up a number of pilot schemes for data collection based upon a provisional specification produced by Eames[14]. By the end of 18 months, the working party had determined a practical compromise on the detail and level of plant reporting and had demonstrated that the stored raw data could be analysed by computer to provide output relevant to reliability analysis. The UKAEA further developed the scheme and continued its operation as part of their overall reliability work. When the Systems Reliability Service (SRS) was formed in April 1970 to apply the reliability technology in other industries the Reliability Data Bank was further elaborated to take account of the requirements of the SRS Associate Members. This culminated in the active and operational system SYREL, first described by Ablitt[15]. SYREL is an end user determined system.

The SYREL system, although refined in detail, has remained in its same basic form up to the present time and is still used by SRS for a large part of its reliability data handling. SYREL comprises two computerized data stores:

1. The Generic Reliability Data Store containing processed reliability data obtained from published information, manufacturers, data collection exercises, etc. It is used primarily to provide data for engineers carrying out system reliability assessments.
2. The Event Data Store, contains raw data from specially structured collection schemes set up in association with plant operators. From this part of the system a detailed operating, maintenance, and test history of each item under surveillance can be obtained. The data are used to provide feedback to the plant operators and as a basis for high quality data used in reliability analysis.

The SYREL Reliability Data Store currently contains some 11 000 entries of summarized reliability data appropriate to a wide range of component parts. The system deals with 800 enquiries a year from among the Associate Members of SRS for specific component reliability data. The growth of data is currently 1500 entries per year.

The overall information flow is shown in Figure 1.9.1-5 and the input data is Figure 1.9.1-6. The basic structure developed in SYREL has been adopted in other Reliability Data Banks[16]. All information is hand coded by engineering staff before it can be entered in the data bank. An example of one of the codes used is given in Figure 1.9.1-7. The 14 digit item code is a two level hierarchic code. At

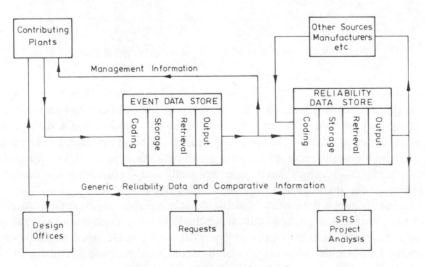

Figure 1.9.1-5 The Syrel data bank

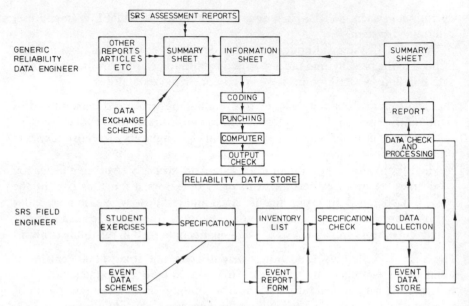

1.9.1-6 Reliability data input-flow diagram

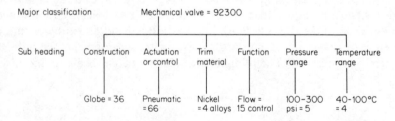

Figure 1.9.1-7 Example Syrel 14-digit item code 92300.36.66.4.15.5.4

the top level the code has five digits thus allowing 99999 different component basic types. Each top level code can have up to six sub-headings and the sub-heading may be unique to that one code. Each sub-heading has a value in the range 1–99 for sub-heads 1, 2, 4 and in the range 1–9 for the remainder. The value 0 indicates a no-entry condition for that sub-heading. Some 15 000 main codes, sub-headings and values have been assigned, and Table 1.9.1-5 indicates a selection of the main codes under which data are available.

Output from the data bank is available at a number of levels from original event lists through to grouped data from the Reliability Data Bank. An example of a single set of Reliability Data is given in Table 1.9.1-6. Grouped data have been published in booklet and microfiche formats for Associate Members of the Systems Reliability Service.

Table 1.9.1-5 List of main headings under which data are held in SYREL (at Dec 1979)

Absorber, radio frequency	Circuit breaker
Accelerometer	Clutch
Accumulator, mechanical	Coil, electromagnetic
Action wheel	Coil, inductance, adjustable
Actuator	Coil, inductance, fixed
Agitator	Coincidence unit
Aileron system	Comminutor
Air conditioning plant or unit	Compass
Air fuelling boom system	Compressor
Alarm system	Compressor bleed system
Altimeter	Compressor unit (motor and compressor)
Ammeter	Computer
Amplifier, electrical, static	Condenser, vapour
Amplifier, electronic	Conductivity indicator
Annunciator	Conductivity sensor
Attenuator	Connector, electrical
Attitude indicator system	Contact pair
Audio device	Contactor
Barometer	Control system
Battery, electrical	Controller, 3 Term
Battery charger	Controller (parameter unspecified)
Bearing	Converter, electrical, rotary, single frame
Bellows	Converter, electronic
Belt	Converter, torque
Blade, turbine	Conveyor
Blower/fan (excl. motor)	Cooling system
Blower/fan (incl. motor)	Cooling tower
Board or panel	Counter
Board, printed circuit	Coupling, drive
Boiler or steam generator	Crane or hoist
Boiler/steam generator tube	Crushing or grinding mill
Bracket	Crystal, electronic application
Brake cylinder unit	Current control unit (instrument use)
Brake shoe or band or block	Cyclone separator
Brake system	Cylinder (piston engine type)
Brushgear, electrical	Data communication terminal (DCT)
Burner	Data logger
Bus-bar	De-aerator
Cabinet or cupboard	De-icing system
Cable, electric, power	Delay, line
Cable, light current and signal	Demineralization plant or unit
Cam	Density indicator
Capacitance probe	Diaphragm
Capacitor, fixed	Die
Capacitor, variable	Diecasting machine
Carburettor	Disc, magnetic storage
Card punching machine	Discriminator
Casing	Drilling machine
Casting ejector (diecasting)	Drum, winding
ghain	Dryer
Chemical analyser	Ducting
Chemical sensor	Ejector
Chimney or stack	Element, heating, electrical
Chuck	Elevator control system, aircraft
Chute	Evaporator

Table 1.9.1-5 (*Continued*)

Explosive device	Landing gear system
Ferrite device (including winding)	Lathe
Filter, mechanical	Lead screw, power transmission
Fire control system	Leak detector
Fire detection system	Level controller
Fire detector	Level indicator
Flame failure detector	Level sensor
Flap system	Level transmitter
Flow controller	Lift
Flow detector	Lifting tackle
Flow indicator	Light intensity meter
Flow recorder	Light source, electrical
Flow sensor	Lighting system, aircraft
Flow transmitter	Lighting unit, non-electrical
Flow trip unit	Line, overhead transmission
Food processing plant	Lining, other than brake lining
Force indicator	Linkage
Fork lift truck	Louvre
Frequency and load controller system	Lubricating system
Frequency meter	Machine tools
Fuel system, aircraft	Magnet
Furnace	Manifold
Fuse, electrical (element)	Memory (not incl disc, drum or tape)
Fuseboard, distribution	Microelectronic circuit, digital
Gear assembly	Microelectronic circuit, linear
Gear wheel or pinion	Mixer
Generator, electrical	Motor, electric
Generator set (combination)	Motor, hydraulic
Gimbals	Motor, pneumatic
Gooseneck (diecasting)	Mounting
Governor	Neutron chopper
Grab	Nozzle
Graph plotter	Nuclear radiation instrument
Gyroscope	Nuclear radiation sensor
Heat exchanger	Nuclear reactor
Heater, electrical	Nuclear reactor control mechanism
Heater, non-electrical	Nuclear reactor fuel element
Heating and ventilating system	Nuclear reactor pressure vessel
Hose	Nuclear reactor refuelling machine
Housing, mechanical	Nut or bolt or washer combination
Humidifier	Orifice, other than measuring
Humidity indicator	Oscillator or signal generator
Hydrogen producing plant	Oscilloscope
Ice maker	Oxygen breathing equipment
Ignition system (other than aircraft)	Pin
Impulse line	Pipe joint and fitting
Insulator, electrical	Piping
Int combust engine, reciprocating/rotary	Piston
Inverter, electrical, rotary	Plunger, piston and non-return valve
Inverter, electrical, static	Pocket, thermocouple
Jacket	Position indicator
Joint, electrical	Position indicator system
Joint, mechanical (not piping or hose)	Power station unit
Ladder	Power supply, electrical, non-rotating

Table 1.9.1-5 (*Continued*)

Power supply, hydraulic	Soldering iron
Power supply, pneumatic	Solenoid
Power supply system, electrical	Spacer or shim
Precipitator, electrostatic	Sparking plug
Pressing machine	Spoiler system, aircraft
Pressure control system	Spring
Pressure controller	Stabilizer system
Pressure indicating system	Starter, motor, electric
Pressure indicator	Starter system internal combustion engine
Pressure sensor	Steering unit
Pressure transmitter	Stem tube, ship
Pressure vessel, unfired	Strain gauge
Printing machine	Structural member
Processor	Superheater
Propeller or impeller	Switch, electrical
Protection system, automatic, aircraft	Switchgear, electrical
Protection system, auto., nuclear plant	Tank
Protective device, mechanical	Tape punching machine
Pulley	Tape reader
Pulse unit, electronic	Tape unit
Pump	Telephone equipment
Pump unit (motor and pump combination)	Teleprinter
Punched card reader	Television equipment
Purging system	Temperature controller
Q meter	Temperature indicator
Radar equipment	Temperature indicating system
Ram	Temperature recorder
Reactivity control element	Temperature sensor, other than thermocouple
Recorder	Temperature transmitter
Refrigeration device	Temperature trip unit
Refrigeration unit or system	Termination, cable/wire, electrical
Register	Thermistor
Relay, other than power protective	Thermocouple
Relay power protective	Timer or programmer
Resistor, fixed	Toggle mechanism (diecasting)
Resistor, variable	Transceiver or receiver or transmitter
Rivet	Transformer
Rope or cable, wire	Transmission unit, mechanical
Rotor, mechanical	Trap
Rudder, system	Trip unit, electrical
Scale	Trip unit, electronic
Scanner	Tube bank, boiler or steam generator
Scrubber	Turbine
Seal	Turbogenerator (combined set)
Semiconductor, diode	Turn and bank indicator
Semiconductor, special type	Typewriter
Semiconductor, transistor	Valve, electronic
Servomechanism	Valve, mech (incl. actuator where applic.)
Shaft	Valveholder, electronic
Shielding, non-radiation	Vehicle, rail transport
Shielding, radiation	Vehicle, road transport
Silencer	Velocity indicator
Slip ring	Voltage regulator
Smoke density indicator	Voltmeter

Table 1.9.1-5 (*Continued*)

Volume indicator	Weighing machine
Water hardness meter	Weighing system
Wattmeter	Winch
Waveguide and microwave equipment	Wiper, windscreen
Wear ring	

Total experience (December 1981) 12 000 000 000 000 hours

Table 1.9.1-6 Example of syrel output

Item description 92300.36.57.0.00.0.0	Value. Mech. (incl. actuator where applic.) Globe. noc Motorized	Data serial no. 1537
Location	/vo2/	
Application	Average (industrial)	
Information type	Field reported	
Data year	1969	
Failure mode	Wontclose	
Failure cause	All	
Number of items	5.30E 01	
Number of failures	2.00E 00	
Failure rate (calender time)		
Mean	4.30E 00 fault/million hours	
Lower limit	5.21E-01 falut/million hours	
Upper limit	1.55E 01 fault/million hours	
Confidence band	95.00 per cent	
Experience (calender time)		
Averate	8.77E 03 hours	
Total	4.65E 05 hours	
Failure rate		
Mean	1.02E–03 fault/cycle	
Lower limit	1.23E–04 fault/cycle	
Upper limit	3.67E–03 fault/cycle	
Confidence band	95.00 per cent	
Experience		
Average	3.72E 01 cycles	
Total	1.97E 03 cycles	
Information source	29	
Comment	Valves, globe, motorized	

The SYREL Data Bank, although still effective, has naturally been under survey for possibilities of further developments and improvements. Specific aspects have been highlighted which it was felt warranted an improved system for Reliability Data.

(a) The Reliability Data Store is a highly structured system of coded and numerical values and although this has reasonably stood the test of time, it lacks flexibility in terms of current standards and requirements.

(b) The SYREL software system, in terms of computer technology, is old, becoming difficult to maintain and lacks the flexibility associated with modern Data Base Management Systems.

(c) SYREL is a batch oriented system, whereas current demands lead to the need in many instances for a fast interactive system of data entry and retrieval.

(d) The SYREL system cannot deal with general descriptive information or new data structures.

In addition to development for Reliability Data, there is an increasing requirement to store risk, hazard, safety, and accident data. In these fields there is no well established user requirement for the type of data retrieval best fitted to the analyst. The proposed improvements to the SYREL system were further reviewed to include the new types of data. A further list of improvements is proposed.

(e) The ability to input data without pre-coding.

(f) On-line interactive definition of data structures, addition of new data values, and association with existing data stocks.

(g) Two orders of magnitude increase in total data storage capability from the current SYREL 30 megabytes to about 3000 megabytes.

(h) Alerting procedures for expert assessors that new data of interest has been entered into the system.

From 1976 to 1980 extensive trials of software systems were undertaken, and pilot systems set up to establish the practicality of developing and operating the expanded SYREL data bank[17].

The new data bank will support the following types of data:

1. Events
    Accident
    Hazard
    Reliability
    Transport
2. Inventories
    Equipment
    Hazardous materials
    Hazardous plant
3. Source Documents
    Expert opinion
    Accident investigations
    Technical reports
    Press cuttings
4. Generic Data
    Derived from events and inventories
    Identified within source documents
    Physical constants
    Hazardous substances and interactions

5. Abstracts
   Technical literature
   Personalities
   Information sources and users

Users will have interactive access to selections of the data controlled by the security processes within the system. For example, a group of users within an organization providing event data would have access to that event data, but others would only have access to the Genetic Data derived from those events combined with events from diverse sources. In this way confidentiality of data can be maintained.

These developments will produce a full range information system in which End User determined data and Original Source data co-reside. New Original Source data are examined by the data bank system for content which matches a profile supplied by a range of expert assessors, called reviewers. The reviewer can then read and research these filtered data and determine the value of the new information. This review process is the start of a gradual migration of data from Original Source towards End User determined requirements.

Just as for codified data bank systems, the reviewer can assign data to a particular structure, either represented as a code value or related by descriptive material added by the reviewer. Thus the existing SYREL structure for Reliability Data can be represented exactly in the new data bank. Moreover, where finer structure to the Item Code becomes necessary, for example to include 10 sub-headings or allow 1000 code values under a sub-heading, then this can be added into the system by a reviewer at the time the revised structure is found to be necessary. This dynamic data structuring is increasingly essentially within the Reliability Data, and where little defined structure exists for risk, hazard, or other data types, allows the on-going development of a single or parallel structure to the data.

Questions 1.9.1

1. What are the basic types of data which are used in safety assessments?
2. Consider a working environment known to you and answer the following:
   (i)   Why, for safety purposes, do we want data and in what form should they be?
   (ii) Describe how you would obtain data appertaining to this working environment.
   (iii) How would you apply these data in your environment to identify those areas in which significant safety related improvements could be made?
3. What are the factors which would lead you to set up your own data bank? Describe the data, types and quantities it would contain and how a user would access the data.

References 1.9.1

1. Stewart, R. M. and Hensley, G. *High Integrity Protective Systems on Hazardous Chemical Plants.*
   European Nuclear Energy Agency, Committee on Reactor Safety Technology, Munich. May 1971.
2. Rasmussen, J. 'Notes on human error analysis and prediction'. In G. Apostolakis, S. Garriba, G. Volta (Eds). *Synthesis and Analysis Methods for Safety and Reliability Studies.* Plenum Press, New York. 1980.
3. Kenton, J. E. *Nuclear Safety Analysis Centre Industry Report*, Sept 1979. Electric Power Research Institute, Palo Alto, California, USA.
4. Green, A. E. and Bourne, A. J. *Reliability Technology.*
   Wiley Interscience, 1972.
5. US Department of Defense *Reliability Prediction of Electronic Equipment.* MIL-HDBK-217C, April 1979.
6. US Department of Defense *Reliability Prediction of Electronic Equipment.* MIL-HDBK-217C, Notice 1, May 1980.
7. Hunns, D. M. and Daniels, B. K. 'Paired comparisons and estimates of failure likelihood.' *Design Engineering*, January 1981. IPC Science and Technology Press Ltd, Guildford, UK.
8. Linstone, H. A. and Turoff, M. *The Delphi Method: Techniques and Applications.* Addison Wesley Inc, MA, 1975.
9. IEEE *Guide to the Collection and Presentation of Electrical, Electronic and Sensing Component Reliability Data for Nuclear-Power Generating Stations.*
   IEEE Std 500-1977, New York. Distributed by Wiley Interscience, John Wiley & Sons Inc., Chichester, UK.
10. Cannon, A. G. 'Reliability – event data retrieval from industrial installations'. National Conference on Reliability, Nottingham, UK. 1977.
11. Negus, A. E. *EURONET Guideline: Standard Commands for Retrieval Systems.* Final report on a study carried out for the Commission of the European Communities. Published by INSPEC, the Institution of Electrical Engineers, London. 1977.
12. Ullman, A. 'The future of EuRa DatA'. In *6th Advances in Reliability Technology Symposium*, Bradford. B. K. Daniels (Ed.). Published as a report NCSR R23 (UKAEA) Warrington, UK. 1980.
13. Mancini, G. 'Feasibility study for the European Reliability Data System'. In *3rd European Reliability Data Bank Seminar*, Bradford. B. K. Daniels (Ed.). Published as a report NCSR R24 (UKAEA) Warrington, UK. 1980.
14. Eames, A. R. *Data Store Requirements Arising out of Reliability Analyses Report.* AHSB(S)R138, UKAEA, Warrington, UK. 1967.
15. Ablitt, J. F. *An Introduction to the SYREL Reliability Data Bank.* SRS Report SRS/GR/14, UKAEA, Warrington, UK. 1972.
16. Bello, G. C. and Bobbio, A. 'A reliability data bank in the petrochemical sector'. In *3rd European Reliability Data Bank Seminar*, Bradford. B. K. Daniels (Ed.). Published as a report NCSR R24, UKAEA, Warrington, UK. 1980.
17. Daniels, B. K. 'Reliability data the LIXIBOSS way'. In *6th Advances in Reliability Technology Symposium*, Bradford. B. K. Daniels (Ed.). Published as a report NCSR R23, UKAEA, Warrington, UK. 1980.

High Risk Safety Technology
Edited by A. E. Green
© 1982 John Wiley & Sons Ltd

SECTION 1.9.2 *Reliability Measurement and Confirmation*

G. L. Crellin and A. M. Smith

Introduction

Formulating an estimate of equipment and/or system reliability is, by definition, the process of reliability measurement. Placing some form of bound on that estimate is the process of confirmation. Reliability measurement is not an easy task; confirmation is more difficult yet. Both are often essentially qualitative in nature and are based on various aspects of engineering judgement and supplemental mathematical analysis.

The problem of measurement and confirmation involves one or more of the following critical steps:

1. Development of a model that accurately portrays the equipment and/or system including both the hardware and operation considerations. In theory, then, this model includes dependencies, interfaces, repair/replacement, operator actions and the like.
2. Development of analytical tools that permit the appropriate mathematical manipulation and interpretation of models and/or data.
3. Establishment of a data-base that provides credible values of failure rates, repair/replacement times, and other pertinent characteristics of the equipment/system under evaluation for use in Nos. 1 and 2 above.

Measurement encompasses pure prediction on the one extreme, and total knowledge on the other extreme. Confirmation, correspondingly, ranges from engineering judgement (a non-mathematical expression of 'confidence' which is often taken as 50%) on the one extreme to some form of classical statistical demonstration at a high confidence level (say 95%) on the other extreme. One always starts at the prediction stage where an attempt is made to model the equipment/system and apply generic data to arrive at 'best estimates'. As the equipment/system in question passes from the design to hardware stage, one attempts to introduce data from that specific equipment/system into the measurement process, and arrive at reliability estimates that now reflect the actual hardware and operational performance situation. It is this latter process that we specifically address in this chapter.

Approaches to Reliability Measurement/Confirmation

We assume, in this section, that the product cycle has now progressed to the point where engineering definition has been completed and the hardware is in some stage of test (e.g. prototype, qualification, acceptance), installation and check-out or perhaps even operation. The question, therefore, is how do we use this actual equipment/system specific data in the reliability measurement and confirmation process. In addressing this question, we recognize that most (if not all) real-world cases are such that sufficient data have *not* been accumulated as yet to permit the straightforward application of classical statistical techniques to arrive at reliability values with associated high confidence levels.

We will thus focus our attention on three methodologies of possible use where classical statistical techniques are of limited value:

1. Trend Plotting
2. Defect Flow Analysis
3. Bayesian Statistics

Trend Plotting

As test information becomes available, it is only reasonable to ask from the very outset just what it is telling us about the initial equipment reliability and where we might be headed in the future (i.e. reliability growth expectation). In 1962, J. T. Duane[1] produced a not-too-widely publicized report which suggested a model for answering these questions. Duane's model stated the following:

$$\theta_\Sigma = \frac{F}{H} = KH^{-\alpha} \qquad (1.9.2-1)$$

where θ_Σ = cumulative failure rate
$\quad H$ = cumulative test hours
$\quad F$ = failure occurrences during H
$\quad K$ = constant determined by initial results
$\quad \alpha$ = reliability growth

Duane's basic postulate was that for a relatively fixed design and a relatively fixed level of test/de-bug/improvement effort, α would remain constant. In practice, the test data (H and F) are plotted on log–log paper and produce a straight line (Figure 1.9.2-1) with slope α and an initial value of $\theta_\Sigma = K$ for the value of H where the first failure occurs.

Duane applied the log–log plotting scheme to several equipment types (e.g. DC motors, gas turbines, steam turbines) and found this straight-line effect to be a consistent pattern (Figure 1.9.2-2).

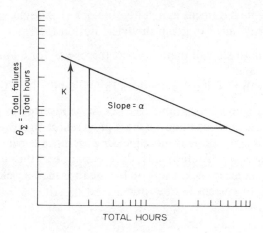

Figure 1.9.2-1 Duane growth model

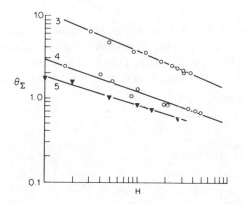

Figure 1.9.2-2 Examples from Duanes original data

Since then, several investigators have continued to confirm the accuracy of the model[2], [3].

The authors have applied this technique, for example, to spacecraft systems, electronic and mechanical components and nuclear plant control rod systems, and have consistently observed the model validity for Duane's original postulate. However, it should be noted that when *significant* design and/or programmatic changes occur, the process must revert to a new beginning for the model to retain its validity. In such cases, a new K always results – and quite often, a new α may result. In this latter regard, it is interesting to note that management often decides to 'back-off' on testing as unit production increases; invariably, α will change for the worse leaving management in a state of shock (I thought we had a mature product!) since they failed to grasp the basic notion that:

1. failure rate is not really all that constant (i.e. $\alpha \neq 0$) for a long time (unless I force it to be), and
2. reliability growth to full maturity does in fact take a long time.

The curves of Figures 1.9.2-1 and 1.9.2-2 can be inverted to plot cumulative MTBF (Mean Time Between Failure); these plots reflect an upward trend (if that is the case) and this is emotionally more pleasing for management visibility (down is 'bad', no matter what!). Additionally, differentiation of the cumulative failure rate with respect to time in the expression for Duane's model yields an expression for instantaneous or current failure rate:

$$\theta_{H_i} = (1 - \alpha)\theta_\Sigma \qquad (1.9.2\text{-}2)$$

where θ_{H_i} = instantaneous failure rate at some time, H_i, of interest,

$$\text{or MTBF}_{H_i} = \frac{1}{(1 - \alpha)} \text{MTBF}_\Sigma \qquad (1.9.2\text{-}3)$$

Thus, the expected θ_{H_i} for the next test is *less* than θ_Σ, or conversely MTBF_{H_i} is *greater* than MTBF_Σ. This is a useful relationship to calculate when questions arise as to the *current* expectations of failure (e.g. when incentive fees may be riding on the next test). The use of MTBF is illustrated on Figure 1.9.2-3.

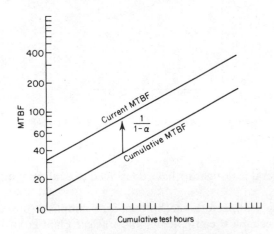

Figure 1.9.2-3 Duane model with MTBF

A special application of the Duane model can be employed for test planning purposes. Assume that some value of θ_{H_i} or MTBF_{H_i} has been set as a goal that must be achieved before the equipment is committed to operation. If K and α have been established, then a unique value of H_i can be determined for the given goal of

θ_{H_i} or MTBF_{H_i}. The test programme must then be planned to achieve H_i – and the programmatic commitment of resources can be clearly established. If these are prohibitive, management can clearly realize the severity of the problem, and may elect to embark upon a significant redesign to effect a dramatic change in α. The authors have witnessed cases where α was either 0 or even negative, and such changes were clearly in order.

In summary, the Duane model is a powerful tool for reliability measurement. While confirmation of the results is essentially a 'best estimate' figure of merit, experience has shown that the results are very descriptive of equipment reliability expectations. As an engineering tool, the Duane model is highly recommended for reliability measurement purposes.

Defect Flow Analysis (DFA)

The DFA technique was originally developed as a method for examining historical failure populations. The basic premise is that succeeding generations of similar equipment will possess like failure characteristics unless something very deliberate is done to preclude such occurrences. This, in turn, necessitates a clear understanding of previous failure histories – what happened and why; what went wrong and what went right. In later applications, DFA was also utilized on a real-time basis for evaluating failure characteristics as they developed. Thus, DFA does not lead to evaluations of failure rate, *per se*. Rather, its objective is to study and understand the nature of the failure process so that appropriate corrective measures can be invoked.

In general, defects are introduced into a product cycle at any one of a thousand places.* Once there, they will stay in the product flow until detected and corrected. The concept of defect flow is a very real phenomenon which occurs whenever a defect is not instantly caught at its point of introduction. Thus, the flow must contain various check points or screens whose job is to deliberately seek out the presence of defects. In developing the DFA strategy, two significant principles evolve. The first, called Basic Quality Level (BQL), is a measure of the number of defects that are introduced during a product development, manufacture, and operation cycle. A high BQL is achieved when a diminishingly small number of defects are introduced. The second called Screen Effectiveness Level (SEL), is a measure of the effectiveness of an organization to systematically search for and eliminate those defects which have been inadvertently introduced.

Thus, a trade-off between BQL and SEL can be developed to allocate resources effectively. If a high BQL can be achieved, the necessity for costly screens diminishes. Conversely, a low BQL necessitates a very effective screening system (a high, and perhaps costly, SEL). The proper mix is very product and organization dependent, but a conscious recognition is required to avoid the

---

* Defect is the underlying cause, problem or error that can potentially lead to some form of equipment malfunction. Failure is simply the manifestation of the defect.

situation where both a low BQL and low SEL occur simultaneously. DFA helps to provide information to management in formulating an effective strategy for this mix.

In using DFA, the analyst takes a file of recorded failures (hence known defects), and systematically evaluates them to determine the population characteristics or dimensions (i.e. a type of Dimensional Analysis) and the manner in which the screen system failed or functioned (i.e. Screen Analysis). Let us illustrate these points with one example each from real life. Several firms have designed and built spacecraft systems. We are about to build a new and improved system. As an aid to developing the test programme, the following questions are posed:

1. What is the minimum level of assembly required to detect defects (Dimensional Analysis)?
2. How strongly can traditional qualification tests be relied upon to qualify the design (Screen Analysis)?

Note that these question are *not* normally the type of information that is captured during the course of the failure analysis process. Thus, the DFA technique requires the further (intuitive) development of information from (*a*) a knowledge of the system design and operation, and (*b*) an evaluation of the recorded facts surrounding the failure. This usually requires the use of a team of two to three people to effect a meaningful analysis. Once formed, the team is limited only by its creativity in developing any number of dimensional and screen questions that may be studied.

For the first question above, we select a group of flight failures from existing experience and categorize each failure point as shown on Figure 1.9.2-4. The population size in Figure 1.9.2-4 is 78 failures, but the analysis has been done with as few as 25 failures, and as many as 200 failures. The resulting message is very significant in that 50% of the failures required a higher level of assembly (subsystem or system) before the defect was even present and available for detection. For our new system, we should thus anticipate that component level test alone is not sufficient; that a significant portion of the potential defect population will reside in the assembled product only! While this may not be welcome news

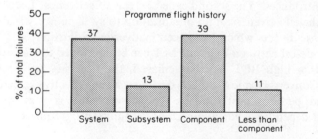

Figure 1.9.2-4 Defect occurence versus level of assembly

(higher level testing is expensive), it is none the less very important to understand this particular dimensional property and plan accordingly. To neglect this finding could lead to serious problems ultimately. (Note that not all products necessarily exhibit this particular dimension, and that primary emphasis on component test only may be fine *if* you have the proper knowledge to guide that decision.)

For the second question, we select a group of system acceptance test failures from an array of various spacecraft systems (here designated A to E). Each recorded failure is evaluated to determine if it could have been detected in the previously conducted qualification test programme. The results in Figure 1.9.2-5 indicate a very disturbing pattern in that sizeable percentages of the acceptance test failures are shown to originate with defects that could/should have been detected in the qualification test. Such a finding should (and did) trigger further analysis to understand the possible deficiencies in the qualification test screens. While not shown here, small sample sizes were a major factor since various problems of variability (e.g. dimensions, material properties, process control, test conditions, etc.) were not permitted to reveal themselves fully. In the larger sample population (i.e. 100%) encountered in acceptance test, these problems were revealed.

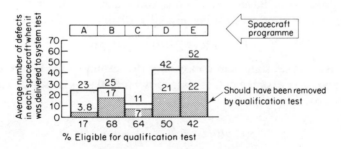

Figure 1.9.2-5 Qualification versus acceptance testing

This analysis thus suggested that conventional qualification tests do not qualify the design; subsequent testing continues to play a major role in equipment qualification and a conscious effort must be put forth to design acceptance tests with this in mind. Failure to do so could well lead to sizeable 'holes' in the screen system for our new project.

This discussion of DFA is admittedly brief. But the basic strategy and means of implementation have been defined. References 4, 5, and 6 present additional information. As an engineering tool in the reliability measurement process, DFA can make a significant contribution to the achievement of high reliability. When used in conjunction with Bayesian Statistics and Trend Plotting, the anatomy of equipment reliability can be thoroughly dissected and understood.

Bayesian Statistics

While Bayes's equation has been utilized in problems of inference for over two centuries, its use in reliability measurement and confirmation has been limited until the past 20 years. Interest in its use as an approach to reliability measurement and confirmation grew from the recognition of the inadequacy of classical statistical methods to cope with limited data availability in high reliability situations. Its major attraction is the acknowledgement that measurement and confirmation have an initialization in predictions as validated by the use of predictions in decisions of system design and test planning.

Briefly stated, Bayes's equation provides a formal analytic basis for updating the confidence one has in achieving a reliability objective as new data and information are received. This is accomplished by first expressing the initial 'confidence' in reliability achievement as a probability density function (called the prior or *a priori*) on the reliability parameter of interest. Commonly, this parameter is the failure rate, and, while other parameters may be employed, our discussion in this section will be limited to failure rate. A revised 'confidence' is then expressed as an updated probability density function (posterior or *a posteriori*) on the same parameter. This revised function is obtained by simple multiplication by the likelihood ratio of the new data.

In equational form, Bayes's equation is nothing more than a reformulation of the multiplicative (product or joint) law of probability theory, which states:

$$p(\theta \cdot D) = p(\theta)p(D/\theta) = p(D)p(\theta/D) \qquad (1.9.2\text{-}4)$$

Thus, if θ is a logical statement concerning (for example) the failure rate value achieved and D is a statement of data observed, then one solves to find the updated probability as

$$p(\theta/D) = \frac{p(\theta)p(D/\theta)}{p(D)} \qquad (1.9.2\text{-}5)$$

Generally, in areas of reliability interest, it is always possible to find $p(D/\theta)$ whenever the conditions of acquiring the data are known and if an underlying stochastic process is adopted as appropriate. For the case of measuring or confirming failure rates, a useful model is the Poisson process. However, caution is encouraged, since $p(\theta/D)$ and $p(\theta)$ generally concern the failure rate to be experienced in the field, and the data may not have been generated from conditions totally representative of the field environment. This usually will not change the stochastic model, but may necessitate consideration of scaling the data appropriately. (That is, one must relate a test hour to a field hour.) In what follows, the data scale is assumed to be unity.

If $p(D/\theta)$ is easily agreed upon as being

$$p(D/\theta) = \frac{(\theta\tau)^k}{k!} \, e^{-\theta\tau} \qquad (1.9.2\text{-}6)$$

where τ = the total unit time
and k = the observed number of failures

then $p(D)$ can be found with little difficulty if one can agree to the prior, $p(\theta)$. This is accomplished by employing the extension rule of probability theory which requires (see, for instance, Reference 7 or 8):

$$p(D) = \int_0^\infty p(D/\theta)p(\theta)\,d\theta \qquad (1.9.2\text{-}7)$$

This leaves the establishment of $p(\theta)$ as the matter of primary interest. If the prior probabilities are known, Bayes's equation becomes a simple, error-free truism. If the prior probabilities are not 'completely known' but are carefully and cautiously selected, estimated, and employed, then Bayes's equation can be successfully utilized to extend inferences of reliability measurement and confirmation. If the prior probabilities, being unknown, are capriciously and arbitrarily adopted, then the results may be inaccurate and misleading. Thus, the key to appropriate and successful application of the Bayesian method is the rational choice of the prior probability distribution.

EXAMPLE 1.9.2-1

Two lots of apparently identical electrical components are in stock. The supplier of one lot assures that his items have a failure rate of $\theta_1 = 0.001\,\mathrm{h}^{-1}$ (that is, on average in the long term, one item fails in 1000 hours) while the supplier of the other lot assures that his items have a failure rate of $\theta_2 = 0.005\,\mathrm{h}^{-1}$. Unfortunately, the identity of which lot is which has been lost.

After randomly selecting one of the lots, a single item is selected from it and is successfully tested for 200 hours. On the basis of this evidence, Bayes's equation allows one to assess the chance that the selected lot contains the components of lower failure rate. (The reader is invited to test his intuition by guessing at this point which lot has been selected.)

Since one of the two indistinguishable lots was picked at random, it is reasonable to assume *a priori* (that is, before testing evidence is available) a probability assignment of $p(\theta_1) = p(\theta_2) = 0.5$. The new information obtained through the test (under the assumption of constant failure rate) is the result of a Poisson process, so the likelihoods are (where k is the number of failures and τ is the test time):

$$p(D|\theta_1) = \frac{(\theta_1\tau)^k}{k!}e^{-\theta_1\tau} = e^{-0.2}$$

and

$$p(D|\theta_2) = \frac{(\theta_2\tau)^k}{k!}e^{-\theta_2\tau} = e^{-1.0}$$

The probability of obtaining the test data is obtained through application of the extension theorem of probability theory:

$$p(D) = p(\theta_1)p(D|\theta_1) + p(\theta_2)p(D|\theta_2) = 0.5\,(e^{-0.2} + e^{-1.0})$$

Finally, all necessary inputs having been obtained, Bayes's equation yields the following posterior results:

$$p(\theta_1|D) = \frac{p(\theta_1)p(D|\theta_1)}{p(D)} = \frac{e^{-0.2}}{e^{-0.2} + e^{-1.0}} = 0.689\,97$$

while

$$p(\theta_2|D) = \frac{p(\theta_2)p(D|\theta_2)}{p(D)} = \frac{e^{-1.0}}{e^{-0.2} + e^{-1.0}} = 0.310\,03$$

Thus, Bayes's equation has provided a rather simple and direct answer to the question of which lot was selected: it is over twice as likely that the better lot has been chosen. Information in this form is suitable for deciding whether to (a) accept, (b) reject, or (c) continue testing.

The reader is invited to confirm that had the component failed at $\tau = 200$ hours, then $p(\theta_1|D') = 0.308\,01$ and $p(\theta_2|D') = 0.691\,99$.

EXAMPLE 1.9.2-2

This example illustrates the application of Bayes's equation to the Bernoulli process. There are two lots of apparently identical pyrotechnic devices in stock. (A pyrotechnic device is one which, when activated, either fires successfully or fails to fire.) One lot is from a supplier that assures his items have a reliability of $R_1 = 0.9$ (that is, in the long term, 9 out of 10 items will fire successfully) while the other lot is from a supplier that assures his items have a reliability of $R_2 = 0.8$. Again, the identity of which lot is which has been lost.

A lot is chosen at random, and a single item is randomly selected for testing. After successfully firing the item, it is desired to assess the chance that the selected lot contains the higher reliability components. Again, this assessment will be based on Bayes's equation.

Since one of the two indistinguishable lots was picked randomly, it is reasonable to assume a prior probability assignment of $p(R_1) = p(R_2) = 0.5$. The new information obtained through the test is usefully represented by a Bernoulli process so the likelihoods are thus:

$$p(D|R_1) = R_1 = 0.9$$

and

$$p(D|R_2) = R_2 = 0.8$$

The probability of obtaining the test result is thus calculated as

$$p(D) = p(R_1)p(D|R_1) + p(R_2)p(D|R_2) = 0.85$$

Finally, Bayes's equation yields the following results:

$$p(R_1|D) = \frac{p(R_1)p(D|R_1)}{p(D)} = \frac{0.45}{0.85} = \frac{9}{17}$$

$$p(R_2|D) = \frac{p(R_2)p(D|R_2)}{p(D)} = \frac{0.40}{0.85} = \frac{8}{17}$$

The chance that the lot is the higher reliability items is only slightly higher. In this case, it might be decided to engage in further testing.

If the test had resulted in a failure, then

$$p(R_1|D') = \frac{p(R_1)p(D'|R_1)}{p(D')} = \frac{0.05}{0.15} = \frac{1}{3}$$

and

$$p(R_2|D') = \frac{p(R_2)p(D'|R_2)}{p(D')} = \frac{0.1}{0.15} = \frac{2}{3}$$

It is twice as likely the lot contains the lower reliability items.

Bayesian Priors and their Construction

Much of the Bayesian literature is filled with examples of what results when priors of particular mathematical form are chosen. These examples are instructive, but the central point should not be lost:

A prior should reflect what is actually known or truly felt about the parameter of interest.

If it does not, then it would be irrational to expect the posterior results to reflect how one feels after the receipt of the data.

The 'non-informative' prior

One way of trying to avoid controversy about the assignment of strong prior belief to the Bayesian process is to attempt to assign 'equal probability' assignments to each competing hypothesis associated with the probability framework under analysis. Simply put, we might say that any failure rate between zero and a value so large as to be almost beyond belief (infinity) is 'equally likely'.

In the case of failure rate measurement, a non-informative prior is described by assigning a uniformly distributed density over the range of possible failure rates. (For this case such a density is considered an 'improper' prior since it will not integrate properly as required by probability theory.) Thus, with $p(\theta) = c\,d\theta$:

$$p(\theta|D) = \frac{c\theta^k e^{-\theta\tau}d\theta}{\displaystyle\int_0^\infty c\theta^k e^{-\theta\tau}\,d\theta} = \frac{\tau(\theta\tau)^k e^{-\theta\tau}\,d\theta}{k!} \tag{1.9.2-8}$$

This posterior density has a mean and variance of

$$E(\theta|D) = \frac{k+1}{\tau} \text{ and } \sigma^2(\theta|D) = \frac{k+1}{\tau^2} \qquad (1.9.2\text{-}9)$$

The mode occurs at $\hat{\theta} = k/\tau$ which identifies with the classical 'maximum likelihood estimate'. Similarly, the upper probability intervals,

$$p(\theta < \theta^*|D) = \int_0^{\theta*} \frac{\tau(\theta\tau)^k e^{-\theta\tau} d\theta}{k!} \qquad (1.9.2\text{-}10)$$

relate directly to the classical confidence intervals formed from the χ-square distribution with $2k + 2$ degrees of freedom.

It has been noted that the posterior distributions reached in the above manipulations have 'best estimate' properties somewhat akin to those of classical maximum likelihood estimates. If one were interested in such properties alone, this type of prior might have some justification – but the counter argument might be: 'Why not simply use "classical" estimates?'

If many data were available, the total posterior would, of course, 'home in' around its true parameter value and numerous such posteriors could then be convolved together to obtain *system parameter* distributions. However, if test data are scarce (as is often the case), the posterior distributions (and their convolutions) remain so diffuse that inferences drawn from them tend to be unsatisfactory and of little utility.

To create a richer inference will then require the definite emplacement of information favouring some competing hypotheses more than others within the structure of the prior. Some of these approaches are discussed in what follows.

Subjective priors

When priors are established subjectively, an individual or group of individuals considered expert and informed concerning the subject matter under analysis is asked to give assessments of values of the parameters(s) of interest:

1. Where a group is being queried a single best estimate (of, for example, an item's failure rate) may be requested from each member and some sort of composite distribution generated from or fitted to the several responses.
2. A 'best estimate' and a 'tail value' may be requested from each respondent: 'the failure rate is probably 0.001 per million hours, but there is a 10% chance that it is greater than 0.005'. A typical technique of fitting the twin values to a distribution of appropriate family (in this case the gamma) is then practised. An example of this is presented below.
3. Each respondent may be requested to give a distributional spectrum of responses: 'There is a 95% chance the failure rate is greater than 0.001, a 75% chance that it is greater than 0.005, a 50% chance it is greater than 0.007, . . . a

5 % chance it is greater than 0.010.' Distributions of discrete form can be fitted to such 'multiple choice' responses with relative ease; it may be difficult to fit continuous distributions of relatively simple family to them.

4. In the second and third cases above, the fitting of the responses of many queried individuals into one overall fitted distribution can be done in a variety of ways: from simple averaging, to weighting in accordance with the 'expertness' of the response, to still more ornate methods.

5. In certain of these approaches, the initial set of responses is collected and shown (without source identification) to the jury members, who then have the opportunity to reconsider their initial judgements. This process is sometimes carried through several iterations.

Some of the more complex methods also permit respondees to view the practical consequences of the distributions fitted to their chosen values. References 9 and 10 should be consulted for further information on these and similar procedures.

Formal difficulties with these approaches are twofold:

(i) The distributional consequences flowing from the judgements made are often 'not really' what the respondent, had in mind. (A fitted tail may be 'too loose'; uncertainty may be too diffuse or too tightly bounded.)

(ii) The behaviour of the fitted prior under the impact of postulated possible subsets of subsequently available test data may yield a posterior distribution (through use of Bayes's theorem) which responsible people would find logically unsatisfying.

Even if judicious choice of the prior parameters could be made, such that the respondents' current belief (and that belief as it might be modified by possible future test data) could be successfully and appropriately modelled, there remains the key matter of convincing neutral to hostile third parties that the subjective belief so given distributional form is somehow 'necessary' and 'true'. (Of course, there is no direct and fully satisfying response to this, since matters are, by definition, 'subjective'.)

An example of fitting subjective prior distributions

The gamma density can be written as:

$$p(\theta) = \frac{\beta(\beta\theta)^{\alpha-1}e^{-\beta\theta}}{\Gamma(\alpha)} \qquad (1.9.2\text{-}11)$$

In reliability terms, it gives the distribution of the exponential failure rate θ as a function of two parameters: β (which can be looked upon as 'pseudo-time') and α (which can be looked upon as the 'number of pseudo-failures').

The gamma distribution can be given quite a variety of shapes by altering the α and β parameters. Figure 1.9.2-6 (with the mean constrained at unity, the mean being given by $E(\theta) = \alpha/\beta$) shows this variety as a function of α.

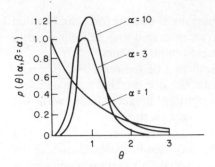

Figure 1.9.2-6 Gamma prior distribution shapes

Increasing β with α constant drives the distributions to the left and sharpens them; alteration of the two parameters together can create a great variety of distributional shapes, one of which may come sufficiently close to meeting the needs as a prior distribution.

The mean of the gamma prior is, as has been stated:

$$E(\theta) = \alpha/\beta \qquad (1.9.2\text{-}12)$$

Its variance and standard deviation are given by:

$$\sigma^2(\theta) = \alpha/\beta^2 \qquad (1.9.2\text{-}13)$$

A design engineer might find himself willing to make some such statement as the following: 'It's a 50–50 proposition that the failure rate of my component is above or below .007. Also, I'm 99% certain it's not higher than .07.'

If we take his statements at face value, a gamma prior could be obtained to fit his specifications somewhat as follows.

His statements amount to the following equations:

$$.50 = \int_0^{.007} \frac{\beta(\beta\theta)^{\alpha-1}e^{-\beta\theta}\,d\theta}{\Gamma(\alpha)} \qquad (1.9.2\text{-}14)$$

$$.99 = \int_0^{.07} \frac{\beta(\beta\theta)^{\alpha-1}e^{-\beta\theta}\,d\theta}{\Gamma(\alpha)} \qquad (1.9.2\text{-}15)$$

Two-parameter gamma densities such as the above are not tabulated; there would be an infinity of them. The standardized or normalized gamma density (where $\theta^* = \beta\theta$ in above usage) is given by:

$$p(\theta^*) = \frac{e^{-\theta^*}\theta^{*\alpha-1}d\theta^*}{\Gamma(\alpha)} \qquad (1.9.2\text{-}16)$$

This distribution is given in *cumulative* form $P(\tilde{\theta}^* < \theta^* | \alpha)$ by indirect use of ordinary Poisson tables (or through use of graphs, such as the one presented by Reference 11).

In the present instance, it is known that:

$$\frac{\theta^*_{.99}}{\theta^*_{.50}} = \frac{.07\,\beta}{.007\,\beta} = 10$$

A final $\alpha \approx .7$ is about right, with $\theta^*_{.99} \approx 4.0$ and $\theta^*_{.50} \approx .40$. Finally, since either $\theta^*_{.99} = .07\,\beta$ or $\theta^*_{.50} = .007\,\beta$, β can be obtained and is equal to:

$$\beta = \frac{4.0}{.07} = \frac{.40}{.007} \approx 57$$

Thus, a prior gamma distribution fitting the engineer's description can be established with parameters $\alpha = .7$, $\beta = 57$.

The posterior results are summarized as:

$$p(\theta|D) = \frac{\theta^{\alpha+k-1}(\beta+\tau)^{\alpha+k}e^{-\theta(\beta+\tau)}d\theta}{\Gamma(\alpha+k)} \qquad (1.9.2\text{-}17)$$

$$E(\theta|D) = \frac{\alpha+k}{\beta+\tau} \qquad (1.9.2\text{-}18)$$

$$\sigma^2(\theta|D) = \frac{\alpha+k}{(\beta+\tau)^2} \qquad (1.9.2\text{-}19)$$

Because the posterior results are similar to the prior (except that α has been replaced by $\alpha + k$ and β has been replaced by $\beta + \tau$), the gamma prior is referred to as a conjugate prior. The interpretation of prior parameters as pseudo-data is obvious, as all results are identical with the classical results, except that one has the prior pseudo-data added to the test results.

Empirical priors

These priors make use of collections of empirically observed failure rates or reliabilities associated with past equipments and typically then statistically fit a distribution of given family to the ensemble of previously observed estimates. Further details concerning these methods can be found in References 12 and 13. Typical arguments against their employment are essentially twofold:

1. The fitted distributions are often quite diffuse in nature.
2. More importantly, it is often logically difficult to show sufficient necessary relationship between a new equipment being developed or procured and an assemblage of past equipments. Such relationships may be logically defensible when we are contemplating 'one more order' of a design essentially long since frozen, but they are much less so in the case of nuclear reactor components and equipments, especially those which are prototypic in nature.

Another means of forming empirical priors is revealed in the work of Green and Bourne[14]. Here it is seen for a large variety of equipments (components through systems) that the ratio of ultimately observed failure rate to initial prediction is very closely fit by a log-normal distribution. This is very similar to a gamma distribution which might be substituted for a prior in order to ease the mathematics.

'Rational' priors

The process of choosing a 'rational' prior is based upon the presumption of an in-depth understanding of the failure phenomena underlying a given failure mode. Specifically, it is based upon possession of a descriptive and operationally useful model which, in conceptual terms, contrasts the innate 'strength' of the designed equipment to the 'stress' impinging upon it from the operational and external environments. Such 'stress versus strength' overlap analysis, in terms of an underlying model of failure phenomenon, is of course not new. What *is* new in the employment described herein is essentially twofold in nature:

1. The modelling of uncertainty (whether empirically observed, or representing gaps in knowledge, or both) is actively incorporated in the analysis.
2. It is observed that when *time* enters the failure model under study (either through the deterioration in time of intrinsic strength or because of time-varying fluctuations in the impinging stresses or both) that the concept of 'failure rate' emerges naturally from the analytic process.

The generation of Bayesian priors in this fashion incorporates prior knowledge in an organized, traceable, and therefore reproducible manner. The starting point is an engineering or scientific model delineating the interaction of variables underlying a potential failure process; uncertainties in these variables can be formally incorporated in the models. Results flow directly from analysis of the model. The prior information is in essence available for step-by-step evaluation and discussion by external auditors or other interested third parties. This can be contrasted with some of the previously discussed methods – where it may be difficult indeed to connect logically the future behaviour of new equipments with the past behaviour of old equipments ('empirical Bayes') or with an expert's subjectively selected choice.

The following is a theoretical approach in which the application of structural reliability concepts leads naturally to a constant failure rate and a distribution on the failure rate. The establishment of this link between constant failure rate processes and structural reliability demonstrates the feasibility of, and provides a guide to, the implementation of the probabilistic physics of failure approach to assigning Bayesian prior distributions on an item's failure rate.

Consider a system of fixed but uncertain capability. The exact capability is unknown, but whatever it is remains unchanged by time or other influences. Thus, the model gives no consideration to degradation or cumulative damage. The uncertainty is expressed as a probability distribution:

$$p(c) = g(c) \tag{1.9.2-20}$$

$$p(< c) = \int_{-\infty}^{c} g(c)dc = G(c) \tag{1.9.2-21}$$

The load upon the system is subject to continuous fluctuation, and is not only uncertain but is not fixed. That is, in each instant of time, a load will apply with the exact value being distributed according to a probability density function. In any fixed interval of time (0 to τ) there will be an infinite sampling of independent instants. Thus, the maximum load occurring in a fixed interval of time, τ, will be distributed according to an extreme value distribution, and the probability of the maximum load being less than L is given by:

$$p(< L|\tau) = F_\tau(L) = \exp[-\exp[-a(L - \mu_\tau)]] \tag{1.9.2-22}$$

where

μ_τ = the expected maximum value

and

$$a \equiv \frac{d\ln(t)}{d\mu_t} = \text{constant} \tag{1.9.2-23}$$

Because each instant in time is an independent sample from the population of loads, the factor, a, is a constant. This is generally true of any phenomenon of engineering interest. As a result, the probability that the maximum load is less than L for any other time, t, may be found:

$$p(< L|t) = F_t(L) = \exp[-\exp[-a(L - \mu_t)]] \tag{1.9.2-24}$$

where from Equation 1.9.2-23

$$\mu_t = \mu_\tau + \frac{1}{a}\ln(t/\tau) \tag{1.9.2-25}$$

Thus, the probability the largest load is less than L in the interval t is

$$p(< L|t) = F_t(L) = \exp[\frac{-t}{\tau}\exp[-a(L - \mu_\tau)]] \tag{1.9.2-26}$$

For a given capability, C, failure will occur in the time interval zero to t whenever $L > C$. When failure occurs in the time interval zero to t, then the time of failure is known to be less than t. Thus, the probability that failure occurs at a time less than t, given the capability, is:

$$p(< t|C) = p(L > C|tC) = 1 - p(L < C|t) \tag{1.9.2-27}$$

Substituting Equation 1.9.2-26 in Equation 1.9.2-27

$$p(<t|C) = 1 - \exp\left\{-\frac{t}{\tau}\exp\left[-a(C-\mu_\tau)\right]\right\} \qquad (1.9.2\text{-}28)$$

The time derivative of Equation (1.9.2-28) yields the probability density function on the exact time at which failure occurs given the capability C:

$$p(t|C) = \frac{1}{\tau}\exp\left\{-a(C-\mu_\tau)-\frac{t}{\tau}\exp\left[-a(C-\mu_\tau)\right]\right\} \qquad (1.9.2\text{-}29)$$

Since the instantaneous hazard rate (failure rate) is defined as

$$h(t) \equiv \frac{p(t|C)}{1-p(<t|C)} \qquad (1.9.2\text{-}30)$$

Substitution of Equations 1.9.2-28 and 1.9.2-29 results in

$$h(t) = \frac{1}{\tau}\exp\left[-a(C-\mu_\tau)\right] = \theta \qquad (1.9.2\text{-}31)$$

The hazard rate is seen to be independent of the passage of time and is thus quite properly a constant failure rate. This failure rate is a direct result of the infinite sampling of independent load instances and is determined by the capability, C, the time interval for which the maximum load distribution is originally defined, and the extreme value distribution parameters a and μ_τ. It should be noted that the failure rate in Equation 1.9.2-31 can take on any value, $0 < \theta < \infty$, depending on the capability, C, and τ, a, and μ_τ.

Generally, τ, a, and μ_τ are known from the observation of loading phenomenon and the subsequent fitting to an extreme value distribution; however, they may also be described by other probability distributions.

As the failure rate is fixed for any given capability, the uncertainty of C leads directly to a probability distribution on the failure rate. Thus:

$$p(\theta) = g\left(C = \mu_\tau - \frac{1}{a}\ln(\theta\tau)\right)\left|\frac{dc}{d\theta}\right| \qquad (1.9.2\text{-}32)$$

$$= g\left(C = \mu_\tau - \frac{1}{a}\ln(\theta\tau)\right)\left|-\frac{1}{a\theta}\right| \qquad (1.9.2\text{-}33)$$

and

$$p(<\theta) = p\left(C > \mu_\tau - \frac{\ln(\theta\tau)}{a}\right) = 1 - G\left(\mu_\tau - \frac{\ln(\theta\tau)}{a}\right) \qquad (1.9.2\text{-}34)$$

One special result is of interest, since it is frequently observed that empirical distributions of failure rates are well fitted by either log-normal distributions or gamma distributions. If the capability is distributed normally, then

$$g(C) = \frac{1}{\sqrt{2\pi}\,\sigma}\exp\left[-\frac{1}{2}\left(\frac{C-\mu_c}{\sigma}\right)^2\right] \qquad (1.9.2\text{-}35)$$

and

$$p(\theta) = \frac{1}{\sqrt{2\pi}\,\sigma a \theta} \exp\left[-\frac{1}{2}\left[\frac{-\ln\theta - \ln\tau + a\mu_\tau - a\mu_c}{\sigma a} \right]^2 \right] \quad (1.9.2\text{-}36)$$

which is a log-normal distribution.

Further information on this approach may be found in Reference 15.

What are the shortcomings of the 'rational prior' method? Two principal ones may be listed:

1. A useful understanding of the failure process is mandatory. That is, the key variables and their interplay must be understood: a useful scientific or engineering model of the phenomenon must exist.
2. Appropriate and accurate interpretation of such a model requires the employment of extensive engineering and scientific manpower for each application. The attempt of engineering generalists to employ such techniques in areas outside their expertise may yield results that are merely simplistic.

Since expertise is required to employ the method, and since this typically requires much time and effort, it is apparent that resources do not exist for broad application of the methodology in any given programme. Applications must, of necessity, be relatively few in number and therefore carefully chosen to yield maximum programme benefit.

Reprise/Conclusion on Bayesian Methodology

Bayesian methods are of interest because they afford procedures by which *all* relevant programme information can be brought to bear on problems of estimation and assessment of reliability and safety parameters. This is of course especially desirable in cases where so-called 'objective' experience (that is, direct test evidence) is and will be scarce in comparison to need. The Bayesian method, by its treatment of *distributions* around parameters, also allows a natural framework for the subsequent convolution of a variety of such distributions to meet the needs of higher level inference.

Prior distributions should not be overly diffuse, since they will then add little to the slight inferential power afforded by the ultimate results of limited test programmes. Neither should prior distributions be too 'tight' in the sense that they over-assert the importance of available information and are therefore not sufficiently sensitive to the possiblity of future test experience running counter to the prior.

Even if a prior is assigned so as to avoid these extremes, it is often difficult to justify to those not directly involved in its construction. An expert may 'know' that his subjectively selected prior is reasonable, but it would be hard indeed for him to demonstrate cogently and satisfactorily to others the *necessity* of his conclusions. The so-called 'rational' prior, flowing as it does from a model of presumable general acceptability and employing distributions on parameters of interest within that model of general defensibility, will allow third-party understanding of the process by which the prior was chosen. Regrettably, it can only be employed (because of the knowledge and resources required) in a limited quantity of cases.

Whether to employ Bayesian methodology throughout a programme or to employ it only on selected aspects of a programme is, of course, an organizational decision. In programmes which raise potential issues of greater than normal sensitivity, such decisions concerning the employment and interpretation of relevant information cannot proceed in a vacuum. Programme contractors are typically subject to the overview of lead contractors, federal authorities and regulatory agencies, and to the public at large (including those members of the public who predictably take adversary stances). The Bayesian method, therefore, will be especially subject to continuing critical review, both in terms of the conclusions it reaches concerning the reliability of safety systems *and* in the methods by which it reaches these conclusions. Any level of programme employment of the Bayesian method should, therefore, be decided upon only after addressing and attempting to assess the impact of the package of methods employed in such a larger context.

Questions 1.9.2

1. As one of several manufacturers of Widgets, I have designed a new Model 700 version to meet my customer's Specification. As one aspect of that Specification, a maximum instantaneous failure rate of $\theta_i = 0.25/\text{yr}$ has been established as an objective for that particular Widget function which has severe safety implications (other functions performed by the Widget are not safety critical). My customer does not have the funds to support a test programme large enough to accomplish a statistical reliability demonstration of the stated objective. But he has asked me to recommend some lesser test programme that will credibly confirm that such a capability does exist. I do have a large quantity of test and field data from the Model 100 to 600 versions, and have found that, when placed on a Duane Plot, using *all* failure incidents recorded, I obtain a straight line with $K = 3.94$ and $\alpha = 0.34$. How could I use these data, plus an appropriate rationale, to recommend a suitable test programme?

2. State Bayes's equation and give the general assumptions which can arise in its use.

3. (a) For Example 1.9.2-1, find the posterior probabilities if the item failed at 200 hours.
   (b) Find the posterior probabilities when two items are tested without failure for a total of 1000 hours.

(c) Find the posterior probabilities when items are tested with one failure in a total of 1000 hours.

4. (a) There is no previous experience with the application of an electronic component. Explain why a uniform prior on failure rate might be appropriate to use in a Bayesian assessment.

   (b) It is is required to show through testing of the electronic component that 0.9 is the value of the probability that the failure rate is less than $0.001 \, h^{-1}$. Assuming a non-informative prior, what is the length of failure-free testing needed.

5. The producers of a component which typically exhibits constant failure rate behaviour have kept very careful records on the failure rates of their production. From these data, they know that the ratio of the observed failure rate (θ_0) to the predicted failure rate (θ_p) is distributed roughly as a gamma distribution with a mean of 0.75 and a variance of 0.1875.

   (a) Determine the parameters for a gamma prior on θ_0 when the new component has a predicted failure rate of $\theta_p = 0.004 \, h^{-1}$.
   (b) What is the mean and variance of this prior?
   (c) What is the prior probability the failure rate is less than the prediction?
   (d) After testing for 1000 hours and observing one failure, what is the posterior mean and variance?
   (e) What is the posterior probability the failure rate is less than the prediction?

6. (a) Extensive computer analysis on a critical structure has indicated it has a capability to withstand an earthquake only up to peak ground accelerations of $0.3 \, g = 294 \, cm/s^2$. Earthquake data for the site of the structure indicate maximum earthquake accelerations in a year are well fitted by an extreme value distribution having $a = 1/31 \, (cm/s^2)^{-1}$ and $\mu_\tau = 17 \, (cm/s^2)$. What is the failure rate of the structure?

   (b) Further analysis shows the capability of the structure is uncertain, but may be described as a normal distribution with mean $\mu_c = 800 \, cm/s^2$ and standard deviation $\sigma_c = 425 \, cm/s^2$. What is the chance the failure rate is less than 10^{-5} failures per year?

   (c) Discuss how the above may be used to create a prior distribution on failure rate.

References 1.9.2

1. Duane, J. T. *TIS Report DF62MD300*, General Electric Company, 1962.
2. Codier, E. O. 'Reliability growth in real life.' *Proceedings of the Annual Reliability and maintainability Symposium*, 1968.
3. Brandt, H. W., Crellin, G. L., Graham, J., Millunzi, A. C., Simpson, D. E., and Smith, A. M. 'Reliability testing of reactor shutdown systems in support of low-risk design.'

ANS/ENS Topical Meeting on Nuclear Power Reactor Safety, Brussels, Belgium, October 1978.
4. Crellin, G. L., Jacobs, I. M., and Smith, A. M. 'A defect analysis program applied to nuclear plant experience data.' *Nuclear Systems Reliability Engineering and Risk Assessment*, SIAM, 1977.
5. Crellin, G. L., Herrmann, C. R., Jacobs, I. M., and Smith, A. M. 'A Defect Analysis Programme Applied to Nuclear Plant Pressure Component Experience Data.' *Proceedings of Reliability Problems of Reactor Pressure Components*, Vol. I, IAEA, Vienna, 1978, pp. 33–47.
6. Thaggert, H. L., Jacobs, I. M., Crellin, G. L., and Smith, A. M. 'Defect flow analysis of control rod drive operational events.' *Nuclear Safety*, Vol. 22, No. 4, July–Aug., 1981, pp. 466–474.
7. Mann, N. R., Schaefer, R. E., and Singpurwalla, N. A. *Methods for Statistical Analysis of Reliability and Life Data*, John Wiley & Sons, New York, 1974.
8. Lindley, A. V. *Introduction to Probability and Statistics From a Bayesian Viewpoint*, Cambridge University Press, 1965.
9. Lin, Chi-Yuan and Schick, G. J. 'On-line (console-aided) assessment of prior distributions for reliability problems.' *Annals of Reliability and Maintainability*, 1970 (*Proceedings of 9th Reliability and Maintainability Conference*), pp. 13–19.
10. *IEEE Guide to the Collection and Presentation of Electrical, Electronic and Sensing Component Reliability Data for Nuclear Power Generating Stations*. Distributed in co-operation with Wiley-Interscience, John Wiley & Sons. IEEE Std 500–1977.
11. Raiffa, H. and Schlaifer, R. *Applied Statistical Decision Theory*, Harvard University, Boston, 1961.
12. Martz, H. F., Jr. and Waller, R. A. *Handbook of Bayesian Reliability Estimation Methods*. Los Alamos Scientific Laboratory Report LA-6572-MS, November, 1976.
13. MacFarland, W. J. *Reliability Manual for Liquid Metal Fast Breeder Reactors*: Section 9, 'Bayesian priors'. General Electric Company, Fast Breeder Reactor Department, Sunnyvale, California, Report SRD-75-064, 1975.
14. Green, A. E. and Bourne, A. J. *Reliability Technology*. John Wiley & Son, London, 1972.
15. Crellin, G. L. and Ingram, G. E. 'Modelling the time-dependent nature of structural reliability.' *Proceedings of Reliability Problems of Reactor Pressure Components*, Vol. I, IAEA, Vienna, 1978, pp. 139–147.

PART 2

SAFETY TECHNOLOGY IN INDUSTRY

High Risk Safety Technology
Edited by A. E. Green
© 1982 John Wiley & Sons Ltd

Chapter 2.1

Chemical

T. A. Kletz and H. G. Lawley

Introduction

In any industry, particularly one that handles large quantities of hazardous materials, it is impossible to remove every risk; sufficient resources are not available and complete safety is usually approached asymptotically. We can always add on more protective equipment, at a cost, and get a little more safety, but the law of diminishing returns soon sets in and we have to decide how far to go.

Often this is a qualitative judgement but sometimes we can make it quantitative. *Hazard Analysis* is the name given to these attempts to apply numerical methods to safety problems, in particular to help us decide whether or not we should take action to reduce the probability that a hazard will occur or to protect people from its consequences. It is called *hazard* analysis, rather than risk analysis, as this latter term is used to describe methods of determining the *commercial* risks of a project[1] and it is called hazard *analysis* because an important stage is analysing the events that lead up to the hazard.

Before we can decide how far to go in removing hazards we must first make sure that we have identified all the hazards. Methods of identifying and assessing hazards are shown in Figure 2.1-1. Some hazards are obvious. The traditional

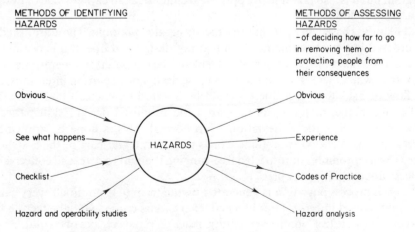

Figure 2.1-1 Methods of identifying and assessing hazards

317

method of identifying hazards was to build the plant and see what happens – 'every dog is allowed one bite'. Until it bit someone, we could say that we did not know it would do so. This method is no longer acceptable now that we keep dogs as big as Flixborough.

Checklists are widely used for identifying hazards but have the disadvantage that items not on the list are not brought forward for consideration and are therefore missed. The more 'open-ended' approach of an operability study, described in the next section, is therefore preferred.

When we come to assess the hazards, in some cases it is obvious that we ought to remove them and there is an obvious way of doing so; in other cases experience or codes of practice tell us what to do. In yet other cases we apply the methods of hazard analysis, though these may range from a five-minute calculation to one taking many weeks.

Operability Studies

Operability studies, also called hazard and operability studies or 'hazops', are the recommended method for identifying hazards.

It cannot be stressed too strongly that a systematic attempt to identify hazards and the ways in which they can arise should precede any attempt to quantify them. It is little use working out the probability and consequences of one hazard if greater hazards have not been identified; it is of little use quantifying one of the routes by which a hazard can arise if we have not identified more important routes.

Operability studies have been described in a number of publications[2]–[7]. The following description is based on the work of Lawley and is particularly suitable for continuous plants. In the example given the method is applied to a line diagram but it is also possible to apply the technique, at an earlier stage of design, to a process flowsheet.

The study is based on a procedure which generates *questions*. These are asked in an ordered but creative manner by a team of design and operations personnel. The method takes the position that a problem – be it in safety or operations – can only arise when there is a deviation from the design or operating intentions, e.g. no flow or backflow when there should be forward flow, etc.

The technique, therefore, is to search the proposed design, systematically looking for *every* process deviation from normal. This is done by applying a carefully chosen checklist of guide words to each integral part of the system in turn. The list promotes unrestricted free-ranging thought to detect all conceivable process abnormalities (Table 2.1-1).

As each process deviation is generated by the team (e.g. no flow), every cause (e.g. valve closed in error, filter blocked, etc.) and its consequential effect on the system as a whole (e.g. pump overheating, reaction runaway, loss of output, etc.) is considered in turn. Potential problems are thereby identified.

Table 2.1-1 Meaning and types of process deviation generated

Guide word	
NONE	No forward flow when there should be, i.e. no flow or reverse flow.
MORE OF	More of any relevant physical property than there should be, e.g. higher flow (rate or total quantity), higher temperature, higher pressure, higher viscosity, etc.
LESS OF	Less of any relevant physical property than there should be, e.g. lower flow (rate or total quantity), lower temperature, lower pressure, etc.
PART OF	Composition of system different from what it should be, e.g. change in ratio of components, component missing, etc.
MORE THAN	More components present in the system than there should be, e.g. extra phase present (vapour, solid), impurities present (air, water, acids, corrosion products), etc.
OTHER	What else can happen apart from normal operation, e.g. start-up, shutdown, uprating, low rate running, alternative operation mode, failure of plant services, maintenance, catalyst change, etc.

The principles on which problem identification is based are summarized in Figure 2.1-2.

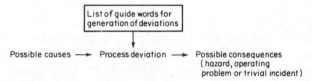

Figure 2.1-2 Flow of steps by which problems are readily identified

For minor hazards or purely operating problems, the need for action is normally decided on the basis of general experience. Taken into account is the likelihood of the event and the seriousness of the consequences. However, for any major risk areas, a full quantitative examination (hazard analysis) should always be carried out. Is the proposed design safe within the limits we consider acceptable? Are further design changes necessary to meet current standards?

The study is normally based on a piping and instrument diagram. This is the earliest stage in the design at which process lines, plant services, instrumentation and control aspects are adequately specified.

The check should begin as soon as possible after these drawings first become available. Thus, modifications can be implemented before design and procurement of major items is too far advanced. This also leaves time for carrying out the normal mandatory reviews, such as layout, electrical area classification, relief and blow-down, etc., on the proposed system.

The study team must be carefully selected. It should represent a wide spread of knowledge, expertise and experience, for all aspects of the study to be properly covered. But the team must also be small enough to be efficient. Team members must be firmly committed to the study. They should be prepared to allocate time. And, they should have authority to make on-the-spot decisions when required.

All studies must be led by a person well trained in the technique. The leader's main responsibilities are to ensure that the full study procedure is followed and that each point, as it arises, is discussed and questioned to the correct level of detail. Pretraining of team members is not necessary. A 10-minute introductory talk and illustration at the start of the first meeting is normally adequate.

For a new grass roots project, the study team would typically consist of the study leader, the project manager or operations-manager designate, the technical project manager, the project engineer and the design engineer. There should be an instrument engineer to advise on the instrumentation and to deal with any control problems. A research chemist is needed if new chemistry is involved.

For plant extensions, the study team would be more works-oriented. It would typically comprise the study leader, the works development manager, the design engineer, the operations manager and the maintenance engineer.

Team meetings should last three hours. Allow one or two days between meetings to give members time to collect information or progress action on the points raised.

At the beginning of the first study meeting, a team member outlines the design intentions for the project as a whole in very broad terms. This ensures that the team has an adequate background knowledge of the process, and of the function of each section within the total system. Thereafter, and just before examining each individual section, a team member states in greater detail the design intention for that section. This covers process conditions, rates and specifications if available.

The study begins by applying the full checklist of guide words in turn to a selected line entering the first vessel. This generates deviations, causes and consequences as already described. The procedure is then repeated for every other line entering the first vessel. When all lines to the vessel have been examined in this way, the vessel itself is studied. For this, only the guide word OTHER is used – not the full checklist. This is because any problem that would be brought to light by use of the first five guide words on the item of equipment should also show up as a problem when these same words are applied to a line joining the item of equipment.

Having examined the first vessel, the procedure is repeated on a line-by-line and vessel-by-vessel basis until the whole section has been studied.

The search for causes and consequences of each process deviation must be carried out thoroughly and systematically, looking for *every* eventuality and its consequences. This includes interactions across the system as a whole. Furthermore, where provision has been made for a contingency, it is important to question whether the provision is adequate. This is not only in terms of suitabilty and reliability, but also in respect of its effectiveness opposite all the possible ways in which the contingency can arise.

Example questions: Will the single check valve safeguard adequately against backflow from the high pressure region in the event of pump failure, or do we need trip protection as well? Is the tank venting adequate for liquid overfilling as well as for normal gas inbreathing and outbreathing? Should the safety valve be trace heated for this particular duty? Is the pressure indicator on the nitrogen supply adequate, or do we need a low pressure alarm as well? Is the trip system sufficiently reliable? Will it be effective opposite all events capable of leading to the unsafe condition? Is the trip system designed to carry out all the necessary trip actions? Will it respond quickly enough? And so on.

As each potential problem is brought to light, the need for action should be evaluated by the team without delay. Problems are given priority on the basis of severity and probability of occurrence. This is facilitated by a rough working knowledge of failure rates for standard items such as pumps, valves, instrumentation, etc., and of fractional dead times for the more common types of protective equipment.

Where action is clearly necessary and the best solution obvious, (e.g. install kickback on pump delivery to prevent overheating when dead-headed) this should be quickly agreed upon by the team and recorded before moving on to the next point. The modification can then be taken immediately into account opposite all deviations on the process line under review and when studying other related parts of the system. This covers any interaction resulting from the change.

On-the-spot agreement to obvious new design requirements which have a bearing on adjacent sections is particularly important. Such agreement avoids the need for an extensive re-study at a later date. It can also be important when several new items are required on the same section of line (e.g. check valve, pump kickback, hydraulic expansion relief, low-point drain). An immediate decision can then be reached on their relative positioning, for all functions to be mutually compatible.

When a satisfactory solution to a problem is not immediately apparent, the point should be recorded for further consideration and evaluation outside the study meetings. Prolonged discussions on how a problem can be solved should be avoided. They interrupt the chain of thought. They are also time-wasting for team members not directly concerned.

Evaluation should also be deferred when agreement on the need for action cannot be quickly reached. It should be deferred when the problem to be evaluated is interrelated with parts of the system still to be studied.

In the case of major risk areas, any obvious design improvements should also be quickly agreed upon. However, for problems falling into this category, a fully quantitative assessment by an expert in hazard analysis may also be required.

Recording of the full study would be very time-consuming. Only points

Table 2.1-2 Operability study of proposed olefine dimerization unit: results for line section from intermediate storage to buffer/setting tank

Guide word	Deviation	Possible causes	Consequences	Action required
NONE	No flow	(1) No hydrocarbon available at intermediate storage.	Loss of feed to reaction section and reduced output. Polymer formed in heat exchanger under no flow conditions.	(a) Ensure good communications with intermediate storage operator.
				(b) Install low level alarm on settling tank LIC.
		(2) J1 pump fails (motor fault, loss of drive, impeller corroded away etc).	As for (1)	Covered by (b)
		(3) Line blockage, isolation valve closed in error, or LCV fails shut.	As for (1) J1 pump overheats	(c) Install kickback on J1 pumps.
				Covered by (b)
				(d) Check design of J1 pump strainers.
		(4) Line fracture	As for (1) Hydrocarbon discharged into area adjacent to public highway.	Covered by (b)
				(e) Institute regular patrolling & inspection of transfer line.
MORE OF	More flow	(5) LCV fails open or LCV bypass open in error.	Settling tank overfills	(f) Install high level alarm on LIC and check sizing of relief opposite liquid overfilling.
				(g) Institute locking off procedure for LCV bypass when not in use.
			Incomplete separation of water phase in tank, leading to problems on reaction section.	(h) Extend J2 pump suction line to 12″ above tank base.

running.

isolated. Check line, FQ and flange ratings and reduce stroking speed of LCV if necessary. Install a PG upstream of LCV and an independent PG on settling tank.

Guide word	Cause	Consequence	Action
More temperature	(7) Thermal expansion in an isolated valved section due to fire or strong sunlight.	Line fracture or flange leak.	(k) Install thermal expansion relief on valved section (relief discharge route to be decided later in study).
	(8) High intermediate storage temperature.	Higher pressure in transfer line and settling tank.	(l) Check whether there is adequate warning of high temperature at intermediate storage. If not, install.
LESS OF Less flow	(9) Leaking flange or valved stub not blanked and leaking.	Material loss adjacent to public highway.	Covered by (e) and the checks in (j).
Less temperature	(10) Winter conditions	Water sump and drain line freeze up	(m) Lag water sump down to drain valve, and steam trace drain valve and drain line downstream.
PART OF High water concentration in stream	(11) High water level in intermediate storage tank.	Water sump fills up more quickly. Increased chance of water phase passing to reaction section.	(n) Arrange for frequent draining off of water from intermediate storage tank. Install high interface level alarm on sump.
High concentration of lower alkanes or alkenes in stream	(12) Disturbance on distillation columns upstream of intermediate storage.	Higher system pressure.	(p) Check that design of settling tank and associated pipework, including relief valve sizing, will cope with sudden ingress of more volatile hydrocarbons.
MORE THAN Organic acids present.	(13) As for (12)	Increased rate of corrosion of tank base, sump and drain line.	(q) Check suitability of materials of construction.
OTHER Maintenance	(14) Equipment failure, flange leak, etc.	Line cannot be completely drained or purged.	(r) Install low-point drain and N_1 purge point downstream of LCV. Also N_2 vent on settling tank.

requiring attention are detailed. These may be actions agreed upon within a meeting or questions requiring further consideration outside the meetings.

A typical study record consists of a list of actions or questions relating to design changes or important operating points as in the 'action required' column of Table 2.1-2. This is written up in sequential order but with no mention of how the need for action came to light. Agreed design changes are also marked up on the line diagram as they arise.

A team member is made responsible for progressing each action or question with the implementing authority (normally the project manager). A review note is eventually circulated to all interested parties. It states the outcome of each point raised in the study.

When the design has to be changed significantly after the study sessions have been completed (due, for example, to deferred evaluation of problems highlighted in the study or to difficulties in procurement of special items) a re-study must always be carried out on the modified and interrelated parts of the system. This is also essential in the event of late precommissioning changes. Indeed, the whole design effort could be negated with disastrous consequences by a single late modification carried out without proper forethought and checking as to its acceptability.

A formal systematic check of the installed system and of the operating manuals is also essential just before commissioning. This check verifies that all equipment modifications and important operating points recommended by the study (or by other mandatory reviews covering specific aspects such as layout, relief and blow-down, electrical area classification, etc.) have been fully implemented.

This precommissioning check should also ensure that no unauthorized changes have been made which could adversely affect safety.

A post-commissioning review after the first 6 or 12 months' operation is wise. For the more hazardous plants, such a review identifies any unexpected shortcomings in the design. It can turn up significant departures from the design concepts or any unanticipated operating difficulties affecting safety. And, it can provide data on plant reliability, equipment and instrument failure rates, etc.

In this way, the original design predictions on plant safety and general operability can be reappraised and updated in the light of running experience. Any areas where changes are required are identified. Any such changes should be operability-studied for acceptability prior to implementation.

After the immediate post-commissioning period, and during the remaining life of the plant, major changes are often required, for example: switch to new catalyst, change in process conditions, alternative operating mode, uprating, etc. Before implementation, all unit operations likely to be affected by the proposed change should be checked by a full hazard and operability study. There should also be other mandatory checks such as relief and blow-down, etc., as appropriate.

The time and effort spent on checking a new design, and on carefully analysing major planned changes throughout the life of a plant, is wasted without attention to detail opposite other more trivial modifications and operating changes required from time to time on most plants. Several serious incidents reported in recent years can be attributed to the day-to-day type of modification carried out for good reason, but without first considering the full implications of the change on the system as a *whole*.

As for major alterations, minor changes to plant, process and procedures should always be checked before implementation. This is to ensure that they will not introduce unforeseen hazards or operating problems either locally or by interaction with other related parts of the system. A technique for doing so is described by Kletz[8].

A simple example illustrating the use of the operability study technique for identifying inadequacies in design and for drawing attention to important operating requirements will be useful. The system considered is the feed section of a proposed olefine dimerization unit, the preliminary design being as shown in Figure 2.1-3 and the process details as follows.

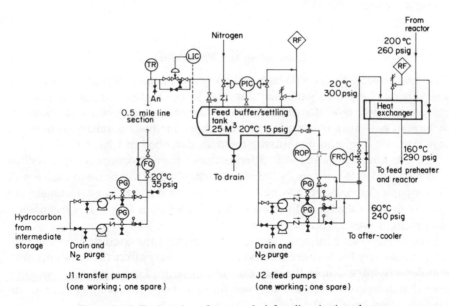

Figure 2.1-3 Feed section of proposed olefine dimerization plant

An alkene/alkane fraction containing small amounts of suspended water is continuously pumped from bulk intermediate storage via a half-mile pipeline section into a buffer/settling tank where residual water is settled out prior to passing via a feed/product heat exchanger and preheater to the reactor section. The water, which has an adverse effect on the dimerization reaction, is run off manually from the settling tank at intervals. Residence time in the reaction section must be held within closely defined limits to ensure adequate conversion of the alkene and to avoid excessive formation of polymer.

Table 2.1-2 summarizes the results for the first line section from intermediate storage up to the buffer tank, and also indicates the manner in which the need for action is recognized. The points which came to light from examination of the next line section up to and including the heat exchanger are summarized[2].

Hazard Analysis

Every hazard analysis, long or short, consists of three steps.

1. Estimating the probability that the hazard will occur.
2. Estimating the consequences to employees, the public, and to the plant.
3. Comparing the results of (1) and (2) with a target or criterion in order to decide if action to reduce the probability of occurrence or minimize the consequences is desirable.

Estimating the Probability

Wherever possible this should be based on past experience. Sometimes we know how often a pump on a particular duty has leaked and how often the leak has fired, or we know how often a tank has overflowed. On many occasions, however, we have to estimate a figure using the recognized methods of reliability analysis, drawing logic trees and quantifying them, as described in Chapter 1.3.

The data for quantification can often be drawn from in-company experience or from data banks (see section 1.9.1). For instruments, data can be freely exchanged; all companies use much the same instruments and changes in environment are unlikely to affect the reliability by a factor of more than four[9]. For mechanical equipment, however, this is not true.

Even though two companies use the same pumps, the conditions of use may vary, and so may the maintenance policy and this may affect the reliability very considerably. For example, the frequency of leaks from flanged pipeline joints in a works handling corrosive chemicals was found to be 10 times higher than in one handling clean hydrocarbons.

The Canvey Report[10] classifies data as follows:

1. A scientifically based figure to which a standard deviation could be attached.
2. Some judgement has been used but the figure could be made definite if more time and trouble were taken.
3. A judgement for which the prospects of reducing the uncertainty are poor. Conclusions should not be drawn from them without testing for sensitivity.

This example should be followed elsewhere.

Estimating the Consequences

Reference should be made to texts on gas dispersion, fires, explosions, etc. for methods of estimating the consequences of a hazard. However, there are a number of general points that should be borne in mind.

1. People often survive fires or explosions simply because they are not on the plant at the time. Observation will show how often a person is near enough to be injured should there be a leak, fire or explosion in or near a piece of equipment. Often people are present for only 5% or 10% of the time.
2. Even if people are present, they may survive the hazard. If a vessel explodes, for example, the solid angle occupied by the fragments is small, and it is unlikely that one person, if present, will occupy this angle. If there is a release of toxic gas, experience shows that people who stay indoors have a very high chance of survival, while others are able to leave the area before the gas reaches them.
3. Actual experience is to be preferred to theory. For example, there are a number of theoretical treatments available for estimating the effects of heat radiation from a fireball on people and these treatments give quite different results. However, from published descriptions of fireballs and their consequences it is possible to make a rough estimate of the probability of death.

Similarly, there have been a number of theoretical treatments[10] of the consequences of a release of toxic gas and very large numbers of casualties have been forecast. In contrast, experience shows that the casualties are much less[11]. Of course, it is always possible that casualties might be higher in unusual weather conditions but the probability of such weather must be taken into account.

Criteria

Risks to plant

This is considered first, not because plant is more important than people, but because the problem is simpler.

If the risk to life is so high that it must be reduced as a matter of priority, then the economic loss is of little concern. If, however, the risk to life is so low that we are not justified in allocating resources to its further reduction, then the economic loss may nevertheless provide a reason for reducing the risk.

The procedure is as follows:

1. The probability that the dangerous occurrence will occur is estimated. Let us suppose it is once in 1000 years.
2. The damage to plant and loss of profit that will result is estimated. Consequential losses should be included as well as the cost of repairing the damage.
   Suppose the loss is £2 million.

3. From (1) and (2), the 'average annual loss' L is calculated. In this case it is £2000.
4. The net cost C (after allowing for tax allowances) of reducing the probability that the dangerous occurrence will occur is calculated. Let us suppose it will cost £15 000 to prevent it happening.
5. The cost C is converted into an annual value (maintenance, depreciation, and interest on capital). In the example considered it will be about £5000.
6. If this annual value is less than L the expenditure of C is justified. In the example, the annual cost is £5000 and the average annual loss L is £2000 so the expenditure is not justified.

Strictly speaking, since most of the 'premium' has to be paid now and the loss, on average, only after some years, the loss should be converted to its 'net present worth'. The data are not, however, usually accurate enough to make this refinement worthwhile.

In calculating the loss, the costs of any injuries or fatalities (compensation, legal costs, costs of investigation, etc.) can be ignored. They are usually negligible in comparison with the costs of the damage and loss of production.

Weighing in the Balance or Target Setting?

The last section provides an example of weighing in the balance – the costs of an accident are compared with the costs of prevention. It would be possible to use this method for risks to people if we had an agreed value for human life. Alternatively, we could set a level of safety which we should not fall below.

When dealing with risks to life the second approach is preferred, for several reasons.
1. Despite many attempts to value life[12][13], there is no generally agreed figure.
2. It is wrong to expose people to a higher level of risk than is generally agreed to be tolerable, just because it is expensive to reduce the risk.
3. If we agree to reduce the risks that are cheap to reduce, but tolerate the risks that are expensive to reduce, then every risk may become expensive to reduce. If, however, we say that all risks above a certain level must be reduced, then experience shows that a reasonably practicable solution can usually be found.
4. Our usual practice in dealing with hazards is to set a level which must be met or a target to aim for; for example, this approach is used in fixing the height of handrails, the concentration of harmful gases or vapours in the atmosphere, the level of noise, and so on.

In the following sections we describe scales for measuring risk to employees and members of the public and we suggest target values that should not be exceeded. The justification of these scales and targets, and indeed of all the methods described in this chapter, is that they help people in real-life situations to allocate their resources better. Experience shows that this is the case, and this is illustrated later by examples.

Risks to Employees

The scale used is the Fatal Accident Frequency Rate (FAFR), sometimes called the Fatal Accident Rate (FAR). It is the number of fatal accidents in a group of 1000 men in a working lifetime-10^8 hours. For the UK chemical industry the FAFR is 4, excluding Flixborough, or about 5 if Flixborough is averaged over 10 years. The FAFR for all premises covered by the UK Factories Act is also 4.

About half the fatal accidents in the chemical industry are unconnected with the materials handled, for example, they are the result of falling downstairs or getting run over, and the other half are the result of fire, explosion, releases of toxic or corrosive chemicals or asphyxiation.

If we can estimate the FAFR for the chemical risks attached to a particular job, we set as our target that the man carrying out the job should not be exposed to an FAFR greater than 2. We will eliminate or reduce any risks that exceed this figure, on new or existing plants.

The justification for this approach is that we should not spend limited resources on reducing the risks to people who are already exposed to below average risks for the industry; instead we should spend our resources on reducing the risks of those who are exposed to above average risks.

Often we have not quantified or are not able to quantify all the risks to which a particular employee is exposed and in these cases we take as our target that any particular risk, considered in isolation, should not expose any employee to an FAFR greater than 0.4. We will reduce any risk that exceeds this figure. We are thus assuming that there are about five significant chemical risks on a typical plant. Other figures can be chosen if considered appropriate for particular plants.

Note that the risks to each person, or the person at highest risk, should be considered and not the average for the plant.

Experience shows that the costs of meeting these targets, though often substantial, are not unbearable.

If a job is manned continuously on shifts then an FAFR of 0.4 is equivalent to one fatality every 30 000 years, or more precisely, to a probability of death from the risk under consideration of 3×10^{-5} per year.

In some cases a simpler target then FAFR can be used.

For example, suppose we intend to use an instrumented protective system instead of a relief valve, perhaps in order to reduce the size of the flare system required. A high-pressure switch will detect a rise in pressure in a vessel and will close a valve which isolates the source of pressure. If the vessel is a distillation column, and the rise in pressure is due to the loss of cooling then the pressure switch would close a valve in the steam line to the reboiler.

How do we decide on the reliability required in the protective system? One possible method would be to estimate the consequences of the protective system failing to work. The vessel would be overpressured. It might fail; we could estimate the probability. Some of the contents could leak out; we could estimate how much. If flammable they might ignite; we could estimate the probability.

Someone might be in the area; we could estimate the probability. He might be killed; we could estimate the probability. Finally we could estimate the FAFR.

A simpler method of calculation is, however, possible. Relief valves are widely used and their reliability, although not 100%, is generally regarded as acceptable; there is no demand for more reliable devices. The protective system used in place of a relief valve should therefore be just as reliable, preferably rather more reliable as we are using something new in place of something well established. Also there is a difference in the mode of failure of relief valves and protective systems; a relief valve which fails to lift at, say, twice the set pressure may still lift at a higher pressure. The same is not true of a protective system. We therefore propose that a protective system used in place of a relief valve should have a failure probability ten times lower [14][15].

Risks to the Public

When we consider risks to the public at large from industry, the level of risk for which we should aim should be much lower than for employees. A man chooses to work for a particular employer or in a particular industry and, unless he chooses a particularly hazardous occupation, the risks he runs are not much greater than if he stayed at home. On the other hand, the public may have risks imposed on them without their permission, and though society as a whole may gain, individuals may not. Not all the people who live near airports wish to travel by air.

Starr[16][17] and Kletz[18][19] have pointed out that we accept voluntarily risks such as driving, flying and smoking, which expose us to a risk of death of 1 in 100 000 or more (sometimes a lot more) per person per year (a FAFR of 0.1). We also accept, with little or no complaint, a number of involuntary risks which expose us to a risk of death of about 1 in 10 million per person per year (a FAFR of 0.001). For example, the risks from lightning, insect bites, falling aircraft, flooding (in Holland), tornadoes and storms (in the US) and earthquakes (in California).

It would be possible to reduce these involuntary risks if there were sufficient pressure from the press and public, but on the whole there is no such pressure. The risk of being struck by lightning or falling aircraft is so small that we accept the occasional death without complaint.

We thus have a basis for assessing risks to the public at large from an industrial activity. If the *average* risk to those exposed in less than 1 in 10 million per person per year resources should not be allocated to its reduction.

Some limitation on the maximum risk of death to which any individual is exposed is desirable and it is suggested that this should be 10^{-5} to 10^{-6} per person per year.

These risks are a good deal lower than those proposed in the Canvey Island report[10], an official report on the risks to the public from the oil refineries and similar industries on an island in the Thames estuary. However, the report, in many people's view, has exaggerated the size of the risks, and so the difference is not as great as it seems at first sight. (The report admits on the last page that it may have exaggerated the risks, as it states, 'Practical people dealing with industrial hazards tend to "feel in their bones" that something is wrong with risk estimates as developed in the body of the report'.)

Nevertheless, the report is a landmark as it shows official acceptance of the view that we cannot do everything possible to avoid every conceivable accident, and that numerical methods should be used to decide what level of risk is so low that resources should not be spent on reducing it further.

We have avoided the use of the term 'acceptable risk' which many people find repugnant, as they say that we should never fail to act when other people's lives are at risk. Of course we should not, but we cannot do everything at once; some things have to be done first, others left until later. Hazard analysis, to repeat what has been said earlier, is concerned with priorities rather than principles.

As with risks to employees, simpler criteria can often be used. For example, suppose we are considering the risk to the public from the transport by pipeline of a material previously transported by road. If there has been no suggestion that road transport is in any way unsafe, then it may be sufficient to show that the new method of transport is no less safe, preferably that it is safer, than the traditional method.

Multiple Casualties

Is any change required in the targets developed, for both risks to employees and risks to the public, if large numbers of people are liable to be killed at one time? On the one hand it can be argued that there is no real difference between an industry that kills 100 people one at a time over 100 years and one that kills 100 people at once every 100 years. On the other hand, the latter event causes a far greater public outcry and it has been suggested that, if N people are killed at a time, the target risk to life should be divided by N or log N or N^2.

There is no logical case for any of these figures and an alternative way of taking public reaction into account has been suggested[20]. The financial loss should be calculated as described previously and it should include an estimate of the cost to the industry of the public reaction to a major hazard – plants might have to be temporarily or permanently closed and many changes made.

Cost per Life Saved as a Secondary Criterion

Although reasons have been given earlier (page 328) for preferring target-setting to arguments based on the cost of saving a life, nevertheless this figure may be useful as a secondary criterion. It is calculated as follows.

Suppose that N people are exposed to a risk of death of 10^{-x} per year and that this risk can be removed at an annual cost C. Then the cost per life saved is

$$\frac{10^x C}{N}$$

This figure is, of course, a notional one. If the money is not spent it is unlikely that anyone will be killed. All we are doing is making a low probability of death even lower.

Sinclair[21] and Kletz[18] have shown that the money spent to save a life varies over a very wide range, from a few hundred pounds for certain forms of medical treatment to £10^7, and that to achieve the standard of safety described above the chemical industry has to be prepared to spend about £10^6 per life saved.

If particular proposals greatly exceed this figure, then the risk should not be accepted but we should look for a cheaper solution. In practice, one can usually be found.

Long-Term Hazards

The criteria described in this chapter have been developed to deal with acute hazards, those that produce their effects quickly such as fires, explosions and large releases of toxic gases.

The chemical industry also has to deal with a number of chronic hazards, those that take a long time to produce their effects, substances such as benzene and asbestos and phenomena such as radiation and noise. It would be desirable to compare the acute and chronic hazards on the same scale so that we know whether or not the allocation of resources between them is correct. This is difficult because, for chronic hazards, in only a few cases do we know with any degree of certainty what effects are produced by various causes. Nevertheless, such evidence as is available suggests that, taking risk to life as the criterion, the present standards for employees exposure to chronic hazards are, within an order of magnitude, comparable with those developed above for acute hazards.

However, it may be questioned whether the risk to life is a suitable criterion for chronic hazards, as each death is preceded by years of illness, or at least loss of the quality of life. An alternative criterion might be loss of expectation of life. This is an area in which further work is required.

EXAMPLE 2.1-1   A Strategic Problem – Strengthened Control Rooms

This example[22] illustrates the application of hazard analysis to a problem where precision is impossible but, nevertheless, numbers may help us to get a better feel for the problem and help us to decide whether or not change is required. The problem is, should we build strengthened control rooms in plants handling large quantities of flashing, flammable liquids that have the potential to produce a Flixborough-type explosion?

Since Flixborough, many companies have reversed their former policies and decided to build blast-resistant control rooms in new plants. There has also been considerable pressure for this to be done by the authorities in some countries. A few companies have even strengthened existing buildings. Is money spent on strengthening existing buildings well spent? Or are we over-reacting to Flixborough?

Bell[23] has estimated that incidents of the Flixborough type have occurred, on average, about once in 1000 yr/site. The authors' own estimates support this if 'site' is taken as equivalent to a large unit such as an ethylene or aromatics plant. If the outside operators have a 50% chance of survival and the control room men no chance (because the building may collapse on them), then the FAFR is 6 for the outside men and 12 for the control room men, clearly far too high.

Averages, however, can be misleading. We hope the industry has learned lessons from the past and that incidents will be fewer in the future. If, by following the best practice in design and operation, we can reduce the average of the past by a factor of 10, then the FAFR becomes 0.6 for the outside men and 1.2 for the control room men. The former figure is close to 0.4, and we have agreed that we should not expend resources on reducing the FAFR below this level. The figure for control room men is higher, but below 2 and they are exposed to fewer risks than outside men; therefore it might also be considered acceptable.

If we could be sure that the chance of a Flixborough-type explosion in our plant is less than one-tenth of the average of the past for the industry as a whole, then the case for a strengthened control room is marginal. But can we be sure that our standards are so high and will remain so? And can we convince others that this is so? If, in addition, we make some allowance for the public reaction to incidents in which a number of people are killed, then we see that we are justified in strengthening control rooms.

If there are demonstrable reasons for believing that the chance of a Flixborough-type incident on a particular plant is well below one-tenth of the industry average of the past, then there is no case for a strengthened control room on that plant. Possible reasons are:

1. The total inventory of flashing flammable liquid is small, say less than 5 or 10 tonnes.
2. The total length of pipeline containing flashing flammable liquids is well below that of a typical unit. We are assuming that vapour cloud explosions occur as the result of pipeline failures rather than vessel failure[24]. If the length of pipeline is small, the chance of an explosion will be correspondingly small.

   Typical petrochemical plants contain 5000 to 50 000 ft of pipeline carrying flashing flammable liquids. If a particular plant contains less than, say, 500 ft, the chance of an explosion may be considered so low that it can be ignored.

3. Even if the length of pipeline is somewhat greater than this figure, there may be no need for a strengthened control room if the chance of a massive leak is reduced by the presence of emergency isolation valves or if the chance of failure is reduced by subjecting pipelines to special procedures of inspection similar to those used for vessels.

EXAMPLE 2.1-2 A Tactical Problem – Pipeline Failure

This example (also from Kletz[22] illustrates the application of hazard analysis to a more detailed problem, the probability of failure of a particular pipeline. It also illustrates the use of a simpler criterion for target setting than the FAFR.

Refrigerated propylene at $-47°C$ has to be pumped through a heater into a 7-mile long pipeline made of ordinary mild steel, Figure 2.1-4. If the heater fails to raise the temperature sufficiently, the pipeline will become brittle and may crack so a low-temperature trip is installed after the heater. The pipeline temperature will also fall due to flashing of the liquid if the pump delivery pressure is lost when there is an open flowpath back to the tank. Is the single trip sufficient to safeguard against both potential causes for low pipeline temperatures, or should a second trip be installed as well?

The propylene is heated by a heat transfer oil that is, in turn, heated by steam. Flow rate from storage is restricted so as not to exceed the propylene heater capacity under the most adverse conditions (during initial pressurization of the 7-mile line at start-up). Propylene is required by the consumer units for practically 100% of the time so intermediate storage capacity is provided at the consumer end of the pipeline to cover occasional interruptions in pumping.

The various possible causes for overcooling are listed in Table 2.1-3, together with an estimate of their frequencies. Potential overcooling due to loss of pipeline pressure via the consumer buffer tank is not included in Table 2.1-3 because it would not be safeguarded against by the export low-temperature trip. It would need to be considered as a further possible cause for pipeline failure when examining the adequacy of the design of the consumer units.

The fractional dead time of the export low-temperature trip is calculated to be 0.0175. Hence, the predicted frequency for low pipeline temperatures due to the causes in Table 2.1-3 is about $0.42/yr \times 0.0175 = 0.0074/yr$, or once in 140 yr on average. Taking a weighted average probability of 40% for pipeline failure when subjected to low temperatures (based on expert opinion), thus gives a failure rate of about 0.003/yr. Note that:

1. 'Negligible' in Table 2.1-3 means negligible in proportion to the other figures quoted.
2. It is assumed that the trip and all the alarms are tested monthly and that the test is thorough.

We now have to decide what failure rate is so low that further action to reduce it is

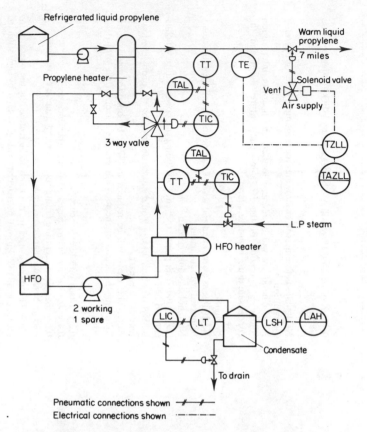

Figure 2.1-4 Schematic of propylene pipeline system.
TT = temperature transmitter (pneumatic); TIC = temperature indicator controller; TAL = temperature alarm (low); TE = temperature element (electric); TZLL = temperature trip (extra low); TAZLL = alarm connected to temperature trip (extra low); LT = level transmitter; LIC = level indicator controller; LSH = level switch (high); LAH = level alarm (high). In valve symbols with arrow pointing toward the controller, valve opens an air failure; and with arrow pointing away from controller, valve closes on air failure

not justified. We might estimate the probability that the gas will ignite, the probability that someone will be in the area at the time, and the probability that he will be killed.

A simpler method, however, is available: the frequency of fracture of the line from this new additional cause should be small compared with the existing probability of fracture. This has already been estimated at 0.001/yr for the 7-mile section as a whole. The additional contribution is thus three times the original and is clearly rather high. Duplication of the trip system was therefore recommended

Table 2.1-3   Possible causes of overcooling

Possible causes of overcooling	Frequency, assuming no protection
Loss of heat to steam/heat transfer oil (HFO) heater due to:	
Loss of steam pressure	0.05/yr
Steam valve fails closed	0.45/yr, of which 0.15/yr will not be detected by HFO low-temperature alarm
Flooding of condensate drum due to failure of level controller	0.5/yr
Fouling of steam/HFO exchanger	Negligible, as fouling will be slow and there is ample time for operator intervention
	Note: The above will be reduced to 0.2/yr by operator intervention following operation of the HFO low-temperature alarm and the condensate drum high-level alarm
Loss of HFO circulation due to:	
Sudden surge of level in HFO surge tank	Negligible
Gradual loss of level in HFO surge tank	Negligible as alarm fitted and several routine inspections in a row could be missed without harm
Line blockage	Negligible, as duty is clean
Pump failure (two simultaneously)	0.07/yr based on plant experience (excluding site power failure which is covered below)
Control valve fails shut	0.55/yr of which 0.15/yr would not be detected by the propylene low-temperature alarm
	Note: The above will be reduced to 0.2/yr by operator intervention following the action of the low-temperature alarm on the propylene stream.
Loss of propylene pump delivery pressure due to:	
Site power failure	0.2/yr based on site experience
Pump failure in service (cable, motor, gearbox or pump faults)	0.4/yr for sudden failure based on plant experience
Pump trip (genuine or spurious)	0.1/yr based on literature failure rate data and general experience
Pump stopped by operator without first isolating pipeline	0.05/yr based on plant experience
	Note: The above causes are not likely to be rectified by the operator in the time available, but will be reduced to about 0.02/yr by operation of the pump non-return valve
	Total demand on the low-temperature trip protection is thus: $0.2 + 0.2 + 0.02 = 0.42$/yr

(fractional dead time then 0.001). This will reduce the frequency of line fracture due to cold embrittlement to about $0.42/\mathrm{yr} \times 0.001$ @ $40\% = 0.000\ 17/\mathrm{yr}$, well below the existing figure.

It is worth noting that this study, including writing up in detail, required 210 minutes work by two men, one familiar with the process and the other experienced in hazard analysis. Note also that the hazard rates calculated will not be achieved in practice unless the trips and alarms are tested regularly. If trip and alarm testing is not the practice, introducing it is much more effective than installing additional trips.

A more detailed hazard analysis on potential fracture of a cross-country pipeline due to cold embrittlement has been published recently for a somewhat similar system handling liquid propane[25].

The derivation of the estimate of the probability of pipeline failure provides another example of how numerical methods can be used in safety studies. A US report[26] had shown that, on average, over a period of many years, pipeline failure had occurred at a rate of $47 \times 10^{-5}/\mathrm{mile/yr}$. The contributions due to each cause (human activity, internal and external corrosion, defect in construction, etc.) were listed. Using these figures as a baseline, a failure rate for the 7-mile pipeline was estimated, taking into account all the factors that could affect human activity, corrosion, etc. For example, it was judged that the probability of a defective field weld was about one-third of the US figure.

EXAMPLE 2.1-3   A Problem in Hazard Analysis with Solution

The problem

If the level in a high pressure distillation column is lost, hydrocarbon vapours would break through via the column bottoms heat interchanger and air cooler into a downstream floating roof storage tank, and so lead to a gross emission of flammable material from the tank roof into the storage area. There is also some chance that the bottoms exchanger or cooler would fracture due to liquid hammer, and thereby lead to a flammable release in the plant area. Thus, both the storage operators and the plant operators would be at considerable risk from a vapour breakthrough incident.

The analysis therefore involves:

1. Determining the maximum allowable frequency (i.e. the target hazard rate) for loss of level in the distillation column base in order to meet current safety standards with respect to both groups of operators.
2. Predicting the expected average frequency for loss of level with the proposed design for comparison with the target.
3. If the target is not achieved, considering possible design or operating changes in order to do so.

Plant system: Proposed design and brief process description (see Figure 2.1-5 for general outline, and Figure 2.1-6 for column bottoms detail)

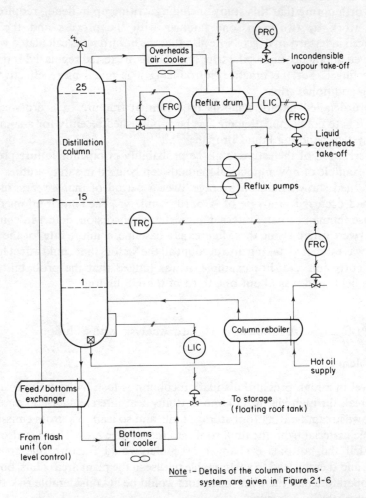

Note :– Details of the column bottoms·
system are given in Figure 2.1-6

Figure 2.1-5 General outline of plant system

A mixed hydrocarbon stream (C_1–C_{10}; 140 t/h) containing traces of dissolved hydrogen is fed on level control from an upstream flash unit operating at 50 °C/265 psig via the shell side of a heat exchanger to a continuous topping column (25 trays) normally operating at base conditions of 240 °C/235 psig. The column bottoms (C_5–C_{10} hydrocarbons; 135 t/h) pass on level control via the tube side of the exchanger and thence an air cooler (3-off unvalved banks in

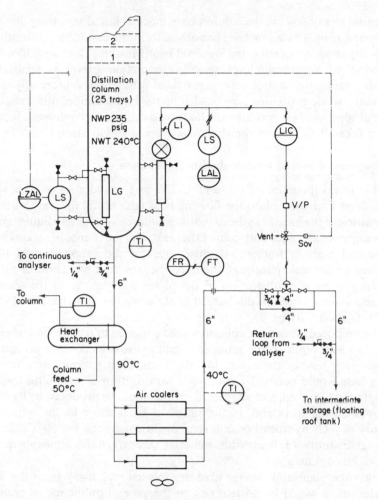

Figure 2.1-6 Original design

parallel served by 3 fans) into a floating roof storage tank (6000 m³). The heat exchanger and cooler are designed to reduce the temperature of the bottoms stream to about 90 °C and 40 °C, respectively. Most of the pressure drop occurs across the control valve, frictional losses via the exchanger and cooler being small by comparison.

Heat input to the column reboiler is on temperature control from round about the column mid-point, and column pressure is controlled by a PRC/PCV located on the 'incondensibles' gas take-off line from the reflux drum (H_2, and C_1–C_2 hydrocarbons; 1 t/h). Column reflux is on flow control, and the liquid overheads take-off (C_3–C_5 hydrocarbons; 4 t/h) is on level control from the reflux drum.

Potential loss of level in the column base (normal liquid inventory 20–25 t) is safeguarded against by a low level pre-alarm from the control loop transmitter, and also by an independent extra low level trip (Mowbrey float switch) which is designed to shut the column level control valve on falling level. To facilitate on-line maintenance, the control valve is provided with a bypass (normally closed).

Normal start-up procedure is to burden up the column under total reflux with the liquid tops and bottoms take-off lines isolated, and to commence tops and bottoms take-off as soon as the streams are up to specification.

Consequences of loss of level in distillation column

If level is lost, vapour initially at $240\,°C/235$ psig would start to pass via the exchanger and air cooler into the floating roof tank at full line rate with some condensation of the heavier hydrocarbons en route causing severe liquid hammer in the equipment and lines upstream of the tank, with the chance of serious impact damage and perhaps fracture. Because most of the vapour would not be condensed on passing up through the tank, the tank roof would lift or tilt, with the chance of buckling/distortion of the plates and damage to the seals, and flammable vapour would be discharged to atmosphere (estimated initial vapour escape rate about 20 t/h).

With the proposed design, the column would continue to receive feed when level is lost, and heat input to the reboiler would increase as column pressure and therefore tray temperature start to fall. Thus, high vapour rates from the column base would persist by continued vaporization of incoming feed until remedial action is taken, and additional vapour could be produced by flashing of hot liquid on the trays and, in the ultimate, by flashing of the reflux drum inventory (C_3–C_5 hydrocarbons; normal drum conditions are $50\,°C/225$ psig). The total quantity of flammable material discharged to atmosphere could therefore be considerable.

Because the plant and storage area are located well away from the Works perimeter, there would be no real risk to the general public in the event of a flammable emission. In the safety context, we are therefore concerned only with the risk to operators in the storage area due to a flammable vapour discharge from the tank, and to operators in the plant area due to a potential emission of vapours in the event of equipment fracture by hammer. Because the two areas are manned by different groups of operators, it is necessary to ensure that acceptable safety standards are met with respect to the man (or men) at greatest risk in each location. Target hazard rates must therefore be derived for both eventualities, and the more demanding taken as the basis for the analysis.

Derivation of the target hazard rate

From a detailed consideration of the area, working habits, wind direction and known sources of ignition, the probability that a storage operator would be killed in the event of loss of column level was estimated at 0.015 (i.e. 1.5% chance).

The probability that a plant operator would be killed in the event of an incident was similarly estimated at 0.006 (0.6% chance).

The risk to the storage operator is thus greater, and the maximum allowable frequency for loss of level in order to meet current safety standards in both locations (not more than 3.0×10^{-5} fatalities per man-year) is given by:

$$3.0 \times 10^{-5}/0.015 = 0.002 \text{ per year}$$

i.e. not more than one incident every 500 years on average or not more than 0.2% probable in any full year.

The average cost of a gross vapour breakthrough incident, calculated on a weighted average severity basis (i.e. with or without fire damage) and including consequential losses, is estimated at £40 000. Thus, if the safety target is achieved, the average annual cost of the incident will not exceed about $40 000 \times 0.002 = £80$, which is the sort of risk we would live with. A target hazard rate of 0.002 per year maximum is thus acceptable on economic grounds.

Prediction of the actual hazard rate for comparison with the target

The procedure for the prediction of actual hazard rate is as follows:

1. A 'fault tree' is first constructed to depict the events or coincidences of events capable of leading to the hazard, allowance being made in the tree for possible remedial operator action in the interval between the onset of the primary event and the time at which the hazard actually occurs, and for possible remedial action by the automatic protection (if any) in this same interval.
2. The completed fault tree is then quantified by incorporating numerical data (frequencies or probabilities) for all inputs to the tree, and processing the mathematics through to give the frequency for the top event.

Construction of the fault tree logic (Figure 2.1-7)

Loss of level can, by definition, occur only when the level control valve or its bypass is further open than it should be for the prevailing column conditions. Thus, the hazard can arise only in the following ways:

(a) Level sensing or measurement faults on the control loop, leading to a high reading in error (either at the time of the fault, e.g. torque tube or transmitter fault, or soon after, e.g. level impulse line blocked or isolated) and thereby causing the control valve to be directed too far open by normal action.
(b) Level control faults or errors causing the control valve to be directed too far open when level measurement is correct, e.g. controller or valve positioner faults, or misdirection of the controller when on manual control.
(c) Control valve faults preventing valve movement or full shut-off on demand.

(d) Control valve bypass too far open when controlling manually on the bypass, or bypass valve left open in error after a period on bypass operation.

Having determined all the events capable of leading to loss of level if they are allowed to exist or continue uninterrupted, allowance must then be made as already mentioned for possible remedial action by the operator or by the automatic protection. This is done by incorporating AND gates for 'operator fails to intervene' and for 'extra low level trip fails to close off column bottoms route to storage on demand'.

Remedial operator action will depend on an awareness of the need to act, on the ease of diagnosing what action to take, on the action required, and on the time available for action. The probability for failure to intervene will therefore be different for many of the primary causes for falling level, and a separate AND gate is required opposite each input.

The automatic extra low level trip protection will be totally ineffective for safeguarding against falling level due to mal-operation of the bypass or due to control valve faults preventing full valve travel and good shut-off (hence no AND gate required), but will normally be effective, i.e. when in proper working order, opposite all the other causes (hence AND gate for 'trip fails to close off bottoms route to storage' required).

This then explains the regrouping of the causes for falling level in Figure 2.1-7.

Quantification of the fault tree (Figure 2.1-7)

The frequencies and probabilities appearing in Figure 2.1-7 are derived as follows (see corresponding note reference in fault tree):

1. Assumes 0.1 fault/year for transmitter failure to high output (typical literature data), together with 0.02 occasion/year for impulse line or torque tube blockage (based on plant experience with the particular fluid, making allowance for the possible detection of developing blockages at routine cross-checks with the level gauge) and 0.03 occasion/year for inadvertent isolation of impulse lines (based on experience).
2. Typical literature data for failure to high output.
3. Based on literature data and plant experience. Covers all faults preventing valve travel/good shut-off on demand, e.g. bent spindle, fractured yoke, spring failure, valve seizure by fouling, damaged seat, fouling of seat, valve plug detached, etc.
4. Implies gross misdirection sufficient to call for remedial operator action based on the pre-alarm, or in the ultimate, operation of the extra low level trip. Gross misdirection would result largely from sudden changes in column conditions, e.g. feed rate, which are not quickly corrected for by the operator – say 0.5 occasion/year based on experience, which takes into account that the controller would be on manual only for a small proportion of total on-line time.

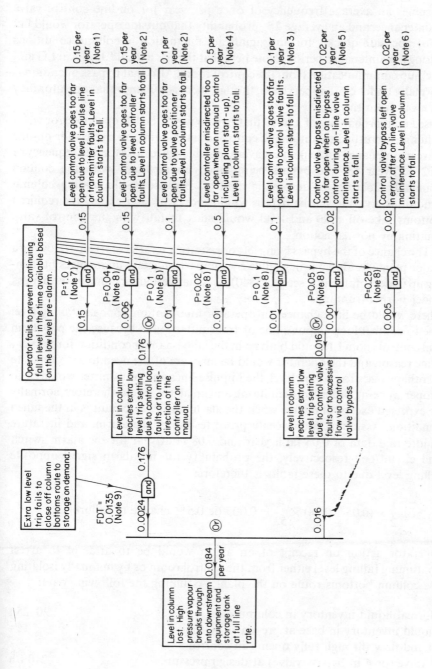

Figure 2.1-7 Fault tree for vapour breakthrough from high pressure distillation column into downstream equipment due to loss of level in distillation column base (original design). See pages 342–346 for reference notes

5. Assumes an average frequency of once per year for on-line control valve maintenance, and an average 2% probability that outside operator would fail to compensate quickly for any significant fluctuation in column conditions during a maintenance period. Note here that there is a local LI, a local LG and back-up communication from the control room. Also that bypass operation is normally under close supervision, though there is a chance that operator may go away from the job.

6. Based on a probability of 1 in 50 that bypass would be left open in error when column is returned to automatic control after running on bypass for on-line valve maintenance (once per year average), which gives an average frequency of 0.02 per year for the bypass being open unknowingly with the column nominally floating on automatic control. Because of fluctuating column conditions, the rate via the bypass would sooner or later exceed the required bottoms take-off rate, and level would start to fall with the control valve shutting by normal action.

   The chance of the bypass being left open during or after start-up following a major plant shutdown is negligible by comparison (start-up involves burdening up under total reflux; see page 340), as is inadvertent opening of the bypass under normal running.

7. There would be little chance for operator intervention opposite falling level due to control loop impulse line or transmitter faults because the pre-alarm and control room LI would both be in the failed-danger condition for the very same reason and the operator would be unaware of the need to act. Opposite all other causes for falling level, the impulse lines and transmitter would be in proper working order (by definition), and an alarm would therefore normally be expected except of course when the alarm switch or circuit is in the failed condition. Assuming four-weekly proof testing of the alarm, and literature failure rate data of 0.05 fault/year and 0.08 fault/year for the alarm switch and circuit/bell respectively, the probability for no alarm signal opposite falling level due to these faults is therefore:

$$\frac{1}{2} \times (0.05 + 0.08) \times \frac{4}{52} = 0.005 \text{ (ie 0.5\% chance of failure)}$$

8. Operator action on receipt of an alarm would be to attempt to arrest continuing falling level either from the control room or by manually isolating the column bottoms route on the plant. Assuming the following data:

Normal liquid inventory in column base	20–25 t
Liquid inventory in base at pre-alarm setting	15 t
Liquid flow through fully open 4 in control valve (or 4 in bypass valve) at design pressure	210 t/h
Normal column bottoms rate	135 t/h

the time available for remedial action from pre-alarm until level is totally lost when bottoms route goes fully open (worst case) under otherwise normal conditions (e.g. column feed continuing) works out at about

$$\frac{15}{210 - 135} \times 60 = 12 \text{ minutes}$$

The estimated probabilities (based on experience) for no effective remedial action on falling level in the 12 minutes available before column level is totally lost when the alarm is in proper working order are given below for the causes of interest, together with reasons for the differences.

Cause of falling level	Probability for no action when alarm sounds		Comments
	%	*	
Gross misdirection of controller when on manual	1.5	(2.0)	Correction possible from control room. Operator normally in close attendance
Controller faults	3.5	(4.0)	Ditto, but conditions more unexpected and operator more likely to be busy elsewhere
Positioner faults or control valve faults	10.0	(10.0)	Manual valve operation on plant required. Request for outside operation would be delayed in order to confirm that alarm is not spurious
Misdirection of bypass when on bypass control	4.5	(5.0)	Higher than for controller faults/misdirection because of communication factor to outside operator who is normally in attendance
Bypass left open in error after valve maintenance	25.0	(25.0)	Unusual fault. Operator would expect to have to close control valve isolations and may overlook bypass

* Figures in parenthesis make allowance for failure of the alarm (0.5 % chance; see Note 7). Thus, for controller faults, overall probability for no action is given by

0.5 %   + 3.5 % of 99.5 %   = c. 4.0 %

(no alarm)   (no action when alarmed)

9. When column level is lost due to causes in this section of the tree, the control valve would be in proper working order and therefore capable of carrying out its trip function. This is because control valve faults have been covered fully as potential causes for loss of level elsewhere in the tree. Thus, valve faults do not enter into the fractional dead time calculations on trip reliability opposite falling level due to the other possible causes. Trip reliability in this context is in fact merely a function of the trip circuit through to and including the solenoid valve, for which the following fail-danger fault rates are assumed:

Blockage of impulse line(s) or float chamber   0.03 f/yr
Random isolation of impulse line(s) in error*   0.02

Float switch faults (contacts break to trip)   0.07
Relay (de-energize to trip)   0.01
Solenoid valve (de-energize to vent)   0.13

* Most likely at a routine cross-check of local LI and LG. Inadvertent isolation at a proof test is covered separately below.

Thus, with four-weekly on-line proof testing, the fractional dead times due to randomly occuring hardware faults only are:
FDT for trip signal asking solenoid to vent

$$= \frac{1}{2}(0.03 + 0.02 + 0.07 + 0.01) \times \frac{4}{52} = 0.005$$

FDT for venting via solenoid valve on receipt of signal

$$= \frac{1}{2} \times 0.13 \times \frac{4}{52} = 0.005$$

To complete the fractional dead time calculations it is necessary to allow for inadvertent isolation of the float chamber from process at a proof test causing the trip to remain dead until the error is detected at the next scheduled test. Say one such error every 300 tests on average, which is equivalent to a dead time of one whole test interval every 300 intervals, i.e. to a FDT of 0.0033.

Finally, allowance must also be made for trip initiator disarmed time whilst testing (say 20 minutes per four-weekly test which is equivalent to $20 \times 13/60 = 4.33$ hours per year, or $4.33/8760 = 0.0005$ of total time). Taking an estimated 60% probability that appropriate remedial action would be taken by the operator or artificer in the event of a live trip demand during the test, gives effectively a disarmed FDT of 0.0005 @ 40% = 0.0002.

By summation, the total probability for failure of the extra low level trip to operate on demand opposite falling level due to causes where it would normally be effective is

(0.005 + 0.0033 + 0.0002)	+ 0.005
(for trip signal to solenoid)	(for solenoid venting)

$$= 0.0085 + 0.005 = 0.0135$$

Comparison of safety standards for the Figure 2.1-6 design with target: Need for design changes

With the Figure 2.1-6 design, the estimated frequency for vapour breakthrough on loss of level in the column base (1.84×10^{-2} per year; Figure 2.1-7) is an order of magnitude greater than target (2.0×10^{-3} per year; see page 341). This stems principally from the inability of the low level trip to safeguard against loss of level due to control valve faults or due to inadvertent flow via the control valve bypass (estimated total frequency for gas breakthrough due to these causes is 1.6×10^{-2} per year). Marginal improvement in the protection against loss of level due to control loop faults/controller misdirection (2.4×10^{-3} per year) is also necessary in order to reach target.

A 10-fold reduction in hazard rate by tackling the causes for the onset of falling level at source is clearly not a practical proposition, in that it would be difficult to increase the reliability of the level control loop or valve substantially, and the control valve bypass must be retained for on-line maintenance.

We are therefore left with possible design changes leading to a higher chance for remedial operator action on falling level and therefore to a lower demand on the trip protection (e.g. installation of an independent low level alarm to assist operator intervention notably against falling level due to control loop impulse line or transmitter faults where there is currently no pre-alarm warning), and/or design changes leading to a more reliable trip system which is effective against all causes for loss of level (e.g. new trip valve outside of the control valve bypass to safeguard against those eventualities where there is currently no trip protection). The latter modification, which is clearly essential in order to stand any chance of meeting target, is the one first considered, with retention of the trip function on the control valve so as to achieve some reduction in the frequency for loss of level due to control loop faults by virtue of the duplicated shut-off.

Evaluation of proposed modified design (Figure 2.1-8)

The proposed change is installation of a new 6 in trip valve, with associated solenoid valve, immediately upstream of the level control valve bypass. New trip valve to close on air failure and new solenoid valve to operate on a de-energize-to-vent basis from the relay of the existing extra low level trip channel. Trip function on control valve also to be retained as in the original design. New trip valve to be tested four-weekly on line using the 3/4 in valved drain on the LCV set to demonstrate good shut-off capability.

Fractional dead time calculations for revised trip system

With the revised design, the demand rate on the trip protection due to all causes will remain as in Figure 2.1-7 (namely 0.176 demands per year due to control loop faults/misdirection, and 0.016 demands per year due to level control valve/bypass faults), but the reliability of the trip protection will be improved as follows:

(a) Loss of column level due to control valve/bypass faults will now be safeguarded against by the new trip valve operating on a 1-out-of-1 basis from the extra low float switch on the column.

(b) Loss of column level due to control loop faults or to gross misdirection of the controller on manual will now be safeguarded against by both the control

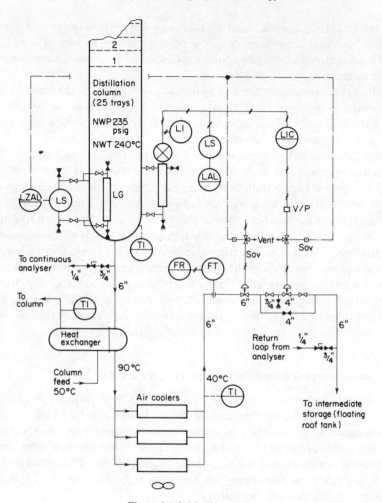

Figure 2.1-8 Modified design

valve and new trip valve operating on a 1-out-of-2 basis from the same float switch on the column.

To assess overall plant safety, it is therefore necessary to derive the fractional dead times of the protection against demands (a) and (b) above. Regular four-weekly proof testing is again assumed.

(a) Fractional dead time for shut-off by new trip valve only.
  Assuming typical literature fail-danger fault rates of 0.13 per year for the solenoid valve and 0.1 per year for the new trip valve (which has of course to be allowed for in the FDT calculations because it is not part of the control

channel), the FDT for trip valve action on receipt of a signal in trip box is:

$$\frac{1}{2}(0.13 + 0.1) \times \frac{4}{52} = 0.009$$

Adding the FDT for trip signal to box (0.0085 total for all causes; see Note 9 on page 345), gives a total FDT for trip action on demand of 0.0175, i.e. 1.75 % chance of failure.

(b) Fractional dead time for shut-off by new trip valve or by level control valve on a 1-out-of-2 basis.

By reference to Note 9, on page 345, the FDT for closure of the control valve on receipt of a trip signal equates with the FDT for solenoid venting (0.005) for all demands in the category of interest. Thus, from (a) above, the FDT for effective trip valve action by control valve or trip valve on receipt of a signal to trip box is:

$$\frac{4}{3} \times 0.005 \times 0.009 = 0.000\,06$$

Adding the FDT for trip signal to box (0.0085 total; see (a) above) gives a total FDT for trip action on demand of about 0.0086, i.e. 0.86 % chance of failure.

Predicted hazard rate with the revised design (*Figure 2.1-8*)

The expected total frequency for vapour breakthrough with the revised design is presented in fault tree Figure 2.1-9, based on the findings in Figure 2.1-7 adjusted for the improvement in trip reliability.

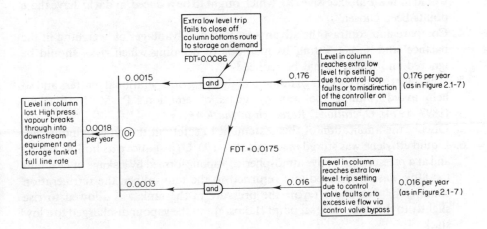

Figure 2.1-9 Fault tree for vapour breakthrough from high pressure distillation column into downstream equipment due to loss of level in distillation column base (Figure 2.1-8 design)

The results indicate that acceptable standards of safety would be achieved opposite potential vapour breakthrough by the proposed modifications (estimated hazard rate is 1.8×10^{-3} per year compared with a target of 2.0×10^{-3} per year).

From literature fail-safe fault rate data, the proposed modifications would be expected to increase the spurious trip frequency from about 0.67 per year (original design) to about 1.0 per year, i.e. by about 50 % or by about one extra trip every three years, which is regarded as acceptable.

Conclusion

The design changes shown in Figure 2.1-8 should be implemented.

Questions 2.1

1. Define what is meant by hazard analysis.
   Choose a hazard in an industry with which you are familiar and list the various steps involved in applying hazard analysis to this hazard.
2. Compare and contrast the advantages and disadvantages of hazard and operability studies and check lists as a means of identifying problems on process plants. Carry out a hazard and operability study on a section of plant with which you are familiar and see if it discloses hazards or operating problems which were previously not apparent.
3. What scales have been used for measuring the risks to which people working in a chemical plant and those living nearby are exposed? What points on the scale would indicate an excessive risk which ought to be reduced, and why have these points been chosen?
4. Compare and contrast the advantages and disadvantages of 'weighing in the balance' and 'target setting' as methods of deciding which risks should be ignored and which should be reduced.
5. 'The scientist can encourage numerical thinking on operational matters and so help avoid running the war by gusts of emotion'. P. M. S. Blackett (1897–1974), *Operational Research in the RAF*.
   Discuss the application of this statement to safety in the process industries.
6. Liquid ethylene was stored in a tank at $-100°C$ (just above its boiling point) and at a pressure just above atmospheric. Vapour formed by leakage of heat into the tank was refrigerated and returned to the tank. When the refrigeration compressor was under repair the pressure in the tank was allowed to rise slightly to the relief valve set point (1.5 psig) and the vapour discharged to a low stack.

   After design was complete and construction had started it was realized that cold ethylene vapour coming out of the stack would fall to the ground where it might be ignited. A late change in design was therefore made and steam was injected into the stack to warm and disperse the ethylene vapour.

Unfortunately the condensate from the steam, running down the stack, met the cold ethylene and froze. The stack (8 in diameter) became completely blocked and the tank was overpressured and split.

If a hazard and operability study had been carried out on the change in design, would this have detected the hazard? If so, how would this have been done?

7(a) A storage tank is fitted with a high level trip which automatically closes a valve in the inlet line when the level exceeds a preset value. The failure rate of the trip is once every 2 years, it is tested every 5 weeks (say 10 times per year) and the demand rate is 1/year

How often will the tank be overfilled?

The time required for testing can be ignored and it can be assumed that the trip is always re-instated correctly after testing.

(b) As (a), but the demand rate is now 100/year.

References 2.1

1. ICI Ltd *Assessing Projects* Book 5, *Risk Analysis*. Methuen, London. 1968.
2. Lawley, H. G. *Chemical Engineering Progress*, **70**(4), 45. 1974.
3. Lawley, H. G. *Hydrocarbon Processing*, **55**(4), 247. 1976.
4. Knowlton, R. E. *R & D Management*, **7**(1), 1. 1976.
5. Anon *Hazard and Operability Studies*, CIA, London. 1977.
6. Rushford, R. *North-East Coast Institution of Engineering Transactions*, **93**, 21 March, p. 117, 1977.
7. Austin, D. G. and Jeffreys, G. V. *The Manufacture of Methyl Ethyl Ketōne from 2-Butanol* Institution of Chemical Engineers, Rugby, Chapter 12. 1979.
8. Kletz, T. A. *Chemical Engineering Progress*, **72**(11), 48. 1976.
9. Lees, F. P. *Symposium Series No. 47*, Institution of Chemical Engineers, Rugby, p. 73. 1976.
10. Health and Safety Executive, *Canvey*, HMSO, London. 1978.
11. Advisory Committee on Major Hazards, *Second Report*, HMSO, London. 1979.
12. Jones-Lee, M. W. *The Value of Life*. Robertson, London. 1976.
13. Mooney, G. H. *The Valuation of Human Life*. MacMillan, London. 1977.
14. Kletz, T. A. *Chemical Processing*, 77. 1974.
15. Kletz, T. A. and Lawley, H. G. *Chemical Engineering*, p. 81, May. 1975.
16. Starr, C. *Science*, **165**, 1232. 1969.
17. Starr, C. *Perspectives in Benefit – Risk Decision Making*. National Academy of Science, Washington DC, USA, p. 17, 1972.
18. Kletz, T. A. *Chemical Engineering in a Changing World*. W. T. Koetsier (Ed.). Elsevier, Amsterdam., p. 397. 1976.
19. Kletz, T. A. *Hydrocarbon Processing*, **56**(5), 297. 1977.
20. Gibson, S. B. *Chemical Engineering Progress*, **72**(2), 59. 1976.
21. Sinclair, C. *Innovation and Human Risk*. Council for the Study of Industrial Innovation, London. 1972.
22. Kletz, T. A. *Chemical Engineering Progress*, **74**(10), 47. 1978.

23. Bell, G. D. *Risks to a Reactor from an Adjacent Oil Refinery.* UK Atomic Energy Authority, Warrington, England. 1974.
24. Kletz, T. A. *Proceedings of the 4th International Conference on Pressure Vessel Technology.* Institution of Mechanical Engineers, London. 1980.
25. Lawley, H. G. *Reliability Engineering* Vol 2, October, 1980.
26. Federal Power Commission. *Safety of Interstate Natural Gas Pipelines.* US Government Printing Office, Washington DC, USA. 1966.

Chapter 2.2

Computer Security

K. K. Wong

Introduction

Security is the protection of the well-being of the organization in conducting its business operations and achieving business objectives.

We regard computer security as the protection of computer facilities and resources to safeguard the proper functioning and survival of the computer department against disruption, destruction, and unauthorized disclosure of computer systems and information processed in the organization.

In this paper we shall review the overall perspective of computer security and highlight areas for management concern. A framework is discussed to determine the control strategy. Finally, ground-rules are laid down for encouraging on-going security awareness and planning for contingency.

Corporate Implications

Figure 2.2-1 shows the typical usage of computers in various business functions in an organization.

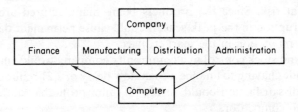

Figure 2.2-1 DP- interaction in an organization

The increased dependence on computing tends to result in a reduction of clerical manning level. In the face of an extended computer disruption, many organizations would have difficulty in coping with the situation by temporarily reverting to clerical back-up. For companies with a long history of computing, the manual back-up procedures may have never been invoked in the past, or tested for acceptance. Worse still, the alternative manual procedures may not be suitable for simple conversion to computer input. This would further aggravate and prolong the recovery operation.

In the meantime, the company would suffer the following consequences:

1. Loss of income through invoicing delays to customers, lost interest on accounts receivable, lost sales and lost future business.
2. Additional cost for extra manning and interest payments on bridging loan.
3. Loss of discount on accounts receivable.
4. Operational inefficiency through lack of control on production and scheduling, and delays in supply and distribution.
5. Legal and contractual penalties for late delivery and potential lawsuits.
6. Loss of goodwill to staff through delays in wage payment, and customers through missing delivery dates.
7. Delays in management reports could lead to inadequate market planning and sales forecasting, and delays in year-end accounts could result in loss of public confidence, and cause a slide in the company's shares.

Business interruption losses are often difficult to ascertain because computer management is rarely provided with the full facts and figures of the company's profit and loss potential.

The following are highlights of some of the security review findings of potential corporate losses arising from an extended computer disruption:

— £1 million of new business per week for a financial institution would be lost throughout the disruption period.
— Serious delays in sending out invoices would cause a publishing house a cash flow of £50 000 per week being held up.
— Between £2 and 3 million of business per week to a manufacturing company would be at risk. Since the products being manufactured are seasonal, a serious disruption in the peak season could cause permanent damage to the company's reputation in the market place.
— Invoicing and stock control in a commodity company would run completely adrift. Besides having to finance a cash flow delay of £21 million per month, customer dissatisfaction could result in some profitable outlets lost permanently to competitors.
— Calculation of wages to hourly paid employees in a brewery would be held up. If disruption lasts for more than two weeks, unless agreement can be reached on alternative pay procedures, union withdrawal of labour would be almost certain.

At the national level, it is now a matter of public record that a prolonged strike of Civil Service computer staff, which halted the collection of VAT, caused a significant increase in the Public Sector Borrowing Requirement and thereby materially contributed to a record 17% Minimum Lending Rate in November 1979.

For a company operating on tight profit margins or high turnover of stock, a major disaster could reduce public confidence in the management and cause a slide in share prices in the stock market. Worse still, the business may never fully recover from such financial losses, culminating in a permanent setback to its operations.

To avoid serious omissions, adequate resources and specialist expertise are needed to build up an assessment of the total possible loss in a large company which is heavily dependent on computing for its normal business activities. Senior internal staff, though perfectly competent at their job, have neither the time nor the experience essential for making such assessments.

Risk Areas

Figure 2.2-2 illustrates the major risk areas of *physical damage* and *system interference* to the computing environment.

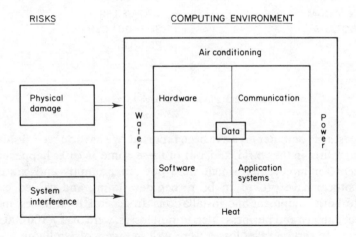

Figure 2.2-2 Risks to the computing environment

Main frame computers require certain basic amenities in the physical environment in order to function properly. Computer systems, on the other hand, rely on the reliability and integrity of input data, central hardware, system software, communications equipment, and application programs to produce meaningful information which should be correct and timely to help management with decision making and business administration.

We shall examine the two risk areas in greater detail in the following two sections.

Physical Damage

Facts and figures

The causes of damage could range from natural hazards of fire, flood, land subsidence, etc. to deliberate acts of arson, explosion, vandalism, and sabotage. Damage may be inflicted on various physical assets ranging from equipment, the physical environment, to storage media, documentation, and people. This is illustrated in Figure 2.2-3.

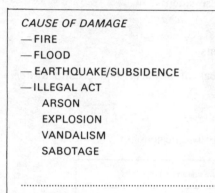

CAUSE OF DAMAGE	*PHYSICAL ASSETS*
— FIRE	EQUIPMENT/NETWORK
— FLOOD	ENVIRONMENT
— EARTHQUAKE/SUBSIDENCE	DOCUMENTATION
— ILLEGAL ACT	INPUT/OUTPUT DEVICES
ARSON	STORAGE MEDIA
EXPLOSION	STATIONERY
VANDALISM	DATA/FILES
SABOTAGE	SUPPORT SERVICE
	PEOPLE

Figure 2.2-3 Physical Damage to Assets

Many cases of computer fire have been reported. We have on record some 60 to 70 computer fires in the world, and, out of these, some 36 cases happened in the UK. Except for minor fires which were put out promptly and caused little damage, others have proved to be rather devastating and caused extensive damage to both equipment and installations. In general, total losses, including both direct damage and business interruption losses, were over £100 000 sterling in each case. In certain cases losses were well in excess of 1 million.

More than 50% of UK fires were started outside the computer suite; 25% were started outside the computer room in the installation; and 16% in the computer room itself. Disgruntled employees, ex-employees, and security guards were responsible for deliberately starting 34% of the fires. Electrical faults caused 30% of the fires.

The above statistics have reinforced the need for adequate fire resistant partitioning both within the installation and from the non-DP functions in the company. Early detection systems should be installed with automatic total flooding systems, in addition to good procedures in the computer room. A contingency plan, which should be tested and proven, ought to be provided to speed up the recovery operation from a major disaster.

Many cases of flooding of computer centres have been recorded in the UK. Many of these were the result of severe winter conditions causing pipe burst. A few were due to heavy rainfall in low-lying areas or a river bursting its banks, or air-conditioning malfunction, causing pipe overflow. The last category resulted in water accumulating in the floor void and could take some time to discover. There was even one case reported of the flooding of an installation located on the twenty-second floor of an office building in London! Most modern computer buildings are now provided with sloping true floors with proper drainage. Some older ones have installed water detectors in the floor voids.

There has been a number of cases reported of fire and explosion caused by such extremist groups as the IRA, and of vandalism by youths. One of the installations suffered two cases of arson within 18 months. Both fires were deliberately started in the same areas, causing serious hold-up in business and inflicting £500 000 sterling of damage on each occasion.

The gloomy economic forecast for the 1980s indicated further depression and higher unemployment. Extremist groups will continue to flourish with occasional bouts of activity. Intellectuals with extremist views in pursuit of ideological goals are particularly dangerous to the community because their schemes of perpetration could be extremely well planned and executed with sophistication. This was exemplified by over 20 cases of bombing of computer centres in Italy, mostly executed by young intellectuals, with calculated speed and callous accuracy. In France several cases have been reported of bombing of computer installations by by an extremist group called CLODO (the Committee to Liquidate or Neutralize Computers). The importance of access control into computer installations and computer rooms cannot be overemphasized, especially outside office hours.

Good recruitment and termination procedures, along with regular progress reviews of staff performance would help to provide early warning of disgruntlement and to initiate preventive measures to pre-empt possible abuse to the installation. Contigency planning plays a key role to ensure the continued survival of the organization.

Contingency planning

Figure 2.2-4 illustrates a typical recovery time scale from a major disaster.

For an installation with a medium to large main frame computer, it takes at least two–four weeks (if you are lucky and stock permitting) to acquire replacement equipment from suppliers, and a further two weeks for installation and commissioning before processing can recommence.

In the meantime, the backlog is gradually building up and requires phased clearance. This can be time consuming and may take up to 12 weeks to complete. Hence at week 16, the installation is back to normal operation. However, on the clerical side the disaster could have more far reaching consequences and another

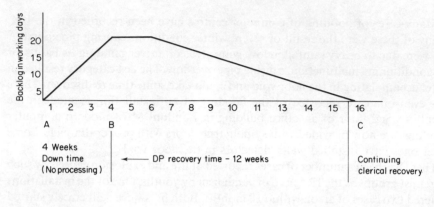

A = Incident
B = Resumption of DP processing
C = DP regained normal production running

Figure 2.2-4 Typical recovery time scale from a major disaster

three months could easily elapse before normal business operations are completely restored.

Such a time scale presupposes that a contingency plan exists and has been proved to work. For a company which has not made any contingency preparations, the delays could be further prolonged. For example, it could take six months to equip new premises with the appropriate power supply, air conditioning, fire protection and access control equipment, plus the installation of transmission lines and communications equipment for on-line working.

There are three aspects to contingency planning:

1. Drawing up a contingency plan
2. Salvage of data and damaged equipment; immediate standby
3. Long-term recovery

In our experience, the longer the interruption, the more severe is the business disruption, and the more difficult it is to return to normal business and computer operation.

To speed up recovery requires the proper coordination of action plans for all key DP areas, as well as making necessary provisions to meet all operational requirements of such plans. In addition, computer users must be involved in the contingency planning preparations to provide additional clerical support to cope with an extended computer disruption.

Considerable resources are needed for the preparation of a detailed action plan. Moreover, various key areas and personnel responsible must be properly coordinated to reduce unnecessary delays and bottlenecks.

The contingency plan should cover the following aspects:

— General strategy; budget authorization; recovery framework.
— Coordination and administration of action plans.
— Detailed action plans for key DP areas.
— Asset inventory.
— Off-site storage requirements.
— Immediate standby arrangements.
— Computer insurance.
— Preparation of reserve site to house replacement equipment.
— Operational requirements and back-up facilities for all key business systems; loss implications.
— Up-dating of action plans to meet future operating requirements.

The up-dating of the plan could pose problems through personnel changes, system upgrades and changes in operating environment, e.g. from batch to on-line processing and distributed network.

The salvage and stock taking of various assets could be helped by the provision of an asset inventory. The damaged or destroyed items would be reordered from suppliers. This includes all main frame units, communications equipment, office equipment and sundry items which are essential for re-establishing the computer service.

If left unattended, contamination from products of combustion could cause permanent damage to the data recorded on magnetic media. The salvage operation should include the copying to new media of all information held on magnetic storage in the disaster location.

Many installations have mutual standby arrangements with others for emergency running. However many such arrangements are rather informal in nature and procedures for invoking standby are rarely tested. Subsequent changes in the operating system, main frame units or storage media (e.g. exchangeable to fixed discs or single to double density discs) could cause havoc arising from slightly incompatible operating environments. The workload of the standby installation may also have increased substantially to render such back-up an impractical proposition in future.

Regular review and testing of standby facilities become more and more urgent if the installation requires immediate back-up at short notice. Appropriate job controls and operating instructions for the back-up installation should be prepared before the event. Discrepancies in accounting or file naming conventions should be resolved when the standby arrangements are first contemplated. There is little room for experimentation or improvisation at a time of extreme emergency. Valuable time and resources could be lost unnecessarily.

The long-term recovery aspects present a new dimension to disaster planning. Various units of equipment may have different lead times for delivery. Most suppliers are sympathetic and would rush through the orders as best they could.

However, it would be foolhardy to assume that immediate delivery could be achieved in every instance. A realistic assessment should be made of the likely delays in delivery for various items and orders for equipment progressed through some network analysis schedule.

If a speedy recovery is vital to the survival of the company's business, it may be prudent to carry a spare stock of those units or components which are hard to come by or could take considerable time to install. For example, this applies to the setting up of a reserve site with the necessary air conditioning, ducting and power supply to accommodate the main frame computer, as well as the installation of modems and communication lines for on-line working.

At present, several commercial concerns are strongly advocating the use of a disaster centre. In essence, a group of companies would subscribe for the provision of a common reserve site which is fully equipped to receive a main frame computer at short notice. A subscriber would be able to use the facilities provided when faced with a major disaster in his installation.

Insurance

Most installations have adequate insurance cover for risks of damage to computer equipment. For leased equipment, the lease agreement normally excludes cover for malicious damage from disgruntled or mentally deranged employees or contractors. Few installations provide cover against denial of access arising from environmental causes, say air conditioning malfunction, or building faults, say collapse of false ceiling or cracking of flooring. In each case the computer service has to be suspended although there may be no damage to computer equipment. However the company may have to provide extra finance to continue the processing of data at another installation. Adequate cover for increased cost of working arising from denial of access and other disasters such as extended hardware failure or computer fire should be provided in order to recover additional costs to maintain a temporary processing capability.

Financial losses incurred by the company arising from a major computer disaster may be recovered through adequate cover for business interruption and loss of profits. If the business operation of the company may be susceptible to computer-related fraud or embezzlement, fidelity guarantee cover should be obtained for both accounting and computer staff.

System Interference

There are two forms of interference to computer systems. The abnormal events include fraud, theft, privacy breaches, industrial action and others, and the *normal* events include, equipment breakdown, system breakdown or malfunction, and errors and omissions.

Abnormal events

Statistics collected of computer abuse in both the US and the UK indicate that the trend is growing. In the UK most computer frauds involved sums of money not exceeding £10 000 sterling. Isolated cases, on the other hand, could well exceed £1 million. Of the 30 odd cases of computer-related fraud collected in our case files the average sum embezzled was £30 000. Schemes of perpetration differ, but mainly exploiting simple loopholes in systems. For example, payment was authorized for goods on order which were never delivered. Special discount facilities were exploited for private gains. The latter involved collusion and detailed knowledge of the sales and accounting system. The most effective safeguards remains with proper segregation of duties, input/output control, proper authorization procedures for system amendments, and to a lesser extent, structured design, and the use of high level languages render the concealment and realization of private gain more difficult. Early detection is helped by the installation of adequate audit trails and monitoring for access violation and unreasonable working habits and lifestyles of employees.

Cases of theft have been reported of sensitive print out, computer equipment and machine time. The most publicized case in Europe was the ransom demand of £250 000 from ICI for the return of magnetic tapes and discs in Rosenberg.

With the growth of personal computing, there is more computing expertise at large to enable potential electronic criminals to perpetrate schemes on the kitchen table. There is wider scope for sophisticated perpetration to interfere with the increased usage of electronic funds transfer systems, where the stakes could be much higher and the abuse more difficult to detect. New technology has brought impetus on business users to exploit the merits of electronic office, word processors, point of sale systems, and electronic mail. The security and protection requirements of these new systems have yet to be defined, and in the meantime could fall prey to potential abuse.

The proliferation of small business systems available on mini- and micro-computers and with such equipment kept in office areas is becoming a growing concern for management. The relatively high value of such systems among other office equipment has brought about a number of reported thefts of portable computer equipment housed in areas with little access control. Moreover manning of such small systems tends to be restricted to one or two individuals with overlapping authority of several normally segregated functions. For example, the person feeding in routine transactions could well be the same individual who runs the equipment as well as maintaining the software. Unless the associated clerical control procedures are adequate, small business systems could be exploited to perpetrate simple frauds and other forms of abuse by disgruntled or dishonest staff in charge of such systems.

Privacy breaches through both accidental and deliberate disclosure of sensitive company and personal information have caused some public concern in the UK.

This is exemplified by the misuse of police computer systems to recruit new clients for gaming clubs and disclosure of sensitive market plans and new product information through exploiting ill-conceived confidential waste disposal procedures in computer installations.

Access control procedures would need tightening up, enforced by secure operating systems, sophisticated access control software and promotion of staff security awareness. The 'need to know' and 'right of access' of confidential information should be reviewed, in some cases, protected by applying encryption in data transmission. Again, early detection would be facilitated by proper audit trails, monitoring for access violation and active involvement of internal auditors in system design.

The UK has been particularly vulnerable to industrial dispute in computer installations in the last few years. Some disputes were caused by non-DP staff in dispute picketing computer installations to prevent staff access. Others were caused by computer staff involved in disputes to work to rule, occupy premises, or going on strike. The most disruptive strikes were those involving the Post Office and the Civil Service in 1979. In each case, several hundred million pounds of cash flow was delayed through the planned withdrawal of a small number of computer operators for an extended period. At the time this led to the Government considering planning to combat future strike action by stiffening industrial relations legislation and reverting to free collective bargaining with total relaxation of pay restraint policy.

Normal events

Equipment breakdown could happen to the central hardware, input/output devices, ancillary equipment, communications equipment and others, with varying degrees of impact on the installation. The technological advances over the last few years have rendered equipment more reliable at the individual unit or component level. However, the general architecture of hardware units and systems is getting more complex. Depending on the importance of a piece of equipment, there could be redundancy features built into the system to alleviate a total breakdown arising from individual unit failure. If necessary, whole systems may be duplicated either at the same site or remotely. Standby arrangements could also be made with hardware manufacturers, another company or a computer bureau. Such arrangements ought to be formalized and regularly reviewed to ensure that standby resources are adequate and available on demand. Good restart and recovery procedures are essential for maintaining the integrity of computer systems.

Most software and application system breakdowns or malfunctions arise as a consequence of poor design, insufficient system testing, and inadequate consultation with users during development. The trend is for more complexity in the systems and software in order to improve performance and to integrate mutually dependent functional areas. The design philosophy and system architecture, on the other hand, tend to be moving towards modularity. There is growing usage of microcodes and middleware which could bend the hardware architecture towards the environment of the user to facilitate his system and programming efforts.

Errors and omissions sometimes contribute to system breakdown and malfunction. For example, in the US a programming error in the design calculation for the nuclear reactor cooling system has caused a structural weakening of cooling pipes and resulted in the evacuation of local residents in Pittsburg. An inexperienced operator working alone on a night shift responded incorrectly to an error misread message from a disc pack and ruined all the three generations of files on various disc packs, in addition to damaging all disc drives on site. In another instance, a telephone subscriber was billed for £2172 for making two local calls.

Major computer manufacturers have made considerable efforts to combat operational errors and omissions in computer installations by providing large operating systems to reduce the extent of operator intervention and to provide more meaningful error diagnostics. Data processing management are channelling more efforts to adopt formal system development methodology, coupled with heavy emphasis on system documentation, structured design, structured walkthrough and system testing. Installations are going for centralized data control and validations such as data directories and data dictionaries. Future privacy legislation would also require installations to pay more attention to the accuracy and timeliness of personal data. There is more emphasis on staff and user training, on the handling of the equipment and systems, as well as the insistence of operation standards and procedures and the provision of comprehensive error diagnostics and restart and recovery procedures.

Risk Management

Thus far, we have examined the two main areas of computer risks, i.e. physical damage and system interference. Management interest areas have been highlighted on the potential business impact arising from an extended computer disruption. Let us now consider how management should work out an acceptable security level for their companies.

In our experience the security awareness and interest of senior executives tend to be fairly low. Proper risk management requires a high level of personal involvement and participation from senior management to play a major role in deciding on the right level of risks which are acceptable to the organization.

This could be achieved by carrying out a *corporate security review* and a *corporate loss assessment* on the company's computer services to users. The corporate security review examines computer processed information as a corporate asset to identify the risks of unavailability, errors and omissions, abuse, and unauthorized disclosure, and to determine their potential implications. Each risk should be assessed in terms of frequency of occurrence and severity of loss or damage incurred. The corporate loss assessment will examine key individual business systems to identify and, where appropriate, quantify losses to the company arising from both short-term and long-term disruption of computer facilities.

Management service executives and data processing managers should be concerned about computer risks and their business implications. As so often happens, by default they have been left with shouldering the sole responsibility for security with little backing from senior management. As a result only lip service is being paid to budget for security to provide for necessary capital spending. Computer managers should acquaint the Board of Directors with the risk exposures relevant to their installations and the likely impact on the organization's operations, and persuade the latter to accept ultimate responsibility for such exposures. The case for senior management attention should be presented with a proper framework of the proposed security strategy for the installation regarding risks, loss implications and costs of protection, in a form that is appropriate for the Board of Directors to appreciate and act to determine implementation priorities and resource scheduling, and to authorize capital expenditure for risk control.

This would be achieved by conducting an *installation review* to examine possible risks in all key areas in the computer installation to assess the various protection and control requirements, and review the extent these have been met with existing control procedures and protection facilities.

Finally, at the operational level, current provisions for standby and back-up facilities at the installation and elsewhere for both short-term and long-term outage of computer power should be reviewed for adequacy in coping with the critical response times required by various key computer systems. If the company is heavily dependent on the use of computer resources in conducting its business operations, business interruption losses could escalate through an extended computer disruption following a major disaster on the installation. In that case, a *contingency plan* should be produced in order to speed up the recovery operation. Adequate insurance cover should also be provided to recover any financial losses.

Conclusion

A systematic and logical approach to reduce computer risks requires the use of proper risk management techniques to identify various risks which are specific to the installation and organization and evaluate their likelihood and consequences. A risk control strategy can then be applied to facilitate early detection and cost effective protection. Total prevention may prove extremely expensive for many risks. In that case a conscious decision should be made to determine an acceptable level of risk taking, through provision of risk reduction measures. Such measures could range from the setting up of new, or improvement of existing, control procedures to the acquisition of access control hardware or software and fire protection equipment.

Guidelines and procedures should be drawn up with detailed action plans to respond to any disaster in order to speed up recovery. This should be augmented by capital expenditure to meet the needs of remote storage of duplicated data, programs and documentation, and standby requirements. Contingency planning is an on-going activity, not a one-off exercise. Contingency plans require regular testing, review, and updating. The major problem facing management in this area remains with that of balanced spending – weighing up potential losses against paying out on protection costs.

Questions 2.2

1. Describe briefly a business computer system you are familiar with. What are its functions?

   Assess the impact on the organization if the system becomes unavailable for processing for two hours, two days, two weeks, and two months. What alternative back-up arrangements can you call on to cope with the organization's requirements in the meantime? Are they effective?

2. Take the same system in Question 1. What would be the impact on the company's business or goodwill to customers if the information produced by the system is fraught with errors and omissions? Can such errors be exploited for private gains? Suggest at least three safeguards which can be built into the system to reduce such errors.

3. Pay a visit to the computer installation which processes the system in Question 1. Make an estimate of the total replacement value of the computer and ancillary equipment.

   Is the installation susceptible to flood, fire caused by flammable materials or processes in neighbouring areas, sabotage or vandalism?

   What would you recommend to the computer manager as counter-measures to improve the installation's physical security? Take into account (*a*) the effectiveness of each counter-measure, (b) cost of installation or implementation, and (*c*) normal operating cost incurred.

4. List three major computer systems used by your organization. What do you regard are the major risks to these systems? Give an account of the risk control measures already in use. Are they effective? If not, can you suggest possible improvements or new controls? What criteria would you use to persuade your management to adopt your recommendations?

Further Reading 2.2

It is difficult to cover the many aspects of computer security in detail in a paper of this size. The following references provide additional sources for further reading:

1. Farr, Chadwick, and Wong. *Security for Computer Systems.* NCC Publication 1972. This book details most of the techniques in hardware, software, and procedures on various aspects of security. The appendix also gives some idea of the cost and effectiveness of each technique.
2. Wong, K. K. *Computer Security – Risk Analysis and Control.* NCC Publication 1977. This book describes in detail the application of risk management techniques in computer security, together with many cases of computer disasters and computer-related fraud discovered in the UK.
3. Wong, K. K. 'Implications of security and privacy in the 80's.' *Information Privacy,* 2 (3), May 1980.
4. Wong, K. K. 'Contingency planning – backing up a distributed system.' *Information Privacy,* 73 (3), May 1981.
5. Wong, K. K. *Security – a matter of reducing the risks.* The Director's Guide to Computing, Director Publications 1977.
6. Wong, K. K. *Computer Related Fraud.* State of the Art Report on Computer System Security. Pergamon Infotech Publications 1981.
7. Wong, K. K. 'Putting a Price on Information.' *Computer Management,* March 1975.
8. Wong, K. K. 'Information – How to analyze its value.' *Computer Management,* April 1975.
9. Wong, K. K. 'Golden rules for buying your first software.' *Computing Europe,* 13 September 1979.

High Risk Safety Technology
Edited by A. E. Green
© 1982 John Wiley & Sons Ltd

Chapter 2.3

Systems Reliability Assessment in Health Care Engineering

A. C. Selman

Introduction

The technological complexity of a hospital is not widely appreciated. While there is a good understanding of the technology involved in modern medicine, for example heart–lung machines, kidney machines, scanners and so on, there is in general little awareness outside of the technical fraternity of the hospital service, of the scale and complexity of engineering installations and equipment in hospitals. These range from basic hotel services such as hot and cold water, gas lighting, heating and electrical power, to more sophisticated services including piped medical gases such as oxygen, nitrous oxide, medical grade compressed air, specialized air conditioning, emergency electrical power supplies, etc. Telecommunication and alarm systems proliferate in hospitals as do large quantities of equipment which, while they do not interface directly with the patient, play a vital part in his treatment. This equipment includes computerized and other analytical devices in laboratories, blood and organ storage refrigeration equipment, centrifuges, various types of autoclaves (sterilizers), etc., as well as the more dramatic life saving equipment already referred to.

Since the primary function of a hospital is to provide patient care it is essential that all the systems associated with that care perform satisfactorily and that all reasonable steps are taken to ensure that failures and outages of the various technological installations of the hospitals do not adversely affect the patient. In the process of eliminating as far as possible technology-based risks, it becomes necessary to identify the relationship between a particular installation or piece of equipment and the treatments available to the patients.

It is fairly clear that with many pieces of medical equipment, particularly when used in a life support situation, the relationship between the equipment and the patient is a very close one and that manoeuvring room in the event of failure is very limited indeed. On the other hand the failure of the public water supply to the hospital is unlikely to be an immediate hazard to the patient, first, because it is generally well removed from the patient interface and, second, because there is usually a second source of supply, i.e. there is redundancy in the system. Not all situations are however quite so clear-cut and it is often dependent on the experience of the engineer to determine the real relationship between a particular system and the patient. Consider for example an air conditioning system. In the normal course of events the primary function of an air conditioning system is to provide a comfortable working or living environment. Not so the air conditioning system of an operating theatre suite. Here the system is the first line of defence against airborne cross-infection and the effectiveness of surgical procedures undertaken, and hence the safety of the patient is largely a function of the integrity of the air conditioning.

It will be readily appreciated that with any engineering installation or equipment there will be a risk of failure or, to put it another way, there is always an element of probability that a particular system will fail to perform in the manner desired. It will be further appreciated that the more critical the performance of a system is in terms of its contribution to the overall capability of a hospital to function the more important it is to quantify the risks associated with system failure. In a hospital this determination can become very complex since criteria are many and variable while the true importance of a potential failure may be obscured by operational circumstances. For example while breakdowns in life support devices are obviously serious and to be avoided at all costs, the failure of such a device may not have the same overall impact as, say, the failure of a pharmacy autoclave (i.e. a sterilizer) which may affect the treatment of a large number of patients, possibly giving rise to a life or death situation as in the case of the more obvious life support equipment failure.

The services and equipment in a hospital may, for the purposes of discussion, be divided into three broad categories, viz. (i) engineering services applicable to the hospital as a whole, (ii) equipment which, while essential to the care of the patient, is not directly involved with the patient's treatment, and (iii) equipment which is directly involved in the treatment of the patient. Techniques of system reliability assessment are applicable to each of these categories and can be of considerable value in identifying areas of design weakness in each of them. It will be useful to continue by examining an example in each of the categories.

Emergency Electrical Supplies

The continuing and rapid advances in the application of high technology in the field of medicine have resulted in an increasing reliance on the permanent engineering installations of the hospital and, in particular, electrical power. The functioning of high risk areas of the hospital such as operating theatres, intensive care units, premature baby units and so on are almost totally dependent on an electrical supply of high integrity. In addition to these high risk patient areas there are many items of equipment in a hospital such as blood banks, human organ storage units, incubators and alarms that are equally critical in terms of the integrity of the electrical supply. Accordingly it is now general practice to equip hospitals with emergency generators to provide standby power in the event of a public utility supply failure. A typical hospital generator set-up is shown in Figure 2.3-1 in simplified diagrammatic form. It is usual for the emergency generator to be sized to provide 40–50% of the normal load of the hospital. The manner in which dependency on electrical power has grown is well demonstrated by comparing this provision with that of some 20 years ago when the electrical standby supply often consisted of no more than batteries capable of supplying just the operating lamps in the theatres for an hour or so. It may also be noted that the increased dependency on electrical supplies is unhappily matched by an increase both in the frequency and duration of electrical outages on the main supply.

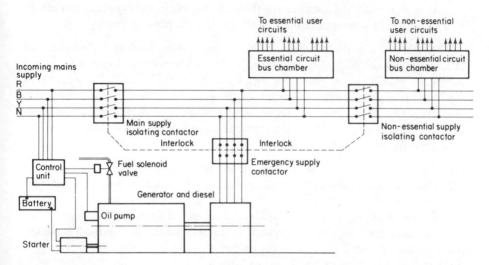

Figure 2.3-1 Schematic system of a typical hospital generator and emergency electrical supply distribution

While emergency generators are hardly novel those used in hospitals are in many cases subject to problems that are perhaps more acute than those found elsewhere. These are as follows:

1. Despite the fact that 45% (say) of the hospital load is met by the generator this still means that the generator is quite small by comparison with, for example, those found in nuclear or petrochemical installations. Typically the emergency generator for a medium size acute hospital will be about 650 kV A. This usually means the purchase of a packaged proprietary set which removes from the purchaser the ability, other than on a limited basis, to vary the machine specification. In a situation where high reliability is sought this constitutes something of a major limitation.

2. It is the exception rather than the rule to find duplexed sets in hospitals. To a large extent this is due to the relatively small size of the sets but it is also due to the fact that duplexing either means the provision of synchronization gear or more complicated bus-bar splitting both of which incur capital cost penalties of an unacceptable order. Either way the reliability improvement that might be afforded by the building of redundancy into the system does not exist.

3. The critical nature of the patient areas covered by emergency supplies often results in a reluctance to permit testing of a generator against its genuine load. This particular difficulty arises from the fact that, on balance, it is subjectively decided that less of a risk is involved in not running the generator against the real load and taking the chance that it will operate correctly when required than in deliberately disconnecting the main supply in the expectation that the generator will start and run correctly for the purpose of testing.

This latter issue itself raises an interesting point, and this concerns the difference between reliability of starting and reliability of running of a generator set (i.e. a diesel powered generator set). Consider first the reliability analysis of starting. Most generators in hospitals of the acute type have automatic starting and loading facilities of varying complexity. Assuming a fairly simple installation a certain sequence of events must take place if the generator is to start and connect its output to the load as follows:

(a) Mains failure must be sensed by the no-volt relay.
(b) Non-essential sections of the distribution system must be disconnected from the bus-bars.
(c) The mains supply incomer must be disconnected from the bus-bar.
(d) The connection between a serviceable battery and the starter must be made.
(e) The starter motor must turn the engine.
(f) The fuel supply must be made available.
(g) Oil pressure must develop.
(h) The contactor between the generator and the essential side of the bus-bar must make.

The failure of any one element in this series of operations will result in the failure of the system as a whole and it is therefore of considerable value to be able to identify weak links in the chain so that action may be taken to reduce the overall failure probability if necessary. Green and Selman[1] have demonstrated a simple approach to the analysis of failure to start probability using generic failure data, as shown in Table 2.3-1.

$$\mu_n = \frac{\theta/10^6 \times t}{2} \tag{2.3-1}$$

where θ is small and μ and θ are exponentially related; t is the period in hours between test runs which, on the basis of a six week interval, is approximately equal

Table 2.3-1 Failure to start generic data

Mode of failure	Failure rate per 10^6 hours	Probability of failure on demand
Batteries	14.0	0.007
Mains failure sensor	0.6	0.0003
Main contactor	2.0	0.001
Starter motor	5.0	0.0025
Fuel solenoid	2.0	0.001
Oil pressure switch	5.0	0.0025
Bendix drive	2.0	0.001
Load contactor	2.0	0.001
Bus-bar isolation contactor	2.0	0.001

to 1000 hours. The overall probability of failure is given by

$$\mu_T = \sum_{n=1}^{n} \mu_n = 0.0163 \qquad (2.3\text{-}2)$$

if all faults are independent. This simple calculation shows that in theory the generator set will fail to start once in every 60 demands. There is an immediate indication that the battery constitutes a weakness in the system and points to a need to consider whether a different type of battery, or indeed a different method of turning over the engine, is desirable.

   The determination of overall failure rates based on generic data is very useful but subject to error which enters because there is no guarantee that the generic data being used is applicable to the specific item of equipment under consideration either by virtue of component design, operating environment, standard of maintenance, etc. The validity of the figures used can only truly be determined by comparing them with field performance data obtained from the installations in question, such data being all the more valuable if the population of installations is large. The National Health Service in England has about 6000 generators of various sizes and has carried out a data collection exercise on the reliability of some 150 sets. In 2417 runs there were 87 failures to start, and the failure rate, θ is given

$$\theta = \frac{n}{Nt} \qquad (2.3\text{-}3)$$

n = number of failures
N = number of tests
t = proof test interval
  = 3 weeks

If, on average, the generator becomes unserviceable half way between proof tests it will be unavailable for $t/2$ hours and the probability of failure on demand μ_A is given by

$$\mu_A = \frac{\theta \times Nt \times t/2}{Nt} = \frac{\theta t}{2} \tag{2.3-4}$$

substituting for θ

$$\mu_A = \frac{n}{2N} \tag{2.3-5}$$

thus the mean unavailability in practice is given by

$$\mu_A = \frac{0.5 \times 87}{2417} = 0.018$$

It is thús the case that in practice, albeit using a shorter proof test period than that hypothesized in the theoretical analysis, the expectation of failure to start is about 1 chance in 55 starts. In fact this figure is subject to slight modification since it is desirable to take into account the repair fault rate and if this is assumed to be 5%, μ_A is given by

$$= \frac{0.5 \times 87}{2417} + \frac{0.05 \times 87}{2 \times 2417} = 0.019$$

It will be seen that the difference between the two values is small. It is evident that there is considerable similarity between the value for μ determined from data bank information and that of μ determined from trial data.

The determination of the running reliability of an emergency generator is more difficult for a number of reasons. First, data are difficult to obtain because, by the very nature of its purpose, the machine does not spend a lot of its time running. Second, those faults observed under test running conditions may often be atypical because of the impracticability of testing the set against a genuine load. Third, the faults occurring during running will be rare compared with the faults to be expected during the start-up sequence. However, during the same trials that give rise to the failure to start statistics observation was also made of running failures occurring. The average running time per test was two hours thus the total accumulated running time was 4834 hours. During this running time nine random failures occurred thus giving a running failure rate of

$$\theta_R = \frac{9}{4834} = 0.019$$

or 2 failures per 1000 hours of running. This compares well with the figure of 2.8 failures per 1000 hours determined Magnon et al[2].

Laboratory Equipment – Sterilizers

In the case of the emergency generator it is of primary importance that the machinery functions immediately on demand and the main value of system reliability analysis in this context lies in its ability to identify weaknesses in the system which might prevent this criterion being met. The priority of immediacy of response is not, of course, the only criterion against which reliability assessment can be made and a case in point is the hospital pharmacy fluids sterilizer[3].

In essence a fluids sterilizer processes a range of fluids which typically includes water, saline solution, dextrose, laevulose, chlorhexidine, glycine, and sodium citrate. These may in some cases be administered intravenously, sometimes used for irrigation and in general be used on, in, and around a patient. The requirement for non-contaminated fluids is obvious and there is a need to ensure that neither bacterial or pyrogenic (non-bacterial) contamination occurs. In the case of bacterial contamination protection is afforded by subjecting the fluid to a sterilizing cycle which consists of heating the fluid to a high temperature by enclosing it in a steam pressure vessel for a prescribed length of time. The bottles containing the fluid are fitted with self-sealing caps which permit ventilation as the fluid is heated but seal the bottle as it begins to cool, thus preventing the ingress of non-sterile air when the pressure vessel is ballasted at the end of the cycle. Contamination of the fluid by pyrogens is prevented by ensuring that the fluid is pure at the commencement of the cycle and that no subsequent opportunity is afforded for such matter to enter the bottles.

A typical sterilizer scheme is shown in Figure 2.3-2. The apparatus consists of a

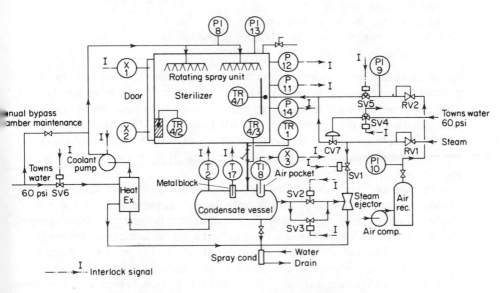

Figure 2.3-2 Schematic system of pharmacy fluids sterilizer

single chamber into which bottles are loaded. Normally the sequence of events is controlled automatically, commencing with a period of steam heating and purging at atmospheric pressure which is continued until the temperature sensed in the pressure vessel drain reaches 100 °C. At this point a drain temperature interlock initiates the pressurized sterilization phase by closing the chamber drain and activating the first of two timers. Expiration of the sterilizing phase starts a cooling cycle by shutting off the steam supply and admitting cooling water and pressurizing air. The second timer is started which controls the duration of the cooling phase and ensures that the chamber door cannot be opened until (a) the correct time has elapsed, (b) the bottles have cooled sufficiently, and (c) the chamber is at atmospheric pressure once again, the temperature being sensed by a thermostat in a dummy bottle, while pressure is sensed directly from the chamber.

Analysis of the system shows that there are basically four ways that a fluids product can be rendered unacceptable by failures of the system. These are as follows:

1. Product has failed to be sterilized for one of the following reasons:
   (a) insufficient sterilizing time;
   (b) inleakage of non-sterile pressurizing air*;
   (c) inleakage of non-sterile cooling water*;
   (d) low sterilizing temperature.
2. Product has been overprocessed as a result of the failure of temperature or time controls
3. Product has been chemically contaminated by:
   (a) oil in pressurizing air*;
   (b) chemicals in the cooling water*.
4. Product is contaminated by particles in the bottle prior to sterilization, e.g. glass particles from the bottle itself.

In the cases marked with an asterisk it will be apparent that not only must the contamination be available to a given load, but there must also be within the load a bottle or bottles with defective cap seals. Given the random nature of bottle selection it follows that in any load there will be less than 100 % faulty bottles and that the most likely situation is one where only part of the load is defective. This is the more dangerous situation since it is much less likely to disclose itself on the instruments of the sterilizer.

Although the fault conditions listed may all, with great advantage, be investigated using systems reliability techniques, it is particularly in the case of this less readily identified situation that the approach is most valuable. Consider the situation obtaining if the probability of part of a load being non-sterile is to be determined. The hazard logic diagram is shown in Figure 2.3-3. Since it is hypothesized that only part of the load is non-sterile it is argued that the criteria of cycle time and temperature were met (otherwise the whole load would be non-sterile). Accordingly it is implicit that for the fault condition to occur either non-sterile coolant must enter a bottle, or, alternatively, non-sterile air. The former can only happen if the chamber floods as the result of either the steam trap on the supply failing to open or the chamber drain line being blocked without the operator noticing the effect that this has on pressure decay at the end of the cycle. The latter situation can only occur if the drain thermostat (TS. 1) or the control system associated with it fails. It will be seen from the hazard logic diagram that the probability of partial non-sterile product per cycle is 4.1×10^{-3}. It also shows that the particular fault is primarily associated with two system components, the steam trap and the thermostat TS. 1. It is also evident that the risk of an operator not noticing the blocked drain outlet is considerable although the risk is largely offset by the low incidence of drain blockage.

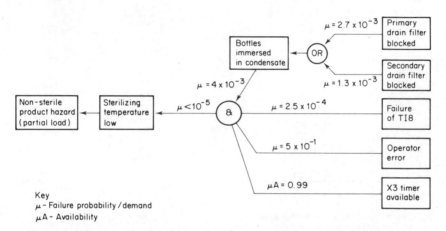

Figure 2.3-3 Hazard logic diagram for the probability of part of a load being non-sterile

   The effect that the human factor has on reliability assessment still requires further investigation to permit better risk quantification but based on current standards of about $\mu = 5.0 \times 10^{-1}$ it is interesting to compare the failure probabilities per demand for the system under automatic control with those when the system is operated manually. Table 2.3-2 gives such a comparison.
   In the case of the sterilizer there is, in addition to the risk from the operator, a risk to the operator. Basically there are two main risks. The first arises from the possibility of the door being opened before the bottles have been properly cooled. This may result in bottles exploding as a result of thermal shock with the operator consequently being subjected to the hazard of flying glass. The second risk arises from the possibility of the door release mechanism being opened while the chamber is still under pressure. The analytical procedure is identical to that previously described and Table 2.3-2 may be extended as in Table 2.3-3. The complete hazard to operator logic diagram is shown in Figure 2.3-4.

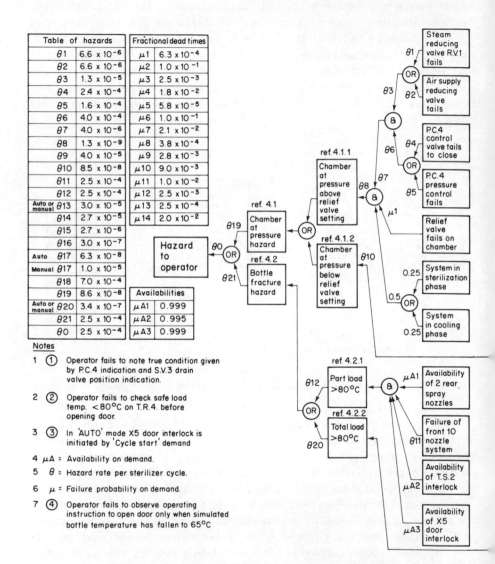

Table of hazards		Fractional dead times	
$\theta 1$	6.6×10^{-6}	$\mu 1$	6.3×10^{-4}
$\theta 2$	6.6×10^{-6}	$\mu 2$	1.0×10^{-1}
$\theta 3$	1.3×10^{-5}	$\mu 3$	2.5×10^{-3}
$\theta 4$	2.4×10^{-4}	$\mu 4$	1.8×10^{-2}
$\theta 5$	1.6×10^{-4}	$\mu 5$	5.8×10^{-5}
$\theta 6$	4.0×10^{-4}	$\mu 6$	1.0×10^{-1}
$\theta 7$	4.0×10^{-6}	$\mu 7$	2.1×10^{-2}
$\theta 8$	1.3×10^{-9}	$\mu 8$	3.8×10^{-4}
$\theta 9$	4.0×10^{-5}	$\mu 9$	2.8×10^{-3}
$\theta 10$	8.5×10^{-8}	$\mu 10$	9.0×10^{-3}
$\theta 11$	2.5×10^{-4}	$\mu 11$	1.0×10^{-2}
$\theta 12$	2.5×10^{-4}	$\mu 12$	2.5×10^{-3}
Auto or manual $\theta 13$	3.0×10^{-5}	$\mu 13$	2.5×10^{-4}
$\theta 14$	2.7×10^{-5}	$\mu 14$	2.0×10^{-2}
$\theta 15$	2.7×10^{-6}		
$\theta 16$	3.0×10^{-7}		
Auto $\theta 17$	6.3×10^{-8}		
Manual $\theta 17$	1.0×10^{-5}		
$\theta 18$	7.0×10^{-4}	Availabilities	
$\theta 19$	8.6×10^{-8}	$\mu A1$	0.999
Auto or manual $\theta 20$	3.4×10^{-7}	$\mu A2$	0.995
$\theta 21$	2.5×10^{-4}	$\mu A3$	0.999
$\theta 0$	2.5×10^{-4}		

Notes

1 ① Operator fails to note true condition given by P.C.4 indication and S.V.3 drain valve position indication.

2 ② Operator fails to check safe load temp. $<80°C$ on T.R.4. before opening door.

3 ③ In 'AUTO' mode X5 door interlock is initiated by 'Cycle start' demand

4 μA = Availability on demand.

5 θ = Hazard rate per sterilizer cycle.

6 μ = Failure probability on demand.

7 ④ Operator fails to observe operating instruction to open door only when simulated bottle temperature has fallen to $65°C$

Figure 2.3-4 Complete hazard logic diagram for hazard to sterilizer operator

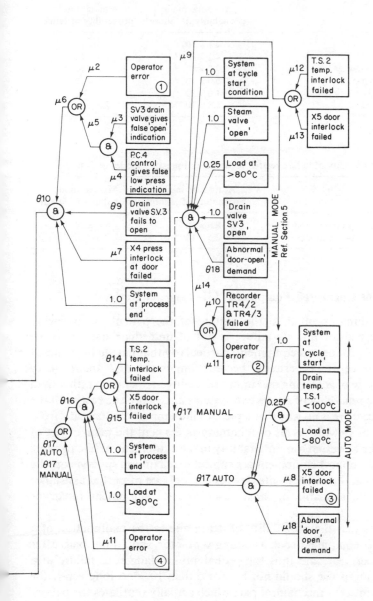

Table 2.3-2 Comparison of automatic and manual control

Hazard	Atomatic mode probability of failure	Manual mode probability of failure
Total load non-sterile	4.1×10^{-2}	5×10^{-2}
Part load non-sterile	4.1×10^{-3}	1.3×10^{-2}
Non-sterile product (bottles/cycle)	7.0×10^{-2}	7×10^{-2}
Particulates in product (bottles/cycle)	6.0×10^{-2}	6×10^{-3}
Chemicals in product (bottles/cycle)	1.0×10^{-3}	1×10^{-3}
Total load overprocessed	1.8×10^{-2}	2.3×10^{-3}

Table 2.3-3 Probability of failure related to bottle and pressure modes of failure

Hazard	Automatic mode probability of failure	Manual mode probability of failure
Hazard to operator/cycle (pressure mode)	8.6×10^{-8}	8.6×10^{-8}
Hazard to operator/cycle (bottle mode)	2.5×10^{-4}	2.5×10^{-4}

Patient Connected Equipment–Life Support Devices

The sterilizer, although essential to the welfare of the patient, is somewhat removed from the bedside and its failure is not of immediate danger to the patient. With patient connected equipment, particularly that having life support functions, different reliability criteria become important and the primary objective becomes one of achieving maximum reliability in operation rather than one of ensuring the reliability of protective systems as is more the case with the sterilizer. In practice there is no way that a piece of life support can 'fail safe' from the patient's point of view. In terms of reliability assessment the main object of any exercise must be to design out unreliability in the equipment. While patients being treated on life support equipment are subject to maximum supervision, the provision of alarms and quasi-fail-safe devices can only be of partial assistance given the short time available in most cases to take equipment repair or rectification action.

Sayers[4] suggests that the probability of death due to the malfunction of a piece of life support equipment such as a lung ventilator, should not be greater than about 10^{-5} per year, and thus the probability of dangerous failure of a ventilator in continuous use should not be worse than 10^{-5}. A lung ventilator may be split down into the mechanical part which actually ventilates the patient and the monitoring part which supervises the performance of the machine and generates alarms if necessary. It has been shown that even the best ventilator monitoring systems have a fail–danger probability slightly worse than 10^{-3} (if the proof period does not exceed one week) and it follows that if the overall fail to danger probability is to be better than 10^{-5} the fail–danger probability of the mechanical part of the machine must be better than 10^{-2}. This figure is difficult to achieve without a measure of redundancy which will (a) add to the cost and (b) add to the complexity of the equipment, but without using some sort of analytical technique it is difficult to identify precisely where improvement might best be made.

Consider the simplified baby incubator shown diagrammatically in Figure 2.3-5. The incubator is seen to consist of a chamber in which the environment is carefully controlled in terms of temperature, humidity, and oxygen concentration. Air is constantly circulated through the chamber by means of a small fan driven by a direct coupled electric motor. The air temperature in the chamber may be raised, in response to a demand from a chamber thermostat, T, by an electric heater in the air input system. Humidity in the chamber may be varied in response to a demand from the chamber humidistat. The performances of the air conditioning systems are monitored by simple independent alarm systems. Oxygen is introduced into the chamber through a manually operated valve and monitoring and adjustment is effected by reference to an oxygen flowmeter.

The failure of any of the four systems will put the infant's life at risk and Figure 2.3-6 shows the hazard logic diagram for the equipment. Based on the probabilities of failure given in Table 2.3-4, and assuming a proof test period of one week, the overall probability of failure on demand μ_T, is 1.7×10^{-4} which is an order worse than would normally be acceptable[5]. It is human interpretation of the alarms that brings the overall failure probability down to the acceptable level of 1.7×10^{-6}. This is dependent on a human failure probability of between 10^{-2} to 10^{-3}, this range being applicable to highly trained personnel in contrast to the figure of 10^{-1} quoted earlier for unskilled personnel operating sterilizers. Although the figure of 1.7×10^{-6} is acceptable it would of course be even better if it were not for the relative unreliability of human beings.

Not all life support equipment possesses monitoring facilities nor indeed does it necessarily operate in the carefully controlled environment of a hospital. A case in point is the implantable heart pacemaker which leaves the hospital in the patient. It will normally only return to the hospital for battery replacement. Pacemakers come in a variety of shapes and sizes but a typical pacemaker weighs about 100 g, is about 50 mm in diameter and is some 25 mm thick. Its object is to stimulate the heart in the event of failure of the natural cardiac pacemaker. Such stimulation may be required continuously or, quite often, only for occasional periods when the natural pacemaker failure is of only a short duration. To meet criteria of satisfactory performance the pacemaker must (i) sense whether the natural pacemaker is operating and (ii) commence artificial stimulation if it is not.

In electronic terms the circuitry of a pacemaker is relatively simple and the electronic reliability, given proper quality of control of the components, is fairly easy to establish. A typical value for the failure rate of a pacemaker circuit is 2.4 per 10^6 hours[5]. This failure rate is largely due to the batteries (one fault per 10^6

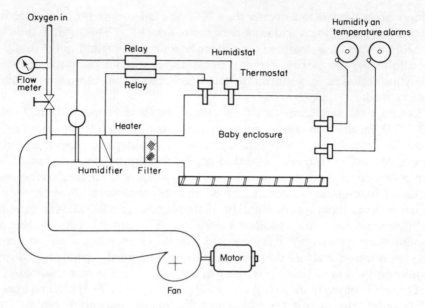

Figure 2.3-5 Schematic system of simple baby incubator

Table 2.3-4 Detailed probability of failure

Component	Probability of failure on demand
Thermostat	1×10^{-3}
Thermostat relay	2×10^{-3}
Heater unit	5×10^{-3}
Heater failure alarm	1×10^{-3}
Air filter	3×10^{-3}
Fan bearings	10×10^{-3}
Fan motor	4×10^{-3}
Air flow failure alarm	1×10^{-3}
Oxygen supply	0.5×10^{-3}
Human factor	3×10^{-3}
Humidistat	2×10^{-3}
Humidifier relay	2×10^{-3}
Humidifier	5×10^{-3}
Humidifier failure alarm	1×10^{-3}

hours) and the large-scale integrated circuits (also one fault per 10^6 hours). This corresponds to a mean time between failures of 47.5 years which means that if the operational life of the pacemaker is three years, 1 unit in 16 will require attention during its lifetime, although it should be noted that such an in-life failure is not normally of great seriousness as long as repair is carried out reasonably quickly.

It is not however the reliability of the electronic circuitry that is dominant in the

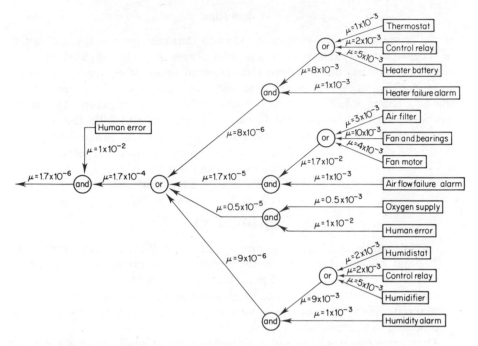

Figure 2.3-6 Hazard logic diagram for baby incubator shown in Figure 2.3-5

overall reliability of the device in its ability to perform satisfactorily. Since the pacemaker is sited in a hostile environment, the human body, its circuitry must be protected from that environment. The manner in which the circuitry is packaged therefore becomes an important element in the reliability equation. A further source of unreliability may be found in the pulse conducting lead(s) between the pacemaker and the heart itself. In some cases bipolar electrodes are used in which two leads are taken from the pacemaker to the heart, while in some cases one electrode will consist of a stainless steel plate which forms one side of the pacemaker with only one lead being taken to the heart. Lead fracture and electrode detachment from the heart are major causes of failure but it is obviously difficult to collect or collate data relating to this type of failure given the somewhat itinerant nature of the device and the fact that, since there is no proof test period, it is extremely difficult if not impossible to determine basic statistical information such as mean time between failure. In any event the selection decision on such aspects as pacemaker type, lead and electrode configuration, etc. is taken essentially on clinical grounds and it is easy to conceive a situation where a particular configuration desirable from a reliability point of view is unacceptable from a clinical one.

Conclusion

It may be seen that in this section consideration has been given to the application of systems reliability assessment to a number of areas in the health care field. They are clearly diverse, ranging from the industrial range of equipment, through scientific/laboratory equipment to equipment used directly on the patient, but all share the common factor of being essential to the care of the patient. The theme has been one of moving closer and closer to the patient until in the last situation consideration was being given to the reliability of the actual patient–equipment interface. It is to be conceded that statistical data relating to this aspect are sparse but it is interesting to conjecture upon whether the next step is to apply reliability technology techniques to the components of the patient himself.

Questions 2.3

1. Consider the hazard logic diagram of Figure 2.3-3. If the control components T18 and X3 were duplexed and the operator was replaced by an automatic lockout system having a failure probability on demand of $\mu = 1 \times 10^{-4}$, draw the resultant hazard logic diagram and calculate the failure probability overall of the sterilizing temperature being low and an unacceptable product being produced in consequence.
2. The suction pump of a medical vacuum system is automatically switched on at 08.00 hours each morning, i.e. 365 times per year. It switches itself off at 20.00 hours each evening. In a given year it fails to start on 5 occasions. On each occasion when it fails to start it takes an average of two hours to repair. Calculate the probability of failure on demand. If the repair fault rate is 10% calculate the amended probability of failure on demand that results.

References 2.3

1. Green, M. R. and Selman A. C. 'Emergency generator systems in hospitals.' *Health Service Engineering*, **38**, 5–12. 1978.
2. Magnon, R., L'Henoret. J., and Reynaud, Y. 'Reliability of standby diesel generator sets in nuclear plants.' *International Conference Proceedings. Reliability of Power Supply Systems*. IEE, London. 1977.
3. Selman, A. C. and Hignett, K. C. 'Systems reliability assessment applied to steam sterlizers.' *Hospital Engineering*, **33(1)**, 6–16. 1979.
4. Sayers, B. *The Use of Quantitative Reliability Techniques in the Design of Medical Equipment*. United Kingdom Atomic Energy Authority, Systems Reliability Service report SRS/GR/34. 1975.
5. Eames, A. R. 'Quantitative reliability techniques – an aid to better medical equipment design.' *I. Mech. E. Journal*, **5**, 31–34. 1976.

High Risk Safety Technology
Edited by A. E. Green
© 1982 John Wiley & Sons Ltd

Chapter 2.4

Coal-Mining

V. M. Thomas

Introduction

To gain an appreciation of mining hazards and the problems in reducing them, it is necessary to sketch, in outline, the layout and activities typical of underground coal mines.

The actual winning or cutting of coal in longwall mining as practised in Europe, and to a much smaller but growing extent in the USA, Australia, and South Africa occurs at the coalface, an excavation of approximately full seam thickness, normally 1 to 2 m in height and 100 to 300 m in length. Figure 2.4-1 depicts such a coalface. Cutter–loader machines travel along the face depositing the mineral on to an armoured conveyor, essentially a series of segmented steel slide plates along which the coal is propelled by chain-driven transverse slats or flights. Access to the coalface for personnel, for transport of the mineral outbye (i.e. outwards from the workings to the shaft and the surface), transport for materials inbye, and for the circulation of ventilating air is provided by two 'gate' roadways, one at each end of the face.

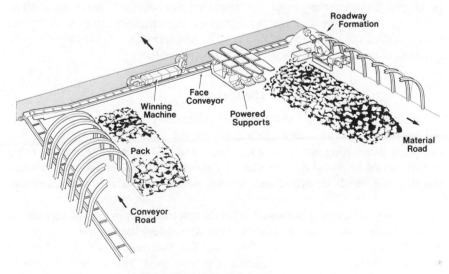

Figure 2.4-1 Longwall advancing coalface

Roof support at the coalface is limited to a strip of a few metres wide, running the full length of the face. It is provided by a line of hydraulic chocks, each consisting of a number of hydraulically extendible legs which support the roof linked to the armoured face conveyor by an hydraulic jack.

Face operations consist of: (i) taking off a strip of coal from the exposed face some 0.5 to 1 m wide, during a traverse of the shearer, a milling-type cutting machine; followed by (ii) advancing the conveyor a corresponding distance by exerting a forward thrust upon the conveyor using the hydraulic jacks attached to the chocks; and (iii) drawing up the chocks, one after the other, by lowering their supporting legs from the roof, applying hydraulic pressure to the ram linking them to the face conveyor to draw it forward, and subsequently reapplying pressure to the hydraulic legs to support the roof once again. After advancing the powered supports one after the other in this way, the unsupported roof soon collapses forming a rubble-filled waste area, or goaf, which subsequently slowly compacts. Due to the strata pressures and yield, the adjacent gate roads providing access to and from the face, are liable to be deformed despite the arched steel supporting 'rings' which provide some resistance. In severe cases roadways may require to be re-ripped or enlarged to ensure an adequate cross-section for access and ventilation purposes.

In total, an established colliery may have 50 to 100 km of roadways consisting of gate roads in the winning districts, of trunk roads for conveying, ventilation and travelling, and auxiliary routes for example in the vicinity of the shafts. A number of collieries also use inclined roadways or drifts, to the surface in place of shafts, especially for mineral transport since there are a number of advantages in a continuous transport system compared with discontinuous shaft winding operations. Services that must be provided for effective working include ventilation; electric power; water; transport for men, material and minerals; and communications. A major factor always to be borne in mind is that the area to be served is continually changing as extraction proceeds.

The Hazards

The hazards in coal mining and their relative severity are summarized in Table 2.4-1 presenting the number of fatalities and the fatal frequency rate per 10^8 hours' exposure. Comparisons with European and US fatalities are made in Table 2.4-2. It should be noted that bracketed figures in the table have been derived from data differently compiled and are not strictly comparable with the other data.

Particularly to be noted in Table 2.4-1 is the low incidence in the UK recently of fatalities from many hazards traditionally associated with mining, i.e. those arising from explosions, fire, and flooding. The two major contributions fall under the headings of 'falls of ground' and transport. The latter is the more important, with over 30% of all underground accidents in the period 1973 to 1977 falling into this category.

Table 2.4-1   Fatalities in UK mining (underground)

Cause	Year 1978	1977	1976	1975	Average 1970 to 1974	Average 1965 to 1969	Average 1960 to 1964
Falls of ground	12	6	14	8	20	52	102
Transport (including shafts)	29	20	20	26	29	48	78
Machinery	4	3	6	6	5	10	10
Explosions (of gas or coal dust)	2	0	0	5	1	7	17
Inrushes of water	0	0	0	0	2	0	0
Fires	0	0	0	0	0	2	0
Electricity	0	0	2	0	1	1	1
Stumbling/falling/slipping	1	1	1	1	1	0	0
Others	0	4	2	9	4	11	18
Total underground	48	34	45	55	63	131	226
Hours $\times 10^8$ (underground)	2.87	2.94	2.87	3.03	(3.20)	(4.94)	(7.03)
Fatal accident Frequency rate (per 10^8 hours)	15	11	16	18	(20)	(27)	(32)
Output tonnes $\times 10^6$	122	121	122	127	(119)	(164)	(188)
Fatal accident frequency rate (per 10^8 tonnes)	35	27	37	43	(53)	(80)	(120)

Note: Some figures prior to 1975 are derived from differently defined statistics and are approximations only: these are bracketted.

Table 2.4-2 Fatal accident frequency rates for UK, European and US coal mining (underground)

Period	Rate per 10^8 hours exposure UK	Germany	France	USA	UK	Rate per 10^8 tonnes mined European Community	USA
1960–1964	32	65	36	106	119	235	54
1965–1969	27	55	40	103	80	151	44
1970–1974	20	42	48	88	54	105	38
1975–1977	15	38	32	45	36	84	39

It is possible to deal here with only a few subjects since the field is wide. In the remaining sections on mining, after an introduction dealing with the general approach to safety only two aspects are dealt with: first, explosions and fires since the methods by which these have been so markedly reduced are clearly of importance; and second, transport, which is the most important single category of current accidents.

The Approach to Safety

The history of mining accidents over the last 130 years, epitomized in Figure 2.4-2 indicates both the great need for improvement, a need that impressed itself

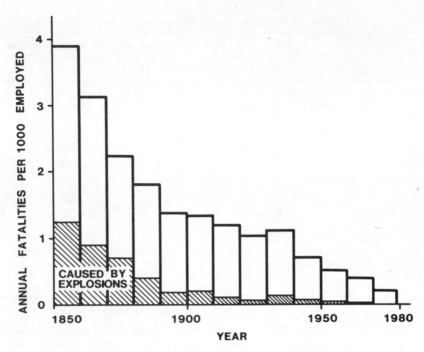

Figure 2.4-2 Accidents in coal mines in Great Britain

upon the industry, public, and Parliament with great force at times of exceptional disasters, and the slow but continuous improvements achieved. The range and magnitude of the dangers have never been in doubt, but the precise causes and the right means to combat them have often been in question.

An early and most notable example undoubtedly is that of mine explosions and the recognition that firedamp or methane, which is given off when coal is worked, was the cause. The name of Sir Humphrey Davy will always be linked with the conclusive demonstration of this fact and with the design of a lamp that could be safely used in such conditions. Much less well-known, however, is the role played by coal dust in propagating and enlarging explosions to disaster proportions that might otherwise have been relatively local ignitions. The understanding of the relevant phenomena, or at least its wide acceptance, took 30 to 40 years, 1870–1910[5]. It also gave a clear demonstration of the importance of scientific research in such work and provided the impetus that led to the creation of the first industrial safety research organization in the UK and probably in the world, the Safety in Mines Research Board (or Establishment as it was subsequently titled).

These are but a couple of examples from many that serve to emphasize that a full scientific understanding of the relevant phenomena and the availability of the measurement and analysis techniques upon which that understanding can be constructed, are essential to an effective attack upon safety problems. In any overall view, however, a range of factors must be recognized:

1. Awareness/recognition of safety problem(s):
   accident and reliability data; reports; managerial attitude.
2. Understanding of the scientific phenomena and engineering issues: measurement; data analysis; identification of the key factors; experimental proof.
3. Appreciation of human and technical environment:
   human error; technical procedures for installation, maintenance; operational management policies; discipline.
4. Development of solutions:
   research and development strategy; engineering development; reliability, safety, and cost analyses; field proving of prospective solutions; improvement of engineering or managerial procedures; standards and codes of practice.
5. Implementation:
   managerial policy; capital and other resource allocation; training; publicity and presentation; feedback on application, difficulties; monitoring of progress; statistical results; legislation.

In coal mining which has had such a well-known history of danger, all recognize the importance of safe practice, from Board Member to young apprentice. As a result of that history, there is a large well established body of regulations and legislation[19]; there are general, specific, and incident review committees; and there are scientific, engineering, and managerial resources committed to the wide range of activities necessary to combat the hazards. Of continuing importance is the magnitude of these resources and their effectiveness, and their relationship to parallel efforts devoted to improving production and productivity. In particular it is the appreciation of and response to change that is significant; change in the methods of winning, transporting, etc. such as the radical changes brought about by mechanization; change in measurement technology that can, for example, allow reliable measurement underground of carbon monoxide in the parts per million range and at a modest cost; change in methods of data transmission, manipulation, and presentation using computers that can permit busy colliery managers to be far more conscious of continually changing environmental conditions in the mine; and change in the form of assessment procedures that can reinforce and extend the engineering and managerial judgement that must continually be used in the design and operation of equipment and systems. Examples of these changes or innovations are quoted later.

The structure and organization of the Safety and Health activities in the National Coal Board have been presented in a recent paper[8]. A number of aspects including the philosophy of safety work, objectives, campaigning, etc. are discussed in papers by Collinson[9]-[2]. The Annual Report of the Chief Inspector of Mines and Quarries[19][20] presents a valuable summary and is a key source of statistical data. A parallel source is the Annual Report of the Safety and Health

Commission for the Mining and Extractive Industries of the European Communities[13]. Many accidents or incidents result in special reports by the Mining Inspectorate[20][21][24].

In the case of the accident at Markham Colliery in 1973, since it was a major accident, a National Committee was set up jointly by the National Coal Board and HM Inspectors of Mines and Quarries to give detailed consideration to the issues raised and to make recommendations[20][22].

Explosions

Explosions of firedamp (methane) or coal dust are not the scourge that they were as Figure 2.4-2 indicates. In the five years 1974 to 1978, there were seven fatalities in the UK or an average of 1.2 per year. There can be no room for complacency, however, since the number of localized ignitions has averaged 20 per year during this period. These arise from a variety of causes – shotfiring, electricity, mechanical friction, but mainly from frictional sparking at the cutter picks of machines.

At first sight the problem appears acutely dangerous. Prior to being worked, coal in the seam has a quantity of methane adsorbed which varies from a few cubic metres per tonne to over $50 \, m^3$/tonne; and a substantial proportion of this is released in the winning operation. Adjacent seams above and below may also release similar quantities due to the strata disturbance and pressure relief caused by the extraction, though this is minimized by draining gas from the adjacent seams using pipes to the surface and a suction fan. Into this environment mechanical cutting machines are introduced with ratings up to 500 kW together with electrical equipment to supply and control them; and shotfiring is frequently essential to the task. Even granting that the electrical hazards did not appear until the early years of this century, it is small wonder that coal mining has had such a history of explosion disasters. Yet today, as a fraction of UK mining fatalities, explosions account for less than 4%.

An ignition requires two conditions to be present simultaneously, a methane–air mixture in the range 5% to 15% (by volume) and an ignition source. The latter can be provided either by an electric spark of quite low energy (in ideal conditions only 0.3 mJ), by an open flame, or by a surface heated to a temperature in excess of 900 °C. Quite small fragments of metal or rock can be heated to incandescence and burn during mechanical cutting especially when intrusions of quartzitic or pyritic rock are present in the coal seam; and where any flame can arise great care is required to ensure its effective containment. The constructional features of the Davy Safety Lamp provide an early example and flameproof enclosure is another. In these cases flame propagation is prevented by controlling the dimensions of all channels and paths connecting the interior volume to the exterior. Once a methane explosion, even of quite small proportions, has been initiated a further danger arises since coal dust alone can support and propagate the explosion throughout large sections of the mine. Reference has already been made to the long delay in identifying coal dust as a secondary cause. Coal dust is such an inevitable accompaniment to the mining process, that dust deposits are unavoidable in the many kilometres of mine roadways and, made airborne in the swirling, turbulence of an explosion, conditions are created for extremely widespread propagation.

Since in practice it is possible neither to eliminate the methane nor to avoid chance ignition sources completely, safety is dependent on a range of activities and controls, which have been learned from the bitter experience of more than 100 years of disasters. There are intended:

(i)   to minimize by ventilation the probability of an explosive atmosphere being present, and to keep to a minimum the volume of any explosive mixture which nevertheless may arise;

(ii) to eliminate as far as practicable all sources of ignition;

(iii) to prevent the propagation of explosions due to coal dust by erecting stone dust or water barriers and by general stone-dusting.

Ventilation

The first requirement to combat the explosion danger is for adequate ventilation to dilute the methane evolved to a non-explosive concentration. Even though the lower explosive limit for methane in air is 5% due to many factors there will be wide variations in the concentration within the winning district and associated roadways and the legal limit[19] for continued working using electric power is much lower at 1.25%. The regulations additionally require that at 2% all men must be withdrawn from the district; and also specify the frequency that measurements of methane, traditionally by sample analyses, must be made and recorded. For example they must be made daily for concentration over 0.8%. Similarly conditions are laid down for the measurement of air quantity. Improvements in the monitoring of methane concentration and of airflow over the last 30 years have been of major importance in increasing colliery awareness of the explosion danger; and in stimulating prompt action to reduce it by improving the drainage, by increasing the airflow or even by stopping production.

Ventilation monitoring   The adequacy of the ventilation system in preventing explosive concentrations of methane arising is checked in two ways, by measuring the methane concentration and, secondly, by measuring airflow.

Methane monitoring   The flame safety lamp has traditionally been used as a crude indicator of the presence of methane and this practice continues to this day. But the first accurate routine methane measurements were based on laboratory chemical analysis of air samples collected from locations underground. J B S Haldane[17] was responsible for the apparatus and measurement techniques introduced in the 1920s. But laboratory sample measurement is inevitably slow, laborious and expensive, and the next important development occurred in the 1950s, when electrical hand methanometers were perfected enabling mining officials to check concentrations on the spot. Over 6000 such instruments are in use in British mines. The basic detection method employed is the oxidation of methane on a heated catalyst[1][14]. The resulting increase in the temperature of the bead on which the catalyst is deposited is measured as a resistive change in an electric bridge circuit, with compensation for ambient temperature variations by means of a similar but non-reactive bead in an opposing arm of the bridge. This technique is now used worldwide (with the exception of some French and East European instruments in which combustion occurs on a platinum filament at a higher temperature). British and European specifications are being formulated to cover the performance of spot-reading hand methanometers.

More recently portable but continuously powered monitors based on this measurement technique have also been introduced. They provide a flashing-light alarm signal automatically when the methane concentration exceeds a set value, for example $1\frac{1}{4}\%$, and give a much increased awareness of any hazard that may be present. They are known in the industry as automatic firedamp detectors and operate for the duration of a working shift before the batteries require recharging on the surface.

A further important development has been the 'fixed'-installation continuous methane monitor (fixed being a relative term in the context of an extractive process continually requiring access to new material). In British mines, the BM1[14] is the instrument in use; and the detector is again of catalytic oxidation type. In addition to providing continuous indication and operating alarm contacts for concentrations exceeding set values it also feeds signals to a local chart recorder or, via standard data transmission systems, to a surface mine monitoring station. The instrument is mains powered but it also has a local battery with a duration of two to three days to guard against an interruption in the normal power supply.

Taken overall, the multiple methane monitoring provisions, i.e. flame safety lamp, local spot-reading hand instruments, portable continuous indicating and alarm instruments, 'fixed' instruments with similar provisions and often with surface computer storage and display facilities and 'tube bundle' back-up monitoring (see page 394), provide a defence in depth. Taking as evidence the two incidents in the last eight years in which loss of life occurred, the present need is to extend available instrumentation to cover working situations more widely, since in both the Goldborne and Houghton Main explosions[19][21] dangerous situations arose that the colliery personnel were not aware of.

Airflow monitoring   Air velocity is measured as a second supporting indicator of a satisfactory ventilation system. Windmill anemometers have been used as portable checking instruments for many years. However, continuous monitoring with local recording facilities or with signal transmission to a central monitoring station has only been introduced in the last five years. A design based on the rotating windmill, designated the BA2 instrument, was initially introduced, but bearing life in dusty humid mine atmospheres was inadequate at four to nine months. An inherently better method of measurement, involving no moving parts was introduced in 1980. The BA4 instrument is based on measuring the frequency of vortices formed in the wake of a strut in the airflow[28]; and early experience of this type of instrument has been very satisfactory.

Supervisory monitoring from a central station The most significant recent development has been the introduction of general supervisory environmental monitoring from a central surface station[3][31]. Using digital transmission systems and computer storage and display facilities, signals from BM1 (catalytic oxidation) transmitting methanometers, BA2 (windmill) or BA4 (vortex counting) anemometers, BM2H (thermal conductivity) high concentration methanometers, and BP1 (diaphragm) pressure and flow instruments for drainage pipe monitoring, an overall picture of the underground atmosphere and the explosion hazard is obtained. One of the most valuable features is the use of visual display units for presenting shift, daily or weekly histograms from which trends can be observed and judgement exercised. Five installations of this type employing the National Coal Board's MINOS computer monitoring and control system[7] have been installed over the last few years. Figure 2.4-3 shows a typical display of methane, airflow, and barometric pressure at one coalface during a night shift at a supervisory monitoring station.

Ignition sources

An ignition source constitutes the other essential constituent for an explosion and great care is taken to design equipment and control operations so that the probability of an occurrence is minimized. After the worst mining disaster in British history, when 396 miners were killed at Senghenydd in South Wales in 1913, due to sparking in a signalling bell circuit, electrical equipment of even quite low power has been recognized as highly dangerous, unless its design conforms to one of several methods of rendering it safe; recognized techniques are those of intrinsic safety, flameproofing and, more recently and to a much more limited extent, 'increased Safety, Type e' and purging of apparatus with an inert gas (BS 5501 Parts 1 to 7, 1977). There are limits however, to the protection that can be afforded. Trailing cables supply electric power to mobile machines and due to the rugged conditions and continuing advance essential in an extractive process, occasional cable damage is inevitable. Flameproof switchgear with highly developed protective systems is employed featuring intrinsically safe pilot remote control to permit control by the operator at the machine of the switchgear some 200 or 300 m away; overload and sensitive earth leakage tripping; sequence

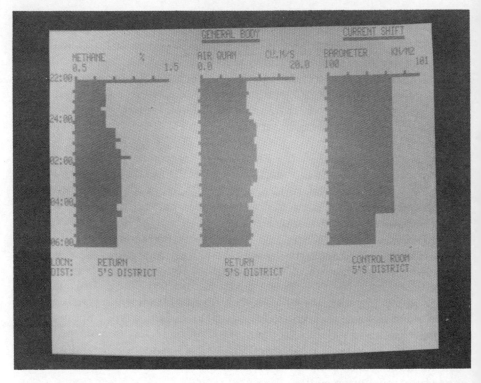

Figure 2.4-3 Typical display of methane, airflow, and barometric pressure

interlocking and timing; and more recently, phase-sensitive short circuit detection[16].

Another source which can hardly be eliminated is frictional ignition at the cutting picks of coal winning and roadway drivage machines. The technique to combat this hazard is good ventilation of the cut; good general ventilation at the face is the first essential but some assistance is provided by a measure aimed primarily at dust suppression – the use of a cutting drum with a hollow shaft ventilator. High pressure water jets entrain air which is carried to the back of the drum and flushes methane from the cut, as well as providing a means of dust suppression.

Suppression of explosions

While all the measures described above minimize the occurrence of ignitions, and reduce the volume of gas involved in the explosion, some ignitions do nevertheless occur. As explained earlier there is a serious danger once initiated of an explosion being reinforced and propagated far more widely due to coal dust. To combat this danger, stone dust or water barriers are installed at strategic locations along roadways to suppress the explosion and prevent propagation. The suppression is the result of the rapid dispersion of a large quantity of an incombustible, finely divided material, normally stone dust, which has the effect of extracting heat from the explosion front and lowering its temperature below that necessary to support propagation. In its usual form the barrier is entirely inert, consisting of heaped stone dust on a series of loosely supported planks of wood. These are dislodged by the shock wave preceding the explosion flame front with a rapid dispersal of the stone dust into the oncoming front. Triggered barriers have also been demonstrated in which radiation from the flame front is detected and used to initiate the rapid dispersions of a large volume of water by means of a compressed gas. As a further precaution stone dust is also liberally distributed along roadways, etc. periodically so that sufficient inert material is also picked up along with the coal dust should an explosion occur.

Fire

Fires underground are more dangerous than in many other environments due to potentially lethal combustion products being carried throughout the entire downwind sections of the mine. The total length of roadways is also a problem – typically 30 km, with much unattended mechanical and electrical plant subject only to periodic inspections. Fortunately continuous monitoring is now increasingly providing a better safeguard.

Underground fires have averaged 50 a year during the 1970s, the principal causes being associated with belt conveyors (40%), electrical equipment (26% and increasing), mechanical equipment (12%), locomotives (8%), and spontaneous combustion (5%). Spontaneous combustion of coal is liable to occur in a number of coal seams with particular characteristics. The waste areas left as extraction proceeds are especially vulnerable. These conditions are obviously very dangerous and if not quickly controlled can lead to the complete loss of a district through the necessity to seal off the affected area to prevent oxygen reaching the burning zone.

Belt conveyor fires can be caused by frictional overheating of conveyor idler rollers and can also result from faulty brake operation, defective bearings, excessive belt slip, etc. at the driveheads. Extensive monitoring is practised to guard against these latter, localized hazards as indicated in Table 2.4-3, but the idler problem (with many millions of idlers in use in some hundreds of kilometres of roadway) and spontaneous combustion and the variety of other plant in use present a distributed hazard which needs to be tackled in a different way. Detection of smoke or the gaseous products of combustion provides such an alternative.

Fire detection, as presently practised in British coal mining is based on smoke detectors of the ionization type[15], and tube bundle systems[6]. In the latter, small bore, 4 to 6 mm, plastic tubes are used to draw samples of the atmosphere from selected locations underground to a surface analysis station. The sample tubes are handled as a composite bundle, like a multi-core cable. At the surface station carbon monoxide in the range 0 to 30 ppm is monitored using a non-dispersive infrared gas analyser as the key indicator of spontaneous combustion; oxygen, methane, and oxides of nitrogen may also be monitored. The main advantages of such a system lie in the use of high sensitivity gas analysis systems at the surface where accessibility, maintenance, etc. is so much better, and in the economical use of such equipment when one or two sets of apparatus are used to monitor samples from many parts of the mine in rotation. It has one important disadvantage – that there can be delays up to several hours in the information becoming available, due to the time the samples take to travel several kilometres to the surface from more remote locations. This is not a major factor in the case of spontaneous combustion however, since this is a very slow process at the incipient stage, typically taking several days to reach significant concentrations. Primarily for this purpose, tube bundle analysis systems have been installed at more than 100 collieries in Britain. In an attempt to minimize the disadvantage due to the transport delay, it is normal to sample and analyse the air in the up-cast shaft. Since the ventilating air circulates through the colliery in 20 to 30 minutes, there is a good chance of identifying quickly the presence of any rapidly developing fire. However, such a warning must serve to initiate rapid follow-up action to locate the cause, since a shaft monitor signal gives no indication of the source location.

The smoke detectors used are of the single chamber type in which ionization is produced by a very small radioactive source. When a potential is applied to two electrodes in the chamber a small current flows due to ion transport. This is markedly reduced when smoke particles are present since ion attachment occurs resulting in lowered ion velocity. Such devices are comparatively cheap, and several thousand are installed. They are of limited effectiveness, however. At distances greater than 30–50 m from a smoke source their sensitivity is seriously reduced, presumably because there is some change in the chemical/electrical properties of the smoke particles affecting ion attachment. They cannot, therefore, be accepted as more than local fire detectors and do not fill the role of a detection instrument for general use.

It is evident that both existing equipments solve only a part of the overall problem. In Germany infrared gas analysers are deployed underground to overcome the transport lag inherent in the tube-bundle system. The problems of maintenance, cost, and reliability are considerable, however, and alternative solutions are being sought. Two new approaches are showing considerable promise; the first is a carbon monoxide detector of electrochemical type, and the second a detector based on conductivity variations of a semiconductor which is extremely sensitive to a number of products of combustion. Figure 2.4-4 shows experimental sensors of these types, as well as an oxygen deficiency sensor also of electrochemical type. Such sensors are small, comparatively cheap and have cell lives of 6 to 12 months. They are scheduled for use in both portable hand instruments and for 'permanently' sited equipment.

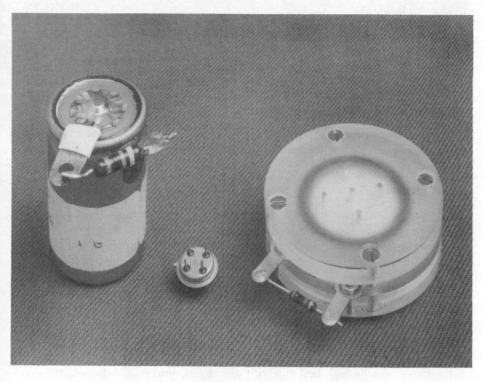

Figure 2.4-4 Experimental sensors for (left) oxygen
(centre) products of combustion
(right) carbon monoxide

Improved performance of ionization type smoke detectors is also possible. Several types of device in which the ionization and detection processes occur in separate chambers have been constructed by different workers and significantly higher sensitivities have been demonstrated[2].

Underground Transport and Winding

The major underground transport task in UK coal mines[34] is summarized by the following figures:

Length of roadways used for transport – 10 000 km.
Daily mineral carried (coal and stone discard) – 700 000 tonnes.
Daily materials (equipment, consumables, etc.) carried – 10 000 tonnes.
Daily number of men travelling – 150 000.
Daily number of journeys – 300 000 (one in; one out).
Average journey distance (shaft to coal face) – 3.6 km.
Number of men in transport services – 40 000.

For comparison with a major surface transport undertaking the London Transport figures are also quoted:

London Transport Buses: daily passengers carried – 143 000; average journey 3.2 km.
London Transport Underground: daily passengers carried – 422 000; average journey 8 km.

There are three principal types of transport in use; belt conveyors and two types of rail transport, locomotive hauled and rope hauled.

Belt conveyors

Coal transport is predominently by conveyor belt, with the majority of conveyor networks remotely controlled from a surface central control station. In the last few years computer control of the conveyor network and associated bunkers has been introduced allowing automatic rather than remote manual control to be exercised from the central station[7]. Approximately 15 % of UK mines had such control in 1981. In both cases of control from the surface, the conveyors, the transfer points linking each conveyor to the next in sequence, and the bunkers, are designed for unattended operation except for mobile conveyor patrolmen who each have responsibility for a few kilometres of roadway. As noted previously unattended conveyor drive equipment with ratings from 50 to 250 kW constitute a significant fire hazard. Belt slip, belt alignment, brake malfunction, transfer chute blockage, overheating of motors, gearboxes, bearings, etc. are all potential sources of danger to men and plant. In consequence extensive safety monitoring of unattended drivehead equipment is practised; as listed in Table 2.4-3. To ensure that personnel along the roadway as well as in the vicinity of the drivehead have adequate awareness of impending conveyor movement and adequate stopping and lockout facilities pull-wire stop and pre-start warning signals are given either by sirens or by roadways loudspeaking telephones mounted at intervals along the entire length of a conveyor.

While the monitoring/tripping arrangements indicated in Table 2.4-3 are not duplicated, some back-up protection is also provided. Failure of the drive to stop when a transducer has called for a power-trip-out is followed by automatic trip-out of a section switch in the case of computer control or by manual section switch tripping in a remote manual control system.

The danger from fire is lessened by the extensive use of non-flammable lubricants to replace conventional mineral oils; and by the use of flame retardant material for the manufacture of conveyor belting. The use of such materials was a consequence of a tragic fire at Creswell Colliery in 1950. After widespread tests of a variety of materials and the development of belting combining both good mechanical and wear properties as well as a high degree of fire resistance, all UK conveyor belting is purchased to an appropriate specification.

Table 2.4-3 Typical monitoring and tripping provisions for remotely/automatically controlled belt conveyor systems

Monitored Function	Obligatory	Optional
Signal key operation (start/stop/lockout)	✓	
Operation in local control mode	✓	
Local Stop	✓	
Local lockout	✓	
Operation in maintenance mode	✓	
Motor start command issued	✓	
Motor running	✓	
Fluid coupling engaged (scoop in)	✓	
Belt running	✓	
Belt slip	✓	
Belt misalignment		✓
Torn belt		✓
Blocked chute		✓
Scoop coupling alarm	✓	
Brake off	✓	
Brake alarm	✓	
Brake overheat		✓
Smoke detected		✓
Water curtain actuated		✓
Man-riding safety gate trip	✓	
Transmission outstation malfunction	✓	

A very extensive testing and development programme has also been carried out by the Mining Research and Development Establishment of the National Coal Board in order to avoid the use of flammable mineral oils underground while still obtaining acceptable wear and economic equipment lifetimes[18][25][26][32].

Fire warning and protection at the drive machinery location is provided by smoke detector(s) of the ionization type and by sprinkler water barriers triggered automatically when the local temperature rises or manually by action from the control station. Fire warning is also provided on a zone basis and overall for the whole mine by means of tube-bundle carbon monoxide monitors. Such zone-based fire detection is a safeguard against the distributed hazard presented by the overheating of conveyor idler rollers. Since temperatures as low as 200 ° C are sufficient, over a period of hours to initiate smouldering of coal dust deposits and more rapidly at higher temperatures, and since coal dust deposits are virtually unavoidable, this hazard is significant. Furthermore, with several million conveyor idlers in use, individual protection has been inhibited by prohibitive cost. Important new developments in fire detection are pending.

Overall, belt conveying systems for mineral transport are a minor source of accidents in comparison with the transport of men and materials. There have been four fatalities in the five years 1973–1977 and 103 serious injuries in the transport of some 500 million tonnes of mineral. It should also be borne in mind that conveyors in the region of the coalface require frequent extension and are subject to ground movement; and that appreciable manriding occurs, both authorized and also to some degree unauthorized, mainly on trunk conveyors; for example, there are approximately 600 conveyors equipped and authorized for manriding.

Rail transport

Locomotive hauled trains, both diesel and battery powered, are used for a variety of transport tasks; for mineral transport in a minority of collieries (570 locomotives were in use for this purpose in 1978); for manriding (170 in 1978); and extensively for materials transport (860 in 1978), i.e. consumables, equipment for installation or withdrawal, spares, etc. The numbers do not imply exclusive use for the particular duty, and many are used for both or all three functions.

Diesel locomotives for use in the restricted ventilation systems of mines and in conditions where methane can be a hazard have to be specially designed and approved[23]. In addition to a regulation of a general nature covering controls, brakes, stability, lights, noise, etc. the air intake has to have a flame trap and air shut off to prevent run-on if appreciable methane is present; the exhaust system has to have a flame and water trap to cool (to 70 °C max.) and scrub the discharged gases; and the combustion conditions must be such as to ensure that not more than 1500 ppm of carbon monoxide and 1000 ppm of oxides of nitrogen are present in the exhaust at any engine loading. Appropriate requirements for electric storage battery locomotives apply also, dealing for example with satisfactory arrangements and tests for hydrogen emission and venting.

The locomotive is confined to trunk roadways where ground stability is good and is not deployed in gate roadways adjacent to the coalface. An unfortunate consequence is that an alternative transport system is required for these sections of the overall journey to the face and to faces under development. Transfer of loads to another (normally endless rope haulage) transport system is necessary and sometimes many transfers are involved. Such transfers are still labour intensive with a corresponding exposure risk. An important objective both for reasons of productivity and safety is the introduction of transport systems which require the absolute minimum of load transfers and handling.

Rope haulage

A rope-haulage installation consists of a winch driving a loop of steel rope installed along the track to which rail vehicles may be attached by rope clips. They are typically 0.5 km in length and operate at speeds from 6 km/h to 15 km/h. For transport of materials it is by far the most common system with over 6000 (i.e. approximately 30 per colliery on average) in use; and, excepting a few developmental systems, they are the sole transport means in the gate roads directly serving the coalface. For collieries dependent solely upon rope haulages for materials transport the transfer problem noted above is even more pronounced and up to 10 separate haulage 'legs' may be involved in the journey from pit bottom to face. This transfer activity together with journeys along poor roads subject to ground movement due to the extraction in the neighbourhood of working faces is, as noted above, a most important cause of accidents.

There are many problems in tackling this largest category of accidents:

1. The track is extensive – 10 000 km, has gradients up to 1 in 2 and in the vicinity of the face is subject to ground movement.
2. The loads are very varied, from simply handled materials to roof supports and machine sections weighing up to 10 tonnes and girders up to 6 m long
3. Mechanical handling is often primitive and labour intensive; and manoeuvring room is frequently minimal.
4. Poor roadway layouts in old collieries (many are 50 to 100 years old) make for inefficient, multiple-leg journeys with too many transfers of loads.

Technical advances which are being introduced include:

— single-pipe braking systems with automatic brake adjustment for manriding trains;
— track (i.e. rail-gripping) brakes to provide an emergency back-up to the wheel-brakes used in normal service;
— toothed rail engagement for steep gradients (> 1 in 15);
— trapped rail systems to prevent derailment;
— monitoring facilities using a specially instrumented vehicle to measure track gauge, alignment, tilt, and roadway clearance.

A recent demonstration of free-steered pneumatically-tyred vehicles using suitably compacted and surface roads in place of rail systems has shown much promise. At a colliery with access through a drift or inclined roadway instead of a shaft, equipment can be loaded into vehicles at the surface using standard industrial mechanical handling gear and transported directly to the face end without a single load transfer, in under one hour. The advantages from a safety viewpoint in the reduced exposure of men are clear.

Shaft winding

Mine winding is an activity which has a public image of considerable danger, and Figure 2.4-5 indicates why this was justified in the past. Since 1960, however, there has been a good record, with the one vital exception of a major accident at Markham Colliery in 1973 with 18 fatalities and 11 serious injuries[20]. This accident resulted in an exhaustive re-examination of the philosophy, design, testing, and maintenance of the entire winding equipment both mechanical and electrical (control, monitoring, and signalling). The work was carried out by a National Committee set up for the purpose with representatives from the Health and Safety Executive, National Coal Board, manufacturers, unions, etc. serving on a variety of sub-committees and examining different aspects[22].

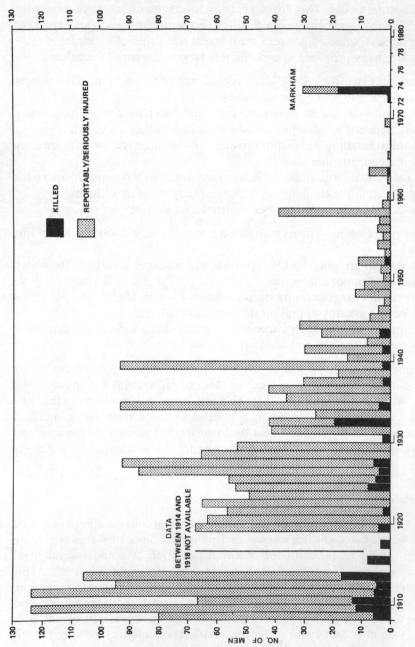

Figure 2.4-5 Number of persons killed and reportably injured due to overwinding in shafts (a) schematic diagram of winder emergency brake application; (b) improved circuit for emergency brake application

The accident resulted from a fatigue failure of a 50 mm diameter steel rod through which a nest of springs applied force to two brake shoes. The rod was subjected to bending stresses due to an unintentional constraint in the free movement at a bearing surface. Two features were of great importance; a complete brake failure resulted from the failure of a single component; and the alternative potential braking means, that of dynamic electrical braking, was removed when the engineman pressed his emergency stop button.

As a result of the National Committee work, the need for redundancy in all braking systems was accepted and a programme to bring all winders into conformity with this requirement was initiated. In respect of the desirability of retaining dynamic electrical braking there were several factors and a variety of a.c. and d.c. winding engines to be considered, some operating with closed loop control but others in open loop systems. An important problem was the danger in some of the installations of compounding, that is of applying electrical braking with simultaneous emergency mechanical braking introducing very high deceleration stresses in a number of components and constituting in itself a hazard to the men riding in the conveyance. Conversely, circumstances could be postulated in which the motor was 'driving through' the mechanical brakes, i.e. applying sufficient torque to overcome the braking effort. The main objective adopted, met in different ways due to the variety of types of equipment, was to retain electrical braking until such time as the mechanical brakes had been 'proved' on. The latter involved the development of sensors that genuinely responded to brake *torque* stresses since, in the presence of contamination of the brake paths by oil or water, checking the thrust of the brake shoe against the brake drum is not a guarantee of effective braking.

These are only two topics from the detailed review of virtually all safety aspects of winding equipment and practices that was carried out. Over 100 recommendations were made; the most important have been implemented, and a programme laid down to complete most of the remainder. A design guide was produced for mechanical brakes; the materials to be used throughout the winding installation were investigated and specifications drawn up; non-destructive testing methods for particular components were recommended; maintenance, testing, and training provisions were throughly examined and revised; comparisons were made with lift practice in other countries; and a variety of control, signalling, and monitoring equipment was either improved or new devices were developed to fulfil identified needs. A brake torque monitor has already been noted above, an electronic supervisory device[33] was also introduced which monitored the winding drum speed and rotation so that the automatic contrivance[29] (a mechanical shadowing device which detects any departure from the expected speed–distance profile of the winder in normal use and is an important item of safety equipment) was itself checked; and a magnetic rope-striping technique developed in order to detect slack rope (in drum winders) or slip (in friction winders) both of which are serious conditions. This technique[27] employs a permanent magnet marker to imprint magnetic stripes every 20 cm along a guide rope (for a drum winder) or suspension rope (friction winder) and a sensitive magnetic field detector and counter to determine travel along the rope. It provides an 'absolute' position of the conveyance in the shaft for comparison with that assumed from rotation of the winder drum; differences exceeding a selected limit are used to give slack (or slip) alarm signals or to trip safety circuits.

A further initiative taken by the National Committee referred to above was to examine safety assessment techniques and to commission such a study for a modern mine winder. The work was undertaken by the Systems Reliability Service of the United Kingdom Atomic Energy Authority[4][30]. The study indicated the overall probability of a hazardous situation arising through random faults of 3×10^{-4} to 8×10^{-4} during any one year period (the dominant contributions to this were from landing platform incidents and rope failure) and corresponds to a fatal accident frequency rate of 3 (number of deaths per 10^8 hours exposure). Such a winder incorporates all the important features identified in the work of the National Committee.

As an example of one element in such an analysis, the operation of the emergency brake system is presented. The mechanical brakes of a winder are spring applied with an hydraulically powered withdrawal system so that in the event of any power failure, or pressure failure the brake is automatically applied. Such an installation is represented schematically in Figure 2.4-6a. The various safety monitoring devices and automatic or manual emergency stop facilities provided operate by releasing one or more solenoid valves which are normally powered in order to permit pressure to be applied to withdraw the brake, but vent the brake cylinder to atmosphere when not powered. Power is supplied to emergency brake solenoid 1 (EBS1) via contacts 1A of safety contactor 1 (SC1) and 2A of safety contactor 2 (SC2), so that the opening of either 1A or 2A results in removal of power from EBS1. In order to minimise the danger arising from a failure of EBS1, a second solenoid EBS2 is used in parallel. It is required to determine the probability of a successful release of pressure when the contactors are de-energized by an emergency trip.

P_1 is the probability that a contactor (either 1 or 2) fails to move to an open position when de-energized.

P_2 is the probability that the contacts of a contactor fail to open when the contactor moves to the open position.

P_3 is the probability that the emergency brake solenoid armature fails to move when de-energized.

Taking first the simpler case of a single solenoid EBS1, Q, the probability of system failure $= (P_1 + P_2)(P_1 + P_2) + P_3$

$$\text{i.e. } Q \simeq P_3 = 1.9 \times 10^{-5}$$

The dominant term is P_3 since the redundancy provided by the combination of

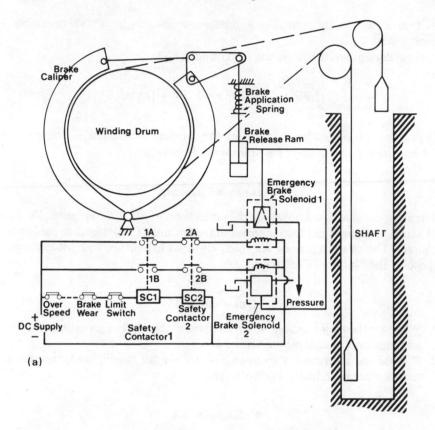

(a)

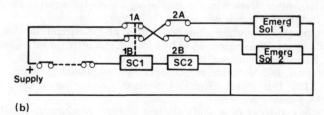

(b)

Figure 2.4-6 Emergency brake systems

Table 2.4-4 Probability values for component failure

	Failure rate faults/yr (a)	Proof check interval (b) year	a × b
P_1	0.001	0.019	1.9×10^{-5}
P_2	0.0015	0.019	2.85×10^{-5}
P_3	0.001	0.019	1.9×10^{-5}

SC1 and SC2 contacts makes a failure to de-energize EBS1 negligible in comparison.

Now taking parallel solenoids EBS1 and EBS2

$$Q^1 = (P_1 + P_2)^2 + P_3^2$$
$$= (1.9 \times 10^{-5} + 2.85 \times 10^{-5})^2 + (1.9 \times 10^{-5})^2$$
$$= 2.6 \times 10^{-9}$$

In practice, it is preferable to cross-link the contacts of SC1 and SC2 as in Figure 2.4-6b, since the redundancy is still more effective.

Acknowledgements

The author is glad to acknowledge the permission of Mr P. G. Tregelles, Director of Mining Research and Development of the National Coal Board, to publish this chapter. The views expressed are solely those of the author and not necessarily those of the National Coal Board.

Questions 2.4

1. What are the major factors which have been observed to give rise to fatalities underground in coal mining in the UK?
2. Discuss the information given in Table 2.4-1 for mining and compare the results with an industry known to you.

References 2.4

1. Baker, A. R. *A Resistance Methanometer Employing a Low-Temperature Catalytic Element.* Safety in Mines Research Report No. 162, 1979.
2. Berry, P. 'The separated ionization chamber.' Paper at the Symposium on Fire Detection for Life Safety, Washington. 31 March 1975. Committee for Fire Research of the National Research Committee, 1975.
3. Bexon, I. and Pargeter, D. J. 'Practical aspects of computers in mine monitoring and control systems.' The Third West Virginia University Conference on Coal Mine Electrotechnology, 4–6 August 1976.
4. Burton, J. E. 'Safety analysis of a shaft-winding system.' Mechanical Reliability Symposium, Bedford College, London. 9 July 1980.
5. Bryan, Sir Andrew. *The Evolution of Health and Safety in Mines.* A Mine and Quarry Publication. 1975.
6. Chamberlain, E. A. C., Donaghue, W. R., Hall D. C., and Scott, C. 'The continuous monitoring of mine gases.' 'The development and use of a tube bundle technique.' *The Mining Engineer*, London, March 1974, pp. 239–251, 1978.
7. Chandler, K. W. 'MINOS a computer system for central control at collieries.' 2nd International Conference on Centralized Control Systems, London, 20 to 23 March 1978.
8. Collinson, J. L. and McLintock, J. S. 'Structure of the National Coal Board.' *Colliery Guardian*, November 1976.

9. Collinson, J. L. 'Making mining safer yet.' *The Mining Engineer*, November 1976.
10. Collinson, J. L. 'Safety: pleas and prophylactics.' *The Mining Engineer*, July 1978.
11. Collinson, J. L. 'Safety–risk rationalization.' *The Mining Engineer*, November 1979.
12. Collinson, J. L. 'Safety–the cost of accidents and their prevention.' *The Mining Engineer*, January 1980.
13. Commission of the European Communities. 16*th Report of the Safety and Health Commission for the Mining and Extractive Industries for the Year* 1978. 1980.
14. Cooper, L. R. 'Transducers for environmental monitoring in British coal mines.' The Third West Virginia University Conference on Coal Mine Electrotechnology, 4–6 August 1976.
15. Cooper, L. R. (1977). 'Transducers for environmental monitoring in British coal mines.' International Conference on Remote Control and Monitoring in Mining, Birmingham, October 1977. Published by the National Coal Board. 1977.
16. Gray, G. W. (1978). 'Phase sensitive short circuit protection.' *Mining Technology*, **60**, 4, 133–136. April 1978.
17. Haldane, J. S. and Graham, J. I. *Methods of Air Analysis*. Charles Griffin & Co. London. 1935.
18. Hall, J. B., Knight, G. C., and Kenny, P. *British Experience with Fire Resistant Fluids in the Mining Industry*. The Institution of Mechanical Engineers, London, February 1974.
19. HMSO, *The Law Relating to Safety and Health in Mines and Quarries*, 1970, Part 2, First Edition, updated 1979 by Health and Safety Executive, 3rd Edition, Part 2 – *Regulations and Orders, Approved Specifications and Procedures Applicable to Mines of Coal, Stratified Ironstone, Shale and Fireclay*.
20. HMSO CMMD 5557. A Report by the Department of Energy on the accident at Markham Colliery, Derbyshire. *Report on the Causes and Circumstances Attending the Overwind which Occurred on 30 July* 1973. 1974.
21. HMSO, Health and Safety Executive. *Explosion at Houghton Main Colliery, Yorkshire*. Report on the Explosion which Occurred on 12 June 1975.
22. HMSO, Health and Safety Executive. *Safe Manriding in Mines*. The First Report of the National Committee for Safety of Manriding in Shafts and Unwalkable Outlets, 1976; Second Report, 1980.
23. HMSO, Health and Safety Executive. *Testing Memorandum No. TM*12, 1977.
24. HMSO, Health and Safety Executive. *The Explosion at Golborne Colliery, Greater Manchester County*, 18 *March* 1979.
25. Knight, G. C., Jones, S. F., and Kenny, P. *Hydraulic Fluid Fires and Fire Resistant Hydraulic Fluids*. The Institution of Mechanical Engineers, London, December 1973.
26. Knight, G. C. *The Application of the Suitability of Hydrostatic Pumps and Motors for Use with Fire Resistant Fluids*. The Institute of Petroleum, London. October 1976.
27. Lewis, D. C. and Ormondroyd, H. 'Magnetic striping of steel ropes.' *Mining Technology*, June 1978.
28. Miller, S. P. 'Instruments for environmental monitoring.' *Mining Technology*, **63**, 9–16. January 1981.
29. Ogden, C. 'Automatic contrivances.' The Association of Mining Electrical and Mechanical Engineers Symposium on the Transportation of Men and Materials in Shafts and Underground. Harrogate, October 1975.
30. Thomas, V. M. and Burgess, H. 'Safety assessment of a modern winder installation.' The Association of Mining Electrical and Mechanical Engineers Symposium on The Transportation of Men and Materials in Shafts and Underground. Harrogate, October 1975.
31. Thomas, V. M. and Cooper, L. R. 'The contribution of monitoring and automation to

health and safety in mines.' Symposium of Health, Safety, and Progress. Harrogate, 27–29 October 1976.
32. Tregelles, P. G. and Knight, G. C. 'Reliability of Mining systems.' Symposium on Productivity through Technology. Harrogate, 18–20 October 1978.
33. Walters, J. S. B. (1975), 'A winder overspeed protection instrument.' The Association of Mining Electrical and Mechanical Engineers Symposium on the Transportation of Men and Materials in Shafts and Underground, Harrogate, October 1975.
34. Watt R. G. (1979). 'Mining transport – short and long term considerations.' *The Mining Engineer*, April 1979.
35. Wood, J. E., Gwatkin, G. H. R., and Whitelam, M. (1975). 'Detection of fire underground by monitoring carbon monoxide in the shaft air.' *The Mining Engineer*, March 1975.

High Risk Technology
Edited by A. E. Green
© 1982 John Wiley & Sons Ltd

Chapter 2.5

Safety Technology for Offshore Oil Platforms

D. Dick

Introduction

The emphasis is what follows will be on safety technology related to the production of oil from offshore platforms in the North Sea[1][2][3]. Only some of the many areas of concern can be mentioned in the space available here. To set the background a brief description of a North Sea production installation will be given to indicate factors which distinguish offshore oil facilities from land-based industrial plants. Basic, mainly engineering methods of achieving safety are described followed by a section on the quantitative techniques which are applied to oil production facilities. The first part of this latter section relates reliability and hazard analysis and shows how reliability analysis can be applied to offshore oil production facilities. An example is given of how the provision of multiple process trains which have extra capacity can increase productivity. Outline considerations of hazard analysis, structural integrity assessment, floatout risks, and blowout hazards are also discussed.

Scenario

Though the processing equipment on an offshore platform is similar to that in use in the land-based chemical industry and though the process in itself is much safer than many onshore, as only physical separation rather than chemical reaction is involved, it will be seen from the following illustrative description that there are circumstances special to such ventures which require particular care to prevent hazard to life, pollution, or indeed costly loss of production.

The crew and plant are in close proximity and there is virtually an unlimited fuel supply which must be controlled and prevented from catching fire. In addition the whole operation takes place in a hostile, remote environment.

The British National Oil Corporation's (BNOC) Thistle platform is located about 130 miles NE of Shetland in 530 ft. of water. The required production equipment, having an operational weight of about 25 000 tons, is supported by a steel jacket piled into the seabed. This is 611 ft. high at production deck level and 967 ft. in total height to the top of the flare tower. The platform area is 240ft × 170ft on which are placed 31 production and drilling modules mainly on three levels, together with living accommodation and a helideck. Full production is expected to be approximately 200 000 barrels a day (there are 35 imp. gal in a barrel). The crude oil and associated gas will come through about 24 wells which are drilled vertically from the platform down 1000 to 3000 ft before deviating outwards as much as 63° to a final vertical depth of up to 10 000 ft.

The crude oil is processed by passing through three stages of separation to remove gas and water before exporting via a pipeline or by direct loading to a tanker from a single point mooring (SALM). Some gas is used for fuel for two gas turbine alternators capable of generating 52 MW of electricity while the remainder is to be compressed and reinjected into the formation. There is a salt water system which filters and treats sea water before injecting it into the reservoir at high pressure (4200 psi) to maintain well pressure and hence oil production rate. The platform has accommodation for 260 persons.

Means used to achieve safety offshore

The basis for safe offshore operation lies in good engineering design. Great thought and effort go into producing a design which not only achieves the required performance as economically as possible but does so in a safe way. Care is taken also in the construction phase to ensure that structure and equipment are built and installed as designed. Inspection, quality assurance and procedures play a large part in ensuring this. Once in operation, maintenance and training play an important part in ensuring safety. Risk analysis (such as will be described later) is being used increasingly to highlight weak areas of design and indicate where design changes could be beneficial, as well as being used to give quantitative assessments of reliability and of chance of remaining hazard free. Ideally, reliability studies should be an on-going concern which impinges on design philosophy and influences design to achieve consistency and adequacy of reliability and safety provision.

An indication of some of the safety provisions incorporated in the engineering design for the Thistle field will be given before moving on to quantitative methods of assessing their adequacy. The structure was designed to withstand the '100 year storm' which is the worst storm likely to recur in a period of 100 years, which for the Thistle location would mean a maximum wave height of 93 ft and maximum wind gust (3 s) of 152 m.p.h.

There is an automatic Fire and Gas detection system designed to detect gas or fire and to take appropriate action to forestall or extinguish a fire. There are some 3000 detectors in the Fire and Gas system – gas, smoke, UV, and heat detectors.

There is an automatic water deluge system and Halon is used in 10 modules as a fast acting fire or explosion suppressant.

The chance of a fire starting is reduced by using intrinsically safe or flameproof equipment and the supervision of any necessary ignition sources by use of permits to work. Fire or gas indicated by a single detector initiates an audible alarm but where two detectors are activated automatic fire suppression takes place by means of water spray deluge or halon discharge together with shutdown of equipment, ventilation, and fire dampers in the module concerned. If the fire or gas release is detected in a hazardous area (zoned as Div 1 or Div 2) then the F & G system is linked to the Emergency Shutdown system which then shuts down wells and production, starts diesel fire pumps and switches the dual-fuel electricity generator to diesel fuel. If this last action fails or if there is a fire in the generator module the auxiliary generators are started. Process vessels are vented to the flare.

Protection against blowout is provided by the installation of four or five tubular casing strings and the blowout preventer stack (BOP) which is installed during the drilling phase and at subsequent workovers of wells. After the well has been drilled the BOP stack is removed and a 'Christmas tree' installed to control the flow of oil and gas. The BOP stack has, typically, an annular preventer, and ram-type preventers, two or three of which are pipe rams which close on the drill pipe and another being a shear ram capable of cutting through it if necessary. A 'Christmas tree' has in line three valves, the bottom one a manually operated master valve and the other two being pneumatically and hydraulically controlled respectively, closing in a fail safe manner. It is supplemented by a hydraulically operated sub-surface safety valve about 1000 ft down the central casing string.

Quantitative risk assessment of production facilities

Two types of risk assessment are distinguished as Reliability studies and Hazard analyses, though the techniques used can be similar for each. For example a reliability study of an oil production system would be made to determine the expected oil producibility in operation.

Such a study is described in detail below. To make the problem tractable the total topside facility is first defined in terms of separate functioning systems, for each of which a fault tree is constructed. This relates logically the plant item failures which have to coincide before the system can fail. Use is then made of the known relations and connections between systems to derive the overall availability and oil producibility.

Some causes of lost production would be due to equipment failure leading to oil leaks. A hazard analysis would be interested in assessing the chance of such leaks leading to fire or explosion. The technique used for this kind of hazard analysis is described later. Other rare event hazards are less amenable to the above approach which develops a logically related fault tree to the items of which realistic probabilities can be applied. The difference is one of degree of randomness. For example, the chance of collision with the platform from shipping may be difficult to assess because of lack of knowledge of ship movements in the North Sea. Some notion of probability can be obtained from statistics on ship collisions, e.g. in the English channel. Lack of precise quantification does not preclude preventive measures being applied; in this instance there is a 500 m 'no go' radius for shipping around platforms. Oil risers bringing oil from the wells to the platforms are positioned where possible within the structure for added protection, rather than on the outside.

Reliability Studies

The method used for assessing the reliability of the Thistle offshore oil production facilities is as follows. Of the 45 systems defined for the Thistle platform 30 impinge directly on the production of oil. For the purpose of the reliability study these 30 systems were modelled as 17 systems to the degree necessary to reflect their impact on producibility (see Figure 2.5-1). The reliability study forms the basis of the evaluation of the fire and explosion hazards involving the same equipment.

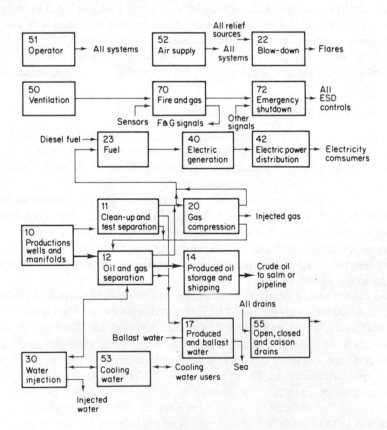

Figure 2.5-1 The platform system

The following paragraphs briefly describe each simulated systems as identified in Figure 2.5-1.

System 10 Wells	Includes the flow of oil from the well to the manifolds leading to the various separation trains.
System 11 Clean-up	Includes the flow of oil through the test separator or the clean-up separator.
System 12 Separation	Includes the flow of oil through the separator trains with oil going to the shipping system and gas to the compression system. The stream from the wells is separated into oil, gas, and water by processing through four parallel and identical separator trains (as many as four trains is not typical of all offshore facilities).
System 14 Shipping	Involves the flow of oil through booster pumps, metering station, and through the shipping pumps into tankers or the pipeline. This system also covers the flow into and out of the platform crude oil storage.
System 17 BS & W	Includes the flow of ballast water and produced water through flotation to separate the residual oil, which is subsequently skimmed and then recycled.
System 20 Compression	Involves the flow of gas from the separator trains through compression, recycling the condensate to the separator trains with the gas going to gas injection wells, to the fuel system, or to the flares.
System 22 Blow-Down	Diagrams the gas flaring system.
System 23 Fuel	Includes the gas from the first stage separator and the knock-out drum through cooling, filtration, and heating – ready for use in the gas turbines and for inert-gas generation. This system also includes the diesel oil flow to the auxiliary diesel power generators and to gas turbine generator set A. The inert-gas generation in the logic description is defined as a common-mode system and identified in System 23.
System 30 Water Injection	Includes the flow of water through the crude oil coolers and subsequent treatment in preparation for water injection. This diagram also includes the storage and distribution of the various chemical additives used.

System 40 Electrical Generation	Includes the electrical power generation system both from the gas turbines and the auxiliary diesel generators.
System 42 Electricity Distribution	Includes the distribution of electricity to the various consumers in priority order.
System 52 Air Supply	Includes the compressed air generation system for plant air and instrument air use.
System 53 Cooling Water	Includes the preparation of the cooling water used in Systems 20, 30 and 40.
System 55 Drains	Includes the drain collector system with the recycle of collected oil and off-loading of residual water.
System 70 F & G and 72 ESD	The emergency shutdown system and fire and gas system are modelled as common systems that affect all other systems. Each individual sensor, alarm, and actuator is not modelled, but rather the aggregate of each type is modelled.
System 73 Inert-Gas Generation	The inert-gas generators are included with System 23.

The systems are interrelated by the flow of materials – crude oil, diesel oil, gas, water, etc. Each individual system was described by a logic diagram or its corresponding fault tree, derived from Engineering Line Drawings (ELDs), cause and effect charts, and process drawings, which showed all the critical pieces of equipment whose failure and subsequent restoration would affect the availability of crude oil. Also the process operating philosophy was incorporated as it affects system capabilities. The modelling of the composite system took account of the separation of common mode systems from the basic oil producibility calculation. A data-base of failure rate and restoration time for each type of equipment being modelled was organized from 16 different sources, possibly the most notable being System Reliability Services (SRS). Equipment was aggregated as far as possible, where the data for the aggregated equipment are the sum of the data for individual components. There is a certain amount of data from worldwide chemical and petroleum process plant and equipment similar to that used offshore and there is an increasing amount of experience being amassed on specifically North Sea plant.

Comparisons of data applicable to the North Sea and to different environments can highlight any adverse environmental effects which might occur from North Sea offshore use. There is need for continuing and increasing efforts' to gather appropriate data from first hand experience of plant used in North Sea oil production. The pooling of such experience for use by all North Sea operators would be of mutual benefit and could be achieved without loss of confidentiality via, e.g. the SRS reliability data bank. The use of computers eases the collection of fault data and build up of equipment histories, indeed automatic fault logging is now possible by means of microprocessors. To obtain all the information necessary to calculate plant availability, cooperation is required between maintenance and operations staff offshore so that operating time as well as failure frequencies and restoration times can be recorded.

The reliability equations were derived from the logic diagram of each system or its corresponding fault tree. The following example will demonstrate this. Each new reliability is expressed in terms of the reliability at the previous node, and the appropriate AND and OR equations (see Figure 2.5-2).

R_9 = reliability of system prior to Figure 2.5-2. For success at node 10 we require to have success to node 9 and either path 1 or path 2 intact. Thus:

$$R_{10} = R_9 \times (1 - q \text{ path } 1 \cdot q \text{ path } 2)$$
$$= R_9 \times (1 - (1 - R_{31} R_{32})(1 - R_{33} R_{34})) \qquad (2.5\text{-}1)$$

and similarly:

$$R_{11} = R_{10} \times (1 - (1 - R_{35} R_{36})(1 - R_{37}))$$
$$\underset{\text{via cooler}}{} \qquad \underset{\text{via bypass}}{} \qquad (2.5\text{-}2)$$

If reliability is defined as the probability of performing adequately for the period of time intended then reliability of component i can be defined as $R_i = e^{-\theta_i t}$ where:

$$\theta_i = \text{failure rate of component, i}$$
$$t = \text{operating time} \qquad (2.5\text{-}3)$$

The failure rates were represented by a triangular distribution bounded by low, mode and high values, and to allow for this randomness, Monte Carlo computer runs were carried out over a range of operating time periods, the samples from each distribution being combined in the above equations to give an overall failure rate curve for the system. This was then curve fitted to give $R = e^{-\theta t}$. The availability of a given system is the expected fraction of time that it is available for service, so:

$$\zeta a = \frac{\text{total time} - \text{dead time}}{\text{total time}} \qquad (2.5\text{-}4)$$

If total time is considered as 1 year and dead time = θT_r.
Where θ = failure rate per year
and T_r = restoration time (converted to years)
than $\zeta_a = 1 - \theta T_r$

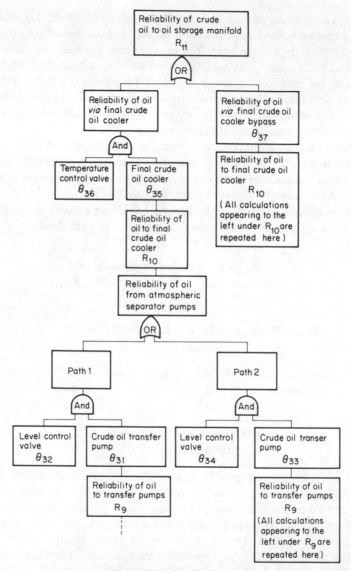

Figure 2.5-2 Partial fault tree

The restoration time for a system is calculated from individual restoration times as (component restoration time) (probability of affecting system restoration time).

The distributions of restoration times for each component were sampled in the same way as were those for failure rates. Thistle has four processing trains with a total capacity in excess of maximum output so the maximum expected

producibility from the trains is calculated from the probability of 4, 3, 2 or 1 trains operating and the expected production for each case. This leads to:
producibility from oil and gas separation trains

$$\zeta = \frac{266.7}{200}\zeta_4 + \frac{200}{300}\zeta_3 + \frac{133.4}{200}\zeta_2 + \frac{66.7}{200}\zeta_1 \qquad (2.5\text{-}5)$$

where ζ_4 = availability of 4 trains

ζ_3 = availability of 3 trains
ζ_2 = availability of 2 trains
ζ_1 = availability of 1 train

200 (thousands of barrels/day) is desired maximum capacity, 266.7 thousand barrels/day is total capacity of all 4 trains.

As by definition one cannot produce more than maximum capacity the first coefficient in the above equation is replaced by unity.

By repeating the Monte Carlo simulations many times a distribution of producibility is obtained.

Besides the processing trains the availability (or producibility where appropriate) of the other 16 systems (see Fig. 2.5-1) is determined in the same way. The simulation process is repeated for the overall system by this time sampling from the producibility distributions of each system. These systems are combined as either directly affecting overall producibility or as common mode failures. A distribution of overall producibility results.

If it is wished to study changes in platform producibility over time it is possible to do a stochastic simulation of the flow of oil, gas, and water through the trains. In this case distributions of component frequency of failure are sampled to determine time to the next failure and distributions of repair time sampled to get clocktime when the component is available again. Calculations are made at each event time where failure will result in lost production and the computer program can log all failures and their effects on oil through the process train, on gas in the process and consequently the loss of oil per year. This method can be used to study the effects of time to shut-down and start-up the process and to switch to another process train.

As an example of how reliability calculations can influence engineering design if carried out early at the conceptual design stage, Figure 2.5-3 shows the effect that multiple process trains and extra processing capacity have on final producibility. If plant availability is known even notionally in this case, it is possible to do a pay-off calculation comparing increased revenue from surer production with the increased capital cost of the extra train(s) and capacity installed to achieve it.

The equations of the graphs in Figure 2.5-3 are obtained as follows: Let k be the proportion of maximum production that one train is capable of processing.

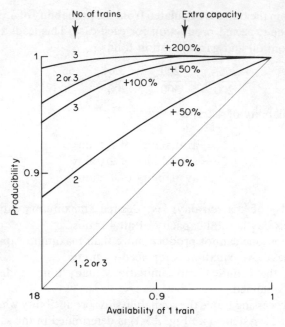

Figure 2.5-3 Producibility as a function of number of
trains and overcapacity

If C = total design capacity

and c = maximum operating capacity

and n = no. of trains

Then $k = \dfrac{C}{nc}$ (2.5-6)

If r trains are in operation the proportion of the total possible

           capacity which will be available $= \dfrac{r \cdot C}{nc}$ (2.5-7)

This may be greater than the proportion that can be used which equals 1. Hence in the following equations if $\dfrac{r \cdot C}{nc} > 1$ it has to be replaced by 1.

(a) 1 train

     producibility $\zeta_T = \zeta$,
     where ζ is availability of 1 train

$$k = \frac{C}{1 \cdot c} \text{ but usable proportion} \leqslant 1$$

$$\therefore \zeta T = \zeta \qquad\qquad\qquad\qquad (2.5\text{-}8)$$

(b) 2 trains

$$\zeta_T = \zeta_2 + k \cdot \zeta_1, \qquad k = \frac{C}{2c}$$

$$= \zeta^2 + k \cdot 2\zeta(1 - \zeta)$$

$$= \zeta^2 + \frac{C}{2c} \cdot 2\zeta(1 - \zeta) \ or \ \zeta^2 + 2\zeta(1 - \zeta) \ \text{if} \ \frac{C}{2c} > 1 \qquad (2.5\text{-}9)$$

Note that if $C = c$, i.e. if there is not surplus capacity then:

(c) 3 trains

$$\zeta_T = \zeta_3 + 2k\zeta_2 + k\zeta_1, \qquad k = \frac{C}{3c}$$

$$= \zeta^3 + 2k \cdot 3\zeta^2(1 - \zeta) + k \cdot 3\zeta(1 - \zeta)^2 \qquad (2.5\text{-}10)$$

where again if $2k > 1$ substitute 1
If $C = c$ this again reduces to $\zeta_T = \zeta$

(d) 4 trains

$$\zeta_T = \zeta^4 + 3k\,4\zeta^3(1 - \zeta) + 2k\,6\zeta^2(1 - \zeta)^2 + k4\zeta(1 - \zeta)^3 \qquad (2.5.11)$$

where the proportion through 3 trains cannot exceed 1. If $C = c$, $\zeta T = \zeta$
It will have been noted that increasing the number of trains only improves producibility if total capacity is also increased. With no extra capacity there is no advantage in duplicating trains if the availability of each train remains the same.

Hazard Analysis

A hazard analysis was done for the production modules. These are designed to be safer than most industrial facilities and the results of the analysis bear this out. Several independent factors would have to occur before oil were released and in addition there is a minimum of ignition sources and mostly the presence of halon to prevent explosion. Figure 2.5-4 shows a fault tree showing how the occurrence of a hydrocarbon release has to coincide with failure of the gas detection and fire detection/suppression equipment within that zone. (Events are shown in rectangles and consequences in hexagons.) To combine failure probabilities in the case of an AND condition the probabilities are multiplied together, while in an OR condition they are added. The latter is an approximation which is justified when the probabilities of failure are very small. Thus for two independent events the range of outcome is given by:

$$(p1 + q1)(p2 + q2) = p1p2 + q1q2 + q1p2 + p1q2$$

where $p1$ is the probability of event 1 happening, or more specifically in this context, plant item 1 failing and $q1$ is chance that it does not fail (so $p1 + q1 = 1$).

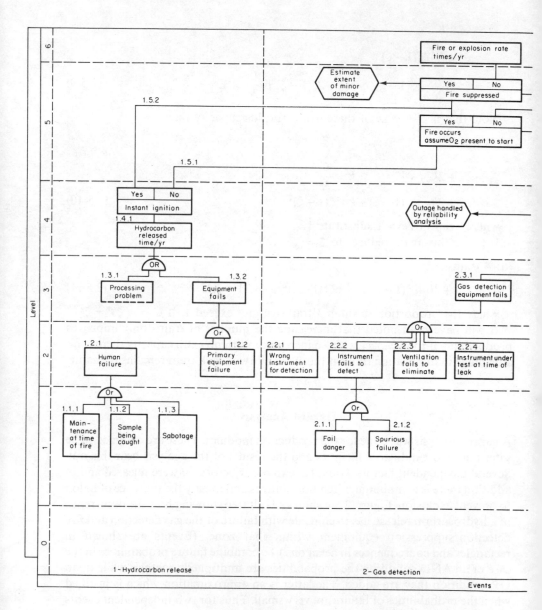

Figure 2.5-4 Cause/consequence fire or explosion diagram

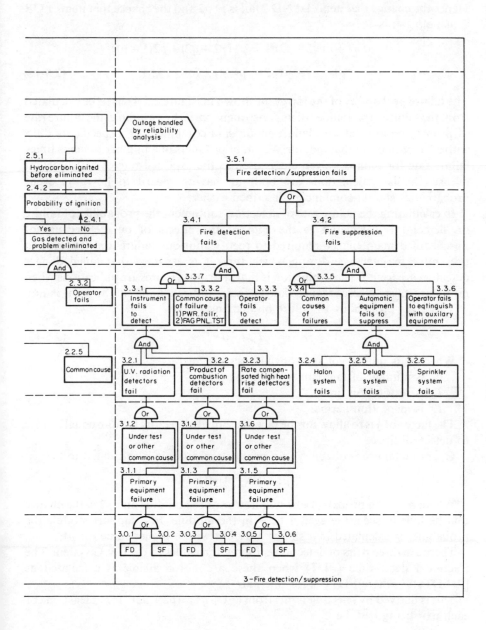

2.5.1
Hydrocarbon ignited
before eliminated

2.4.2
Probability of ignition

2.4.1

| Yes | No |
Gas detected and
problem eliminated

And

2.3.2
Operator
fails

2.2.5
Common cause

Outage handled
by reliability
analysis

3.5.1
Fire detection / suppression fails

Or

3.4.1
Fire detection
fails

3.4.2
Fire suppression
fails

And

And

Or 3.3.7

3.3.1
Instrument
fails
to
detect

3.3.2
Common cause
of failure
1)PWR. failr.
2)FAG PNL.TST

3.3.3
Operator
fails
to
detect

Or 3.3.5
3.34

3.3.4
Common
causes
of
failures

3.3.5
Automatic
equipment
fails to
suppress

3.3.6
Operator fails
to eqtinguish
with auxilary
equipment

And

And

3.2.1
U.V. radiation
detectors
fail

3.2.2
Product of
combustion
detectors
fail

3.2.3
Rate compen-
sated high heat
rise detectors
fail

3.2.4
Halon
system
fails

3.2.5
Deluge
system
fails

3.2.6
Sprinkler
system
fails

Or

Or

Or

3.1.2
Under test
or other
common cause

3.1.4
Under test
or other
common cause

3.1.6
Under test
or other
common cause

3.1.1
Primary
equipment
failure

3.1.3
Primary
equipment
failure

3.1.5
Primary
equipment
failure

Or

Or

Or

3.0.1   3.0.2   3.0.3   3.04   3.0.5   3.0.6

| FD | SF | FD | SF | FD | SF |

3 — Fire detection / suppression

Hence the change that items 1 AND 2 fail is $p1\,p2$ and the chance that items 1 OR 2 or both fail is

$$p1\,p2 + p1\,q2 + p2\,q1 = p1\,(p2 + q2) + p2\,(1 - p1) =$$

$$p1 + p2 - p1\,p2 = p1 + p2 \text{ for } p1 \ll 1. \qquad (2.5\text{-}12)$$

The failure probability of the top event shown in Figure 2.5-4 has to be calculated from the failure probabilities of its constituent parts. For example, the failure rate of primary equipment uses failure mode data combined for a specific module, with a list of items in that module. A sum of products of a number of items times failure rate for pumps, valves, etc. is used as the expected rate of hydrocarbon release. Similar calculations are built up having regard to how the event probabilities should combine as described earlier.

In calculating the failure rate of detection equipment the probability of failure on demand is calculated as the sum of the effects of outage because of unscheduled maintenance required to rectify spurious failure and that due to preventive maintenance during which time it is assumed that failed danger detectors are found. The distinction is that spurious failures are indications when no failure exists whereas failed danger faults indicate all is well when it is not.

This gives probability of failure on demand:

$$= \theta_s T_R + \tfrac{1}{2}\theta_D T_i \qquad (2.5\text{-}13)$$

Where θ_s is number of spurious failures per year:
   θ_D is number of fail danger events per year
   T_R is restoration time
   T_i is inspection period.
The factor of $\tfrac{1}{2}$ is to allow for the fact that not all detectors have an equal chance to detect all fires.

Operator failure probability in a production module is calculated to be:

$$p1 + (1 - p1)p2$$

Where $p1$ is the probability he will not be in the module and $p2$ is the chance that he fails to see a fire even if he is in the module. In a support module the probability is calculated in a similar way but with a higher value for $p1$.

There are three pairs of detection/suppression cables to the F & G system. The fractional dead time (FDT) when these are not available is calculated as $[2(\text{FDT}_1)]^3$ where FDT_1 is for one pair.

This is derived by selecting terms from $(p1 + q1)^3 (p2 + q2)^3$ for at least one of each pair being bad, i.e.

$$(p1 + q1\,p2)^3 = (2p1)^3. \qquad (2.5\text{-}14)$$

The probability of ignition used is that where no obvious ignition source exists. The consequence analysis of the monetary impact of hazards in each module is estimated in terms of the percentage effect of a fire on a module times the cost of the module. This gives the module replacement cost. The lead time is used to calculate loss of oil revenue.

Structural integrity

The risk of a structure failing

$$P = P(R \leqslant S) = \int_0^\infty \left[\left(\int_0^s f_R(r)dr \right) f_s(s) \right] ds = \int_0^\infty F_R(s)f_s(s)ds \quad (2.5\text{-}15)$$

where S = randomly distributed loading on structure
R = structural resistance to the loading.

$f_R(r)$ and $f_s(s)$ are the probability density functions (PDF) of resistance and loading respectively and $F_R(s)$ is the cumulative distribution function of R. There is still a dearth of data to define distributions of loading and strength parameters and a lack of proof and fatigue tests for large welded tubular structures. Such difficulties in quantification result in the use of factors of safety and predominant reliance on engineering judgement.

To calculate the risk of a catastrophic structural failure of a jacket during a 25 year lifetime the following approach can be used: The loading distribution is got by analysing wave data to get an exceedance curve giving the number of occasions per year that wave height can be expected to exceed any given value.

If n = number of exceedance cycles of wave height H in 1 year.
and N = total number of waves in 1 year average, then the probability of any single wave exceeding H is:

$$P = \frac{n}{N}$$

Probability if no wave exceeding H in one year

$$= \left(1 - \frac{n}{N} \right)^N \quad (2.5\text{-}16)$$

Probability of no wave exceeding H in 25 years, $P = \left(1 - \frac{n}{N} \right)^{25N}$

$$\therefore p \simeq e^{-25N}$$

from which the PDF can be determined.

There is an accepted relationship between wave height H (ft) and load factor R (i.e. wave loading expressed as a fraction of design load). This is $H^2 = 98.5 R^2$.

Hence the PDF for loading is obtained. It is next necessary to compute the ultimate strength of the jacket. This is done by considering the high level of redundancy which exists in the structure as well as the level of confidence which is attributed to computed member and joint loads. The redundancy is achieved by the arrangement of bracings between horizontal levels on all four planes. These provide alternative load paths by diagonal tension in the event of the compression bracings buckling. By considering the magnitude of member and joint interaction factors achieved during the 50 year storm case it is possible to show that only a small proportion of structural elements are loaded to full capacity. Also the order in which members and joints would fail can be assessed. It is found that many bracing members and joints would have to fail before structural collapse could take place.

To obtain the distribution of structural resistance to load, use is made of experimentally determined distributions of variation in material properties for the worst of the relevant steels. The properties involved are yield stress, ultimate tensile stress, and Charpy V impact energy. Used in conjunction with a static strength assessment of critical joints and members these stresses are converted to reserve factors to give the PDF for resistance which is used to calculate the probability of failure of the jacket from Equation 2.5-1. Pile failure is assessed by comparison with the chance of jacket structure collapse. It is considered that complete failure of the piling is unlikely before this happened. Factors leading to a lower probability include low stress levels for extreme storm, comparative insensitivity of pile loading to changes in environmental loading, because the major load is gravitational, and in addition the redundancy in pile groups and grouted shear connections. Another consideration is the chance of foundation failure. Detailed soil tests are done before foundation design. During the carefully controlled piling operation further measurements are used to ensure that the foundation profile is as assumed during design.

Yet another risk assessment is made to determine whether a number of fatigue failures could result in structural collapse during the life of the platform. A typical assessment of the distribution of fatigue lives shows all to be greater than 25 years and ranging up to 250 years. Because of uncertainties in fatigue data it is found that predicted stress ranges give rise to an 85 % surival level for computed fatigue lives. Nevertheless it is considered that regular inspection of the structure makes it unlikely that fatigue failure would occur and lead to catastrophic failure.

Float out risks

It can take about a week to tow out the steel jacket structure to its required position in the North Sea. The structure is floated out on its side if it is a self-floater and is towed by tugs which are capable of holding the unit in any likely gale. Once on-site, the structure has to be up-ended and flooded down to its correct position on the seabed. The integrity of the structure under tow is checked to ensure that it will withstand the 100 year storm applicable for the summer months of the tow.The assessment includes an analysis of the stresses induced by out-of-balance wave loads. The structure is also designed to be able to survive one major compartment being floaded during tow. Where the structure has unstable phases a crush dive might be required but usually the up-ending is done stage by stage with checks between each. Positioning, after up-ending, can be done slowly by a combination of winching, towing, and possibly ballasting. Three separate control methods are available for up-ending by umbilical cable, by radio or manually. Back-up valves are independently operated.

A piled structure's own weight may not be sufficient to resist the overturning moment of large waves. Summer waves are unlikely to cause overturning but could cause rocking at seabed level with possible damage or uneven penetration, so piling operations are conducted speedily until the first set are in place.

Conventional fault tree analyses and failure mode and effect analyses can be used for assessing, for example, the flood and de-ballast system which is required for up-ending the structure. Another technique that has been used is sneak circuit analysis which is a method of identifying unintentional connections and interaction in complex electrical circuitry. Sneak circuits are identified by computer analysis of topological network trees produced from manufacturing detail.

Blowout hazard

Engineering provisions, as previously described, are such as to make an uncontrolled blowout very unlikely; however there are instances, especially during workovers, where operator error has led to a blowout. A description of the procedure and what went wrong in one such case (not Thistle) might be of interest for the light it throws on human error type failures. Ten thousand feet of production tubing were to be pulled from a well and to do this the 'Christmas tree' had to be removed and a blowout preventer stack (BOP) installed. In the time between these operations the well is open. Two measures were employed to reduce the risk during this time, viz. before the tree was removed the well was killed with mud. To do this a column of mud is pumped down the well, the mud's density being such as to balance the well pressure. Though the mud used in this case appeared adequate it was of lower density than specified and because a sliding sleeve wouldn't open there was no circulation. As the second measure a back pressure valve should have been installed in the production tubing but this was not done as it was thought that the tubing hanger located on the production deck was of the wrong type to accommodate it. It was decided to substitute an alternative type of downwhole safety valve (DHSV) but seven attempts (over a 30 hour period) to install such a valve, 500 ft down in the tubing failed though in fact the last attempt was thought to be successful.

It was not possible to check from below that the DHSV could withstand the required pressure because of the sliding sleeve being closed. The BOP (17 tonnes in weight) was still in two parts after the 'Christmas tree' was removed instead of being assembled and in position for immediate use. Mud was observed coming from the well but the significance of this was not appreciated and the BOP was assembled (in 3 hours) and installation started with the lower section upside down. Blowout occurred before the BOP had been bolted completely and the platform was evacuated within 15 minutes. Fortunately no one was hurt, the blowout did not catch fire and the resulting oil slick had dispersed before reaching the coasts. The blowout was controlled after $7\frac{1}{2}$ days.

A lot has been learned from an analysis of this accident and measures have been introduced to forestall a recurrence. These measures concentrate on good procedural practice and training. Should a blowout catch fire the consequent risks to life and of damage to the topsides and structure are mitigated by good evacuation facilities and procedures, a good fire water system (FWS), and the proximity of a maintenance support vessel (MSV) for further and subsequent fire fighting capability. The platform FWS will continue in operation until the diesel oil supply is exhausted after about 24 hours.

Calculations of the effects of heat radiation from a burning blowout on the topside facilities, in particular the living quarters (which often have a heat shield on the side facing the wells), can be made. These results can be used to assess the consequences of such a hazard. The accuracy of such calculations depends on knowledge about flame temperature and emissivity and estimates of these are made from such test work as is available.

Questions 2.5

1. Compare the hazards associated with an offshore platform and a land-based plant. Discuss those factors which are specially connected with hazardous consequences which arise principally from the fact that the platform is not 'land-based'.
2. What are the basic means used to achieve safety offshore?
3. Discuss the techniques for undertaking a quantitative risk assessment of an offshore oil production facility. Describe the basic principles involved in showing the reliability of the facility and its constituent parts.

References 2.5

1. *Offshore Installations: Guidance on Design and Construction.* Department of Energy. Her Majesty's Stationery Office, London. July 1977.
2. 'Oil from deep water'. *Proceedings of the offshore Scotland Conference, Aberdeen.* Published by Offshore Services, 1973.
3. BOSS Conference, London. Papers on General Safety. 1979.

High Risk Safety Technology
Edited by A. E. Green
© 1982 John Wiley & Sons Ltd

Chapter 2.6

Pharmaceutical

G. T. Dickson and F. W. Teather

Introduction

The pharmaceutical industry is almost unique in creating products deliberately designed to have a significant and specific effect upon one or more of the body systems of man or upon organisms infecting those systems. The special features of these products are well covered by the definition of 'medicinal product' in the United Kingdom Medicines Act of 1968. This can be summarized as a substance or article which is administered for a medicinal purpose; 'a medicinal purpose' meaning treating, preventing or diagnosing disease, contraception, inducing anaesthesia, or 'otherwise preventing or interfering with the normal operation of a physiological function, whether permanently or temporarily . . .'. It is this latter phrase which provides an indication of the potential for harm which might be possessed by medicinal products. Ideally a medicinal product should produce its beneficial effects without other unwanted effects. Much of the effort of the pharmaceutical industry in producing new drugs is channelled into identifying and quantifying possible unwanted effects so that a risk–benefit assessment may be made.

Thus the consideration of risk in the pharmaceutical industry falls into two categories:

1. Hazards associated with the various chemical and pharmaceutical processes – that is, risks to operators in the factory or laboratory environment;
2. Hazards associated with use of the product – that is risks to patients. The major part of this chapter deals with this aspect.

Before considering this aspect however, a different set of hazards will be considered – the hazards of manufacture. A crude analogy may be drawn from vehicle manufacture, where, on the one hand, much thought is given to designing motor cars which will be reliable and safe on the roads; but their manufacture involves processes of high risk like the handling of molten steel, the rolling and pressing of steel sheet, and the casting and machining of engine parts. Similarly operators in the pharmaceutical industry, in making safe drugs, are exposed to many chemical risks from the processes used.

Hazards to Operators

Manufacture of pharmaceuticals takes place in two phases: (*a*) Primary

Manufacture – the creation of the pharmacologically active ingredient; and (b) Secondary Manufacture – the compounding of that active ingredient for ultimate use making sterile preparations for injection, tablets or capsules for ingestion, aerosols for inhalation, or creams or ointments for external application. See Figures 2.6-1, 2.6-2 and 2.6-3.

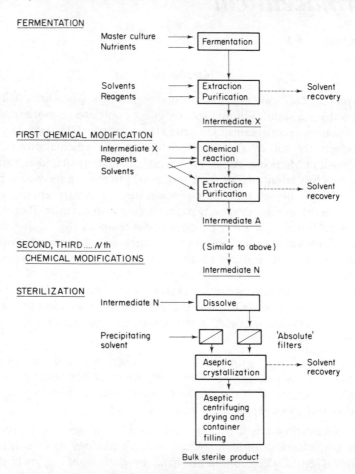

Figure 2.6-1 Simplified flow diagram of typical fermentation-based primary pharmaceutical manufacture

Hazards to operators in primary manufacture

The hazards of primary manufacture are largely those of the chemical industry. Many primary processes are in fact chemical syntheses, others may commence with biological processes, for example the culture and 'fermentation' of micro-organisms on a vast scale to produce antibiotics, but are then followed by

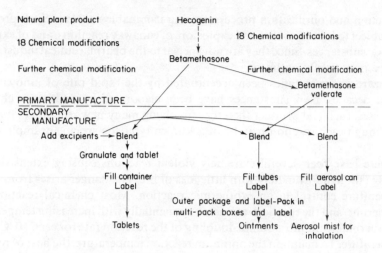

Figure 2.6-2 Example of pharmaceutical production (corticosteroids)

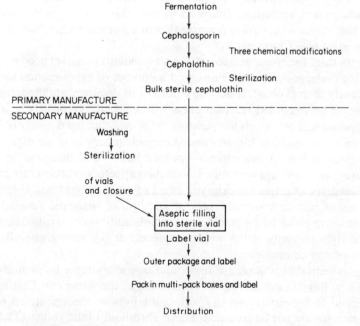

Figure 2.6-3 Example of pharmaceutical production (antibiotics)

extraction and purification processes using flammable solvents. Such processes are subject to hazards of fire and explosion, of runaway reactions, and of exposure to toxic substances. Since they are not special to the pharmaceutical industry they will be only briefly outlined.

Hazards have at times been accentuated by the rapid rate of innovation in recent decades. New substances have been elaborated by multistage chemical syntheses, each end product therefore involving many new reactions – that is to say though each reaction may be of a known type, the specific example will be new.

There have been several extremely violent reactions causing extensive plant damage though fortunately with little loss of life. The danger arises from loss of temperature control in an exothermic reaction. Most chemical reactions are exothermic, and the reaction rate rises exponentially with increasing temperature: a useful rule-of-thumb assumes doubling of the reaction rate for each 10 °C rise in temperature. To maintain the optimum reaction temperature, the heat of reaction is removed by suitably controlled cooling. If for any reason the rate of heat generation exceeds the rate of cooling the scene is set for disaster: the excess heat raises the temperature, accelerates the reaction producing heat even faster until the temperature 'runs away'. At higher temperatures, secondary decomposition reactions may ensue, or the solvent may be rapidly boiled off, or gaseous by-products may be evolved causing a rapid rise of pressure. It is such circumstances which have led to virtually explosive destruction of reaction vessels and surrounding plant, sometimes followed by fire. This scenario is mathematically comparable to the accelerating neutron flux in a nuclear reactor which is going out of control (see Chapter 1.3).

It is important to foresee and forestall such incidents on novel processes before embarking on large-scale operations, and a number of experimental techniques have recently been evolved. It may be sufficient to demonstrate the temperature above which an uncooled reaction mass will self-heat exponentially, and then plan to operate well below this temperature[1]. On the other hand it may be essential to determine the reaction kinetics and thermochemistry in some detail and for this purpose elaborate calorimeter systems have been designed to control temperature and measure heat flux[2]. Variables affecting reaction rate must also be examined; for example the catalytic effect of unusual but feasible impurities; the ingress of rust or water; the rate of addition of reagents. Possibilities for human error may also be evaluated, for example addition of erroneous reagents, or of the right reagents in the wrong sequence, or the wrong quantity, or the omission of one component.

Many industrial chemicals are toxic, and operators must be protected from their effects. For toxic materials in common use, the American Conference of Government Industrial Hygienists[3] annually publish recommended exposure limits which should not be exceeded. These Threshold Limit Values (TLV) are of two types: (*a*) the time-weighted average concentration (TWA) for a normal eight hour work day to which nearly all workers can repeatedly be exposed day after day without adverse effect; and (b) the Short-Term Exposure Limit (STEL) which is the maximal concentration to which workers can be exposed for 15 minutes continuously without ill-effects: for example, TLVs for methanol are TWA 260 and STEL 310 mg/m³ in the 1979 edition. These recommendations have received wide international acceptance.

There are special problems in primary pharmaceutical production arising from the use of unusual or even unique chemical reagents for which no human epidemiology is available. For such reagents it is necessary either (a) to make arbitrary extrapolations from animal toxicology or (b) arbitrarily to assume that the new reagent must be assumed to be toxic until proved safe. It is not difficult to assess and extrapolate likely acute toxic effects from acute animal toxicology, but long-term chronic effects in man are difficult to predict, and human allergy virtually impossible to predict from animal experiments. In practice, operators are protected if there is any doubt about a process material's safety.

Hazards to operators from exposure to active products

Since the end products are highly active pharmacologically, it is usually necessary to protect operators from the product itself. Their exposure is of a type different from exposure of a patient to whom the drug is administered in controlled amounts under medical supervision and in many cases only for a limited period. In contrast, the operators in a pharmaceutical plant could be subjected to regularly repeated random doses every working day of their lives. It follows that more stringent protection is necessary. For example a drug designed to prevent blood clotting in the treatment of thrombosis may, after repeated tiny doses, prevent the healing of wounds through continual bleeding which is no longer stopped by formation of a blood clot.

A special class of risks relates to the natural hormones or hormone analogues. For example, in the manufacture of oestrogens, widely used for contraceptive preparations, signs of exposure include development of breast tissue in male operators, and vaginal haemorrhage in women through response to the female hormone. A different hormone effect is illustrated by operator response to repeated small doses of cortical hormones or analogues which would cause the pituitary controlling mechanism to shut down the normal output of the adrenal cortex, so that the body would have to rely on the 'external' dose supplied regularly at work: it would then become hazardous for the operator to be absent from work, e.g. following an accident, since it would take several weeks for normal cortical function to return when the 'external' dose had been cut off. The effective 'external' dose may be fractions of a microgram per cubic metre of air breathed; or absorption of milligrams through the skin. This hazard is overcome by avoiding skin contact with the product and avoiding inhalation of product dust.

A large number of pharmacologically active substances are capable of causing allergy in a susceptible individual. Common examples are the penicillins (natural and semi-synthetic). Although at first, sizeable doses may be required to initiate an allergic response, with repeated exposure the triggering dose becomes progressively less until the subject may have an asthmatic attack or violent skin eruption if a trace is brought into the same room, on slightly contaminated clothing. It is obviously preferable not to elicit such responses in the first place, by total protection of the whole workforce, but in practice this desirable aim cannot always be achieved. From time to time individual operators will show signs of allergy, and the only remedy is to remove them to other work where the substance is not used.

Operations may be on quite a small scale. Production batches smaller than 1000 litres in volume and containing only about 100 kg of material may suffice for the demand. On such a scale, unit operations such as filtration and product drying may have to be carried out by hand. In that case total protection of the operators may be the only safe procedure, ideally by total enclosure of the process, more usually by protection of operators in one-piece plastic air suits. Stringent precautions include detergent scrubbing and showering before such air suits are removed so that contamination on the outside of the suit is not transferred either to the operator's hands or to the inside of the suit for later contamination of skin or underclothes. Complete sets of disposable work wear may be provided, or sets reserved for work may need to be laundered after every working shift, and regular personal showering may be desirable so that small amounts of drug do not contaminate the skin or take-home clothing.

Hazards to operators in secondary manufacture

In secondary manufacture, the hazards relate almost entirely to the processing of the active ingredients since the other substances used (fillers, disintegrating agents and lubricants in tablets, waxes and emulsifying agents for creams, etc.) are chosen because they are innocuous.

The problems arise in the many dust-producing operations such as particle size reduction, granulation, compressing, and filling operations. Protective methods include enclosure of the equipment, effective dust extraction, and the use of special helmets incorporating a battery-operated pump recirculating air through a dust filter, to provide dust-free air in the operator's breathing zone.

Hazards to Patients

Definition of hazards

Medicinal products are administered to patients with the intention of:

— re-establishing normality in a person with disease or controlling the disease state (a therapeutic effect);
— preventing the development of a disease state or condition in a person likely to be at risk to that disease or condition (a prophylactic effect);
— providing an indication of the integrity of a particular system (a diagnostic effect).

Obvious hazards are those which arise as a result of *unwanted effects* being produced by the medicinal product. Perhaps less obvious but of no less importance are those which may arise as a result of the medicinal product failing to produce its anticipated effect, i.e. *lack of efficacy*. In a patient suffering from a disease a lack of effect could result in the disease deteriorating. Even for a prophylactic drug although the immediate consequence of lack of efficacy may not be obvious, the individual is unknowingly left unprotected from the disease from which he is seeking protection. A recent example of this occurred with the issue of sub-potent Rubella vaccine. This vaccine is administered primarily to adolescent girls and occasionally to women of child-bearing age to provide protection from 'German measles' which can cause foetal abnormalities in pregnant women. Individuals who had been injected with the material had to be contacted to be given the opportunity to be revaccinated.[4]

Assessment of risk of inadequate activity

The primary objective of producing a new medicinal product is to obtain a product which is effective for the particular medicinal purpose for which it is being developed. The procedures for assessing its efficacy will at the same time evaluate the risk of inadequate activity. At various stages during these procedures, tests and evaluations are also carried out to assess the risk of unwanted effects. The investigations from which an assessment of the risk of inadequate activity can be made will be considered first.

Early screening studies will select compounds for further study on the basis of their activity in some classical pharmacological model. An anaesthetic is selected on the basis of its ability to cause loss of the righting reflex in small animals. Antibacterial compounds are selected on the basis of their activity in *in vitro* microbiological studies.

Thereafter the compounds are subjected to more rigorous testing to further differentiate compounds which might be expected to have the greater potential. For instance, an anaesthetic is evaluated in larger animals to determine the duration of anaesthesia, recovery time, and effects of repeat doses. Antibiotics selected initially on the basis of activity against a wide range of organisms are tested to show resistance to degradation by bacterial enzymes and levels achieved in blood after dosing (taking account of both peak concentration and duration).

Further animal studies are then undertaken to determine whether the activities of the selected compounds are likely to be affected by metabolic changes or by other compounds commonly administered with medicinal products of the class under development. With an anaesthetic further selectivity is achieved by observing the effects of the anaesthetic on other systems – for example vasomotor effects could produce unwanted twitching or there may be undue excitation during the recovery phase. The final stages in this preliminary selection assess the effects of other drugs administered as adjuncts during anaesthesia to ensure compatibility.

After appropriate toxicological studies the drug is then studied in human volunteers to determine absorption, distribution, metabolism, and excretion after dosing by the proposed route of administration. These studies must show that the compound produces the required profile in man. Thus if an antibiotic required for oral administration is found to be absorbed to only a limited extent, there is little point in proceeding to the next stage in evaluation. Similarly, if because of rapid metabolism, it fails to produce adequate levels in blood or urine it is unlikely to have much potential in clinical practice.

The final steps in the evaluation of efficacy take place in patients. Initially studies are carried out to determine the appropriate dosing schedule and thereafter full-scale clinical trials are undertaken to assess the efficacy of the product during normal use in patients.

These studies are relatively simple to interpret since the investigator is measuring a positive effect, i.e. activity, and unless the compound showed good activity it would not be progressed to the market.

Assessment of risks of unwanted effects

Tests to evaluate unwanted effects are much more difficult to interpret than those described above for evaluating efficacy. In these tests one is trying to predict to what extent the type of effects produced by exaggerated dosage of the compound in animals will occur after therapeutic doses in man. Further development of the medicinal product could be abandoned on the basis of results of these tests. Those which are cleared for progression would not be expected to exhibit a significant number of unwanted effects during clinical trials. The following studies are carried out:

1. *Pharmacodynamic studies* are carried out on the body systems which are not the specific target of the new drug in order to identify possible side effects. For example, a new drug being developed for the action on the cardiovascular system (e.g. a beta blocker) would be tested for its effect on the central nervous system; an anaesthetic would be tested for its effects on the cardiovascular system and an antibiotic would be tested for its effect on both the cardiovascular and central nervous systems since one would not want such a compound to produce effects on those systems.
2. *Acute toxicity studies* are single dose studies carried out on a range of species including mouse, rat, dog, rabbit, cat, and monkey. Usually a single dose is

given and the animals observed for 7 to 14 days. These tests show what the acute effects of overdosage might be, and indicate whether there are any species-related differences in toxicity which might suggest different mechanisms of metabolism.

An additional acute toxicity test in which the animals are killed two days after dosing may also be carried out to identify target organs for the toxic effects of the drug. These tests can then be expanded to study interactions with other drugs which are also known to affect the same organ. For instance certain antibiotics may produce adverse effects on the kidney at high doses. Similarly certain potent diuretics produce adverse effects on the kidney. The two drugs would therefore be given together to evaluate the effect of such combined dosage and to indicate whether special attention should be drawn to any particular hazard of combined use in patients.

3. *Repeat dose toxicity studies* are undertaken to assess the effects of repeated dosage. The duration of such studies will be determined by the type of study in man for which they are designed to provide clearance. Thus a 2 week study in a rodent and dog would provide clearance for single dose studies in human volunteers and, in addition, would serve to establish the correct dosage schedule for later studies. The relationship between the proposed period of treatment in man and the duration of dosing in the repeat dose toxicity studies is as follows:

1 or several doses within one day – 2 weeks' toxicity study
7 days' dosing in man   – 4 weeks' toxicity study
Up to 30 days' dosing in man   – 3 months' toxicity study
More than 30 days' dosing in man– 6 months' toxicity study

During these studies animals are dosed at three dose levels, the lowest being related on a body weight basis to the maximum proposed daily dose in man, the highest producing some toxic effects and the intermediate dose being logarithmically spaced between these.

Animals are observed daily and haematological and biochemical parameters are measured at intervals. Finally after dosing for the requisite period the animals, with the exception of a small group on the top dose and a similar group of controls, are killed and subjected to extensive post-mortem examination to determine organ toxicity. The surviving groups are observed to determine which, if any, of the drug-related changes seen in the animals are reversible on stopping treatment.

4. *Pharmacokinetic studies* are undertaken in several species to determine whether the absorption, distribution, metabolism, and excretion of the drug are similar in the animals being used for toxicity studies and in man. If these are shown to be different the results obtained in animals are unlikely to have predictive value for man.

5. *Oncogenicity studies* are studies to determine whether the compound is likely to be tumorogenic in man. Such studies are usually undertaken for medicinal

products which are to be administered to man for periods in excess of six months; if the active ingredient has a close chemical analogy to a known carcinogen or co-carcinogen; or if suspicious changes occur during repeat dose toxicity studies.

These studies are usually carried out in the mouse and the rat and involve dosing for 18 months and two years respectively. As in the repeat dose toxicity studies three dose levels and a control are used. A positive result in such tests will almost inevitably result in termination of development of a product. If positive effects were seen after marketing, it is likely that the drug would be withdrawn, e.g. the withdrawal in the United Kingdom of products containing the antihistamine methapyrilene after a report of a carcinogenic effect of high doses of the drug in rats[5].

6. *Reproduction studies* – Three types of test are included in these studies to reveal the presence of any drug-related effect which might result in foetal abnormality, foetal loss, or damage to the offspring in later life.

   — *A fertility study*, in which males and females are dosed before conception and dosing of females is continued throughout pregnancy, is carried out in one species, usually the rat. In half the females litters are delivered by Caesarian section and examined for foetal abnormalities, while the remainder are allowed to litter normally and rear their progeny. One male and one female from each litter are allowed to breed and produce one litter as a further test of reproductive function.

   — *A teratology study*, in which pregnant females are dosed during the period of organogenesis, i.e. that period of pregnancy during which organs are formed, is carried out to determine the effect of the drug on the developing foetus. This type of study is usually undertaken in the rat and the rabbit. This test was introduced to predict the type of effects produced by Thalidomide, which produced malformation of the foetus when taken by pregnant women[6]. It was this clinical disaster which gave a great impetus to more rigorous safety testing of drugs and to the wider introduction of governmental control over the marketing of new drugs.

   — *A perinatal study*, in which pregnant females are dosed after the period of organogenesis and throughout lactation up to weaning, is carried out to assess effects on future generations. The progeny are then allowed to litter to assess their reproductive capacity.

7. *Mutagenicity studies* – The predictive value of these studies is still being assessed. They differ from the animal studies described in the earlier sections in that they are usually *in vitro* studies carried out on micro-organisms (bacteria or yeast) or mammalian cells. The tests rely upon an effect being manifest among a very large number of organisms/cells which are individually not very sensitive. There are two group of tests: one group evaluates genetic damage and the other group chromosomal damage.

8. *Clinical studies* – During the course of clinical trials, although careful note is made of all adverse effects, special care is taken to look for those adverse effects

which have been predicted by animal tests. Thus those patients with metabolic defects which might affect the activity of the product or functional defects which might be aggravated by the product are observed particularly carefully. Only those effects which occur with a relatively high frequency are likely to be picked up during clinical trials. For instance during clinical studies on Meliamide, a histamine H2 receptor antagonist, seven patients developed agranulocytosis, that is, the patients stopped producing certain white blood-cells which are essential in fighting infection. As a result of this the clinical trials of this drug were stopped[7].

9. *Post-marketing surveillance* – An extension of these observations might also be achieved by devising a scheme of 'monitored release' when the product is first marketed. Under such a scheme doctors using the product are provided with special record forms to record at specific intervals the reactions of their patients to the new product. Several variations of such schemes are under discussion, but each of them fundamentally requires that it shall be possible to identify and obtain records for patients who have received the product. Support for the value of such schemes is obtained from the number of occasions on which adverse effects of drugs are picked up only after widescale clinical use. An example of this was the recognition of the oculocutaneous syndrome due to Practolol[8]. This syndrome includes dry eye and scarring, fibrosis, and metaplasia of the conjunctiva; rashes which are often psoriasi-form; pleural and pericardial reactions; fibrinous peritonitis; and serious otitis media. The drug was withdrawn from general use after it had been making considerable contribution to the control of blood pressure and heart disease. It has been known for many years that long-term use of phenacetin is associated with nephropathy and, finally in September 1979, after review by the Committee on Review of Medicines, the general use of phenacetin in human and veterinary medicine was banned by an Order promulgated under the UK Medicines Act of 1968. (The Medicines (Phenacetin Prohibition) Order 1979 SI No. 1181.)

Standardisation of the product and process

Standardization of chemical purity

The control of quality in food and drug manufacture is of obvious importance, and has been the subject of regulation for many years, both by government agencies, and by the medical and pharmaceutical professions. The self-regulatory standards of the industry are also very strict. No major mishaps are on record in recent times, but the inherent possibilities are illustrated by a recent analogous disaster in the USA where an animal feed compounder made up many tonnes of feed containing a long acting poison (polybrominated biphenyl) in place of a mineral supplement (magnesium oxide), the substitution going unnoticed because of the similar trade names and physical appearance. Hundreds of farm animals were affected and the poison spread via milk, eggs, and meat to the human population[12].

Meticulous control is essential in final compounding procedures – any contaminant introduced will eventually reach the patient. As an example the protocol enforced by the US Food and Drug Administration and the Medicines Division of the UK Department of Health and Social Security requires for each raw material:

1. Purchase only from approved suppliers with high quality control standards.
2. In-house checking of identity and purity.
3. Storage of data on the subsequent use of each batch of material.
4. Segregation of approved and under-test materials.
   and for each lot of final product:
5. The 'family-tree' of all input materials to allow recall of lots with a common ancestry.

Similar quality standards apply to materials used for packing the products.

Identical control criteria are applied in primary manufacture, though the sense in applying these criteria to each reagent at the start of a multistage synthesis must be regarded as lacking in logical rigour – the probability that contaminants will be carried over into the end products is negligible. The control protocol is none the less meticulously applied.

What is of great value in multistage primary manufacture is to assess the identity and purity of each intermediate substance and to standardize all reaction variables. Changes in process are introduced with great caution, the product being quarantined at later stages until demonstrated to be of immaculate quality: even then, major changes in method must be approved in advance by the regulatory authority before the final product is released for sale.

Standardization of process in vaccine manufacture

In vaccine manufacture, the hazard to patients is most dramatically exemplified by an incident in the USA. In 1955 a chemically inactivated virus in an extensive field trial was proved to be a safe and effective vaccine for poliomyelitis. In anticipation of its release for sale, six pharmaceutical companies had manufactured stocks sufficient for 10 million doses. In the first two weeks after release 4 million doses were administered mainly to children in order to anticipate and protect from the annual summer epidemic of the disease. A significant number of inoculated people developed paralytic poliomyelitis, which seemed to be related to vaccine from one manufacturer, who was immediately asked to recall all vaccine. In a subsequent inquiry[13] it was shown that 94 persons had developed poliomyelitis following vaccination and a further 166 contacts developed poliomyelitis which must have arisen through their infection by vaccinated people who (though symptomless) must have had an active infection following vaccination. The vaccine incriminated by these studies came from one production blend from one manufacturer, this blend being subsequently shown to contain low but detectable amounts of live virus.

Stringent precautions must be taken to avoid such mishaps in regular manufacture. Vaccine processes rely on the creation of modified virus material which retains its ability to trigger the body's defensive antibodies, while losing its capacity to proliferate or to cause morbid effects. Such modification may be genetic or chemical. Taking chemical processes as an example, the first process is to provide host cells, usually by tissue culture, which are then infected with live virus. The latter invades the host cells to provide a highly infectious culture which must be totally contained to protect the operators and the environment. The infective culture is then subjected to controlled chemical treatment and shown by exhaustive assay to be safe. At this stage the modified (safe) vaccine must be totally protected from potential contact with the live infective culture. Geographical segregation of live, under test, and passed cultures, and meticulous surveillance are essential.

Effective sterilization of products for injection

Injection by-passes the body's natural defensive barriers against infection, so that contaminating organisms have a chance to proliferate without first having to pass such barriers. Accordingly products for injection must be free from live micro-organisms. Though, at first sight, it might be supposed that antibiotics would be self-sterilizing this is far from true – no antibiotic is of universal efficacy against pathogenic organisms.

Sterilization by heat

The damage caused by infection will depend on the site of injection. There is a chance of localization of an intramuscular infection, while an intravenous infection has much more chance of spreading through the bloodstream, and an infection of the cerebrospinal fluid can be rapidly fatal being inaccessible to attack by the defensive leucocytes which cannot cross the blood/brain barrier. The simplest and most effective method of sterilization is to sterilize the final sealed container, for example by heating sealed ampoules. In this way subsequent contamination is avoided. The hazard of inadequate sterilization of fluids for intravenous infusion was highlighted by an incident in Plymouth, UK[14]. Solutions of glucose (allegedly sterile) used for intravenous infusion, being suspected of causing untoward reactions in patients, were found to be infected with *Klebsiella aerogenes*. It was subsequently shown that the heat treatment cycle during autoclaving had been inadequate. Strict guidelines for the prevention of microbial contamination of medicinal products were subsequently drawn up by the UK Medicines Commission[15].

Sterilization by filtration

Heat treatment may cause decomposition of the active ingredient. This indeed happens with most of the common antibiotics such as the penicillins and cephalosporins. With such substances sterility is achieved by passing a solution through an 'absolute' membrane filter; that is one whose pores are so small (*c.* 0.2 μm) that all cells and spores are held back. The sterile solution is then treated with a similarly sterilized precipitating reagent. For example a sterile solution of a soluble penicillin salt in acetone if treated with a sterile solution of potassium acetate will cause the potassium salt of penicillin to crystallize. Alternatively the product may be precipitated by adding a suitable solvent.

The entire process must be carried out under aseptic conditions. Before the start of the process the entire plant must be cleaned and sterilized usually by steam or superheated water. Thereafter asepsis is maintained. The process stream may be maintained at a positive pressure relative to the outer environment using sterile air or nitrogen. In addition it is customary to surround the plant with a very clean environment whose standards of cleanliness considerably exceed hospital operating-room standards. Some procedures such as container filling and sampling are carried out under the protection of laminar flows of sterile air, sterilized by absolute filtration.

Problems of sampling arise especially in attempting to prove the sterility of ampoules. Regulatory authorities stipulate the number of ampoules to be sampled, e.g. 20 in every lot of 1000 to 50 000, in a randomized pattern[16]. Statistically such a sampling frequency means that relatively 'light' infections could escape detection, as has been demonstrated statistically[17]. It follows that sterility testing, by itself, will not be an adequate safeguard, but will only indicate a serious trend. In practice, reliance must be placed on meticulous plant operating procedures.

A further risk in injectable drugs is contamination with the artefacts of bacterial contamination. Such bacteria can be killed by heat, but the dead bacteria release water soluble macromolecules termed pyrogens, which can cause allergic reactions and in some cases may even precipitate anaphylactic shock when injected. Water supplies are particularly at risk, and must be kept sterile and be tested for such pyrogenic impurities if used for manufacture of products for injection.

Standardization of formulation

Having undertaken an extensive assessment of the potential risks associated with a product it is essential to standardize to the particular formulation that has been tested. Pharmaceutical studies must therefore be undertaken on the formulation to show that on storage there is neither unacceptable reduction in potency nor the production of toxic degradation products. Changes to the formulation should only be made after careful consideration of the effects that such changes might have on the bioavailability of the active ingredient. If necessary pharmacokinetic studies may be required to show bioequivalence.

The effects of certain changes are relatively predictable – for example, changing the particle size of a poorly soluble active ingredient will affect bioavailability. A typical example of this effect was seen on reducing the particle size of griseofulvin which allowed a reduction of virtually 50 % in the dosage required to produce the desired therapeutic effect[9]. Other effects may be unexpected as in the case of changes to the method of manufacture of digoxin tablets or the change of inert excipient used in phenytoin capsules. In the former example a change in the method of mixing during production resulted in a halving of the bioavailability[10]. An increase in toxicity of capsules of phenytoin sodium in Australia was attributed to the change in inert diluent from calcium sulphate to lactose[11].

Standardization of labelling and product information

Regulatory authorities grant approval to market medicinal products after assessing the results of the extensive programme of testing undertaken by the manufacturer to demonstrate that the product has an acceptable degree of safety and efficacy in the treatment of certain conditions. The marketing authorization sets out details of the proposed uses and dosage together with the warnings and precautions relevant to such use. It is therefore not surprising that the regulatory authority seeks to ensure that appropriate information relevant to the approved use is conveyed to the doctor, the patient, and others concerned with the use of the medicinal product. This is achieved by promulgation of regulations which control the content of labelling and promotional material issued in connection with the product.

These regulations control the final stage in the production of a medicinal product to ensure that use of the product is limited to the uses approved by the regulatory authorities.

Risk–benefit ratio

The benefit provided by a medicinal product varies over a very wide range. A product on unrestricted sale is likely to provide relatively minor benefit such as the relief of cold symptoms; a product available on prescription only from a General Practitioner will exert considerable benefit such as control of blood pressure; the other extreme of the spectrum is a product used by specialized medical staff in hospitals for cancer chemotherapy.

The risk which is acceptable for the product available over the counter for self-medication is obviously much smaller than that which would be acceptable from a product used in cancer chemotherapy. The ailment being treated by the over-the-counter product is likely to be of such a relatively minor nature that recovery would occur without treatment, and thus the treatment must produce virtually no risk of ill-effects to the patient. On the other hand the condition of a patient with cancer is very serious and one is thus prepared to accept a very much greater risk of ill-effect from cancer chemotherapy to obtain some benefit. Such a balance has to be taken into account when proposing the marketing of a new product.

Attitudes of regulatory authorities

Most regulatory authorities follow the general principles set out in the earlier sections. However differences occur between authorities which, although probably of minor significance from the point of view of impact on safety, can result in studies which are acceptable to one authority being unacceptable to another. For instance the requirements for toxicity testing may differ in respect of the number of animals per group; the frequency of taking samples; the duration of observation after dosing, etc. This leads to difficulties for international drug development.

These difficulties can also be exaggerated for the company developing a new drug where it is not clear whether some of these apparent minor differences are based on national attitudes or are in fact specific requirements of the regulatory authority.

General Comments

Regulatory control of new drug development has been exercised for various periods of time in different countries. Sweden was probably one of the first countries to introduce legislation and the United States was another country which introduced legislation during the late 1930s. The great impetus to the spread of control was the Thalidomide disaster in the late 1950s. It would be incorrect to give the impression that before the introduction of such controls by government agencies the industry did not carry out safety testing. However a greater awareness of the need for more sophisticated studies was evoked by the Thalidomide disaster.

The extent of the testing in animals and of the carefully controlled clinical studies in man to which new drugs are now subjected ensures that much is known about a new drug before it is marketed. Even this cannot ensure that a new drug will be free from the risk of adverse effects. In seeking greater efficacy from more potent drugs, a certain degree of risk has to be accepted in order to attain that benefit. Total safety is probably unattainable.

Thus while it is important to continue to undertake testing to assess risk, it is equally important that the demands of such tests do not become so great that new medicines are delayed. The loss of benefit from such medicines might be less acceptable than the hazard being prevented. This effect is already seen to some extent in relation to drugs used in the treatment of diseases with a low incidence. The potential sales of a drug to treat such a disease are unlikely to be sufficiently high to cover the very heavy costs which are now incurred in the development of a new drug, and in consequence their development is not commercially viable.

Finally, it is important to keep in perspective any demands for increasing the requirements placed on the pharmaceutical industry to assess these risks. Caution is indeed to ensure that there is not over-regulation with consequent stifling of activities aimed at producing new medicines which can make a beneficial contribution to mankind. An article somewhat provocatively entitled 'The British way of death'[18] in which death rates for various diseases in 1975 are compared, helps to put in perspective the risks associated with medicines. A particularly dramatic comparison is that in which the number of deaths from oral contraceptives (approximately 100) is compared with the numbers of deaths from smoking (100 000), traffic accidents (6000), alcohol (approximately 5000), and murder (500). Not surprisingly the author comments that 'preventing smoking (or somehow reducing the toxicity of cigarettes) is the single most important public health measure for developed countries'. The article concludes that the present trend in death rate suggests that vigorous expensive social or political action against aspects of modern life such as pollution or food additives (and presumably one could add control of the development of new medicines) is unlikely to have the impact on health and mortality that a strong attack on smoking could have.

Questions 2.6

1. What are the hazards which might occur as a result of lack of efficacy of a medicinal product? Describe the methods employed progressively during the development of a medicinal product to demonstrate that it is likely to be effective.
2. List the types of tests normally carried out to determine unwanted effects of a medicinal product and indicate the purpose of each test.
3. Having assessed the safety and efficacy of a medicinal product, what steps are taken to ensure that these properties do not change during production and marketing?
4. Outline some of the problems in applying a risk–benefit criterion to the development of a new medicinal product.
5. What hazards to plant operators may occur during primary pharmaceutical or other fine chemical manufacture and what practical steps may be taken to minimize them?
6. What risks to the ultimate user of the medicinal products prepared for injection, may arise during processing and how may they be avoided? Exemplify from the fields of vaccine and antibiotics manufacture.

References 2.6

1. Coates, C. F. and Riddell, W. 'Assessment of thermal hazards of batch processing.' *Chemistry and Industry*, 7 Feb 1981, p. 4. Society of Chemical Industry London.
2. Sikarex: An example of commercially available equipment is a safety calorimeter manufactured by System-Technik AG. CH-8803 Ruschilikon Switzerland who provide technical literature on request.
3. American Conference of Government Industrial Hygienists PO Box 1937 Cincinnati, Ohio 45201, USA.
   Note: By agreement the UK Health and Safety Executive publish a British Version, e.g. Threshold Limit Values for 1979.
   HM Stationery Office, London, 1980. ISBN 0 11 883193 3
4. *Pharm. J.* 'Faulty rubella vaccine questioned.' *Pharm. J.*, 1 December **1979**, 568.
5. *Pharm. J.* 'Methapyrilene to be withdrawn from UK market.' *Pharm. J.*, 14 July **1979**, 49.
6. McBride, W. G. 'Thalidomide and congenital abnormalities.' *Lancet*, **2**, 1358. 1961.
7. Burland, W. L., Sharpe P. C., Colin-Jones D. G., Turnbull P. R. G., and Bowskill P. 'Reversal of Metiamide-induced agranulocytosis during treatment with Cimetidine.' *Lancet*, **2**, 1085. 1975.
8. Marshall, A. J., Baddeley, H., Barritt, D. W., Davies, J. D., Lee, R. E. J., Low-Beer, T. S., and Read, A. E. '*Practolol peritonitis.*' *Quart. J. Med.*, **46**, 135. 1977.
9. Atkinson, R. M., Bedford, C., Child, K. J., and Tomich, E. G. 'Effects of particle size on blood griseofulvin-levels in Man.' *Nature*, **193**, 188. 1962.
10. Hamer, J. and Grahame-smith D. G. 'Bioavailability of Digoxin.' *Lancet*, **2**, 325. 1972.
11. Tyrer, J. H., Eadie, M. J., Sutherland, J. M., and Hooper, W. D. 'Outbreak of anticonvulsant intoxication in an Australian city.' BMJ, **4**, 271. 1970.
12. Kay, K. *Environmental Research*, **13**, 74. 1977.
13. Nathanson, N. and Longmuir, A. D. *Amer J Hygiene*, **78(1)**, 16. 1963.
14. Clothier, C. M. *Report of the Committee appointed to inquire into the circumstances, including the production, which led to the use of contaminated infusion fluids in Devonport Section of Plymouth General Hospital*. Chairman C. M. Clothier, HM Stationery Office London. Cmnd 5036 (1972). SBN 10 150350 4.
15. Rosenheim: *Report on Prevention of Microbial Contamination of Medical Products*. Chairman Lord Rosenhein, HM Stationery Office, London 1973 SBN 11 320492 2.
16. World Health Organization Technical Report Series No. 200. *Requirements for Biological Substances*. 6. General Requirements for the Sterility of Biological Substances. Geneva 1960.
17. Greenberg, L. 'Statistical significance of sterility test results, as affected by sampling procedures – a controller's viewpoint.' Paper presented at Round Table Conference on Sterility Testing held by the International Association of Microbiological Societies, London, England. 18, 19, and 20 November 1963.
18. McGinty, L. 'The British way of death' New Scientist, 30 August 1979, 649.

High Risk Safety Technology
Edited by A. E. Green
© 1982 John Wiley & Sons Ltd

Chapter 2.7

Electric Power

B. J. Garrick and S. Kaplan

Principal Types of Power Plants

Large central station power plants (400 MW(E) or greater) represent the foundation of most of the industrial nations' electric power systems. These systems involve numerous types of electric power generation plants, usually defined by the fuel or energy source. Plants in operation today include coal, oil, gas, nuclear, geothermal, and hydro. The types of plants under development are based on advanced coal and nuclear technologies and renewable energy sources[1][2].

Examples of advanced coal systems are fluidized-bed combustion, coal modification processes (e.g. desulphurization through solvent extraction and coal gasification), and combinations that result in both a gas and steam cycle to enhance efficiency. Renewable energy sources for central station electricity are still mostly in the exploratory stage. They include direct and indirect solar energy as well as tidal and geothermal sources. Direct solar concepts are thermal, e.g. the 'power tower', and photovoltaic. Examples of indirect solar concepts are wind energy and energy from ocean wave action. Other renewable energy sources for producing electricity are ocean thermal energy conversion (OTEC), biomass, and tidal power.

Research and development activities have been dramatically increased on all of the above advanced technologies directly impacting the future production of electricity. In the near term (the next 20 to 30 years), there will continue to be heavy dependence on the types of plants currently in operation, although improvements are expected with respect to efficiency, environmental impact, and safety. The thrust of the remaining discussion on electric power and risk will be with respect to plants currently in operation.

For most plants currently in operation a boiler is used to produce steam that drives a turbine. The process is seen in simplified form in Figure 2.7-1 where the steam-driven turbine drives the generator that produces the electricity[3]. After the steam has been used to impel turbine blades that rotate a shaft driving the generator the steam is condensed into feedwater and returned to the boiler for recycling. In the case of the geothermal plant, the boiler is provided by mother nature's naturally existing hot brine which is cycled from the ground formation to exchange heat with a low temperature steam conversion fluid such as isobutane

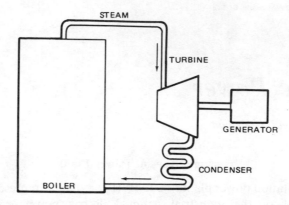

Figure 2.7-1 Schematic power plant cycle

before it is recycled back to the formation. Some geothermal plants use water recycled through a hot rock formation to form steam. The hydro-type plant differs in principle from the others in that water with a high kinetic energy due to its elevation impels directly on turbine blades without a steam cycle and drives the turbine which in turn drives the electricity-producing generator. Combinations of these different plant types are often used by electric utility companies to meet their variable power supply needs on an hourly basis. The mix depends on the fuel resources most available to the utility and the cost effectiveness of each of the plant types.

Consider the different boilers for coal, oil, gas, and nuclear plants. Figure 2.7-2 shows a typical coal burning boiler. This particular type uses a cyclone furnace and has superheaters for producing the steam. A typical size of this type of plant being built today is 800 to 1000 MW(E). Oil- and gas-fired boilers are similar to coal burning boilers in principle except for the fuel feed mechanisms[4].

In a nuclear plant the heat is generated by enhanced fissioning of uranium atoms. There are several types of nuclear steam supply systems in operation today. The majority of these are the light water reactor (LWR) type of which there are two basic designs, the pressurized water reactor (PWR) and the boiling water reactor (BWR) seen in Figure 2.7-3[5]. The pressurized heavy water moderated and cooled reactor such as the CANDU is another type which, unlike the other reactors, uses uranium that is not enriched. An attractive feature of CANDU is on-line refuelling. The CANDU is the only existing foreign reactor that, so far, has been able to compete in the world market against the light water reactors first introduced by the Americans. The first country to launch a commercial nuclear power programme was the United Kingdom in the mid-1950s. The first generation plants were natural uranium fuelled, gas-cooled reactors with graphic moderation – known as Magnox reactors. More recent UK plants involve advance gas-cooled reactors fuelled with enriched uranium oxide.

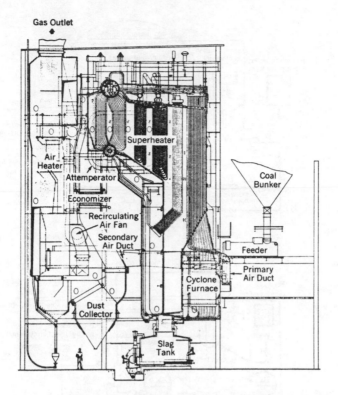

Figure 2.7-2 Typical coal burning boiler

In principle, the nuclear reactor core comprised of fuel assemblies becomes a heat source as a result of the nuclear fission process. A fluid (demineralized water in a LWR) acting as the coolant passes through the core and removes the heat generated by the reactor. In a PWR, pressurized high temperature water is cycled through a steam generator which produces steam on its secondary side. That steam is transmitted to the turbine where it impacts on the turbine blades. In a BWR, the steam produced in the reactor is separated from the heated water inside the reactor vessel after which it is transmitted directly to the turbine.

Before discussing the safety issues involved in electric power projects, it is important to point out briefly that the plant is merely one part of a fuel cycle, regardless of plant type. Typically a fuel cycle consists of fuel exploration and extraction, fuel processing, transportation to and from a power plant, power generation, and waste disposal. Certainly hydro and geothermal power systems do not necessarily involve all of these phases in their fuel cycles, but most other types do. When evaluating safety in the fuel cycle, all phases need to be considered including facilities construction and power transmission.

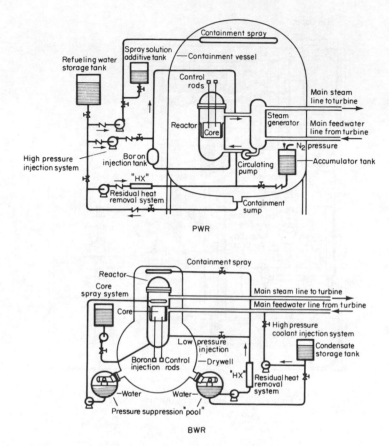

PWR

BWR

Figure 2.7-3 Nuclear power reactors

Safety and Availability Requirements

Not unlike many industries, power demands create two somewhat conflicting goals. On the one hand there is the need for electric energy to supply industries which provide the necessities and conveniences of life. In order to minimize the cost of power, one or another of the fuels is the most economical energy source in any given location because of the combination of available resource and lower net costs for extraction, transportation, generation plant, fuel, and operations. Whenever a particular power source is lower in total net cost for an area, it becomes the most highly desirable energy source to meet that utility's baseload service demand. The incentive is then to maximize the availability of energy generated by that energy source and to rely on the other energy sources only to provide the difference between baseload and peak demands, usually at increased cost. On the other hand power plants cannot be operated in a way that maximizes their availability at the expense of the safety of plant personnel or of the public.

Looking at it another way, these goals of safety and plant availability are not really conflicting. The combination of safety interests and regulatory controls requires that the plants either be operated with a high level of safety, or they are otherwise de-rated or shut down, thereby forcibly reducing the amount of power which is available. So it is in the power industry's best interest to maximize plant availability consistent with safety. As will be seen shortly, the techniques employed in maximizing safety are similar to those used in maximizing cost effective plant availability.

Safety issues in power plants are associated with both normal operations and accident conditions. Accidents may derive from equipment failures, natural catastrophies, or human error or procedures. Safety may also be construed to include an issue of environmental impact during normal operation. In a sense, some safety concerns can be found in every phase of the fuel cycle and for every type of power concept. It is necessary to sort through these issues, determine which ones significantly contribute to risk, and resolve what might be done to reduce the risk. For example, what are some of the safety considerations in the different fuel cycles and what has been done to ameliorate them?

Safety Issues and Protective Measures

Studies[6] have been performed in which occupational and public hazards and the environmental impact of coal, oil, gas, and nuclear fuel cycles have been examined. The mining of coal and the impact of its transport represent the largest occupational and public hazards of those studied. Coal mines present explosion hazards when ventilation is inadequate. The greater impact of strip mining that denudes the landscape and of ash storage at the plant site impacts on land use. The large amount of air contamination in the coal fuel cycle also stands out as a significant hazard. The main issue in a coal-fired plant is its environmental impact under normal operating conditions. The slag tank, dust collectors, and gas outlets are all potential contributors to air pollution. Scrubbers installed in the gas outlet path have been required to reduce substantially sulphate emissions and particulates.

Although the outlet gases in oil plants do not contain fly ash, they carry a considerable amount of sulphates and nitrates. Oil plants also have some non-gas residue after burning, but the volume is not great. Gas plants are substantially cleaner than either coal or oil plants, producing no fly ash and less of the noxious gases.

The main operational safety issues in hydro plants come from the extensive land use for water storage, the usurping of natural habitats when the dam is first filled, and changes in regional water distribution. Under accident conditions dam failure is of great concern[7]. Retirement or reinforcement of unstable earth dams and lowering of water levels are corrective actions being taken for reducing risk of existing dam failures. Greater care in seismic design is being demanded for new dams, and re-evaluation of the seismic risk from existing dams has also taken place.

The main safety issue in nuclear power plants is the possible release of radioactive material following an accident. Under normal operating conditions radiation is emitted in the nuclear fission process within the reactor and from the plant support areas where fluids and solids containing nuclear materials are collected or stored. Plant personnel are protected from this radiation by thick concrete walls. The public is also protected by radiation shields, primary and secondary containment, and by the establishment of minimum distances between the plant and populated areas. Small quantities of radioactive gas are discharged to the environment but at a rate which results in great dilution and essentially zero risk.

Nuclear plants are designed so that when abnormal conditions occur, the nuclear fission process in the core is automatically shut down. Redundant and diversified safeguards systems are included in nuclear plants as seen in Figure 2.7-3. Strategically placed and redundant sensors pick up abnormal conditions such as loss of coolant or unacceptable flow rates, high radiation levels, or unacceptable temperatures and pressure. When these unacceptable conditions occur, safeguard systems are automatically initiated to safely shut down the plant. These include control rod insertion into the core to upset immediately the fission process, high and low pressure injection systems to introduce neutron absorbing material to interrupt the fission process, and auxiliary heat removal systems.

In addition to the shutdown systems for accident conditions, nuclear plants have concrete and steel containment structures around the reactor vessel designed to contain pressures and radioactivity that might be released in the event plant shutdown and safeguard systems were to fail and steam is released in the plant. Containment spray systems and, in the case of the BWR, a pressure suppression pool are included to condense any steam that might be released into the containment and which might contain fission products. Spray systems will reduce steam pressure in the containment and collect contaminants. In addition, all penetrations into the containment are sealed and isolation valves are included in piping penetrations. These containment systems and design features reduce the risk to occupants of the plant site and to the surrounding public.

Redundancy and diversity are provided in the operational and safety systems of nuclear plants in order to reduce the negative impact of possible system or equipment failures. Redundant components are placed at different elevations and in different areas of the buildings to reduce the impact of common failure causes, such as from fire, flood, or earthquake. In addition, operating and maintenance procedures are provided to backcheck critical actions taken by plant personnel on important operational and safety equipment in the plant.

It is clear that all plant types have hazards. The truth is, there are always hazards, i.e. sources of danger – not just with plants and industrial complexes, but with the very process of human existence. Hazards cannot be eliminated but they can be controlled. The extent to which hazards can be controlled is what is meant by 'risk'. The balance of the discussion addresses this notion of risk and how risk analysis is performed.

Reliability, Safety, and Risk Analysis

Fundamentally, the approach to quantifying risk should be the same regardless of the type of power plant or its fuel cycle. For that matter, the same basic approach to risk analysis should apply to any industry – power, chemical, petroleum, hazardous materials, etc. One approach that meets this requirement and has been presented in detail elsewhere[8] is based on the 'set of triplets idea' where the risk R is given by

$$R = \{ \langle s_i, p_i(\phi_i), q_i(X_i) \rangle \} \tag{2.7-1}$$

where,

$s_i = $ a scenario identification or description

$p_i(\phi_i) = $ the probability p_i that the scenario s_i will have a frequency of occurrence of ϕ_i

$q_i(x_i) = $ the probability q_i that the level of damage as a result of s_i is X_i.

Equation 2.7-1 captures the three essential questions of any risk analysis, namely: (1) what can go wrong (s)? (2) with what frequency (ϕ)? and (3) what consequences (X)? The p's and q's are simply expressions of states of confidence (probabilities) in ϕ and X, respectively. The braces of Equation 2.7-1 denote the 'set of' and, in particular, Equation 2.7-1 in tabular form is as shown in Table 2.7-1.

Table 2.7-1   Risk table

Scenario	Frequency	Consequence
S_1	$p_1(\phi_1)$	$q_1(X_1)$
S_2	$p_2(\phi_2)$	$q_2(X_2)$
S_3	$p_3(\phi_3)$	$q_3(X_3)$
$\vdots$	$\vdots$	$\vdots$
S_n	$p_n(\phi_n)$	$q_n(X_n)$
S_{n+1}	$p_{n+1}(\phi_{n+1})$	$q_{n+1}(X_{n+1})$

The $n+1$ scenario is that category of scenarios representing the 'other' category. The 'other' category completes the list of scenarios as it represents all those not otherwise on the list. More is said about this in other work[8].

Finally, another and more contemporary way of presenting Equation 2.7-1 is in the form of a risk curve. For example, if Table 2.7-1 is arranged in order of increasing severity of damage and the probabilities are appropriately accumulated, the result is Figure 2.7-4. Figure 2.7-4 is a risk curve in probability of frequency format. For example, to use this diagram, enter at a specific damage level, say X, and choose the appropriate P curve, say $P_4 = 0.90$. The ordinate of this curve, $\phi_{0.90}(X)$, is then the 90th percentile frequency of X. In particular, there is 90% confidence that the frequency with which damage level X or greater occurs is not larger than $\phi_{0.90}(X)$. Probably the best known examples of such curves were published in the *Reactor Safety Study*[9].

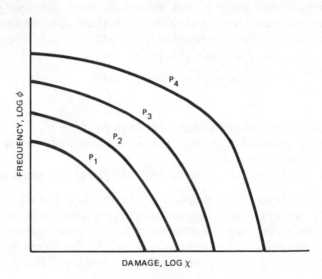

Figure 2.7-4 Risk curve

The process of obtaining risk curves of the form of Figure 2.7-4 is logically divided into two concurrent activities (Figure 2.7-5): the activity preceding the release of radioisotopes (or other contaminants if not a nuclear plant), i.e. the establishment of the 'source' condition, and the activity subsequent to release which describes the movement of the contaminants and the resulting damage. The output of the first activity is a risk curve in probability of frequency format with the abscissa being a measure of the type of release. Releases associated with individual scenarios are grouped into 'release categories' p. The second activity produces a set of conditional risk curves which for each release category ρ gives for each damage level X, the fraction, F, of times that level X is exceeded. The major source of variability in these curves is the weather. The output of the two

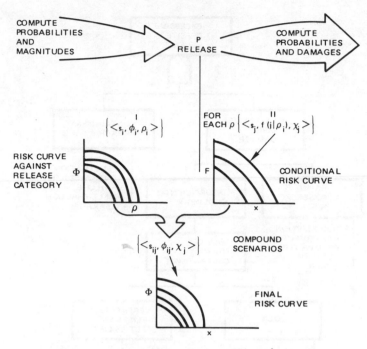

Figure 2.7-5 Structuring the scenario list basic pinch point

activities is combined into final risk curves giving the frequency, including uncertainty, of different levels of damage. Each of the two activities is discussed below.

Release frequencies

To establish release frequencies, it is necessary to construct a set of scenarios. The first step in structuring scenarios is to identify the categories or classes of events that could initiate an accident scenario or sequence. Such events are identified as initiating events. One approach to identifying initiating events is to use a logic diagram such as Figure 2.7-6. Such a diagram will be referred to as a Master Logic Diagram since it represents a summary overview of the relationship of initiating event categories and plant states. For example, the initiating event category 'small' pipe break may lead to a 'loss-of-coolant-accident' (LOCA) state for the plant. In many cases, the initiating event categories are subdivided into more detailed events depending on their importance and the ability to define them. For example, Figure 2.7-6 shows 21 initiating events and the number usually varies between 40 and 60 in typical nuclear power plant probabilistic risk assessments.

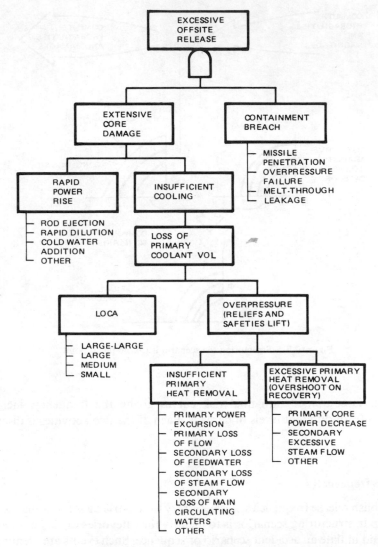

Figure 2.7-6 Master Logic Diagram

The format often adopted for scenario construction is the event tree. The event tree displays the success and failure states of systems (hardware or functional) that might impact the course of events following an initiating event. Figure 2.7-7 illustrates the basic structure of an event tree having eight scenarios or sequences. The number of event trees involved is dictated by the number of initiating events or initiating event categories. Typical numbers of event trees for a nuclear plant

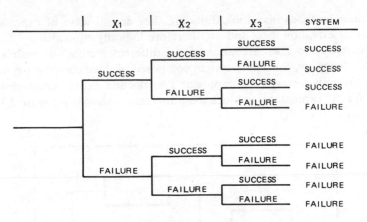

Figure 2.7-7 Event tree

risk assessment may be 7 to 14. The actual number of scenarios involved may be in the millions depending on the number of interacting systems or functions. However, experience indicates that relatively few scenarios dominate the risk – usually less than ten and often three or four.

Each of the system states noted in the event tree must be assigned a probability of occurrence to obtain scenario probabilities. The scenarios from the event trees are eventually grouped into release categories. The event tree expresses the relationship between systems and the state of the plant (the plant model). There is now a need for a relationship or 'system function' between the state of a system and the set of states of its components (the system model). To formalize the idea, let X_i be a variable characterizing the state of the *i*th component. For most work, it is adequate to take X_i as a two-state variable, that is, either the component 'succeeds' or 'fails'. It is emphasized that, while this two-state treatment is customary, it is neither essential nor universal. It should be regarded as a discretization, i.e. as a finite modelling of a continuous, infinite reality. Let

$$Q = Q(X_1, X_2, \ldots, X_n) \qquad (2.7\text{-}2)$$

represent the system function. If the set of the X_i variables constitutes a mathematical space, the component state space, S_c, and if the system variables, Q, constitute the system state space, S_s, then Equation 2.7-2 can be thought of as a mapping from S_c to S_s, i.e. from the component state space to the system state space. Here again the system space may contain two states: 'works' or 'fails', or it may contain several, as in the release categories of the Reactor Safety Study.

There are numerous methods in common use for expressing or describing system functions. Among these are fault trees, event trees, block diagrams, Boolean expressions, signature diagrams, truth tables, minimal failure sets, etc.

While these methods may look different, they are all ways of expressing the mapping of Equation 2.7-2 and are therefore logically equivalent.

The best way to illustrate some of the different methods is with a simple example. Therefore, consider a system composed of three components arranged so that the system works if component 1 works and either component 2 or 3 works. We can express this in block diagram form as shown in Figure 2.7-8 or in

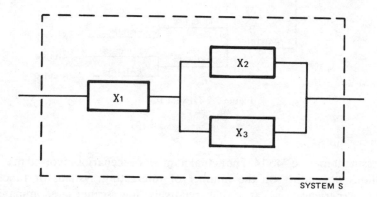

Figure 2.7-8 Systems *S* block diagram model

Boolean form:

$$S = X_1 \wedge (X_2 \vee X_3)$$

(where $\wedge$ and $\vee$ are the Boolean symbols for 'and' and 'or', respectively) or in the form of a 'success tree' (see Figure 2.7-9) where $\sqcap$ and $\sqcap$ are the tree symbols for 'and' and 'or'.

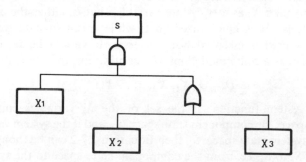

Figure 2.7-9 Success tree

Alternatively, we could use a fault tree (see Figure 2.7-10) which says that the system fails if either component 1 fails or c_2 and c_3 fail. In Boolean form, this

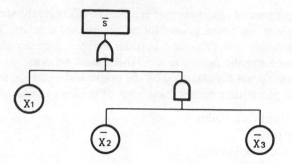

Figure 2.7-10 Fault tree

statement is:

$$\overline{S} = \overline{X}_1 \vee (\overline{X}_2 \wedge \overline{X}_3)$$

Table 2.7-2 shows truth table form

Table 2.7-2   Truth table (Components X_1, X_2, X_3)

X_1	X_2	X_3	Q
0	0	0	0
0	0	1	0
0	1	0	0
0	1	1	1
1	0	0	1
1	0	1	1
1	1	0	1
1	1	1	1

where 0 represents 'success' and 1 represents 'failure'.

In event tree form, the same logical relationship is expressed as shown in Figure 2.7-11.

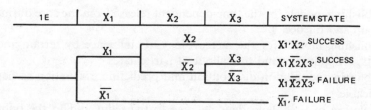

Figure 2.7-11 Event tree; logical interrelationship (component X_1, X_2, X_3)

It is such component functions that are used to calculate the state probabilities of systems given the state probabilities of components or, more precisely, component failure modes. Of course, in order to input, for example, the frequency of occurrence of specific failure modes, there must be cause data.

Cause data are generally classified by the origin and nature of faults. There are hardware and procedural causes. Examples of causes of failures are:

— Random Hardware Failures
— Human Error
— Maintenance and Testing
— Environmental Factors
— Combinations of Causes
— Common Causes

Cause data are obtained from a variety of sources. Among the more widely used sources are:

— Reactor Safety Study[9]
— Systems Reliability Service Data Bank (UKAEA)
— Nuclear Plant Reliability Data System (NPRDS)
— IEEE Std-500[10],
— LER Data[11][12]
— Human Error Rates[13]
— North American Electric Reliability Council (NERC)
— EPRI Reports

For the case of operating plants, an important source of data, of course, is the plant itself. Bayes's theorem is the basis for combining population data and plant-specific data for risk assessment work.

Consequence analysis

The second activity of a probabilistic risk assessment is the evaluation of the consequences following a release. A conditional consequence analysis can be performed concurrently with the release frequency analysis. The two analyses are then combined to determine the final consequences. In particular, the consequence analysis steps for a nuclear power plant are as follows:

1. Establishing models for the time-dependent release from the core during core heat-up and for decay.
2. Establishing models for the meteorology; its influence by terrain and lake effects, if applicable; and the fallout of particulates.
3. Establishing population distribution and predicting evacuation scenarios if applicable.
4. Calculation of time-dependent damage to the plant and to the public (as measured by deaths, injury, or property damage).

Computerized models for decay and time-dependent release from the core during core heat-up are developed for nuclear plants. Fission product decay and deposition within containment are analysed as well as gas accumulation, hydrogen build-up, and containment pressure. Models for structural response of the containment to pressure transients are developed and analysed for determination of containment leakage or failure.

The release from containment of fission products whose characteristics change with time makes it necessary to develop an environmental model which depicts the possible environmental distribution of the released material with time. This considers atmospheric stability, wind direction and speed, as well as the likelihood of precipitation during the release period. Computer programs such as CRAC (used in WASH-1400) examine the distribution of released materials in a plume on the assumption the plume and evacuees move radially outward from the release point. Other programs such as CRACIT[14], developed by Woodard and Potter, consider site-specific characteristics which could complicate the distribution. For example, it includes the effect of significant terrain variations or nearby lake effects on weather patterns, meteorologic data from more than one station, and special evacuation patterns and land uses. An example of the plume overlay on a ground grid at a particular point in time is seen in Figure 2.7-12, indicating the path and distribution of the plume and of the evacuees or residents. Where plume and evacuees or residents cross, the computer program calculates the dose received by occupants in the grid element, considering the dispersion and

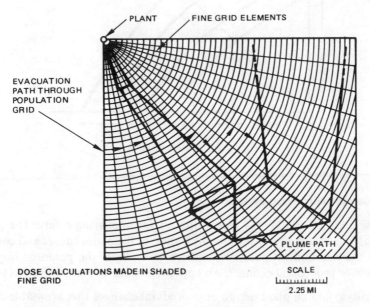

Figure 2.7-12 Illustration of plume and evacuation paths on fine grid

attenuation of radiation with time. Thousands of scenarios might be examined to establish probabilities of wind patterns and evacuation paths. Dose calculations are made for the permanent and transient inhabitants in the area under investigation.

The types of consequence effects usually selected for nuclear studies include early fatalities and injuries, latent cancers, and land and crop interdiction estimates. The values are determined from the accumulation of these effects' parameters from all the grid elements of the model. The final data might be presented as a family of conditional risk curves in which, given a release, the frequency of occurrence of different levels of damage would be depicted. A family of curves reflects the range in uncertainty of the results. These conditional risk curves are integrated with the curves developed for annual frequency of release to provide a final risk diagram in probability of frequency format as seen in Figure 2.7-13.

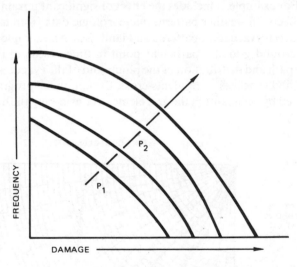

Figure 2.7-13 Risk diagram

Decisions for corrective actions

The risk analysis results may reveal options for improving safety. The process previously discussed permits the identification of the major causes and contributors to the risk and the tools with which to evaluate the potential impact of hardware or procedural changes. An approach might be summarized as follows:

1. Identify potential plant design or procedural changes that are viable options for reducing risk.

2. Predict potential costs and benefits for each option of all these contributors and determine their net benefit.
3. Define logical combinations of these options, present their net benefits on a probabilistic basis, and select optimum solutions.

The identification of the options for reducing risk is accomplished with the aid of the analysis model – often system fault trees and event trees. By selecting alternatives, it is possible to reflect the proposed modification in the models; and, by using failure frequency data applicable to the changes, the potential impact in system failure frequency can be calculated as well as the release frequency for each event tree reflecting the system. Options which do not significantly reduce release frequencies would not be considered further. A cost/benefit evaluation would be made for the others.

Cost can be based on dollars while benefits can be based on reduced damage converted to an acceptable dollar worth in order to determine a net worth for each option. By addressing the uncertainty of each parameter contributing to cost and benefit, the net benefit of each option can be expressed in a probabilistic format. In this way the range in, and likelihood of, various net benefits is seen before selecting the better options. This method will be explored further in the next discussion with respect to an availability example.

Availability Analysis

Many of the principles used to assess the safety and risk of plant operations can also be applied to plant availability improvement programmes. These programmes are concerned with the cost effective maximization of plant availability. The methodology frequently used in the power plant industry can be applied to other industies as well. As in the case of safety considerations, availability analysis is appropriate for both the design and operating phases of a power plant project.

The main steps in determining cost effective improvements to plant design and procedures are:

1. Identifying the largest contributors to plant unavailability and establishing the most critical items.
2. Modelling the plant and the systems contributing most to unavailability.
3. Identifying and analysing options for improving plant availability.
4. Recommending the most likely cost effective combination of improvements that will increase plant availability.

Contributors to plant unavailability

For operating plants it is necessary to establish a data gathering system which will help to pinpoint the main sources of production losses. In the case of power plants these might be generally categorized as:

1. Unplanned outages
   (a) Equipment failures
   (b) Administrative or procedural causes
   (c) Regulatory restrictions
2. Planned outages
   (a) Refuelling
   (b) Overhauls
   (c) Test and maintenance

The cause and nature of outages must be known to determine potential corrective actions. A common parameter for assessing the magnitude of unavailability is to accrue unavailability data in terms of lost megawatthours as a result of equipment or system failure. In some cases an outage might occur at a time when there is no demand for the production, in which case there is no allocation of lost megawatt hours to the failure or cause of lost power.

Periodically a review of the data enables the prioritizing of plant unavailability contributors. However, the ranking cannot be made only on the basis of lost production. It must also consider whether the loss is even controllable and the extent to which failures or occurrences might have been reduced. For example, if a piece of equipment has been found to require overhaul, say every 18 months, and the overhaul requires about 21 days each time, it is not possible to eliminate the overhaul and thereby completely eliminate the lost production. However, consideration could be given to changing the frequency of the overhaul or to take steps to reduce its duration, or perhaps both. By examining the data from operating plants, it is possible to establish a baseline or reference plant in terms of availability performance. The difference between the potential improvement in availability and the reference plant represents a basis for ranking candidate items for improving availability. Those items which appear to offer the largest reduction in plant unavailability warrant further analysis to determine their costs and to better assess the potential benefits. This ranked list of problems might be termed a critical problems list.

Plants which are in the design phase do not of course have accumulations of plant unavailability data with which to focus so easily on potential improvements for increased availability. However, as will be discussed shortly, there are other approaches that can be taken.

Modelling the plant

In order to assess the potential impact of plant design improvements it is necessary to determine more precisely the relationship between the reliability of a component with the reliability of its system, and the reliability of production-related systems with the availability of the plant. This is accomplished by first constructing the plant model to identify all the systems on which plant production depends and their impact on plant production if any one of them were to fail. Detailed system availability models would be developed for each of those systems found to be high on the critical problems list for operating plants. These system models would take the form discussed in the section on Release Frequencies.

Identifying and analysing improvement options[15]

Whether for operating plants or for plants in the design phase, and whether for availability improvement or safety improvement, the technique for assessing potential improvements is similar. The system models are the basis for evaluating the possible outcome of improvements; that is the possible savings from reduced lost production (or safety hazards).

For plants in design, the models are the basis for first predicting the plant availability of the reference design and then for examining the impact of potential improvements if the predicted availability of the reference plant is unacceptable. In that case, the reference plant unavailability is determined by applying generic equipment failure rates and repair times as obtainable from the information sources given in the section on Release Frequencies. After calculating the resultant plant unavailability, it is possible to identify the potential major contributors through examination of the models for the plant in design and to construct a critical problems list from that information.

For each critical problem a number of corrective options would normally be identified for reducing equipment failure rates, their repair or replacement time, or for building in redundancies that would alleviate the plant unavailability. By predicting the change in component and system unavailability, through re-analysis using the models it is possible to determine the potential reduction in plant unavailability. The cost of the improvement, whether it be equipment replacement, system redesign, or changes in operating, maintenance, or test procedures, can be determined also.

Implemented changes to plant design or operations for plant availability improvement should be determined from the most cost effective combination of actions which might be taken. A limited budget for capital or administrative improvements further emphasizes the need for obtaining the most improvement for the investment. Therefore, the various options that are available for solving each problem should be logically combined with other critical problem solution options. These combinations can be placed in an increasing order of annualized benefit coupled with an increase in annualized cost. Figure 2.7-14 indicates how these various combinations might be presented graphically indicating the predicted resulting outage hours (at an equivalent full capacity) per year. The reference design would be the existing design of an operating plant or the existing design of a plant in the design phase for which improvements in plant availability

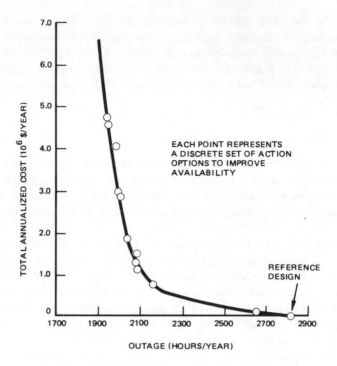

Figure 2.7-14 Minimum cost corrective actions

are desired. Any point to the right of the curve is theoretically achievable.

By next converting the outage hours saved by corrective actions to benefit dollars (in this case $500 000/outage day for replacement power) a benefit curve can be plotted as seen in Figure 2.7-15. Subtracting the cost curve from the benefit curve yields a net benefit curve which in this case is maximized at about 2000 outage hours/year. Referring to Figure 2.7-15, it is possible to choose the combination of option actions which permit achievement of the maximum net benefit.

It is possible to increase the merit of an availability decision, however, by considering the uncertainty of the knowledge about each parameter against which costs and benefits are determined. This can be accomplished by assigning probability distributions to the costs and benefits of the various individual options considered. Usually, the distributions take the form of histograms such as Figure 2.7-16. What remains is probability arithmetic. The cumulative probability distribution of the net benefit values is obtained and a net benefit probability density curve finally developed. By superimposing each option or group of options as seen in Figure 2.7-17, there results a visual display of the net benefits for each option and a measure of uncertainty in that prediction. On this

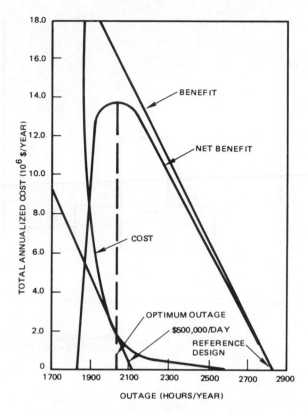

Figure 2.7-15 Net benefit versus outage reduction

basis, options which have higher probabilities of smaller net benefits might be more desirable than one having a higher benefit but lower probability.

Summary

Formal safety analysis is more prevalent in the electric power industry than other heavy industries. This situation has been driven by the nuclear power field where safety analysis and, more recently, probabilistic risk assessments have been developed to a high level of sophistication. The methods employed are basic and hence applicable to the risk assessment of any hazardous system. The modern approach is probabilistic which means a quantification of uncertainty. The format is a probability of frequency format of various levels and kinds of damage. The structure not only quantifies risk but provides the framework for evaluating the impact on risk of design and procedural changes. The same techniques apply to availability analysis and allow setting in place a procedure for optimizing and controlling plant reliability and availability.

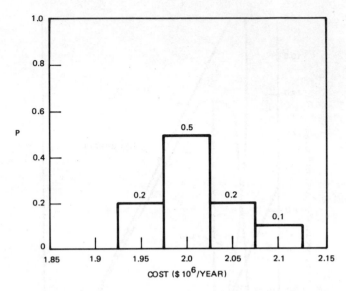

Figure 2.7-16 Spare rotor and diagnostic instrumentation

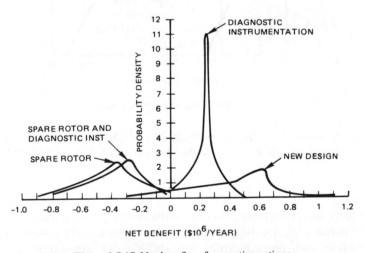

Figure 2.7-17 Net benefits of corrective actions

Acknowledgements

The authors wish to acknowledge the assistance of Mardyros Kazarians and Harold F. Perla, both of Pickard, Lowe and Garrick, Inc., for their valuable research and editorial assistance.

Questions 2.7

1. List the different types of power plants which can be used for producing electrical energy. Compare their advantages and disadvantages from a safety point of view.
2. What are the main safety issues which give rise to protective measures limiting the use of the different plants considered in Question 1?
3. For a power plant outline the basic methods of dealing with the following items:
   (a) consequence analysis;
   (b) availability analysis;
   (c) identifying and analysing improvement options.
4. List separately, from Figure 2.7-3.
   (a) the 'redundant' safeguards
   (b) the 'diversified' safeguards
   for each (PWR, BWR) system.
   Can any safeguard be both redundent and diversified or vice versa; and disregarding differences in technical detail in each system, does either system lack diversified or redundant safeguards which are present in the other?

References 2.7

1. Landsberg, Hans H. *Energy: The next Twenty Years.* Ballinger Publishing Co, Massachusetts. 1979.
2. Schurr, Sam H. *Energy in America's Future.* The Johns Hopkins University Press, Baltimore and London. 1979.
3. Loftness, R. L. *Engineering Handbook.* Van Nostrand Reinhold Company, New York. 1979.
4. Babcock and Wilcox.*Steam/Its Generation and Use.* 36th Edition. (Undated).
5. Granger, M. *Energy and Man: Technical and Social Aspects of Energy.* Morgan (Ed.). IEEE Press, New York. 1978.
6. United States Atomic Energy Commission, *Comparative Risk, Cost, Benefit Study of Alternative Sources of Electrical Energy.* WASH-1224. Dec. 1974.
7. Okrent, D. 'Comment on societal risk.' *Science*, **208**, 372–375. 1980.
8. Kaplan, S. and Garrick, B. J. 'The quantitative definition of risk.' *Risk Analysis*, 1 (1). 1981.
9. United States Regulatory Commission. *Reactor Safety Study*, WASH-1400, (NUREG-75/014). Oct. 1975.
10. Institute of Electrical and Electronics Engineers. *IEEE Guide to the Collection and Presentation of Electrical, Electronic and Sensing Component Reliability Data for Nuclear Power Generating Stations.* Report No. IEE Std-500. 1977.
11. Poloski, J. P. and Sullivan, W. H. *Data Summaries of Licensee Event Reports of Pumps at US Commercial Nuclear Power Plants.* 1 January 1976 to 31 December 1978. Report No. NUREG/CR-1362. 1978.
12. Sullivan, W. H. and Poloski, J. P. *Data Summaries of Licensee Event Reports of Pumps at US Commercial Nuclear Power Plants.* 1 January 1972 to 30 April 1978.' Report No. NUREG/CR-1205. 1980.

13. Swain, A. D. and Guttman, A. G. *Handbook of Human Reliability Analysis with Emphasis on Nuclear Power Plant Applications.* Draft Report NUREG/CR-1278. 1980.
14. Garrick, B. J., Kaplan, S., Bieniarz, P. P., Woodard, K. W. *et al, OPSA, Oyster Creek Probabilistic Safety Analysis.* Draft Report PLG-0100. 1979.
15. Garrick, B. J. 'Availability decision-making under uncertainty.' *34th Annual Technical Conference Transactions*, American Society for Quality Control. 1980.

High Risk Safety Technology
Edited by A. E. Green
© 1982 John Wiley & Sons Ltd

Chapter 2.8

Reprocessing Irradiated Nuclear Fuel

J. A. Williams

Introduction

The title of this book deals with high risk and it is appropriate to consider the nature of the risk of reprocessing irradiated nuclear fuel. What is different about nuclear fuel? We know that exposure to electromagnetic radiation or to sub-atomic particles of high energy can cause cancers of various types or with very high exposure fairly rapid death. The exposure can be external or by ingestion and internal exposure. The workers on the plant can be exposed in the course of their occupation, the public can be exposed by accidental release or by misjudged effluent discharge. Radiation can in addition cause genetic changes so that low level widely dispersed exposure must be a factor which is considered.

Some radioactive constituents of irradiated nuclear fuel can accumulate in the body or can enter food chains which are later absorbed into the human body. Some radioactive materials have long half-lives and so will remain a hazard for very long periods. This can be said about many substances outside the nuclear industry; arsenic, lead, cadmium, enter food chains or are retained in the body or will contaminate the environment if discharged accidentally or deliberately. There are however some properties of radioactive materials which are significant and different.

Radioactivity can operate at a distance and cannot be detected by the senses. The tolerance level for some of the radioactive substances which the body accumulates is very small. Radioactive materials can cause radiolytic decomposition of water with the generation of hydrogen and oxygen and a consequent explosion hazard. Radioactive materials can be self-heating. Exposure of structural materials to intense radioactivity can cause changes and be a source of weakness especially with organic structural or containing materials.

Radioactive materials possess one major advantage; by the use of instruments they can be rapidly detected at very low levels, much lower than those of other toxic or dangerous materials.

Reprocessing presents a potential hazard. In 30 years of reprocessing there has never been a Flixborough or a Seveso incident. Operators of reprocessing plants are aware of the hazards and precautions are taken to ensure safety. Provided the

467

vigilance and the standards of operation hitherto are maintained the probability of a major accident is very low indeed but we must take care that with the miser in *Silas Marner*,

> the lapse of time during which a given event has not happened is, in the logic of habit, constantly alleged as a reason why the event should never happen, even when the lapse of time is precisely the added condition which makes the event imminent.

A Short History of The Reprocessing of Irradiated Nuclear Fuel[13]-[18]

Uranium is used as fuel in a nuclear reactor, the uranium-235 is fissioned by neutrons giving rise to energy. The atom splits to form fission products ranging from mass number 72 to mass number 155. Some neutrons are captured by the predominant uranium-238 atoms forming plutonium-239 atoms which are themselves fissionable by neutrons. Eventually so much of the uranium-235 is used that it will no longer fully sustain the chain reaction, the fuel may also be at the limit of its mechanical strength, it is taken out of the reactor and can be reprocessed to recover the useful materials and prepare the fission product wastes for disposal.

The reprocessing of irradiated fuel started in the United States more than 30 years ago. The objective was to recover plutonium from uranium metal irradiated in the Hanford piles, no attempt was made to recover or recycle uranium. The process used at Hanford was based on carrier precipitation of plutonium with bismuth phosphate. Batch centrifuges removed the precipitate from the large volumes of active solution.

The next step in the evolution of reprocessing was the introduction of solvent extraction using 'Hexone' to extract uranium and plutonium from a nitric acid solution, leaving the fission products behind in the nitric acid. Separation of uranium and plutonium involved the reduction and oxidation of plutonium, hence the process was known as 'Redox'. Aluminium nitrate was added to the nitric acid to increase the partition of uranium and plutonium into the solvent and this gave rise to large volumes of fission product containing waste for storage.

The British at Chalk River in Canada and at Harwell in the UK, selected dibutyl carbitol (Butex) as their solvent. Butex did not need the salt to extract uranium and plutonium. The Butex process was used for the first UK recovery process at Windscale, Cumbria, which started operation in 1952.

Work had meantime progressed in the United States on a new solvent tributyl phosphate (TBP), like Butex it has the advantage of not needing salts in the nitric acid to give high partition of uranium and plutonium into the solvent. In addition it gives better separation from fission products and is chemically more stable. A process using TBP was called 'PUREX' and was made the basis of plants at Savannah River in 1954 and Hanford in 1956.

In France 'PUREX' was used as the basis for the Marcoule plant in 1958 and later at La Hague in 1967. In the UK two plants using TBP were built at Dounreay in Scotland, the first for Materials Testing Reactor Fuel using an aluminium alloy with uranium highly enriched in the ^{235}U isotope and the second for the enriched uranium metal fuelled Dounreay Fast Reactor. Then a large plant of about 1200 tonnes capacity was built to reprocess natural uranium metal fuel for the UK 'Magnox' reactor programme, this plant started in 1964.

Other plants based on 'PUREX' were the Eurochemic plant at Mol in Belgium (an experimental plant operated by a group of European countries), the WAK plant at Karlsruhe in Germany, the NFS plant in New York State, the Trombay plant in India, the EUREX plant at Sallugia in Italy, the ITREC plant at Rotandella in southern Italy. A small plant has recently been commissioned at Tokai Mura in Japan. A large plant was built near Savannah River, the Barnwell plant, for US power reactor fuel, the plant was completed but has not operated as a matter of US policy. A further plant at Morris, Illinois was based on the 'Aquafluor' process, in which the uranium is dissolved in nitric acid. The fission products and plutonium are separated using TBP, and the uranium while still contaminated with fission products was converted to UF_6 for recycle to a diffusion plant. The plant has technical difficulty during commissioning and operation was abandoned.

A complete list of plants is given in Table 2.8-1 and an outline flowsheet for Purex in Figure 2.8-1. Other processes have been developed but have not been used commercially. A process for remelting metal fuel and slagging of fission products followed by remote fabrication was developed at Idaho in the USA. Another process developed in the USA produced gaseous UF_6 and PuF_6 directly from solid UO_2 from the fuel. Work has also been carried out using ion exchange.

Plant Design

Because of the radioactivity of irradiated fuel, a large part of the process must be conducted behind heavy concrete shielding, combined with effective ventilation control. Once the fission products and hence the gamma radiation is removed it is necessary to prevent escape of plutonium because of its toxicity and the consequent low air tolerance and body burden levels, so extremely tight containment is needed. Two types of plant have been developed to cope with these difficulties: remote maintenance and direct maintenance plants. Direct maintenance calls for extreme reliability and very high construction standards, achieved by avoiding moving parts wherever possible. To carry out major maintenance the plant needs to be cleaned out very effectively, this may mean washing out for periods of months with corrosive chemicals. Nevertheless, the plants at Windscale, Dounreay, and in France which use this design have been maintained satisfactorily over many years. Major modifications have been successfully carried out.

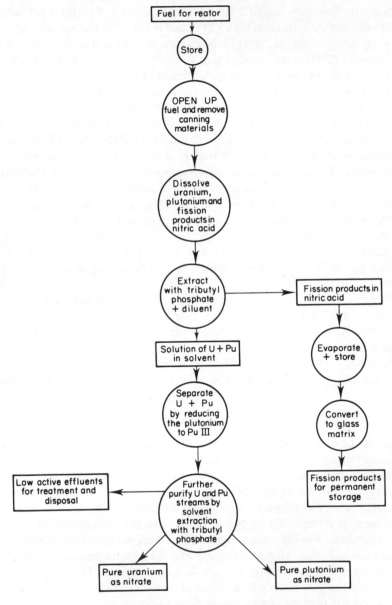

Figure 2.8-1 The 'Purex' process

Table 2.8-1   The world's reprocessing plants

Name*	Site	Start Up	Notes	Present position
Savannah River	Georgia, USA	1954	Military plant	—
Hanford	Washington State, USA	1966	Purex (military plant)	Operating
		1950	Hexone process	
Marcoule	Nr. Avignon, France	1956	Purex process (military plant)	Operating
Idaho	Idaho, USA	1960	Purex process	Operating
			Purex process for MTR and later zirconium alloy fuel	
Windscale	Cumbria, UK	1952	Butex process (military plant)	Operating
		1964	Purex process for civil power programme	Operating
Dounreay	Caithness, UK	1958	Purex process for materials testing reactor fuel	Operating
		1959	Purex process for metal fast reactor fuel	
		1980	Modified for oxide fast reactor fuel	Commissioning (1980)
Trombay	India	1965	Purex plant for oxide fuel	Operating
West Valley	NY State, USA	1966	Purex oxide fuel	Closed 1972
La Hague	Nr. Cherbourg, France	1966	Purex plant for metal	Operating
		1975	Modified to take oxide fuel	
Eurochemic	Mol, Belgium	1966	Purex plant for oxide fuel	Closed for commercial reasons 1972
Tokai Mura	Ibaraki, Japan	1977	Purex pilot plant (~ 200 tonnes)	Operating
Barnwell	USA	1976	Purex plant for oxide fuel	Commissioning postponed indefinitely by government policy
Morris	Illinois, USA	1975	'Aquafluor process'	Closed following commissioning troubles
Karlsruhe	German Federal Republic	1970	Purex pilot plant for oxide and fast reactor fuel	Operating
Itrec	Rotandella, Italy	Built for	Built for thorium fuel reprocessing	Abandoned
Eurex	Saluggia, Italy	1975	Built for MTR fuel, later modified for Purex oxide fuel reprocessing	Closed indefinitely

* Pilot plants have been built in Canada (1950s), Spain, Sweden and Taiwan. France also has a fast reactor reprocessing pilot plant at 'La Hague'.

Remote maintenance is the basis of US design. Plant items can be removed and replaced remotely. A special pipe joint, the 'jumper', which allows pipes to be broken and remade remotely is a vital part of this technique. Hanford, Savannah River, and Idaho have all used this design of plant. Both systems are satisfactory and there is a balance between complication in the remote maintenance case and the possibility of prolonged outage plus difficulty of making changes with direct maintenance.

The major potential hazard which might be caused by a large radioactive release means that plant must be designed to withstand natural hazards such as earthquakes, floods, high winds, and aircraft impact. In all modern reprocessing plants, design ensures that in the event of the most extreme possible conditions the radioactivity will be contained.

Radioactivity

There are a number of types of radiation which have to be taken into consideration. Alpha, beta, and gamma rays occur naturally and in addition X-rays and neutrons are produced by artificial means such as reactor irradiation or from elements produced during reactor irradiation. Radiations are emitted with energy, related to the specific source.

Alpha radiation

Alpha particles are helium nuclei which are emitted in the energy range of 4–9MeV by elements of high atomic weight, e.g. uranium-238, plutonium-239, americium-241. The range of an alpha particle in air is only a few centimetres.

Beta emitters

Beta particles are electrons emitted at high speed from the nucleus of the radioactive atom. The particles have ranges up to many metres in air and will penetrate a few millimetres of body tissue. Most of the fission products formed in a nuclear reactor give rise to beta emissions.

Gamma emitters

Gamma rays are electromagnetic radiations of very short wavelength. The rays often accompany alpha and beta particle emission. The rays are very penetrating.

Neutron emitters

Neutrons are uncharged particles of weight similar to a proton. Neutrons are emitted when certain heavy elements undergo spontaneous fission and also by the interaction of alpha particles with light elements such as aluminium, oxygen or fluorine.

Safety Problems Associated with Radiation

1. *Containment*
The need to confine radioactive materials under both normal operation and accident conditions.
2. *Shielding*
Operators require protection from external effects of radioactivity.
3. *Criticality*
Fissile material in sufficient quantity can give rise to a fission chain reaction with the consequent generation of a high neutron flux and considerable heat.
4. *Long-term plant reliability*
Because of the radioactivity normal maintenance may be difficult or even impossible.

Containment[1]

To prevent the release of radioactivity into the working environment or the public domain. The permissible levels of release are derived from the permissible levels of dose and ingestion recommended by the International Commission on Radiation Protection (ICRP). These recommendations are adopted in legislation by national bodies responsible.

Table 2.8-2

Nuclide	Max. permissible concentration in air $\mu Ci/cm^3$	Body burden μCi	μgm
^{239}Pu	2×10^{-12}	0.04	0.65
Natural uranium	7×10^{-11}	0.005	7400

The primary containment is usually the process equipment, the equipment is designed with this in mind. Where the process or equipment might give rise to a suspension of activity in a gas phase, for example in a steam ejector, then means of removal of the activity is included in the equipment. The secondary containment is often a cell in which the plant is placed and where only limited access or no access is permitted. The construction of the secondary containment will depend on the material to be processed, short-range activity, alpha emission, can be contained within transparent plastic cells with integral flexible gloves (glove boxes). For γ emitting material a thick concrete wall may be required.

The ventilation of active plants is important and is described in detail in Regulatory Guides. The confinement of activity is by means of multiple zones, each zone is bounded by barriers, the equipment, the cell walls, building walls, etc. Zone 1 will be the plant cell or glove box. Zone 2 will be an area surrounding the cell and where entry into the cell or removal of materials may give rise to a break in the zone 1 containment. Zone 3 will be the operating area. Pressure differentials should be maintained between building containment zones and the outside atmosphere to ensure that airflow is from less to more active areas.

Before release to the atmosphere the air is passed through efficient filters, e.g. a HEPA (high efficiency particulate) filter and the level of activity monitored before release to atmosphere.

Shielding[2][3]

The energy range of interest for gamma radiation from irradiated fuel extends from the low keV range to several MeV. The three main processes that contribute to attentuation by shielding are the photoelectric effect, Compton Scattering, and pair production. Each process results in energy of the incident photon being deposited, and the last two cause emission at a lower energy in a new direction. High density, high atomic number materials such as lead make the most efficient gamma shields.

Neutron reactions are with nuclei. Such reactions are more complicated, and in many cases not well understood. One of the most important interactions is elastic scattering of the neutron. Since the neutron can lose more energy in such an event if the nuclei of a light element are involved, the most efficient neutron shields have a large hydrogenous content. Materials such as water or concrete are good neutron shields.

Criticality[4]–[7]

Certain heavy fissile nuclei split apart following the absorption of a neutron, to give fission product fragments and further neutrons. Each of these nuclear fissions liberates roughly 200 MeV of kinetic energy, the binding energy previously required to hold the fragments together, and two or three 'fast' high energy neutrons. If one of these secondary neutrons happens to collide with another fissile nucleus, a further fission may occur, liberating yet more neutrons giving perhaps yet more fissions. A self-sustaining 'chain reaction' may therefore be set up which may 'diverge' if the number of fissions increases with time. If the number of fissions remains constant, i.e. just enough neutrons collide with fissile nuclei to maintain the chain reaction, the system is 'critical'.

The conditions required to bring about a critical system are numerous and interdependent. Under ideal conditions, less than 500 g of plutonium may be critical, whereas under other conditions tonnes of plutonium may be safely kept in the same store. It is clear that if anything more than trace quantities of fissile nuclei are present in a system, then the possibility of criticality exists.

Methods of criticality control

Methods of ensuring that criticality does not occur are as follows:

1. *Geometry control*
   It may be possible to design the geometry of the system so that there is sufficient neutron leakage for criticality to be impossible. This type of control leads to 'ever safe geometry' designs, e.g. slab tanks, columns.
2. *Mass control*
   If the system is isolated and restricted to less than the known minimum critical mass under optimum conditions of moderation and reflection, there is sufficient neutron leakage to guarantee that criticality cannot occur.
3. *Composition/concentration control*
   This requires the material composition to be controlled so that the fissile concentration is always low enough to guarantee sufficient neutron leakage or parasitic absorption in other materials.
4. *Moderation control*
   The guaranteed exclusion of moderating materials will considerably increase the minimum critical mass.
5. *Deliberate poisoning*
   The addition of a material with a large parasitic absorption cross-section can be very effective at reducing the system reactivity. Boron and gadolinium are typical poisons which can be used as solids or in solution.

The energy yield from an excursion crucially depends on the rate of approach to critical, maximum super-critical reactivity achieved, and total time at critical. Total yields from observed excursions normally fall in the range 10^{17} to 10^{19} fissions, and the energy release from 10^{18} fissions is roughly equivalent to the energy release from exploding about 10 kg of TNT. The disruptive effects of an excursion of this magnitude will not necessarily be so severe however, depending on the size of the initial 'power burst', which in turn depends on the maximum super-reactivity achieved. For a very slow approach to critical followed by a delayed critical excursion there may be no initial power burst, and the energy release may be spread out over many seconds or even minutes. The major hazard therefore will not usually be from explosion, but from the very large radiation doses associated with critical fission rates. Reference 6 devotes a chapter to historic accidental excursions.

Instrumentation and Control

Complete protection against hazards arising from incidents generated within the plant is based on the plant being operated at the designed conditions. To achieve this, as far as is possible, control and instrumentation equipment must be provided to monitor and control the various plant parameters, and to warn the operator of unacceptable divergencies from normal operating conditions. Failure of this equipment must be considered in the overall fault analysis for the plant but it is worth while outlining the design philosophy which might be applied in relation to a nuclear reprocessing plant.

The instrumentation may be considered in two broad categories:

1. That which monitors the performance of the plant for the purposes of plant control which is exercised either by the operator or by automatic equipment linked to instrumentation.
2. That which is provided solely for personnel protection, generally, this warns the operator when an abnormal condition has been reached.

In addition to the normal process control instrumentation there is certain instrumentation which is specific to nuclear plants. This is reviewed below.

Critical incident protection

For plutonium, the concentration of the fissile material in solution can be monitored by means of scintillation counters, measuring the emitted alpha particles or 384 KeV gamma photons, or by neutron counters using boron trifluoride counting tubes. The neutron counters make use of the alpha/neutron reaction which occurs in the plutonium solution. Because large numbers of counting tubes are used, scanning systems and automatic data processing are employed.

The response of alpha particle counters to background gamma radiation must be taken into account, as well as the possibility of other alpha-emitting elements being present in the plutonium bearing stream.

Critical incident detection

Despite the safeguards incorporated in the plant design to prevent a criticality incident the possibility of such event cannot be ruled out. To enable the plant operators to evacuate the building as quickly as possible, and thus to minimize the exposure to harmful ionizing radiations, it is essential for an unmistakable warning to be broadcast throughout the plant in less than a second of the incident. Detectors are used which respond to gamma radiation.

Airborne Radioactive Contamination

The plant ventilation system is designed to ensure that under normal operating conditions the direction of airflow in the plant is always from inactive areas. With a continuous flow of air into the plant it is necessary for air to be continuously exhausted to the environment usually via a gas clean-up system and a stack.

The effluent air must be continuously monitored for the level of alpha and beta particulate activity.

Within the plant it is necessary to measure the level of airborne particulate activity to warn the operators of any possible hazard arising from a breach in the plant containment. This type of monitoring is usually achieved by trapping particulate matter on a filter which is then exposed to a radiation detector.

Contaminated Liquid Effluents

All liquid effluents from the plant must be monitored to assess the quantity of radioactivity discharged. In addition it may also be necessary to measure continuously the specific activity of the effluent so that action can be initiated if it exceeds a predetermined level.

Fault Analysis[8]–[12]

To demonstrate, as far as is possible at the design stage, that unplanned releases of radioactive material caused by plant faults, and by external hazards such as earthquakes, gas cloud explosions, missile impacts, floods, high winds, etc., do not expose the operators or the public to an unacceptable level of risk it is necessary to perform a detailed fault analysis.

The object of such analyses is to determine the fault sequences which could occur, their probabilities, and the resulting consequences. Clearly, when considering the spectrum of fault sequences which could occur, as the consequences increase in severity the probability of occurrence must decrease. This concept, which was proposed by Farmer in a paper on nuclear reactor siting policy[11] is equally applicable to nuclear extraction plants or to non-nuclear chemical plants. In the identification and analysis of fault conditions increasing use is being made of event and fault tree techniques pioneered by the aerospace industry and developed further in the study on the safety of water reactors in the United States headed by Norman C Rasmussen[12].

Non-Nuclear Hazards

There are hazards present which are not directly due to nuclear radiation. Some examples are:

1. The presence of 'Zirconium' fines. Light water fuel is canned in Zircaloy and

can give rise to the presence of a proportion of fine metal when the fuel is broken up and dissolved. Zircaloy will react violently with water or nitric acid especially when it is damp.

2. Radiolytic decomposition of water with the potential evolution of hydrogen and oxygen has already been referred to.
3. Nitrated organic materials. The TBP and its diluents can be nitrated. This is especially true when the fission product wastes from a reprocessing plant are evaporated to reduce their volume. Violent reactions have been known to occur. These are overcome by operating at reduced temperature in plant where solvent may be present.
4. Self-heating. The radioactive decay of fission products and plutonium produces heat and in a storage tank for fission products this may be large, of the order of 1 MW. Cooling must be provided and of such reliability and redundancy that cooling failure becomes virtually impossible.

Conclusion

Reprocessing of irradiated nuclear fuel is a potentially hazardous process, nevertheless in 30 years of operation of a large number of plants there has been no major release which has affected the public or the workforce significantly. This is no doubt due to the precautions taken, the extreme care with plant design, high standards of operation and maintenance, and a real effort to search out and eliminate failure and fault routes before they give rise to risks.[19] In some respects the treatment of safety when reprocessing fuel is more akin to that adopted in the aircraft industry than the chemical industry. Perhaps this indicates the future for many high technology industries which are potentially hazardous but necessary to the advance of modern living.

Questions 2.8

1. Discuss the history of reprocessing irradiated nuclear fuel. Give an outline flowsheet for the 'PUREX' process.
2. In the design of the plant for reprocessing irradiated nuclear fuel what are the safety problems which have to be considered arising from radiation?
3. What are the factors involved in ensuring safe conditions to operators in the type of plant in Question 2 with reference to the following:
   (a) Containment
   (b) Shielding
   (c) Criticality?
4. What part does instrumentation play in the methods of criticality control? In principle what types of radiation are detected and in which way?
5. Discuss the types of non-nuclear hazards which can arise on a plant for reprocessing irradiated nuclear fuel.

References 2.8

Containment

1. Burchsted, C. A., Fuller, A. B., and Kahn, J. E. *Nuclear Air Cleaning Handbook*. ERDA 76–21.

Shielding

2. *Handbook of Radiation Protection*, Part 1. HMSO, 1971.
3. Jaeger, R. G. (Ed.). *Engineering Compendium on Radiation Shielding*. Vol. 1. Springer-Verlag. 1968.

Criticality

4. Carter, R. D. *et al. ARH*-600 *Criticality Handbook*. Vols. I–III. Atlantic Richfield Hanford. 1968.
5. *Chemical Processing of Reactor Fuels*. J. F. Flagg (Ed.). Chapter IX.
6. *OECD Book – Criticality Control in Nuclear Processing Plant*.
7. Thomas, J. T. *Nuclear Safety Guide*. TID-7016 Revision 2. U.S. Atomic Energy Commission. 1978.

Legislation and Safety

8. *Regulatory Guide* 3.26. ISNRC Guide Series, US Nuclear Regulatory Commission.
9. *Individual Risk – A Compilation of Recent British Data*.
   Safety and Reliability Directorate, United Kingdom Atomic Energy Authority.
10. *Annals of the ICRP*. ICRP Publication 26.
11. Farmer, F. R. *Siting Criteria*. A new approach. Conference on Periodic Inspection of Pressure Vessels 9–11 May 1972.
12. *Reactor Safety Study*. An assessment of accident risks is US commericial nuclear power plants. ReportNo. WASH-1400.

Reprocessing Plants

13. Orth and McKnibben *Trans Amer. Nucl. Society* **12,** 28. 1969.
14. Stevenson, R. L. and Bradley, B. G. HW 19170. 1961.
15. Geien, R. G. HW 49542. 1957.
16. Howells, G. R. *et al. Prog. in Nuclear Energy Series III*. Pergamon. 1961.
17. Warner, B. F. *et al.* 3rd Int. Conf. Peaceful Uses of Atomic Energy, Geneva **10,** 224. 1964.
18. Birch, C. *et al.* 2nd Int. Conf. Peaceful Uses of Atomic Energy, Geneva **17,** 23. 1958.

General

19. Benedict and Pigford, *Nuclear Chemical Engineering*. McGraw-Hill.

PART 3

SPECIFIC HAZARDS

High Risk Safety Technology
Edited by A. E. Green
© 1982 John Wiley & Sons Ltd

Chapter 3.1

External Hazards

G. I. Schuëller

Introduction

External hazards to which most systems and structures are subjected, may be either due to environmental and/or man-made conditions. Experience has shown that their effects have to be considered extremely important for design. *Environmental hazards* are caused by various natural mechanisms generated by flood, wind, earthquake, and wave action, etc. A general treatment of this subject is presented in Kates[1]. *Man-made hazards* are the result of accidents of technical equipment developed by men. As an example one might consider the crash of an aeroplane on a nuclear power or chemical plant, or the effect of a gas explosion in the vicinity of such structures.

There is sufficient statistical evidence to believe that their occurrence is random with respect to time and space, i.e. frequency of occurrence and magnitude. As a consequence of this fact it is necessary – for the purpose of modelling their properties – to utilize probabilistic methods. The results of such a probabilistic analysis provide then the pertinent information needed as input for a global risk analysis of technological systems. They are considered to consist of electronic, mechanical, and structural systems. Although the loading generated by the external hazards will generally be applied directly to the structural system, it will, in an interactive way, also affect the mechanical and the electronic systems. As an example the effect of an environmental hazard such as an earthquake occurrence on the primary piping of a Light Water Reactor which, is part of a mechanical system, might be considered. The forces generated by earthquakes are applied through the bottom plate of the containment structure – which acts as a filter – in terms of displacements to the primary piping. Due to the earthquake, excitation malfunctions of the electronic protective system can be caused as well by which additional loading of the primary piping may be caused. If, due to this sequence a loss of coolant accident occurs, the containment structure itself experiences a loading caused by this accident.

The purpose of this example is to show that external hazards are not just of concern for the design of nuclear structures, dams, tall buildings, bridges, offshore structures, TV towers, and other types of structures but also for mechanical as well as electronic systems contained by these structures.

Statistical Properties of External Hazards

Historical data of various hazard occurrences reveal their statistical characteristics already by visual observation. As a representative example historical observations of storm data are shown in Figure 3.1-1. In this figure the number of events occurring within a particular time interval (i.e. one year) are plotted along a time axis. The events are defined as those exceeding a particular threshold (exceedance) of the wind velocity. If such records are not available one can infer the data from damage records. In cases where data over a longer time range are needed, i.e. when for example a statistically stable estimate of a mean rate of occurrence of events is sought, one might resort to this type of method of data collection. However, the lower quality, particularly that of the older data should be kept in mind. A similar type of randomness has been claimed and observed for earthquake occurrences[3], and floods[4] which has proved to be valid for other types of environmental hazards as well.

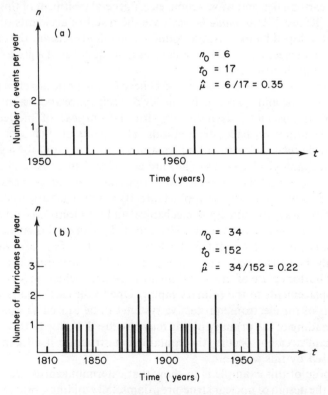

Figure 3.1-1 Historical data of severe storms. (a) Occurrence of severe storms in the southern North Sea region, Hamburg site. (b) Occurrence of hurricanes in the Gulf of Mexico region, Mustang Island site[2]

In the case of wind loading or floods the annual maximum values are recorded in many instances, but mostly over a relatively short time range only. As an example the annual maximum gust velocities which were recorded in Düsseldorf between 1951 and 1967 are listed in Table 3.1-1 and will be analysed later. Obtaining this type of information for earthquake accelerations or wind generated water wave heights presents more difficulties.

Table 3.1-1   Annual maximum gust velocities, Düsseldorf site (reference height $z = 10$ m)

Year	Gust velocities (m/s)	Year	Gust velocities (m/s)
1951	25.1	1960	26.0
1952	23.1	1961	28.2
1953	33.0	1962	27.0
1954	31.0	1963	22.1
1955	24.3	1964	28.5
1956	28.8	1965	30.0
1957	23.3	1966	29.0
1958	26.5	1967	31.7
1959	22.9		

For cases of man-made hazards much less statistical evidence is available. However, for a number of these types of hazards, the statistics in terms of mean rate of occurrences are available. For example it is reported in the Deutsche Risikostudie Kernkraftwerke[5], that the average rate of occurrence of aeroplane crashes on nuclear power plants is 10^{-6}/yr (for military planes). For the pressure wave due to external explosion a range of rates between 10^{-5}/yr and 5×10^{-7}/yr has been estimated. It is interesting to note that the same study reports an average rate of 3×10^{-1}/yr of the reactor building being hit by lightning.

Probabilistic Models for Prediction of External Hazards

General

Statistical evidence is a necessary but not sufficient piece of information to forecast frequency of occurrence and magnitude which is needed for design purposes. Statistical data provide information about the characteristics of the events which have already occurred. Since structures and systems have to be designed for loads which have *not* yet occurred, probabilistic models have to be developed in order to predict beyond the data range.

Depending on the prospective problem and the type of data available, the appropriate method for prediction of the frequency of occurrence and the magnitude of the events to be discussed in the following will be utilized.

The return period concept

This concept is the most simple and also the most widely used probabilistic model to forecast external and in particular environmental hazards. It describes the *average* time between particular statistically independent events, i.e.

$$\bar{R}(x) = \frac{1}{1 - F(x)} \qquad (3.1\text{-}1)$$

In the equation above $\bar{R}(x)$ denotes the return period and $F(x)$ the cumulative distribution function of the event x. The parameters of this distribution may be estimated from statistical data. The generally used statistical procedures are to be applied for this purpose. Before discussing the properties of $F(x)$ in more detail a few remarks with respect to the interpretation of the return period are in order. With reference to the definition of Equation 3.1-1, an event $X > x$ with a return period of 100 years is associated with the probability of 10^{-2} (or 1%) exceeding x, i.e. $\bar{R} = 100 = 1/(1 - 0.99)$. It also implies that $\bar{R}(x)$ has a distribution itself and there is a probability that a 100-year event may occur again only one year after its previous occurrence. The probability of this event is small but not zero. It will be shown below that the probability, that the event $X > x$ occurs before the mean value $\bar{R}(x)$ is 64%. Since structures are generally designed for a finite life, the determination of a proper return period is based on the fact that it is a multiple of the design life. The definite relation between return period and design life T can be described by the following exceedance probability expression:

$$E(x) = 1 - [F(x)]^T \qquad (3.1\text{-}2)$$

Inserting Equation 3.1-1 in the above equation yields

$$E(x) = 1 - \left[1 - \frac{1}{\bar{R}(x)}\right]^T \qquad (3.1\text{-}3)$$

Using this equation one can show that there is for example a 64% chance that the 100-year event occurs within a design life of the same span. If the design life is limited to 50 years this chance decreases to $\sim 40\%$. If, for instance, within those 50 years an exceedance probability of only 2% is acceptable, a return period of 2475 years has to be chosen. It has been mentioned before that in order to determine the time between events, the distribution $F(x)$ must be referenced to a certain time scale. If one decides to choose for example a scale of one year, one also bases the analysis on an extreme part of the population of all possible realizations of x. In other words extremes, such as the annual maximum values, are analysed. This implies that in the case of the previously used example of wind velocities, the annual maximum value is one representative value for all possibly occurring wind velocities during that year. Under these circumstances only a limited number of models can be used to describe $F(x)$, in particular only the class of the extreme value distributions applies. In practice the distributions of the asymptotic type

(Gumbel)[6] are generally used. For physical reasons the following two distributions have received particular attention. The extreme values of a particular phenomenon x can be predicted either by the so-called *Gumbel* distribution where the cumulative distribution function is defined by the following relation

$$F(x) = \exp\left(-e^{-\alpha(x-u)}\right) \qquad -\infty < x < \infty \qquad (3.1\text{-}4)$$

where u denotes the mode and α the shape parameter of the distribution, or by the so-called *Fréchet* distribution which is defined by

$$F(x) = \exp -\left(\frac{v}{x}\right)^k \qquad 0 < x < \infty \qquad (3.1\text{-}5)$$

where v is again the mode and k the shape parameter. (Note the difference in units as compared to α.)

The utilization of the extreme value distributions has, among others, the advantage of not being directly dependent on the entire distribution of population which is – due to insufficient statistical information – not known in most cases. To determine which type of the asymptotic distribution applies, only the functional expression of the tail of interest of the distribution of the population has to be known. In those cases in which this tail is of exponential shape, the *Gumbel* distribution applies. In cases in which the tail follows a potential law the *Fréchet* distribution will be appropriate. However, it should be mentioned in this context, that the extreme value distributions in general do not depend too strongly on the underlying information concerning the distribution of the population. If there is no information concerning these distributions available, it is sufficient to utilize physical reasoning and procedures of statistical inference to arrive at the appropriate asymptotic distribution. In Figures 3.1-2 and 3.1-3 examples of the annual extreme values of wind velocities and floods are shown. *Gumbel* as well as *Fréchet* distributions are fitted to the data. The return periods of various magnitudes are also indicated.

Referring back to Equation 3.1-1 it is important to note, that the return period $\overline{R}(x)$ is an average value which depends with respect to its confidence on the sample statistics of $F(x)$. By considering the wide spread of this confidence bound for a longer range of extrapolation one should exercise care by reviewing the results. For example, the analysis of the flood data shows for a flow rate of 5668 m^3/s, a return period of 100 years. Within the 10% calculated confidence bounds this rate could vary between 5163 m^3/s and 6374 m^3/s. It is therefore sometimes advisable to turn to more sophisticated models such as the one discussed in the following section.

Application of stochastic processes

In general terms, a stochastic process is a time-dependent random process. In engineering application the so-called *Gaussian* and *Poisson* processes are of

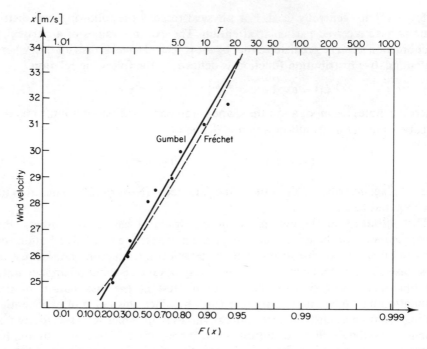

Figure 3.1-2 Distribution of maximum annual gust velocities listed in Table 3.1-1;
Gumbel probability paper

predominant importance. An in-depth discussion concerning their properties is given by Parzen[7]. In context with the prediction of *extreme* events and the type of long-range data available the *Poisson* process proved to be an extremely useful and in many cases, based on statistical tests, also an appropriate model. In addition, there is a strong physical argument to be explained in the following which also supports the hypothesis of the *Poisson* model.

If one considers only two types of magnitudes of loads, i.e. a load with a magnitude $> x$ (event A) and a load with a magnitude $\leqslant x$ (Figure 3.1-4), the probability p that the event A occurs exactly r times in n independent trials can be predicted using the well known binomial distribution:

$$p(r) = \binom{n}{r} p^r (1 - p)^{n-r} \qquad (r = 0, 1, \ldots n) \qquad (3.1\text{-}6)$$

If, moreover, one considers very large magnitudes – which may be considered as rare events – the number of trials n must be obviously increased to exceed the threshold magnitude x. In turn the probability of occurrence of this event decreases.

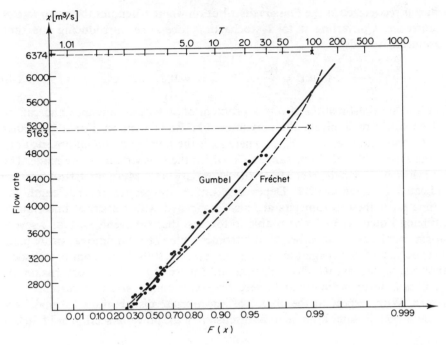

Figure 3.1-3 Distribution of maximum annual flow rates (Neuhaus, Inn-River; 1914–1976)

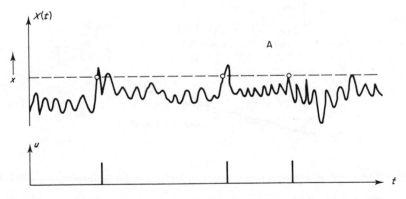

Figure 3.1-4 Exceedance of a particular threshold x by a random process $X(t)$

If now the number of trials $n \to \infty$ and the occurrence probability $p \to 0$ such that the product $np \to \mu$, it can be shown[8][9] that Equation 3.1-6 can be transformed into the following expression:

$$p(r\,|\,\mu) = \frac{\mu^r}{r!}e^{-\mu} \qquad\qquad r = 0, 1, \ldots, \infty \qquad\qquad (3.1\text{-}7)$$

which is recognized as the *Poisson* distribution where μ denotes the mean rate of occurrence. Converting it to a stochastic process, i.e. introducing the time parameter it reads

$$p(r|\mu,t) = \frac{(\mu t)^r}{r!}e^{-\mu t} \qquad r = 0,1,\ldots,\infty \qquad (3.1\text{-}8)$$

The statistical estimation of μ is a problem of statistical inference and can be solved easily by applying the maximum likelihood method. For a particular sample this estimate is $\hat{\mu} = n_0/t_0$ where n_0 is the number of actual observations of the event within the time range t_0 in which these observations are made. The estimates of $\hat{\mu}$ for the records of the Hamburg and Mustang Island site are indicated on Figure 3.1-1. Depending on the respective record length the estimates of these parameters are also associated with statistical uncertainty. Utilizing Equation 3.1-8 one is able to forecast discrete events such as a severe storm, earthquake, wave height occurrences, etc. The conditional probability distribution of the magnitude given the event results then from a statistical analysis of the events. For example in Figure 3.1-5 data from maximum significant wave heights which were observed during storm occurrences are shown. With respect to the choice of appropriate distributions to model the phenomena, the same arguments as discussed in the previous section hold. In this

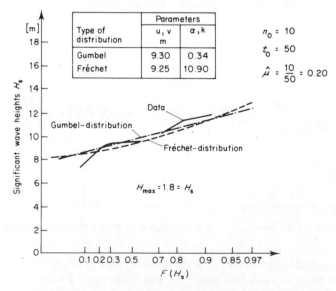

Figure 3.1-5 Fit of asymptotic distributions to maximum significant wave heights (H_s) resulting from 10 storm events observed within 50 years. (Source: Jahns and Wheeler[10]; Gumbel probability paper)

case, however, the distributions refer to a particular, i.e. discrete event; in the previous case the reference has been made to events occurring in a particular time interval, i.e. one year. From Figure 3.1-5 it can be seen that both the *Gumbel* and the *Fréchet* distributions show a good fit to the data. The mean rate of storm occurrence, $\hat{\mu} = 0.20$, is taken for the example. As an additional example in Figure 3.1-6 the conditional probability distributions of earthquake-induced ground accelerations for two different locations are shown on *Gumbel* probability paper.

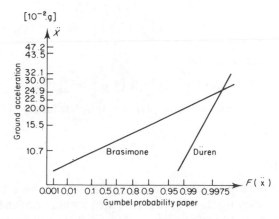

Figure 3.1-6 Conditional probability distributions of earthquake-induced ground accelerations for two different locations

In order to test the goodness of fit of the stochastic model, in this case the *Poisson* process, several statistical methods such as the χ^2-test, etc. are available. In this context the test utilizing the distribution of waiting time between events deserves particular attention. The probability density function of the waiting time between *Poisson* events is $f(t) = \hat{\mu}\exp(-\hat{\mu}t)$. Utilizing the complementary cumulative distribution function, which is of exponential type, i.e. $Q(t) = \exp(-\hat{\mu}t)$, the goodness of fit can be determined by simply plotting the data on exponential paper. If the *Poisson* process is the appropriate model for the underlying mechanism, i.e. records, the data have to be distributed along a straight line. In most rare event type cases of environmental and man-made origin, the *Poisson* process has proved to be a very useful model. This model has now to be combined with the model used to forecast the magnitude of the given event.

As mentioned before, this distribution, which is also modelled using an extreme value distribution, is referenced to the particular events and not to a particular time range as is the case when using the concept of the return period. The probability that a particular magnitude is not exceeded – given events with the distribution defined by $F(x)$ – can be calculated by $[F(x)]^r$, where r is the number of occurring events and $F(x)$ the distribution of the load magnitude given events such as storm waves (Figure 3.1-5), earthquakes (Figure 3.1-6), etc. The probability of occurrence of r events during a particular design life is to be predicted by the *Poisson* model (Equation 3.1-8). Its mean rate of occurrence μ is estimated using historical data. Under the assumption of independence between frequency of occurrence and magnitude of the event the rule of multiplication may be applied. After summing up all possible events, i.e. for $r = 0, 1, \ldots, \infty$, the following expression is obtained

$$Z_T(t) = \sum_{r=0}^{\infty} \left(\frac{\mu t^r}{r!} e^{-\mu t} \right) \cdot [F(x)]^r \qquad (3.1-9)$$

This expression reduces to

$$Z_T(t) = \exp\{-\mu t [1 - F(x)]\} \qquad (3.1-10)$$

The exceedance probability is then $F_T(t) = 1 - Z_T(t)$. This development is now exemplified as follows.

EXAMPLE 3.1-1

Given the data of a wave height distribution as shown in Figure 3.1-5, what is the probability that a wave height of 25 m is exceeded within a design life of 15 years? Utilizing Equation 3.1-10 and assuming a *Fréchet* distribution to model the data one obtains after calculating the mean rate of occurrence of $\hat{\mu} = 10/50 = 0.2$ from historical data.

$$Z_T(t) = \exp\{-0.2 \cdot 15 \cdot (1 - 0.99)\}$$
$$= 0.965$$

This result implies that there is a 3.5% chance that this wave height will be exceeded. If this risk is considered too high, the design wave height of 25 m has to be increased accordingly.

The example as shown here is of utmost importance for the determination of the design wave height for offshore or dam structures. It can also be applied to determine the design wind velocity or design earthquake acceleration based on a given risk of exceeding the set criteria.

To determine the actual failure probability of a structure under external hazards the distribution of the structural resistance is needed as well and it is used in terms of a convolution integral in Equation 3.1-10. However, a more detailed treatment of this aspect is given in Reference 11.

Spectral representation of external hazards

So far the magnitudes of the external hazards have been described only by the distribution of their extreme values. However, particularly in view of the dynamic properties of a large class of structures, stochastic external loading, such as loads due to storm, wave, and earthquake action, etc., are also represented by power spectral densities. These spectra allow the evaluation of the energy inherent in a stochastic process, also, the energy distribution at the various frequencies present in the process.

It is generally known that the power spectral density of a stationary stochastic process is obtained by *Fourier* transform of its autocorrelation function. As an example, the power spectrum of the well known El Centro earthquake record is shown in Figure 3.1-7. Ample literature is available in which numerical methods for obtaining these spectral densities are described[12]. Another important feature of the spectral densities is the fact that an integration over all frequencies results in the variance of the process. For example, if $X(t)$ represents the stochastic

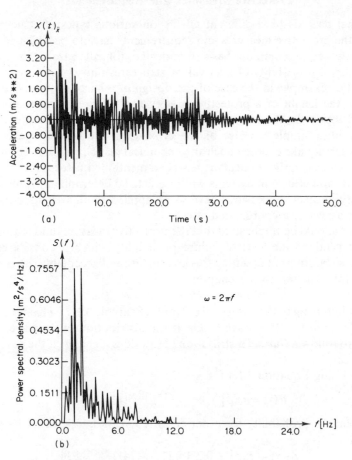

Figure 3.1-7 Californian earthquake recorded at El Centro 18 May 1940. (a) Time history. (b) Power spectral density

process, as shown in Figure 3.1-7(a), then its variance $\sigma_{\ddot{x}}^2 = \mathrm{Var}\,(X(t))$ $= \int_0^\infty S(\omega)\,d\omega$. For a stationary *Gaussian* process the distribution of the extreme values of the process follows a *Rayleigh* distribution[13], i.e. in terms of the probability density function

$$f(\ddot{x}) = \frac{\ddot{x}}{\sigma_{\ddot{x}}^2}\exp\left(-\frac{\ddot{x}^2}{2\sigma_{\ddot{x}}^2}\right) \qquad (3.1\text{-}11)$$

The forecasts obtained by the above equation yield comparable results as obtained by using Equations 3.1-4 and 3.1-5 respectively.

Protective Measures and Requirements

As external hazards have different effects on various types of structures and systems, the protective measures and requirements have to be designed accordingly. However, a common basis for design for all hazards is a certain predetermined probability of survival of structures and systems against these hazards. For example, in the case of the design of a harbour it may be easy to determine the height of a protective breakwater for protecting against storm generated waves. In the case of wind exposed structures one can control the reliability by a simple increase of the cross-sectional area. However, if one considers earthquake exposed nuclear or chemical plants, etc., the determination of the measures is rather difficult as has been mentioned before since structural, mechanical, and electronic systems will interact. To determine the total survival probability, protective measures have to be applied to all systems, particularly under the aspect of a coordinated effect.

Finally it should be emphasized that the protective measures and requirements directly depend on the level of failure probability which is considered to be acceptable. The criterion of setting this level in terms of acceptable risk is referred to in the Introduction to this chapter.

Problem: Referring to the numerical example (Example 3.1-1), what is the chance of exceedance of the 25 m wave height if the distribution of the wave height is assumed to follow a *Gumbel* distribution? How does it change if the design life is only 10 years?

Solution: Using Equation 3.1.4 for $x = 25\,\mathrm{m}$

$$F(x) = \exp\left[-e^{-0.34(25-16.74)}\right] = 0.941$$

The exceedance probability for a period of 15 years is calculated by utilizing Equation 3.1.10

$$Z_T(t) = \exp\left[-0.2\cdot15\cdot(1-0.941)\right] = 0.834$$

i.e. the chance of exceedance of a 25 m wave is 16.6%.

For 10 years the result is

$$Z_T(t) = \exp[-0.2 \cdot 10 \cdot (1 - 0.941)] = 0.889$$

This means for 10 years the chance decreases to 11.1%.

Questions 3.1

1. Discuss what is meant by the phrase 'Environmental Hazards and Their Causes'. Consider how an earthquake may affect a system known to you.
2. Consider the range of external hazards to which a chemical plant may be subjected and the statistical properties related to these hazards.
3. Discuss 'Return Period' and how the concept is used to forecast external and environmental hazards.
4. Explain how extreme events can be used to provide probabilistic data essential to the design of potentially hazardous structures.
5. Referring to Example 3.1-1, what is the chance of exceedance of the 25 m wave height if the distribution of the wave height is assumed to follow a Gumbel distribution? How does it change if the design life is only 10 years?

Notation 3.1

$\bar{R}(x)$	return period
$F(x)$	cumulative distribution function
$f(x)$	probability density function
T	design life
$X(t)$	random process
u, v	modal values
α, k	shape parameters
p	probability of occurrence of event
n	number of trials
r	number of occurrences
$\binom{n}{r}$	binomial coefficient
μ	mean rate of occurrences
$Z(t)$	reliability
$F_T(t)$	failure, i.e. exceedance probability
$\ddot{x}$	horizontal acceleration of earthquake
σ	standard deviation
H_s	significant wave height
H_{max}	maximum wave height
n_0	number of observations within time range t_0
t_0	time range of observation
$Q(t)$	complementary cumulative distribution function
$S(\omega)$	power spectral density

References 3.1

1. Kates, R. W. *Risk Assessment of Environmental Hazard.* John Wiley & Sons, New York, 1978.
2. Russell, L. R. and Schüeller, G. I. 'Probabilistic models for Texas Gulf Coast hurricane occurrences'. *J. Petroleum Techn.*, March 1974. S. 279–288.
3. Benjamin, J. R. 'Probabilistic models for seismic force design.' *J. Struct. Div.*, ASCE, Vol. 94, No. ST5, Paper 5950. 1968.
4. Kite, G. W. *Frequency and Risk Analysis in Hydrology.* Water Resources Publication, Fort Collins, Colarado 1977.
5. Deutsche Risikostudie-Kernkraftwerke, Verlag TUV Rheinland, Köln, 1979.
6. Gumbel, E. J. *Statistics of Extremes.* Columbia University Press, New York, 1958.
7. Parzen, E. *Stochastic Processes.* Holden Day, San Francisco, 1967.
8. Benjamin, J. R. and Cornell, C. A. *Probability, Statistics, and Decision for Civil Engineers.* McGraw-Hill, New York, 1970.
9. Schüeller, G. I. *Einführung in die Sicherheit und Zuverlässigkeit von Tragwerken*, W. Ernst u. Sohn Verlag, Berlin, 1981.
10. Jahns, H. O. and Wheeler, J. D. 'Long term wave probabilities based on hindcasting of severe storms.' Preprints, Offshore Technology Conf. Paper No. 1590 May 1972.
11. Freudenthal, A. M. and Schüeller, G. I. 'Risikoanalyse von Ingenieurtragwerken.' Reports, Konstr. Ingenieurbau, Hrsg. W. Zerna, Report No. 25, Vulkan-Verlag, Essen, Aug. 1976.
12. Bendat, J. S. and Piersol, A. G. *Random Data: Analysis and Measurement Procedures.* John Wiley & Sons, Inc., New York, 1971.
13. Rice, S. O. 'Mathematical analysis of random noise.' In *Noise-Stochastic Processes.* Dover Publ., New York, 1954, also in *Bell Tech., J.*, **23**, 282, 1944 u, **24**, 46, 1945.

High Risk Safety Technology
Edited by A. E. Green
© 1982 John Wiley & Sons Ltd

Chapter 3.2

Fire and Explosion

J. G. Marshall and P. V. Rutledge

Fire and Flame–Definition and Data

Combustion is the process of chemical reaction involving the release of energy in the form of heat and light. The chemical reaction is usually an oxidative one.

Combustion is usually associated with the oxidation of a fuel by atmospheric oxygen, though the process can occur in other atmospheres.

Fire occurs in the gaseous state. Solids and liquids either vaporize or yield combustible vapours by degradation, which burn when ignited. Ignition results if the vapour is mixed with air or oxygen in the appropriate proportion and this mixture meets an ignition source of sufficient energy.

Table 3.2-1 Comparison of fuel oxidant systems and their flame temperatures

Fuel	Oxidant	Approximate flame temperature (K)
Saturated hydrocarbon	Air	2200
Saturated hydrocarbon	Oxygen	3100
Saturated hydrocarbon	Hydrogen peroxide	3000
Methane	Chlorine	3725
Hydrogen	Chlorine	2400

Flame is the physical consequence of this combustion. Flame temperatures range from 600 to 3000 °C, though mostly they will be in the range 800 to 1200 °C.

The thermal energy of open fire is dissipated by convection, conduction, and radiation. The proportion of combustion energy which is radiated depends on the flame emissivity, its thickness and its temperature, but as a general rule it does not exceed about one-third.

Combustion is the most energetic of all chemical reactions. Heats of combustion are quoted in the literature and in chemical data books[1][2]. The numerical values cover a wide range dependent on the material concerned. Hydrocarbons in general have the highest heats of combustion mostly in the range of 40–45 MJ/kg, while nylon is 33 MJ/kg and cellulose in the form of paper, wood or cotton is approximately 16 MJ/kg.

The change from combustible to incombustible is not clearly defined. Conventional tests define as combustible those materials with heats of combustion in excess of 1.5 MJ/kg of fuel. This figure is approximately 10% of the calorific value of wood[3].

Types of Flame and Fire

Flames are classified into pre-mixed and diffusion according to whether the air for combustion is mixed with fuel upstream of the flame or enters via the flame envelope. Controlled combustion is usually pre-mixed with some diffusion while uncontrolled combustion is predominantly diffusive in nature.

The rate of spread of fire is largely controlled by the physical state of the material being burnt. Flame can travel through flammable gas/air mixtures at metres per second, across many combustible liquid surfaces at centimetres per second, and across solid surfaces at rates as low as centimetres per hour. In the case of solid materials the rate is very dependent on physical layout and the direction of spread.

Liquid pool fires can be a consequence of a major catastrophe on chemical plants, refineries, and ships. The rate of combustion of liquid fuels as pools is related to the pool diameter for pools up to 1 m across. It is generally accepted that, for such fires, the flame height is approximately twice the pool diameter at least up to several metres across. This approximation does not hold for very large liquid pool fires as the air for combustion cannot reach the pool centre. Large pools of burning hydrocarbon will consume up to 4 mm/min of fuel and will radiate heat at up to 130 kW/m² of radiation from flame surfaces[4][5].

The radiation received at any point can be calculated and compared with known values of radiation effects. For example, solar radiation can be up to 0.8 kW/m², the threshold of pain in man is 1.5 kW/m², 2.0 kW/m² can be tolerated for up to a minute, and combustible materials are ignited at intensities of 20 to 30 kW/m²[6].

At the opposite end of the scale, the smouldering of solid fuels represents one of the slowest rates of combustion (within the definition of fire). As a general rule, smouldering fires occur in low density cellulosic materials, typically sawdust. Rates of smouldering are usually controlled by air accessibility in the presence of excess fuel.

The most familiar form of smouldering combustion is the lit cigarette, which smoulders at 0.001 g/s.

Ignition

The processes whereby fires (and explosions) initiate are collectively called ignition. Ignition results when a small volume of a fuel/air mixture comes into contact with a spark, hot surface or pre-existing flame. It presupposes that the fuel/air mixture is of the correct proportions to sustain combustion.

The most common ignition sources can be classified according to their energy source. These are as follows:

Chemical The reaction of one energetic chemical with another until flame is produced. Matches are the most familiar example.

Electrical The temperature of an electrical spark can attain thousands of degrees centigrade, sufficient to ignite all gases and vapours. Prolonged or high intensity electrical discharges will ignite solid materials.

Hot surface Surfaces whose temperatures exceed 250 °C can cause ignition of many common materials.

Self-heating This is an uncommon form of ignition in which a material reacts with its surroundings (usually air) until it erupts into flaming combustion. Self-heating may start through microbiological action, cease when the temperature kills the bacteria, but may be continued by oxidative processes depending on the fuel. The theory of self-heating has been described by Thomas[3].

Friction The mechanical rubbing of one surface on another can produce either hot surfaces or sparks. Both of these can cause ignition.

Spontaneous or self-ignition Almost all combustible materials will ignite if heated sufficiently in air. Ignition temperatures are not sharply defined like flash points (see below) and are dependent on the type of equipment in which the test is carried out.

The ignition property of a combustible liquid which is of considerable technical importance is its flash point. The flash point is the lowest temperature at which the vapour from that fuel can be ignited by a small source of ignition, usually either a small flame or a spark. It is possible to ignite liquids, and even solids, at temperatures well below their flash point if they are suitably absorbed on to a substrate which forms a wick.

The Results of Uncontrolled Fire

There are about 350 000 fires per year attended by the Fire Brigades in England Wales, and Scotland. Of these, fires in buildings account for about 100 000. These fires result in the death of 600–900 persons annually in the UK. (The United States with a population five times that of the UK has 10 000 fire fatalities per annum.) The ratio of fatalities to injured is approximately 1:10 (UK Fire Statistics)[7].

Building fires give rise to significant levels of toxic combustion products in which the structure may assist their retention. The proportion of carbon monoxide in the combustion products largely determines its toxicity though living beings have little resistance to high environmental temperatures. Carbon monoxide rapidly produces dizziness and collapse. The presence of trace quantities of aldehydes, cyanides, and halides, coupled with the thermal stress and smoke, will have a weakening effect on those exposed. It has been suggested that in fire, people succumb to lower levels of toxic combustion products than indicated by laboratory tests[8][9].

Fire results in enormous damage to property and equipment. Direct insured losses exceeded £1 million per day in the UK in 1980. Consequential loss from fire affecting the means of production is several times the direct insured loss, while uninsured losses are also considerable.

The course of a fire can be broken down into three distinct phases, that of initiation and growth, followed by fully developed fire, and then decay. The growth of fire from small sources has been shown to be exponential in rate. Growth will occur provided that fuel, air, and heat are available to it, the balance of these factors controlling the behaviour of the fire. The study of fire growth from small sources is of importance since virtually all uncontrolled fires originate from small sources.

The thermal output of a fire in an enclosure is approximately as follows:

65% lost by convection from openings;
25% lost to the enclosure walls;
10% lost by radiation.

A feature of building fires of great importance to fire fighters is the phenomenon known as 'flash over'. A slow build up of a compartment fire may heat the surfaces of combustible materials. These may then evolve flammable gases of decomposition. The sudden admission of air, due to say window failure or human action, may allow these gases to be burnt with great rapidity. The resulting 'flash over' may cause burn injuries to fire fighters.

Any large fire in a building may well involve the burning of considerable masses of wood and similar combustible materials. To assess the fire hazard posed by a substantial quantity of combustibles, use is made of the fire load parameter. This is the theoretical heat output per unit area of floor. The weight of combustibles multiplied by their heats of combustion is divided by the total floor area of the compartment. The result is expressed in megajoules per square metre. As a guide, an average living room will have a fire load of 200 MJ/m^3, an office approximately 500 MJ/m^2, and a high racked warehouse 10 000 MJ/m^2 (NFPA Handbook, 1976)[10].

Fire Prevention

Good organization and management of staff, plant, and resources can greatly help to prevent ignition and stop escalation to damaging fire. People and their carelessness are a major cause of fire. Close control of all personnel, whether directly employed or the employees of others or visitors, will go a long way towards eliminating fire. The careless use of smoking materials is a common cause of fire. Smoking is not permitted in high hazard chemical plants and should not be allowed in a wide range of industrial operations. Good housekeeping is of great importance in reducing the chance that a small ignition source, however it may arise, does not escalate into an uncontrollable fire.

As 25% to 30% of all fires are electrical in origin, high quality electrical equipment, well maintained, will greatly reduce the risk of electrical fires.

Chemical plants are divided into hazard areas according to the risk of combustible liquids and gases being present. The electrical equipment selected must match the hazard. This matching of equipment against the hazard could well be applied in other high risk situations.

Passive Fire Defence

Passive fire protection is the term used to describe fire defence of a building which can be considered as part of the structure.

Active fire protection involves the provision of automatic or manual fire fighting systems in buildings or structures. Installed hose reels and sprinkler systems are typical of manual and automatic fire defence systems.

Passive fire protection may involve the choice of the most suitable material of construction for buildings or plant. Wood is relatively cheap, easily constructed, and maintenance costs are low in dry conditions. Light sections are readily consumed in fire but, in substantial sections, it may withstand prolonged fire attack where steel would fail. Steel has great cold strength but fails quickly in substantial fire. Reinforced concrete, when properly designed, has good fire resistance but the structure cannot easily be modified.

Active Fire Defence

Fire must be detected before fire defence measures can be initiated. At the very earliest stages of fire, intelligent human observation is the best detection system. The observer can decide what action stands the best chance of defeating the fire. Where human surveillance is too costly or not physically possible, e.g. inside closed equipment, there is a requirement for automatic fire detection, of which there are a number of different types.

Rate of temperature rise detectors are widely used for fire detection, as are the fixed temperature type. They are robust, do not give false alarms and, with modern circuitry, alarm if a fault develops.

The most widely used smoke detector is the ion chamber type which compares the current flow in two similar chambers containing air ionized by a microscopic amount of a radioisotope. Modern versions of this device are said not to respond to cigarette smoke. Historically they have a high false alarm rate.

Optical photoelectric smoke detectors rely for their effect on the light scattering property of a light beam containing smoke particles. A photocell at an angle to the beam detects the scatter.

Radiation detectors respond to infrared or ultraviolet radiation from flames. They may be designed to accept a signal lasting seconds before triggering, to avoid false alarms.

Another variation of the optical detector is the beam or laser beam detector. They are used to cover large open span areas. They can be designed to respond to thermal signals generated at the early stages of a fire.

The Extinction of Fire

There are a number of processes whereby flame, once initiated, can be extinguished. These are by cooling, oxygen dilution and exclusion, fuel elimination and chemical flame inhibition.

In most fire fighting activities, several of these processes are involved at the same time. In a well-developed fire the extinguishing water cools the fuel and, on conversion to steam, blankets the fire to reduce the oxygen concentration. Some extinguishing agents act by breaking the chain carrying reactions which are a major part of the molecular mechanism of flaming combustion.

The cooling action of fire extinction is most prevalent when the water is in excess in relation to the volume of fire. Sprinkler and deluge systems are effective because they normally operate at the early stages of fire.

Oxygen in the atmosphere can be diluted or excluded by foam, steam, the products of combustion, carbon dioxide, vaporizing liquids or inert gas systems. It can be excluded by physical methods such as sand and foam.

Fixed Extinguishing Systems

These may be classified as manual and automatic, though manual extinguishment is not necessarily static. Provision of portable fire extinguishers is normally a legal or insurance requirement. Extinguishers can be of the water, foam, carbon dioxide, dry powder, or vaporizing liquid type. Selection of type must be according to the potential hazard. Just as important as providing portable extinguishers is adequate and proper training in their use.

Run-out hose reels would be essential in most hazardous areas to deal with small outbreaks of fire on a first aid basis. They must be supplied with water of sufficient pressure and have the necessary reach to all potential targets. Hose reels must be sited where adequate protection for fire fighters could be provided by plant, buildings or entrances. To do this implies a foreknowledge of the potential fire risk positions.

Explosion Characteristics

The essential characteristic of an explosion is the rapid release of the expansion energy of gas at high pressure, manifested in the form of blast waves in the surrounding air or of missile fragments. Explosions also usually involve the release of flames and hot gases since they are most frequently due to the rapid evolution of heat by chemical reaction which creates the high pressure in the reaction products. However the bursting of a vessel filled with gas or liquefied gas (which is unreactive) at elevated pressure can also be considered an explosion. In order to create the initial high pressure the system must usually be initially confined but this is not essential in the case of condensed phase high explosive compostions or that of very large volumes of flammable gas/air mixtures where UVCEs, i.e. unconfined vapour cloud explosions, can arise (see Chapter 3.3, Gas Clouds).

Materials which Present an Explosion Risk

Reactive materials

The principal class of materials which present an explosion risk comprise those mixtures or single substances which are capable of rapid exothermic reaction resulting in a large relative increase in the volume of products when measured at atmospheric pressure. Such materials can exist in a variety of physical states and a classification on this basis with examples is shown in Table 3.2-2.

Table 3.2-2 Classification of reactive materials presenting an explosion risk with examples

Class and physical state		Examples
(A)	Single component	
	(i) Gas	Vapours of exothermic compounds such as ethylene oxide, hydrogen peroxide
	(ii) Liquid	Nitropropane, organic peroxy compounds, liquid monopropellants and explosives, e.g. nitroglycerine
	(iii) Solid	Solid propellants and explosives
(B)	Reactive mixture	
	(i) Gas	Combustible gas/air Combustible gas/oxygen Hydrogen/chlorine
	(ii) Liquid/gas	Combustible liquid mist/air
	(iii) Liquid	Liquid propellants and explosives Polymerization process mixtures Oxidation process mixtures
	(iv) Solid/gas	Combustible dust/air
	(v) Solid/liquid	Ammonium nitrate/fuel oil explosives
	(vi) Solid	Solid propellants and explosives

The explosive properties of any two-component reactive system will diminish as the ratio of the reactant concentrations departs from the stoichiometric ratio. There will be a limited range of compositions which are explosive; for fuel/oxidant systems the limits are often referred to as the lower and upper explosive limits (LEL and UEL) corresponding to the lowest and highest fuel concentrations respectively. Reactive single-component or two-component systems can be rendered non-explosive by the addition of sufficient inert material.

There are three distinct mechanisms by which reactive materials can produce an explosion, i.e. thermal explosion or spontaneous ignition, deflagration, and detonation. The first differs from the latter two in that reaction proceeds simultaneously throughout the volume of material whereas in the others a reaction zone propagates through the volume from the point of initiation. The principal characteristics of the three mechanisms are discussed in turn.

For a given volume of a composition capable of undergoing rapid exothermic reaction there will be a critical external temperature determining whether a thermal explosion occurs. Below this temperature the heat transfer characteristics of the system and the kinetics and heat of reaction are such that a stationary state is possible in which the heat loss equals the heat evolved and the system reaches a constant temperature slightly elevated above those of its surroundings. If the external temperature is increased, then the rate of heat evolution will increase more rapidly than the rate of heat loss to the surroundings and, above a critical ignition temperature, a constant system temperature will no longer be possible. Instead there follows a rapidly increasing rate of reaction or, in other words, an explosion. The theoretical relationships regarding the critical conditions are discussed by Frank-Kamenetskii[29]. A thermal explosion is the most common mechanism by which a liquid reaction mixture, e.g. one undergoing polymerization, oxidation or other exothermic reaction, can give rise to an explosion. For a gaseous reactive material the critical temperature is sometimes called the spontaneous ignition temperature (SIT) or auto-ignition temperature (AIT); it can be measured by rapidly introducing the gas into a vessel at a controlled temperature and determining the minimum value at which a violent pressure rise occurs. It will be appreciated that the critical temperature is not an absolute value for a particular material since it depends also on the volume and heat transfer characteristics of the container; the larger the vessel the lower the critical temperature.

Deflagration is characterized by a reaction zone propagating through the material from the source of ignition at a subsonic velocity; the reaction zone is narrow in thickness and effectively separates completely reacted material from completely unreacted material. The zone propagates by conduction of heat and diffusion of radicals or other active species from the reaction zone to the unreacted material. In the case of a deflagrating explosive gas mixture such a reaction zone is called a flame; for typical flammable gases in air the maximum flame velocity with respect to unburnt gas is in the range 0.3–3 m/s. Liquid and solid propellants also 'burn' by means of such a mechanism.

Detonation also involves a reaction zone propagating through the material but in this case at supersonic velocity. The reaction zone is associated with a shock wave; the arrival of the latter in unreacted material heats it by compression effectively instantaneously and adiabatically and so brings about reaction. Detonation is a characteristic of liquid and solid high explosives with typical propagation velocities of up to 8000 m/s but can also occur with gaseous systems, where velocities of 1700–3000 m/s are usual.

The general characteristics of these phenomena are described in the literature as follows: gas deflagration and detonation[11]–[14], dust explosions[15][16], condensed phase deflagration[17], and condensed phase detonation[18].

Fluids initially at high pressure

Fluids which possess significant quantities of pressure energy are usually either gases or liquefied gases under pressure at temperatures above their atmospheric boiling point. Classic examples of this type are high pressure steam boilers failing because of excessive pressure or weakening of the boiler and cylinders containing gas or liquefied gas rupturing when involved in a fire.

Initiation of Explosions in Reactive Materials

For a thermal explosion to occur the relationship between the factors determining the rate of heat evolution and those determining the rate of heat loss has to exceed a critical value. Such explosions can therefore result from the following:

1. Changing the composition or temperature of the system to increase the rate of reaction, e.g. by addition of catalysts or by rapid compression in the case of a gas.
2. Increasing the external temperature or in other ways reducing the rate of heat loss from the system, e.g. by introducing the mixture to a hot vessel.

Whatever the circumstances a thermal explosion involves the whole of the reactant mixture increasing in temperature homogeneously.

The initiation or ignition of a deflagration, on the other hand, involves raising a relatively small volume of the explosive medium to a temperature at which rapid reaction occurs; the energy input has to be such that the reaction zone thus created is large enough to initiate propagation through the remainder of the medium. The methods of ignition commonly encountered are:

— Flames
— Hot surfaces
— Friction or impact sparks created with hard materials, e.g. some metals, stones, etc.[19]

— Electric arcs or sparks
— Electrostatic discharges
— Pyrophoric or catalytic materials
— The thermite reaction[20]

The minimum energy required to initiate an explosion can be conveniently measured in the case of gas and dust explosions by use of electric sparks of known energy. Typically, the most readily ignitable hydrogen/air mixture has a minimum ignition energy of about 0.02 mJ, hydrocarbon/air mixtures about 0.3 mJ, combustible metal dust suspensions in air about 15 mJ, and combustible organic dust suspensions 10–100 mJ.

Detonation can be initiated directly in a material by means of a shock wave from another source, e.g. from a detonator, exploding wire or by a very high energy spark. In other cases a deflagration may under certain conditions convert to a detonation. Circumstances favouring such a translation are a high energy release and rate of propagation, a large mass of material and a high degree of confinement and, in the case of gas mixtures, the development of a high level of turbulence from initial gas movement or from the accelerating flame movement, the presence of obstacles, bends, etc. in the containing vessel.

The Consequences of an Explosion

For an explosion occurring within a vessel or other enclosure the most important question is whether the explosion will give rise to a maximum pressure which exceeds that which the vessel or enclosure is capable of containing without failure.

In the case of a thermal explosion or a deflagration the maximum pressure is determined by the thermochemistry of the processes that can occur and, more particularly, since we are dealing with constant volume processes, with the internal energy change ΔE of the processes. It is therefore possible, if the final composition of the reaction mixture can be determined from a knowledge of equilibrium constants, etc., to predict the maximum final pressure that could be achieved in a sufficiently strong enclosure. For gaseous fuel/air or oxygen mixtures the theoretical maximum pressures that can be achieved will depend on how closely the mixture composition approximates to stoichiometric decreasing as the composition tends to the limits of flammable concentration. Typical theoretical and experimental maximum pressures for stoicheiometric hydrocarbon/air mixtures originally at atmospheric pressure are in the range 800–1000 kN/m² absolute. For compositions initially in the condensed phase the theoretical maximum pressures will usually be much greater in magnitude and incapable of being withstood by a vessel.

In the case of compositions which detonate the maximum instantaneous pressure which can be developed can be calculated from the hydrodynamic theory of detonation, as can the maximum temperature and the velocity with which the detonation wave propagates through the composition. For gaseous compositions the maximum theoretical pressures that can be developed are approximately twice those which can arise from the deflagration or thermal explosion of the same mixture, e.g. for stoichiometric hydrocarbon/air mixtures originally at atmospheric pressure they are of the order of $1700-2000 \, kN/m^2$ absolute. For condensed phase high explosives the peak pressure can exceed 10^5 bar. The pressure rapidly falls after the passage of the detonation wave.

If a deflagration or thermal explosion can give rise to a pressure greater than that which the vessel or enclosure can withstand, then clearly the latter will rupture. In these circumstances the damage that can result will be due to (a) the emission of flames and hot reaction products which can initiate fires and cause injury or damage as a direct result of their high temperature and (b) the air blast and the high velocity missile fragments of the enclosure.

The maximum energy that theoretically could be available to create air blast and to accelerate vessel fragments is the isentropic expansion energy of the reaction products expanding from the pressure at which the vessel fails down to atmospheric pressure. This applies equally to the failure of a vessel containing compressed gas or a liquefied gas under pressure. The isentropic expansion energy can be estimated from tables of internal energy and entropy when these are available. In the case of a gas it can be estimated from the relation:

$$\Delta E = \frac{P_{max} V}{\gamma - 1} \left\{ 1 - \left(\frac{P_{max}}{P_a} \right)^{\frac{1-\gamma}{\gamma}} \right\} \tag{3.2-1}$$

where ΔE is the isentropic expansion energy   J
        V is the volume of the enclosure m^3
        P_{max} is the maximum pressure attained N/m^2
        P_a is the atmospheric pressure N/m^2
        γ is the specific heat ratio of the reaction
        products

This energy will be significantly less than the enthalpy or internal energy change of the reaction since a significant proportion of this is still retained in the hot reaction products as thermal energy after expansion.

The division of the expansion energy between air blast and fragment acceleration will depend on the mode of failure of the vessel and it is only possible to give a rough guide. It has been suggested that the proportion going to create the shock wave is normally between 40 % and 80 % of the total[21].

To estimate the characteristics of the air blast wave that is created, it is recommended to convert the appropriate fraction of the expansion energy to an equivalent quantity of TNT by use of the value of 3.72 MJ/kg for the isentropic expansion energy of TNT. It is then possible to estimate the characteristics of the blast wave, i.e. peak overpressure, duration of overpressure, etc. by reference to the well established data for TNT.[22][23]

If it is possible to predict how much of the vessel or enclosure will be projected as missiles, then their initial velocity can be estimated by equating the appropriate fraction of the expansion energy to the initial kinetic energy of the missiles. There are other empirical methods for estimating the initial velocity. From this the maximum range of missiles can be estimated by standard methods of ballistics.

The Avoidance of Explosions and Minimization of their Consequences

The preferred way to avoid accidental explosions is to prevent the formation of a composition and/or the maintenance of conditions permitting an explosion. In the case of multireactant systems, e.g. fuel and oxidant, it may be possible to maintain the composition outside the range of explosive compositions. For any system the use of an inert diluent can often serve the same purpose. Thus explosive gas compositions can be rendered non-explosive by the addition of gases such as nitrogen, water vapour, and carbon dioxide (in order of increasing effectiveness) while dust explosions can be avoided either by addition of such gases to the air or by mixing inert inorganic materials with the combustible material. These diluents exert their inerting effect primarily by reducing the maximum temperature that can be developed by reaction.

Where explosive compositions cannot be avoided completely, steps must be taken to exclude ignition sources. The following are appropriate steps:

1. Adequate separation distances from equipment containing flames such as direct fired heaters and the banning of matches, lighters, and smoking materials.
2. The banning of hot work, e.g. cutting and welding, except under carefully controlled conditions.
3. The use of electrical equipment which is incapable of igniting the explosive medium, e.g. intrinsically safe, flameproof, etc.[24].
4. The banning of other sources of electric sparks, e.g. spark ignition engines.
5. The prevention of electrostatic sparks, particularly by avoiding large insulated conductors where charge can accumulate and subsequently discharge to give an energetic spark[25].
6. The avoidance of hot or catalytic surfaces capable of igniting the media that may be present.
7. The prevention of stones and other hard material entering moving equipment where incendive frictional sparks might be created.

Where, as is not infrequently the case, it is not possible to reduce the probability of an explosive medium occurring together with a source of ignition to an acceptably low level, it is necessary to accept that an explosion may occur and to take steps to minimize the consequences. One method of doing this in the case of gas or dust explosions is to suppress the explosion sufficiently rapidly that no mechanical damage occurs. The normal method is to employ sensitive pressure rise switches to detect the explosion at a very early stage and to trigger the rapid dispersion of suppressing agents, either a fluorinated hydrocarbon such as chlorobromomethane or an inorganic powder such as ammonium phosphate[26]. In this way it is possible to stop the propagation of the explosion.

Another means of minimizing the consequences is to limit the spread of an explosion and prevent it propagating throughout the extent of the explosive medium by compartmenting the equipment and providing barriers to flame propagation. In the case of gases and vapours flame arresters, which provide a multiplicity of small diameter channels which quench the flame, can act in this way. With solids handling systems which present a dust explosion hazard, such barriers can be provided by a rotary valve or a horizontal screw conveyor from which one flight has been removed.

Wherever it is concluded that an explosion may occur in a plant capable of rupturing the plant, it is necessary to provide explosion relief. This will consist of devices such as bursting discs or relief panels which rupture or blow off at a relatively low pressure and prevent the maximum pressure inside the plant from reaching a level at which the plant itself would fail. The area, operating pressure, and inertia of the relief have to be chosen carefully in the light of the explosive characteristics of the plant contents[27][28]. It is also important that the materials emitted are directed away from any area where they could cause injury or damage.

Finally consideration must be given to the possibility of an explosion leading to secondary explosions. By the latter are meant explosions involving, for example, combustible dusts which are not initially dispersed in air but which can become so and then ignited as a result of the primary explosion. Layers of combustible dust, e.g. grain dust in the buildings containing grain handling plant, are a classic example of this hazard. They can be avoided by adequate safe relief of the plant itself and adequate regular cleaning of all floors and horizontal surfaces. If this is not practicable, the building itself should be provided with relief.

Questions 3.2

1. Define the meaning of fire and flame. Give examples of the different types of fire and flame.
2. What is ignition and classify some common ignition sources?
3. Discuss the results of fire becoming uncontrolled together with the consequences in situations known to you involving significant safety issues.
4. What basic principles are involved in fire prevention?
5. Discuss passive and active fire defence and the means of extinguishing fires.
6. What are the characteristics of an explosion and how do they differ from those of a detonation?

7. Discuss materials which present an explosion risk and how the initiation of explosions may occur in reactive materials.
8. Consider an industrial process known to you involving severe consequences if an explosion were to arise. Investigate the possibilities of an explosion occurring, and the steps you would undertake to avoid accidental explosion.

References 3.2

1. Weast, Robert C. (Editor) *Handbook of Chemistry and Physics*. CRC Press, Inc. 1963.
2. Dean, John A. (Editor) *Lange's Handbook of Chemistry*. McGraw-Hill. 1967.
3. Thomas, P. H. *Self-heating and Thermal Ignition—A Guide to its Theory and Application in Ignition Heat Release and Non-combustibility of Materials*. ASTM, STP 502, pp. 56-82. 1972.
4. Hall, A. R. *Pool Burning. A Review*. Rocket Propulsion Establishment Technical Report. 1972.
5. Stark, G. W. V. 'Liquid spillage fires.' 4th Symposium on Chemical Process Hazards. I. Chem. E. Symposium, Series 33. 1972.
6. Craven, A. D. 'Thermal radiation hazards from the ignition of emergency vents.' 4th Symposium on Chemical Process Hazards. I. Chem. E. Symposium, Series 33. 1972.
7. *UK Fire Statistics*, Home Office, HMSO.
8. Kimmerle, Georg. 'Aspects and methodology for the evaluation of toxicological parameters during fire exposure.' *Journal Fire & Flammability Combustion Toxicology*, **1**, 4. 1974.
9. Birky Merritt, M. *Hazard Characteristics of Combustion Products in Fires—The State of the Art*: Review NASA, PB 267828. 1977.
10. NFPA, Boston, Mass. *Fire Protection Handbook*. 14th Edition. 1976.
11. Lewis, B. and von Elbe, G. *Combustion Flames and Explosions of Gases*. Academic Press. 1961.
12. Jost, W. *Explosion and Combustion Processes in Gases*. McGraw-Hill. 1946.
13. Zabetakis, M. G. 'Flammability characteristics of combustible gases and vapours.' US Bureau of Mines Bulletin No. 627. 1965.
14. Coward, H. F. and Jones, G. W. 'Limits of flammability of gases and vapours.' US Bureau of Mines Bulletin No. 503. 1952.
15. Palmer, K. N. *Dust Explosions and Fires*. Chapman & Hall. 1973.
16. Burgoyne, J. H. 'The testing and assessment of materials liable to dust explosion or fire.' *Chemistry & Industry*, 81–87. 1978.
17. Taylor, J. *Solid Propellant and Exothermic Compositions*. George Newnes. 1959.
18. Taylor, J. *Detonation in Condensed Explosives*. Oxford. 1952.
19. Powell, F. "Ignition of Gases and Vapours", *Ind. Eng. Chem.*, **61**(12), 29–37. 1969.
20. Gibson, N., Lloyd, F. C., and Perry, G. R. 'Fire hazards in chemical plant from friction sparks involving the thermite reaction.' I. Chem. E. Symposium Series No. 25, 26–35. 1968.
21. High Pressure Technology Association, *High Pressure Safety Code*. 1975.
22. Glasstone, S. (Ed). *The Effects of Nuclear Weapons*. USAEC, Washington. 1962.
23. Advisory Committee on Major Hazards. Second Report HMSO, 1979.
24. Hall, A. 'Explosion protected electrical equipment.' *Protection*, 10–14. 1978.
25. Strawson, H. 'Electrostatic fires and explosions.' *The Chartered Mechanical Engineer*, **20**, 91–96. 1973.

26. Maisey, H. R. 'Explosion suppression and automatic explosion control.' I. Chem. E. Symposium Series No. 58, 171–191. 1980.
27. Anthony, E. J. 'The use of venting formulae in the design and protection of building and industrial plant from damage by gas or vapour explosions.' *J. Haz. Mat.*, **2**, 23–49. 1977/78.
28. Maisey, H. R. 'Gaseous and dust explosion venting.' *Chem & Process Eng.*, **46**, 527, 662. 1965.
29. Frank-Kamenetskii, D. A. 'Diffusion and heat exchange in chemical kinetics.' Princeton Univ. Press (Trans. N. Thon). 1955.

High Risk Safety Technology
Edited by A. E. Green
© 1982 John Wiley & Sons Ltd

Chapter 3.3

Gas Clouds

G. D. Kaiser

As has been mentioned in Section 1.3.5 on consequence assessment, the analysis of the safety of advanced technological installations often requires the calculation of the atmospheric dispersion of radiotoxic, chemically toxic or flammable gases. It is towards the understanding of this process that this chapter is directed. It is convenient to consider three different modes of atmospheric dispersion – passive, buoyant, and heavy.

Passive Dispersion

The modelling of passive dispersion (the behaviour of gases that do not modify the ambient turbulence) has already been discussed in Section 1.3.5. It is particularly useful in the modelling of radiotoxic releases or of slow releases of most materials. No further discussion will be devoted to this phenomenon here.

Buoyant Releases

It is sometimes the case that a release of radioactive material is expected to be hot. For example, if there should be an accident in a nuclear reactor then the plume of radionuclides will emerge from a region in which there may have been nuclear excursions, fires or the failure to remove decay heating. The Rasmussen study identifies certain, albeit highly improbable, accident sequences in which the rate of heat release can be as high as about 150 MW, see Table 1.3.6-1. In the chemical industry, many processes are carried out at high temperatures. If a hot release should take place, *plume rise* may occur. This is a complex and fascinating phenomenon and it can only be briefly described here. For those wishing to learn more, there is an article by Briggs[1] that is highly to be recommended.

Figure 3.3-1 contains an 'artistic' impression of a typical history of plume rise. First, the plume is emitted from the reactor building (or chemical plant structure) and interacts with the turbulent building wake. In general, this wake causes considerable dilution and it is first necessary to consider whether the plume will rise at all – that is, whether it will 'lift-off'. The interaction of buoyant plumes and building wakes is still, at the time of writing of this book, the subject of experimental and theoretical research. The tentative indications from present

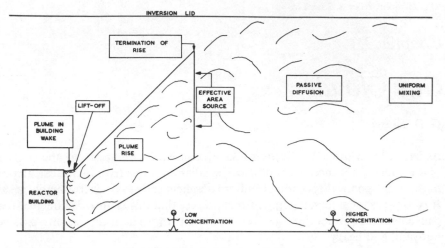

Figure 3.3-1 Typical history of plume rise

studies are that even high rates of heat release – say 100 MW – may not be sufficient to ensure clean lift-off in the wake of a building 100 m in width. Assuming, however, that the plume does begin to rise, its trajectory and radius may be calculated using one of a host of standard theories that are expressions of basic physical conservation laws:

1. Conservation of mass
2. Conservation of momentum (i.e. Newton's second law of motion)
3. Conservation of enthalpy

To these must be added (4) an entrainment hypothesis, for which it is generally assumed that the entrainment velocity U_e is proportional to the rate of rise of the plume. The entrainment velocity gives the volumetric rate at which air is pulled into the plume across unit area of its boundary due to the action of internally generated turbulence. During this stage of the plume's rise, the material within it is generally confined by sharp boundaries and the ground level concentrations below it are *extremely* small.

Sooner or later, the plume must stop rising and this is brought about by interaction with the atmosphere. As it rises, the 'vigour' of the turbulence within it decreases. A measure of this vigour is sometimes taken to be the turbulence energy dissipation rate, which is the rate at which turbulent energy is converted into heat by the action of viscous forces. When this quantity is the same both inside and outside the plume it is plausible to argue that the plume can no longer be distinguished from the atmosphere. Subsequently, it disperses as would a passive plume and eventually spreads throughout the boundary layer until, far enough downwind, it has 'forgotten' that there was an initial rise. The rise may also cease as the plume rises into a temperature inversion. The inversion may begin at the ground, in which case the weather category is stable, or it may form a 'lid' on the atmospheric boundary layer. Very buoyant plumes may punch through this lid into a less stable region above, in which case they escape from the atmospheric boundary layer and may travel large distances before any of the material within the plume returns to earth.

Heavy Releases

These are vapour clouds that are denser than the surrounding atmosphere; most chemically toxic or flammable gases come into this category. To begin with, there is a host of gases that are heavy because their molecular weight exceeds that of air. Chlorine is an example of a dense toxic vapour, while the petrochemical industry produces and transports a range of heavy, flammable hydrocarbons such as propane, butane, and propylene. Next, methane (or natural gas consisting largely of methane) has a molecular weight smaller than that of air, but is dense if released at its boiling point of $-161\,°C$. Natural gas is often transported as a refrigerated liquid and cold, dense vapour can be generated if LNG should be spilled on to land or water. Even a gas such as ammonia, which, as a pure vapour, is buoyant at its boiling point of $-33\,°C$, can form part of denser-than-air mixture, as can be seen on Figure 3.3-2, which shows the path of an ammonia cloud released after a train derailment in Pensacola, Florida in 1977[2]. Some 40 t of anhydrous ammonia escaped quickly to the atmosphere and was picked up on the radar at the nearby airport, which was directly upwind. After five minutes the cloud was observed to be 'about a mile across'. Subsequently, the cloud was tracked for nine miles over land and water during which time it did not lift off the ground. The reasons for this behaviour will become clear later in this chapter.

Finally, hydrogen fluoride, a toxic gas used by the nuclear industry during the fuel manufacturing process and by the chemical industry as a catalyst in alkylation plants, is an interesting curiosity. Its molecular weight is nominally 20 but, as a pure vapour, it is highly associated and consists of a mixture of hexamer and monomer with an effective molecular weight of about 70. Dissociation takes place as the hydrogen fluoride is diluted with air, but the heat required to do this is extracted from the mixture and this keeps it denser than the surrounding atmosphere. It is fair to point out, however, that, in contrast to the case of the ammonia described above, there is no known example of a large-scale accidental release in which this density effect has been observed.

At the time of writing there are considerable uncertainties in the calculation of the atmospheric dispersion of dense gases. These will only be removed as better

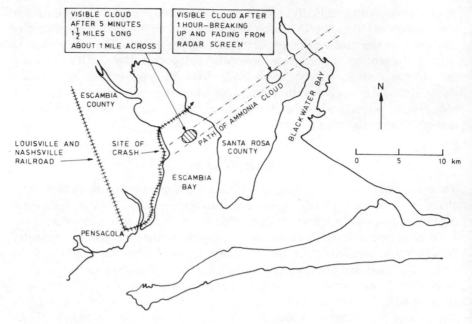

Figure 3.3-2 Path of ammonia cloud, Pensacola, Florida, 1977

experimental data become available during the next few years. Meanwhile, any model of dense vapour dispersion should contain the following four elements.

1. Specification of the source term for atmospheric dispersion
2. Gravitational slumping
3. Entrainment of air
4. Heating of the cloud

Source term

The nature of the source has a profound effect on the subsequent atmospheric dispersion. For example, if a gas is kept liquefied under pressure in a vessel which for some reason develops a small leak in the vapour space, a pure vapour jet results. If, on the other hand, the failure is catastrophic, bulk boiling results and the whole of the contents of the vessel may be thrown into the air as a mixture of vapour and fine liquid droplets. At the same time, much turbulence is generated and air is entrained, which evaporates the liquid droplets and is itself cooled as it does so. In the case of ammonia, this process can produce a mixture of ammonia and air at a temperature of $-33\,^{\circ}$C and with a density exceeding that of air – this is the explanation of the behaviour of the ammonia cloud observed after the Pensacola derailment. If the pressurized vessel should spring a leak in the liquid space, or in a pipe leading from the liquid space, then high velocity jets of liquid or liquid/vapour mixtures will emerge into the atmosphere and 'flash', entraining air as they do so. If the gas is kept as a refrigerated liquid, then it will boil if spilled on to land or water. If it is an unbunded spill, the evaporating pool could spread to become very broad. If the gas escapes into the turbulent wake of a building, it may be thoroughly mixed throughout this wake, leading to a substantial dilution before it begins to travel downwind. This description of source terms only begins to scratch the surface of a fascinating subject. It is discussed more thoroughly in Reference 3 in the context of ammonia.

Gravitational slumping

If a puff of dense vapour is formed as a result (say) of the catastrophic failure of a pressurized vessel, there is a useful rule of thumb which shows that it may be approximated by a cylinder. Subsequently, it slumps as would a column of liquid. Figure 3.3-3 shows the behaviour of a puff of Refrigerant-12 released from a collapsible 'tent' during recent open air experiments at Porton[4]. The 'slumping' behaviour may be modelled by a liquid column analogy in which the radius R_a increases with time as follows

$$\frac{dR_a}{dt} = \kappa \sqrt{(\rho - \rho_a)g H_a/\rho_a} \qquad (3.3\text{-}1)$$

where g is the acceleration due to gravity, H_a is the height of the puff, ρ is the density, ρ_a is the density of the surrounding atmosphere and κ is an empirical constant taking a value in the range 1–2.

Entrainment of air

This is the area of greatest uncertainty in the modelling of heavy vapour dispersion and this section should be regarded as indicative of present thinking rather than as a definitive treatment. The entrainment of air occurs in the presence of turbulence, which is generated mechanically in the presence of strong velocity gradients and convectively in the presence of heat fluxes.

The gravitational slumping generates high velocities and velocity gradients, particularly at the edge of the cloud. The rate at which the mass of air m_a entrained into the cloud increases is simply assumed to be proportional to the product of the circumference of the cylinder and its radial velocity

$$\frac{dm_e}{dt} = 2\rho_a \pi R_a H_a \alpha^* \frac{dR_a}{dt} \qquad (3.3\text{-}2)$$

where α^* is an empirical constant between 0 and 1. Scientific opinion is at present

Figure 3.3-3 Experiments at Porton. (a) View of source tent being filled with Refrigerant — 12. (b) Collapse of tent — release of denser-than-air puff. (c) Collapse of puff to a height of less than a metre when there is no wind. Crown Copyright

divided about the importance of edge entrainment. Another significant source of turbulence is the atmosphere itself, which acts mainly over the top surface. The usual way of treating entrainment is to define an entrainment velocity U_e (for example, $U_e = \alpha^* dR_a/dt$ in Equation 3.3-2). For the action of ambient turbulence over the top surface, U_e should be dependent on some form of the Richardson number R_n,

where
$$R_n = \frac{gl_s(\rho - \rho_a)}{U_1^2 \rho_a} \tag{3.3-3}$$

Here l_s is a length scale – which may be simply the height of the puff, or some more sophisticated measure of the turbulence length scale in the atmosphere – and U_1 is the longitudinal turbulence velocity which, loosely speaking, is a measure of the rate of spreading of the passive plume in the atmosphere. R_n is therefore the square of the ratio of a gravitationally induced spreading velocity

$$\sqrt{(gl_s(\rho - \rho_a))\rho_a}$$

(see Equation (3.3-1)) and a velocity characterizing the state of atmospheric turbulence.

Intuitively, it is to be expected that turbulent entrainment of air is suppressed in the presence of a stabilizing density gradient. A simple guess at the form of U_e would be

$$U_e = \alpha' U_1 R_n^{-1} \tag{3.3-4}$$

and, indeed, there is evidence from experiments carried out in water tunnels, with water advancing over a layer of salt, that this is a suitable form for U_e. α' is an empirical constant which is usually assigned a value of about a half. Equation 3.3-2 now becomes

$$\frac{dm_a}{dt} = \rho_a (\pi R_a^2) U_e + 2\rho_a \pi R_a H_a \alpha^* \frac{dR_a}{dt} \tag{3.3-5}$$

Heating of the cloud

In principle, it is necessary to take into account the ways in which the cloud can be heated – by the sun, for example, or by heat generated by turbulent natural convection, driven by a temperature difference ΔT_g between the cloud and the ground. The classical expression for the rate of increase of cloud temperature T_e is given by

$$\frac{dT_e}{dt} = \frac{\alpha^+ (\pi R_a^2) \Delta T_g^{4/3}}{m_a C_{pa} + m_g C_{pg}} \tag{3.3-6}$$

where ΔT_g is the temperature difference between the cloud and the ground, m_g is the mass of toxic or flammable gas, C_{pa} is the specific heat of air at constant pressure and C_{pg} is the specific heat of the gas, also at constant pressure. α^+ is a constant depending on the thermodynamic properties of the gas–air mixture[5].

As warmer air is entrained into a cold cloud, this also causes a temperature rise and Equation 3.3-5 becomes

$$\frac{dT_e}{dt} = \frac{\dfrac{dm_a}{dt}C_{pa}\Delta T_a + \alpha^+ (\pi R_a^2)\Delta T_g^{4/3}}{m_a C_{pa} + m_g C_{pg}} \tag{3.3-7}$$

where ΔT_a is the temperature difference between the cloud and the surrounding atmosphere.

Equations 3.3-1, 3.3-4 and 3.3-7 form three coupled linear differential equations in the independent variables R_a, m_a, and T_e and may be solved numerically. It is usual to assume that, as U_e approaches U_1, or as $(\rho - \rho_a)/\rho_a$ becomes sufficiently small, say 0.001, the cloud becomes passive.

The model presented here is grossly simplified, but none the less displays the four basic elements – source term specification, gravitational slumping, entrainment of air, and cloud heating – which are common to any approach, no matter how sophisticated it may be.

EXAMPLE 3.3-1

A pressure vessel containing 40 t of anhydrous ammonia as a liquid at 20 °C fails catastrophically. Assuming that all of the contents of the vessel become airborne, and that the turbulence generated by the failure causes the entrainment of 20 times as much mass of air as there is of ammonia, calculate the density and radius of the source for the atmospheric dispersion modelling, assuming that this source is cylindrical with height equal to radius. If the source begins to slump, what is its radius after five minutes assuming that, during this time, entrainment of air and cloud heating may be neglected?

Suppose that, as the ammonia cools, a fraction f is vaporized. The heat released by cooling 1 g of liquid anhydrous ammonia from 20 °C to -33 °C is 65.7 calories. In this range, the latent heat of vaporization is about 320 cal g^{-1}. The ratio of these two numbers gives $f = 0.2$. If 20 g of air are now added and cooled from $+20$ °C to -33 °C, 250 calories are liberated. This will evaporate 250/320 $= 0.8$ g of liquid ammonia – that is, it will evaporate the remaining ammonia droplets leaving a mixture which is mainly air at -33 °C. It is left as an exercise for the reader to prove that the density of this mixture is 1.42 kg m^{-3}.

Altogether, there are 40 t of ammonia and 800 t of air, a total mass of 8.4×10^5 kg. This occupies a volume of $8.4 \times 10^5/1.42 = 5.9 \times 10^5$ m^3. If the volume is cylindrical with radius and height equal, $\pi R_{ao}^3 = 5.9 \times 10^5$ from which $R_{ao} = 57$ m.

Neglecting entrainment of air and cloud heating, the volume V_o of the puff remains constant and equation 3.3-1 becomes

$$R_a^2 - R_{ao}^2 = 2\kappa t \sqrt{g(\rho - \rho_a)V_o/\rho_a \pi}$$

Taking $\kappa = 1$, $(\rho - \rho_a)/\rho_a = 0.18$, $g = 9.8$ m s^{-2}, $V_o = 5.9 \times 10^5$ m^3, $R_{oa} = 57$ m and $t = 5$ minutes $= 300$ seconds, then $R_a = 780$ m and the cloud is about 1.5 km across. This ties in well with the observed dimensions of the cloud during the Pensacola incident. Such a cloud width could not have been generated unless there had been gravitational slumping. Numerical calculations show that the predictions for the rate of growth of radius are not sensitive to the entrainment and heating assumptions, at least for the first few minutes, so that the good agreement with observation still stands.

Effects of Gas Clouds

Some of the potentially harmful effects of gas clouds have already been discussed in Section on 1.3.5, Consequence Assessment, in which considerable attention was devoted to radiotoxic releases into the atmosphere. As has been seen, such releases may lead to the development of cancers and genetic defects in the surrounding population and, if large enough, may also cause early deaths and morbidities.

The case of chemically toxic materials has also been considered in Section 1.3.5 where it was shown that the important quantities are the average airborne concentration χ_A during the duration of cloud passage τ_p. Once this pair of values is known it is in principle possible to consult toxicological data and determine by inspection whether the calculated concentration may prove fatal, cause severe illness, lead to distress, or whatever level of consequence is being assessed.

For flammable gas clouds, the problem is more complicated. There are two consequences that need to be considered. The first is the effect of any pressure waves that may be generated by an explosion, should the cloud be ignited. The second is the duration and intensity of radiant heat from the burning cloud. In order to calculate both of these effects, the concentration profile of flammable material within the cloud is required as a function of time: that is, instantaneous concentration distributions are required. The problem of turbulent fluctuations becomes important – the concentration at any particular point may be below the LFL when averaged over the duration of passage of the cloud but, from time to time, may exceed the LFL for short periods. The importance of this effect is still the subject of experimental research. Meanwhile, many workers in the field use half the LFL -2.5% by volume in the case of methane – when calculating hazard ranges, but this is an *ad hoc* assumption with little experimental or theoretical backing.

Once a cloud has ignited, the question arises, how damaging are the pressure waves generated? In short, will the cloud explode? The answer to this question is uncertain. An explosion in air is an event in which energy is deposited in a sufficiently small volume at a sufficiently rapid rate to generate discontinuous pressure waves. If an amount E of energy is deposited in a volume V over time-period t, important parameters are the energy density E/V and the energy deposition rate E/t. For many explosive substances (e.g. TNT) E/V and E/t are independent of the total quantity W of explosive and there is a simple scaling law which states that the distance from the centre of the explosion to the point at which a given overpressure is felt is proportional to $W^{1/3}$.

For a vapour cloud, however, the energy deposition rate E/t depends on the flame speed. The initial flame speed following successful ignition depends on the ignition energy density and the rate of ignition energy deposition. A flame or a hot surface is a weak ignition source and, for the common hydrocarbons of average reactivity such as propane and butane, flame speeds in excess of 10 ms^{-1} and overpressure levels in excess of 0.01 bar have not been observed. Highly reactive fuels such as acetylene or ethylene oxide may exhibit somewhat higher flame speeds, however.

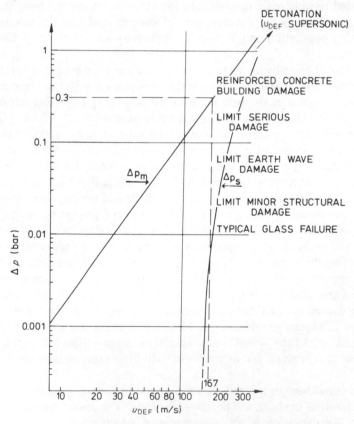

Figure 3.3-4 Variation of overpressure at the leading shock (Δp_s) and at the flame front (Δp_m) with deflagration wave speed (u_{DEF}). (Reproduced by permission of W. Geiger)

High flame speeds may be obtained by the use of strong ignition sources such as a 1 kg charge of TNT, but the simultaneous occurrence of a flammable gas cloud and a strong ignition source is improbable. The analysis of damage done in major incidents such as that at Flixborough[6] suggests that flame speeds greater than 100 m s^{-1} were obtained. This observation introduces one of the major outstanding problems in the study of unconfined gas cloud explosions – that of *flame acceleration*. Figure 3.3-4 shows how the overpressure at the leading shock wave and just behind the flame front varies with flame speed. This is for spherically symmetrical combustion wave propagation at a constant velocity in an infinite flammable gas cloud, but the qualitative features of the flow patterns are typical of all deflagrations and detonations. The flame front and the leading shock wave merge as the transition is made from deflagration to detonation. It is clear that flame speeds in excess of 200 m s^{-1} are required for the leading shock wave to produce significant damage outside the burning vapour cloud. At present, the cause of flame acceleration in previous incidents is unknown. Several mechanisms have been postulated, including continuous flame acceleration by some mechanism such as flame-induced turbulence in the unburnt gas, or discontinuous flame acceleration following flame propagation into a confined space such as narrow pipework. Meanwhile, *ad hoc* methods are required in order to predict the frequency with which a severe explosion will occur, given the formation of a flammable gas cloud, and the variation of peak overpressure and peak dynamic pressure with distance. A review of current information on the causes and effects of explosions in unconfined vapour clouds has been given in Reference 7.

Questions 3.3

1. If l_s in Equation 3.3-3 is set equal to H_a, the cloud height, and if the puff is at ambient temperature and contains a mass m_g of heavy gas with pure vapour density ρ_g, prove that Equations 3.3-1 and 3.3-4 have the following solutions:

(i) $R_a^2 = R_{ao}^2 + 2\gamma t$

where

$$\gamma^2 = \kappa^2 g m_g (1 - \rho_a/\rho_g)/\pi \rho_a$$

(ii) $V = \lambda R_a^6 + \left[\lambda R_{ao}^2 H_o^a - R_o^6\right] \left[\dfrac{R_a}{R_{a_o}}\right]^{2\alpha *}$

where

$$\lambda = \frac{\pi \alpha' \kappa U_1^{\frac{3}{4}}}{(6 - 2\alpha *)\gamma^3}$$

2. What are the essential elements of a 'model' of heavy vapour dispersion and give reasons?
3. Choose a heavy toxic or flammable gas which is widely used in the chemical industry and from the essential elements derived in Question 2 illustrate how they would apply to this gas.

Notation 3.3

g	acceleration due to gravity
H_a	height of gas cloud
H_{ao}	initial height
l_s	length scale
m_a	mass of air
m_g	mass of denser-than-air gas
R_a	radius
R_{ao}	initial radius
R_n	Richardson number
T_e	cloud temperature
U_1	longitudinal turbulence velocity
U_e	entrainment velocity
V	volume of puff
V_o	volume of puff at time $t = 0$
ΔT_a	temperature difference between the cloud and the surrounding atmosphere
ΔT_g	temperature difference between the cloud and the ground
κ	constant in gravitational slumping formula
ρ	density of a vapour cloud
ρ_a	density of air
ρ_g	density of gas
τ_p	duration of vapour cloud passage
χ_A	Average airborne concentration

References 3.3

1. Briggs, G. A. 'Plume rise predictions'. In *'lectures on Air Pollution and Environmental Impact Analysis'*. American Meteorological Society, Boston, Massachussets. 1975.
2. NTSB, *Railroad Accident Report: Louisville and Nashville Railroad Company Freight Train Derailment and Puncture of Anhydrous Ammonia Tank Cars at Pensacola, Florida, 9 November* 1977. US National Transportation Safety Board Report No. NTSB-RAR-78-4. 1978.
3. Griffiths, R. F. and Kaiser, G. D. *The Accidental Release of Anhydrous Ammonia to the Atmosphere – A Systematic Study of Factors Influencing Cloud Density and Dispersion.* Journal of the Air Pollution Control Association, **32**, 1982, 66–71.
4. Picknett, R. G. *Field Experiments on the Behaviour of Dense Clouds*. Report Ptn IL 1154/78/1. Chemical Defence Establishment, Porton Down, Wiltshire. Contract to the Health and Safety Executive, Sheffield. 1978.

5. McAdams, W. H. *Heat Transmission*. McGraw-Hill, New York. 1954.
6. HMSO. *Report of the Court of Inquiry into the Flixborough Disaster*. Her Majesty's Stationery Office, London. 1975.
7. Briscoe, F. 'A review of current information on the causes and effects of explosions of unconfirmed vapour clouds'. Appendix 1 of the Canvey Island Report. HMSO, 1978.

High Risk Safety Technology
Edited by A. E. Green
© 1982 John Wiley & Sons Ltd

Chapter 3.4

Radiation

SECTION 3.4.1 Non-Ionizing Radiation

A. F. McKinlay

Introduction

The electromagnetic spectrum can be divided into various named regions as illustrated in Figure 3.4.1-1. The boundaries of these regions have been developed by usage and adopted by convention. The demarcation between non-ionizing radiation and ionizing radiation is dependent on the absorbing medium. Conventionally non-ionizing radiation has wavelengths longer than 100 nm, i.e. photon energies less than 12 eV[1]. The actual value taken has philosophical rather than practical significance, and ultraviolet radiation with wavelengths shorter than 200 nm is unlikely to constitute any direct hazard.

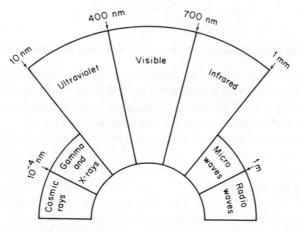

Figure 3.4.1-1 The electromagnetic spectrum

There are two basic methods of generating electromagnetic radiation:

1. The acceleration and deceleration of charged particles. This method is associated particularly with the production of X-radiation and, for non-ionizing radiation, of radiofrequency radiation.

527

2. The excitation of atoms and molecules, and subsequent de-excitation with the emission of radiant energy. This process describes most sources of ultraviolet, visible, and infrared radiations.

Table 3.4.1-1 gives examples of some of the artificial sources of non-ionizing radiation which are a common feature of modern industrial life.

Table 3.4.1-1   Some artificial sources of non-ionizing radiation

Spectral region (wavelength range)	Sources
Ultraviolet (200–400 nm)	Bactericidal lamps Photo reactor lamps (UV curing) Spectrographic arcs Black lights Welding and plasma arcs
Visible and infrared (400 nm–1 mm)	Photo processing lamps Spectrographic arcs Welding and plasma arcs Floodlights Furnaces Lasers
Microwave and radiofrequency	Microwave ovens RF heating coils Heat sealers Communication equipment Radar

Hazardous Biological Effects of Non-Ionizing Radiation

It is a reasonable thesis that exposure to artificial sources of non-ionizing radiation will not constitute a significant hazard if the resultant exposure is less than that which one would *normally* experience from natural sources.

The natural radiation environment that mankind, in common with other diurnal animals, has adapted to is dominated by that part of the sun's radiation which passes through the atmosphere to reach the earth's surface. This is broadband non-coherent radiation and, depending on the sun's elevation, the peak irradiance may approach 10^3 W m^{-2}, with 99.9% contained within the wavelength limits of 295 nm and 4 μm. The effects of atmospheric filtration are most marked on the shorter wavelength ultraviolet radiations which constitute 5–10% of the total. Visible radiation contributes about 40% and infrared radiation about 50–55%. The wavelength limits 295 nm to 4 μm represent a 'window' in the filtration of the atmosphere and there is another, used by radioastronomers, in the microwave and radiofrequency region. The exposure irradiance of man from these radiations is, however, only some 10^{-9} W m^{-2}. The more energetic X-radiations also reach the surface of the earth but at an even lower irradiance – about 10^{-13} W m^{-2}.

Non-ionizing radiation with wavelengths shorter than a few centimeters has only limited penetration into tissue so that the effects are generally limited to the surface of the body, viz. the skin and eyes. Penetration is greater at longer wavelengths and continues to rise as wavelengths lengthen, and radiant energy may then be deposited in deeper tissues and organs. The biological effects of non-ionizing radiation vary greatly with spectral region and are summarized in Table 3.4.1-2.

Table 3.4.1-2 Injuries to the eye and skin produced by non-ionizing radiation

Spectral region	Eye	Skin
ULTRAVIOLET		
200–315 nm	Photochemical corneal injury (keratitis)	Erythema, skin ageing, skin cancer
315–400 nm	Photochemical cataract Lens yellowing	Photosensitivity
VISIBLE	Photochemical, thermal	Skin burn
400–700 nm	and thermomechanical retinal damage	Photosensitivity
INFRARED		
700–1400 nm	Thermal and thermo-mechanical retinal damage	Skin burn
	Infrared cataract	
1400 nm–1 mm	Corneal burn	Skin burn
MICROWAVE AND RADIOFREQUENCY	Corneal burn lenticular	Skin burn
1 nm–30 km	cataract	Overheating

The known harmful effects can be categorized into three types: thermomechanical, thermal, and photochemical.

Thermomechanical damage occurs when the rate of thermal expansion of tissue is explosively high and causes mechanical disruption. It requires very high rates of heat input and is important for short ($< 1 \mu s$) pulses of laser radiation focused by the eye to form small retinal images.

Thermal damage occurs when tissue temperatures are raised by more than a few degrees. Denaturation of protein is the predominant effect, and consequential cell death is a rapid function of both temperature and time.

Photochemical injury depends on the total energy absorbed before repair mechanisms are effective. It is the limiting effect for bare skin and eye exposure over the range of wavelengths 200–315 nm, and for ocular exposures of longer than a second or so over the wavelength range 400–500 nm.

There is no reason to believe that line spectra are more harmfully effective than continua, or that the properties of coherence as such are biologically significant.

Ultraviolet Radiation

The non-ionizing component of the ultraviolet spectrum extends over the range 100–400 nm. Ultraviolet radiation with wavelengths shorter than 200 nm is rapidly absorbed in air and does not constitute a direct biological hazard; it is, however, responsible for ozone generation. Radiation with wavelengths between 200 and 315 nm is most effective is causing injury. In this range injury is a consequence of primary photochemical interactions and there is a latent period between exposure and the onset of symptoms. For acute injury, which is likely to show within a few hours, the total exposure required is independent of whether this is acquired in a few seconds at a high rate or at a lower rate over a period of hours.

The principal acute effect on the eyes is the painful but usually reversible condition photokeratitis, known variously as 'arc eye', 'welder's flash', and 'snow blindness'. The symptoms are severe pain, a sensation of grit in the eyes, and an aversion to bright light. The cornea and the conjunctiva may become inflamed. For a given exposure the maximum effect occurs with radiation of wavelength 270 nm. At this wavelength a radiant exposure of $40 \, \text{J m}^{-2}$ will produce a threshold effect[2][3]. The chronic effects on the eye include yellowing of the lens due to absorption of ultraviolet A (315–400 nm) radiation by lenticular pigments. Ultraviolet radiation of wavelengths longer than 290 nm may penetrate the aqueous humour, the iris and lens, and evidence furnished by animal experiments indicates that ultraviolet A and ultraviolet B (280–315 nm) radiation can induce cataracts[4].

The main acute effect on the skin is erythema, known to and experienced by most people as sunburn. The redness produced is caused by dilation of blood vessels; in white skin it is most efficiently produced by ultraviolet radiation within the range 240–315 nm[2][3]. The erythemal sensitivity of individuals to ultraviolet radiation varies greatly, depending largely on skin pigmentation and thickness.

The chronic effects of repeated exposure include premature skin ageing and the induction of skin cancers. Skin ageing is a natural process and there is evidence that prolonged and repeated exposure to ultraviolet radiation accelerates it[4][7].

It is now generally accepted that there is a causal link between exposure to solar ultraviolet radiation and the incidence of skin cancers[8]. However, the susceptibility of any group of individuals can vary greatly and depends not only on total exposure but also on cultural behaviour, and racial and genetic factors. The action spectrum for ultraviolet carcinogenesis in humans has not been determined but there is evidence that it is similar to that for erythema.

Visible and Infrared Radiation

In the visible and infrared A range of wavelengths, 400–1400 nm, the ocular media are mainly transparent. Radiation entering the eye is refracted primarily by the cornea, and focused to a small image on the retina[9]. Adaptation has produced aversion responses to bright lights which normally preclude staring at high luminance (i.e. visibly bright) sources such as the sun for longer than about 0.25 s. If the aversion response is delibrately overcome then retinal injury is likely. Similarly prolonged exposure to arcs and arc lamps can cause retinal burns. However few devices, apart from lasers, will cause injury within the reflex time of the aversion responses.

Infrared A radiation 700–1400 nm may also cause some damage to the anterior regions of the eye. In particular it could contribute to the formation of lenticular cataracts. In the infrared B region, 1.4 to 3 μm, both lenticular and corneal damage may occur, i.e. cataracts and corneal burns. In the infrared C, 3 μm to 1 mm, the effects are limited to the cornea, and mostly to its outer epithelial layer. The main hazard to the skin from visible and infrared radiations is thermal burning.

Microwave and Radiofrequency Radiation

The main effect of microwave and radiofrequency radiation on people is heating of their tissues. When applied locally this can result in burns, while for whole body exposure there will be thermal stress. Temperature excursions will be limited by the body's normal capacity for temperature stabilization but this mechanism can be overloaded. These effects have been shown in animal experiments and form the basis of the Western view to microwave and radiofrequency radiation as a hazard, as first pronounced about 25 years ago.

The lens of the eye is particularly at risk from thermal overload as an effectively cooling blood supply is lacking. However, the irradiance of microwave and radiofrequency radiation required to produce cataractous changes observed in experimental animals is relatively high ($> 10^3$ W m^{-2})[10].

The Eastern European approach is that stress and behavioural changes manifest themselves at lower levels than overt injury and must also be guarded against. No fully plausible mechanisms have been suggested, but theoretical and experimental studies, especially over the last decade, have shown that the radiant energy can be deposited very unevenly[11]. This may lead to localized overheating or to local temperature differentials. There is no suggestion of serious injury resulting, but that functional changes will result in people feeling unwell.

Exposure Standards

The most comprehensive occupational exposure standard for protection against ultra-violet radiation is that of the American Conference of Governmental Industrial Hygienists (ACGIH, 1980)[12]. The recommended maximum permissible exposure levels for ultraviolet radiation are shown in Figure 3.4.1-2.

At the time of writing drafts of various laser exposure standards are being discussed by interested organizations; these drafts include the American National

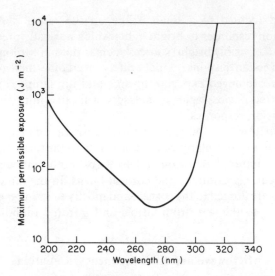

Figure 3.4.1-2 Occupational maximum permissible exposure levels (MPEs) for monochromatic ultraviolet radiation. In the wavelength range 200 to 315 nm the radiant exposure on the unprotected eyes and skin should not exceed, within any eight hour period, the values shown. In the wavelength range 315–400 nm the total radiant exposure should not exceed 10^4 J m^{-2} for exposure periods less than 10^3, and the irradiance should not exceed 10 Wm^{-2} for exposure periods greater than 10^3. For polychromatic ultraviolet radiation the maximum permissible exposure is calculated by summing the relative spectral contributions to the irradiance, each being weighted by a relative spectral effectiveness factor corresponding to the inverse of the maximum permissible exposure at each wavelength and normalized at 270 nm[12]

Standards Institute Standard which will replace ANSI-Z136.1. 1976 (ANSI, 1976)[13]; the British Standards Institution Standard which will replace BS 4803. 1972 (BSI, 1972)[14]; and an International Electrotechnical Committee Standard. The maximum permissible exposures at the cornea for direct ocular exposure (intra-beam viewing) to laser radiation recommended in the Draft British Standards Institution Standard are shown in Figure 3.4.1-3. In Western countries the continuous exposure standard for microwave and radiofrequency radiation is 100 W m^{-2} (e.g. ANSI, 1974[15]; HO, 1960[16]; MRC, 1970[17]). The heat stress on the body resulting from a continuous exposure at this level is small compared with that caused by normal physical effort even under thermally adverse environmental conditions. However there is some concern that over the frequency range 10 MHz to 1 GHz, where resonance absorptions may occur in

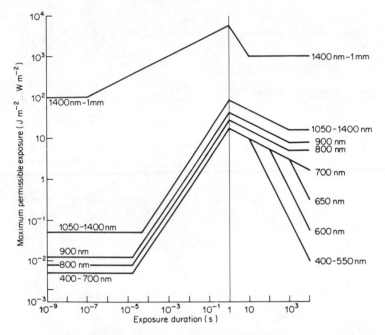

Figure 3.4.1-3 Maximum permissible exposure levels (MPEs) at the cornea for direct ocular exposure (intra beam viewing) to laser radiations in the wavelength range 400 nm to 1 mm. The maximum permissible exposure levels are expressed in units of radiant exposure $J\,m^{-2}$ for exposures shorter than 1 s, and in units of irradiance $W\,m^{-2}$ for exposures of 1 s or longer. (Adapted from the maximum permissible exposure levels contained in the 8th Draft (May 1980) of the British Standards Institution Standard BS 4803)

humans, safety factors may have been overestimated. The marked differences in current maximum permissible exposure levels for microwave and radiofrequency radiations among Western and Eastern European countries are illustrated in Figure 3.4.1-4.

Protection Principles

The most serious risk in working with non-ionizing radiation is damage to sight and this is much more likely from a single gross overexposure that from repeated small overexposures. Although instrumental assessment of hazard is often difficult, it is usually easy to provide adequate protection. This may be achieved by a combination of the following:

1. Engineering controls
2. Administrative controls
3. Personal protection

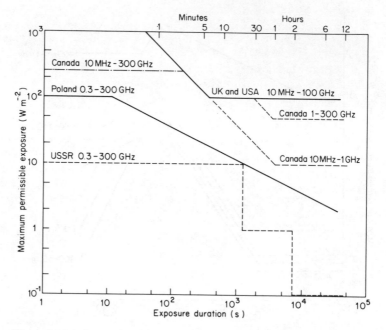

Figure 3.4.1-4 Occupational maximum permissible exposure levels (MPEs) for microwave and radiofrequency radiations in the frequency range 10 MHz to 300 GHz, as a function of exposure time, adopted by the USA, UK, Canada, Poland, and the USSR

Wherever reasonably practicable the source should be enclosed and leakage minimized. A prime example is the modern microwave oven. Ovens conforming to the emission standard in the UK, $50\,\mathrm{W\,m^{-2}}$ at $5\,\mathrm{cm}$[18][19]; and in the USA, $10\,\mathrm{W\,m^{-2}}$ when the oven is new and $50\,\mathrm{W\,m^{-2}}$ thereafter[20], will give a time averaged emission irradiance of order $10^{-2}\,\mathrm{W\,m^{-2}}$ and would therefore be acceptable even under the restrictive Eastern European exposure regulations. Where there is significant emission of non-ionizing radiation warning signs should be used and staff should be informed about possible dangers. Access should be controlled. Regulations require the provision of personal protection for the eyes in some processes and a wide selection of eyewear and clothing is available covering much of the non-ionizing radiation frequency spectrum. However, the preferred method of limiting exposure is by engineering controls, followed by administrative controls.

Questions 3.4.1

1. What distinguishes non-ionizing radiation from ionizing radiation? List different types of non-ionizing radiation known to you.

2. What are the hazardous biological effects which can be produced by non-ionizing radiation? Discuss the effects of different wavelengths.
3. Discuss ultraviolet radiation and the effects it can have on the eyes or skin of a human being. Does the wavelength of the radiation have a significance in the severity of the effects and how would this be manifested?
4. Give examples of visible and infrared radiations and state the main effects of microwave and radiofrequency radiations with reference to the tissues of people.
5. What standards and protection principles are applied to prevent serious risk to people working in an environment with non-ionizing radiation?

References 3.4.1

1. Matelsky, I. *Industrial Hygiene Highlights.* Volume 1. Industrial Hygiene Foundation of America, Pittsburg. 1968.
2. Pitts, D. G. and Tredici, T. J. 'The effects of ultra violet on the eye.' *Am. Ind. Hyg. Assoc. J.*, **32**, 235–246. 1971.
3. Pitts, D. G. 'The human ultra violet action spectrum.' *Am. J. Optom. Physiol. Opt.*, **51**, 946–960. 1974.
4. Parrish, J. A., Alderson, R. R., Urbach, F., and Pitts, D. G. *UVA Biologic Effects of Ultra Violet Radiation with Emphasis on Human Responses to Longwave Ultra Violet.* Wiley, Chichester. 1978.
5. Everett, M. A., Olson, R. L., and Sayre, R. M. 'Ultra violet erythema.' *Arch. Dermat.* **92**, 713–719. 1965.
6. Freeman, R. G., Owens, D. W., Knox, J. M., and Hudson, H. T. 'Relative energy requirements for an erythemal response of skin to monochromatic wavelengths of ultra violet present in the solar spectrum.' *J. Invest. Dermatol.*, **47**, 586–592. 1966.
7. Magnus, I. A. *Dermatological Photobiology.* Blackwells, Oxford. 1976.
8. Emmett, E. A. 'Ultra violet radiation as a cause of skin tumours.' *CRC Crit. Rev. Toxicol.*, **2**, 211–255. 1973.
9. Sliney, D. H. and Freasier, B. C. 'Evaluation of optical radiation hazards.' *Appl. Opt.*, **12**, 1. 1973.
10. Cleary, F. H. 'Microwave cataractogenesis' *Proc. of the IEEE.*, **68**, 49–55. 1980
11. Tell, R. A. and Harlen, F. 'A review of selected biological effects and dosimetric data useful for development of radiofrequency safety standards for human exposure.' *J. Microwave Power*, **14**, 405–424. 1979.
12. ACGIH, *Threshold Limit Values for Chemical Substances and Physical Agents in the Workroom Environment.* American Conference of Governmental Industrial Hygienists, Cincinatti. 1980.
13. *Standard for the Safe Use of Lasers Z136.1.* American National Standards Institute. New York. ANSI 1976.
14. *Guide on Protection of Personnel Against Hazards from Laser Radiation BS4803.* British Standards Institution, London. BSI 1972.
15. *Safety Levels of Electromagnetic Radiation with Respect to Personnel ANSI–C9 5.1.* American National Standards Institute, New York. ANSI 1974.
16. *Safety Precautions Relating to Intense Radiofrequency Radiation.* Her Majesty's Stationery Office, London. HO 1960.
17. *Exposure to Microwave and Radio-frequency Radiations*, MRC Press Notice 70/1314. MRC, London. MRC 1970.

18. *Safety of Household Electrical Appliances · Microwave Ovens. BS 3456.* British Standards Institution, London. BSI 1976.
19. *Safety of Commercial Electrical Appliances Using Microwave Energy for Heating Foodstuffs, BS 5175.* British Standards Institution, London. BSI 1976.
20. *Regulations for the Administration and Enforcement of the Radiation Control of Health and Safety Act of 1968.* HEW Publication (FDA) 798035. Department of Health Education and Welfare, Washington. FDA 1979.

High Risk Safety Technology
Edited by A. E. Green
© 1982 John Wiley & Sons Ltd

SECTION 3.4.2 *Ionizing Radiation*

J. A. Dennis

Characteristics of Ionizing Radiations

There are two types of ionizing radiation. One type consists of particles possessing electric charge and is called *directly ionizing radiation*; the other type does not possess electric charge and is called *indirectly ionizing radiation*. The particles of practical importance are:

Directly ionizing:
   alpha particles (helium nuclei)
   beta particles (negatively charged electrons)
   positrons (positively charged electrons)

Indirectly ionizing:
   gamma rays (photons)
   X-rays (photons)
   neutrons

Directly ionizing particles, because of their electric charge, lose energy in a continuous fashion to the electrons in the matter through which they pass and as a result have a finite range. In human tissues a typical alpha particle with an energy of about 5 MeV has a range of about $50 \mu m$, and a beta particle with an energy of a few MeV has a range of 1 or 2 cm.

Indirectly ionizing particles do not lose energy by a continuous process. The photon radiations, gamma and X-rays, may react with some statistical probability with electrons to lose all or part of their energy, and in this process produce directly ionizing electrons. Neutrons have a statistical probability of reacting with atomic nuclei. The result may be radioactive atoms, possibly recoiling nuclei which are directly ionizing particles or nuclear fission with the production of more neutrons. Beams of photons or neutrons are attenuated exponentially as they pass through matter. This attenuation is characterized by a relaxation length or mean free path which is the distance in which the particle fluence decreases to $1/e$ or 37.8 % of its original value. This distance is about 14 cm for 1 MeV gamma rays in tissue and about 2 cm for 1 MeV neutrons. At lower energies the mean free paths become smaller.

537

Stable atomic nuclei contain very roughly equal numbers of neutrons and protons but, owing to the electrostatic repulsive force between protons, rather fewer protons than neutrons can be accommodated in stable nuclei as the total mass increases: the larger nuclei tend to be less stable generally. Unstable nuclei which occur naturally or are created artificially are subject to radioactive decay in which gamma rays, X-rays, positrons, beta or alpha particles may be emitted. Some very large nuclei which have been created artificially emit neutrons. It is not possible to predict when a particular radioactive nucleus will decay, but the decay of a large number of identical radioactive nuclei can be characterized by a 'half-life' during which half of the original number will have decayed.

Table 3.4.2-1 lists some of the more well-known radioactive nuclei and their half-lives.

Table 3.4.2-1   Some radioactive nuclei of interest

Nucleus	Half-life	Major radiations emitted
Phosphorus-32	14.28 days	beta particles
Potassium-40	1.26×10^9 years	beta particles, positrons
Cobalt-60	5.26 years	beta particles, gamma rays
Strontium-90	27.7 years	beta particles
Iodine-125	60.2 days	X-rays
Cesium-137	30.0 years	beta particles, X-rays
Radium-226	1602 years	alpha particles, gamma rays
Uranium-235	7.1×10^8 years	alpha particles, X-rays
Uranium-238	4.51×10^9 years	alpha particles, X-rays
Plutonium-239	24 390 years	alpha particles, X-rays
Californium-252	2.646 years	alpha particles, X-rays, fission fragments, neutrons

Biological Effects

The biological effects of ionizing radiations are to a significant extent determined by the amount of energy absorbed per unit mass of tissue. This is called the *absorbed dose* (D) which is expressed in the unit of the gray (Gy) that is equivalent to an energy absorbtion of 1 joule per kilogram (an older unit, the 'rad', is equivalent to 100 ergs per gram). Some radiations, notably neutrons and alpha particles, are nevertheless more damaging per unit of absorbed dose than others, such as electrons, gamma rays, and X-rays. For radiological protection purposes therefore another quantity has been defined and is known as the *dose equivalent* (H) which may be considered as the weighted absorbed dose. The weighting is known as the *Quality Factor* (Q). For electrons and photon radiations it is specified as unity. Its maximum value is 20 for alpha particles and other heavy particles. The formal relationship between dose equivalent, absorbed dose, and the Quality Factor can be expressed as

$$H = D \times Q \qquad (3.4.2\text{-}1)$$

The unit of dose equivalent corresponding to the absorbed dose of 1 gray with unit Quality Factor is the sievert (Sv). The older unit was the rem and the conversion is

$$1 \text{ sievert} \equiv 100 \text{ rem}$$

In the remainder of this section the term 'dose' will be used as synonymous with 'dose equivalent'.

After exposure to high doses of radiation there may be direct damage to the tissues of the body which impairs their proper function, possibly to the point of death of the individual. Such damage requires a 'threshold dose' to be exceeded in a comparatively short time, i.e. in less than a few weeks or days; when the dose is delivered over much longer times the recovery powers of the tissues reduce or prevent direct damage. When the dose delivered in a very short time exceeds 3.5 to 4.5 Sv (350–450 rem) to the whole body, damage to the bone marrow may ultimately cause death by depletion of the white blood cells. Higher doses cause death by damage to the gastrointestinal tract or the central nervous system. High doses confined to small areas of the body can cause radiation burns and in the case of the eyes may result in cataract either in the longer or shorter term.

The effect on man of exposure to low doses of radiation is an increased probability of developing cancer and an increased probability of hereditary defects in any subsequent children of the irradiated individuals and their further descendants. The probability of such consequences is assumed to increase proportionally to the radiation dose without any threshold. The size of the radiation dose does not influence either the nature or the seriousness of the effect. The assumption of direct proportionality between the dose and the probability is questionable, and whether this assumption results in an under- or over-estimate of the risks is controversial, although most informed authorities believe that the risks may be overestimated.

The International Commission on Radiological Protection (ICRP), drawing upon sources of information such as those summarized by the United Nations Scientific Committee on the Effects of Atomic Radiation[1], has published estimates of the sensitivities of different organs and tissues of the body for the induction of fatal cancer and of the gonads for the induction of hereditary defects that might be expressed in the children or grandchildren of exposed persons[2]. These absolute and relative sensitivities are shown in Table 3.4.2-2. They are the average sensitivities that may be taken to apply to a typical working population, the sensitivities are both sex and age dependent and those of young persons and women may be greater than the average values to the extent that their overall risk is about 50 % greater; the risk to older persons may be about 80 % less than the average.

Table 3.4.2-2   Absolute and relative organ sensitivities adopted by the ICRP for the induction of cancer and hereditary defects

Tissue or organ	Sensitivities	
	Absolute probability per sievert	Relative
Gonads	40×10^{-4}	0.25
Breast	25×10^{-4}	0.15
Red bone marrow	20×10^{-4}	0.12
Lung	20×10^{-4}	0.12
Thyroid	5×10^{-4}	0.03
Bone surfaces	5×10^{-4}	0.03
All other tissues together	50×10^{-4}	0.30*
Total for whole body	165×10^{-4}	1.00

* In calculating effective dose equivalent the five other tissues receiving the highest doses are taken to have a relative sensitivity of 0.06 each.

Comparison of Radiation and other Industrial Risks

A simple comparison between the risk of death from exposure to radiation and the risk of death by accident in industries not involving exposure to radiation is unsatisfactory, because there is usually a fairly long period between exposure to radiation and a consequential death from cancer; the loss of life expectancy may therefore be comparatively small whereas the loss due to a fatal accident occurring at the same age as the exposure to radiation is much greater. The ICRP has published a suggested method for comparing accidental and radiation risks by means of an Index of Harm[3], which takes into account the loss of life expectancy. Using this Index it is possible to compare the average annual risks of different industries in the United Kingdom, Table 3.4.2-3, with the risks arising from occupational exposures to radiation, Table 3.4.2-4. This comparison

Table 3.4.2-3   Fatal accident rates in UK industries during 1977 (Health and Safety Executive, 1978)[7]

Industry	Fatal accident rate per 10^6 at risk	Index of Harm
Vehicle manufacture	9	0.41
Paper, printing, and publishing	22	0.89
Textiles	23	0.93
All manufacturing industries	34	1.3
Food, drink, tobacco	40	1.5
Chemicals and allied industries	46	1.7
Ship-building and marine engineering	104	3.7
Metal manufacture	122	4.3
Construction	131	4.6
Coal and petroleum products	195	6.7

Table 3.4.2-4   The distribution of radiation risks by occupation in the UK (Taylor and Webb, 1978)[8]

Occupation	Number of workers	Annual average exposure (mSv)	Index of Harm
Nuclear fuel cycle			
Fuel fabrication	2 200	2.5	0.6
Fuel enrichment	600	0.7	0.2
Nuclear reactors:			
CEGB/SSEB	6 500	2.7	0.6
BNFL	800	9.5	2.2
Fuel reprocessing	4 400	11.0	2.5
Research			
Nuclear	9 600	4.0	0.9
Teaching institutions	10 000	2.5	0.6
Industry			
Industrial radiography			
factory	5 000	9.0	2.1
site	2 000	27.0	6.2
Luminizing	130	8.0	1.8
Radioisotope preparation	900	7.8	1.8
Aircraft crew	20 000	1.2	0.3
Others	11 000	4.3	1.0
Medical			
General	33 000	2.1	0.5
Dental	20 000	1.6	0.4
Veterinary	4 000	0.6	0.1
Others	2 500	3.1	0.7
Total	133 000	3.3	0.8

1 mSv $\equiv$ 0.1 rem

indicates that the majority of the average exposures to radiation result in risk levels that are equivalent to the average fatality risks in the safer industries. The most obvious exception is that of site radiography. Radiation risks are, of course, additional to the other accidental industrial risks.

Principles of Protection

The ICRP recommends a system of dose limitation with three main principles which expressed as in Reference 2 are
1. No practice shall be adopted unless its introduction produces a positive net benefit.
2. All exposures shall be kept as low as reasonably achievable, economic and social factors being taken into account.
3. The dose equivalent to individuals shall not exceed the limits recommended for the appropriate circumstances by the Commission.

Cost–Benefit Analysis

The first principle implies that an explicit cost–benefit analysis should be carried out to determine the justification for any practice involving the exposure of people to radiation. The difficulties of doing this objectively are recognized, particularly when those suffering the possible detriment are not necessarily those in full receipt of the benefits.

The second principle implies that even when a practice is judged to be beneficial it is necessary to carry out a differential cost–benefit analysis to optimize the net benefit. This is achieved when the increase in the cost of protection per unit dose equivalent is equal to the decrease in the cost of the detriment per unit dose equivalent.

Usually the quantity of interest in cost–benefit studies is the *collective dose equivalent*, S, arising from the practice, and which may be defined by

$$S = \int_0^\infty H P(H) \, dH \qquad (3.4.2\text{-}2)$$

where $P(H) \, dH$ is the number of individuals who will receive a dose equivalent in the range H to $H + dH$.

Dose Limits

Primary limits

The ICRP has recommended dose limits with the aim of limiting the risks to individuals from cancer and hereditary effects to acceptable levels and of completely preventing the occurrence of other types of damage to any tissue of the body. In attempting to keep the incidence of cancer and hereditary effects to acceptable levels the principle that doses should be kept *as low as reasonably achievable* (known by the mnemonic ALARA), economic and social factors taken into account, must be kept in mind, and the ICRP does not believe that any person should be exposed continuously throughout his life to the annual dose limits which it recommends. In the use of the dose limits it is essential to recognize that they do not represent a sudden transition from safety to danger. Although from regulatory viewpoint the exceeding of a dose limit should be an offence and invoke appropriate sanctions, unless the overexposure is excessive there will be no attributable or immediate medical or biological consequences. Managements and regulatory bodies may adopt flexible attitudes in cases where there have been minor infringments of dose limits and attempt to meet the general objective of improving the working situation.

The basic dose limit recommended by the ICRP for uniform whole body of radiation workers is 50 mSv (5 rem) per year. To deal with situations in which the body is irradiated non-uniformly either from external radiation or more usually from the intake of radioactivity into the body the concept of *effective dose equivalent* is introduced which makes use of the relative organ sensitivities given in Table 3.4.2-2. The effective dose equivalent, H_E, is defined as

$$H_E = \Sigma_T w_T \times H_T \qquad (3.4.2-3)$$

where w_T is the weighting factor or relative sensitivity for organ or tissue T and H_T is the dose equivalent in that tissue. The effective dose equivalent derived by the use of the formula indicates the same risk of cancer or a hereditary defect as that given by uniform irradiation of the whole body to the same dose equivalent value. However, if the formula is used to derive the doses to individual organs (assuming the unlikely possibility that they could be irradiated in isolation from the rest of the body) which would result in the same overall risk as the dose limit of 50 mSv (5 rem) to the whole body, it allows doses to insensitivie organs such as the thyroid which would be so high that there would be an appreciable risk of direct damage. For this reason the ICRP imposes an additional dose limit of 500 mSv (50 rem) on any organ in one year. Except for the eye this limit is less than the actual threshold for direct damage. The limit for the eye is set at 150 mSv (15 rem).

Since the very young, the very old and the infirm may be at greater risk from radiation than those who are fit to work, and since members of the public may be exposed for the whole of their lives, the dose limits for members of the public are set at one-tenth of those for workers. The resulting level of risk is judged by ICRP to be no greater than that which the public generally accept in the context of everyday life. However no member of the public is expected to be exposed to the dose limits continuously throughout his life.

Secondary limits

To assist in controlling and recording the exposure arising from radioactive substances which are taken into the body the ICRP has recommended a set of secondary limits known as *annual limits of intake* (ALI)[4]. These limits are obtained from calculations based on models for the metabolsim of different radioactive nuclides by the body and on the route of intake whether by ingestion or by inhalation. An ALI is that quantity of a radioactive nuclide which if taken into the body in one year would result in risks of fatal cancer or hereditary effects that do not exceed those of 50mSv (5 rem) of uniform, whole body exposure, with the additional limitation that no organ or tissue exceeds 500 mSv (50 rem). The actual doses from intakes of radioactivity are received over a period of time after the intake, and for nuclides with long half-lives and which are retained in the body for long periods this may extend over the whole of the subsequent life. In the calculation of ALIs the period is taken to be 50 years for radiation workers and 70 years for members of the public.

Derived limits

For the control of environmental levels of radioactive substances it is necessary to derive additional limits for permissible concentrations of radioactive materials in air, water, foodstuffs, and on surfaces. These are obtained from the ALIs with knowledge or assumptions about pathways which could lead to the intake of the radioactivity by human beings. While such derived limits are extremely useful for an immediate comparison of measurements with radiation protection standards, they must be used with an understanding of the assumptions underlying their derivation.

The Control and Monitoring of Radiation

The exposure of workers is best controlled by monitoring the workplaces. An effective monitoring programme is aided by the adoption of the ICRP recommendations[9] for classifying working areas as either *controlled* or *supervised*. A *controlled area* is one in which a worker, if he worked continuously, might be exposed in excess of three-tenths of the annual limit, i.e. to an effective dose equivalent in excess of 15 mSv (1.5 rem). Access to such an area will normally be limited by physical means. A *supervised area* is one in which workers may exceed one-tenth of the annual dose limit, i.e. the limit for members of the public.

By appropriate instrumentation it is possible to measure the radiations from sources and to measure the degree of contamination with radioactivity of the air and surfaces. *Individual monitoring* of the exposure of workers to external radiations may be carried out by means of personal dosemeters based on photographic film or thermoluminescent materials. The exposure of individuals to radioactive substances which have been taken into the body is not so easily assessed and in general reliance has to be placed either on the direct measurement of the radioactivity within the body by external radiation detectors (the usual instrument for this is a 'whole body counter'), or on the measurement of the radioactivity excreted in the urine and faeces. The assessment may also have to rely on the measurements of the radioactivity, usually the concentration in the air, within the exposed individual's workplace.

The ICRP recommends that individual monitoring should always be carried out when the effective dose equivalent is likely to exceed 15 mSv (1.5 rem) in a year and that records of the assessed doses should be kept for at least 30 years. While this implies that only workers entering controlled areas need be monitored it does not necessarily require all such workers to be monitored since their doses will depend on the actual conditions of their work and the time they spend in the area.

In addition to the determination of the individual doses the results of monitoring a group of workers may form part of the programme for monitoring the workplace. Routine monitoring of the workplace is known as *environmental monitoring* and is generally carried out by means of portable or fixed instruments depending upon the type of work and the nature of the radiation hazard. Where radioactive materials are handled it may be necessary to monitor workers for any contamination on their clothing or extremities as they leave a controlled area. It may also be necessary to monitor items of equipment and rubbish which are removed from the area. Such monitoring will be in addition to measurements of surface and air contamination in the area itself. A comprehensive environmental survey will usually be carried out when there is a substantial change in the working arrangements or conditions. When the situation is not likely to change the amount of monitoring can be reduced and designed to give a warning of any increase in the radiation hazards.

Often when carrying out a procedure involving high dose rates it is advisable to perform *operational monitoring*. In these circumstances the results of the monitoring may govern the continuing conduct of the work. The monitoring will usually be continuous and the workers may carry personal dosemeters that are capable of rapid assessment or which give an audible or visual warning of excessive doses or dose rates.

Special monitoring involving more instruments or measurements than required for either environmental or operational monitoring should be performed when a new procedure involving high dose rates or large amounts of radioactivity is carried out for the first time, or when workers are suspected of having been exposed accidentally to excessive doses of external radiation or significant amounts of radioactivity.

Reference Levels

In a monitoring programme it is often helpful to compare measurements not with the basic or derived limits but with established 'reference levels' which can be used to decide further action. The action might range from a simple investigation as to the cause of the observed measurement, through a formal consideration as to whether the ALARA principle has been achieved, to a major emergency requiring active intervention to restrict the exposure of individuals and minimize the consequences.

Emergency Reference Levels

In major emergencies it is generally accepted that the potential risks to those directly involved will be higher than those of normal circumstances. The ICRP recommended that counter measures should be taken only when their social cost and risk is less than that which would result from further exposure[2]. This implies that this balance should be made in the context of the actual circumstances prevailing at the time of the emergency. However, advance guidance is extremely helpful and in the United Kingdom has been given in the form of Emergency

Reference Levels by the Medical Research Council (1975), Table 3.4.2-5[5]. These are levels of exposure for members of the public below which counter measures are unlikely to be justified. In the case of exposure to radioactivity they are the *committed dose equivalents*, i.e. the dose equivalents to the body or organ which would accumulate over the 50 years following an intake. Recent thinking has suggested that such a table with its implied simple distinction between action and inaction should be modified according to local circumstances with the emergency reference levels lying in the range 5 mSv (0.5 rem) to 500 mSv (50 rem) for the effective dose equivalent and 500 mSv (50 rem) for individual organs and tissues[6]. Whatever the values of the emergency reference levels they should be used for guidance rather than for the rigid application of counter measures, although their very existence is likely to influence the judgement of those responsible for control of the emergency. One of their uses is to define the geographical limits of counter measure once it has been decided to take action. As with the basic dose limits it may be convenient to use the emergency reference levels to derive other levels related directly to environmental measurements.

Table 3.4.2-5 Emergency Reference Levels of dose currently in use in the UK (Medical Research Council, 1975)[5]

Organ or tissue	Emergency Reference Level of dose (mSv)
Whole body	100
Thyroid	300
Lung	300
Bone	
Endosteal tissue	300
Marrow	100
Gonads	100
Superficial tissues irradiated by beta particles	600
Any other organ or tissue	300

1 mSv ≡ 0.1 rem

Questions 3.4.2

1. Distinguish between directly and indirectly ionizing radiations. What is the range of half-lives over which the well-known radioactive nuclei extend?
2. Discuss the biological effects of ionizing radiations with reference to human tissues. Can the risk of death from exposure to ionizing radiations be compared with that arising by accident in industries not involving ionizing radiations and what are the results?

3. What principles of protection should be adopted for people involved in potentially significant exposures to ionizing radiations? How would the dose limits be set and what doses would be considered as low as reasonably achievable?
4. Discuss the methods of controlling and monitoring ionizing radiation in a controlled or supervised area where people may be working. What part does instrumentation play in the role of a monitoring programme?

References 3.4.2

1. *Sources and Effects of Ionizing Radiation.* Report of the United Nations Scientific Committee on the Effects of Atomic Radiations to the General Assembly. UN, New York, UNSCEAR 1977.
2. ICRP Publication 26, *Recommendations of the International Commission on Radiological Protection.* Pergamon Press, Oxford. ICRP. 1977.
3. ICRP Publication 27, *Problems Involved in Developing an Index of Harm.* Pergamon Press, Oxford. ICRP. 1977.
4. ICRP Publication 30, *Limits for Intakes of Radionuclides by Workers.* Pergamon Press, Oxford. ICRP. 1979.
5. Medical Research Council, *Criteria for Controlling Radiation Doses to the Public after Accidental Escape of Radioactive Material.* HMSO, London. 1975.
6. Clarke, R. H. 'Radiological protection criteria for controlling doses to the public in the event of unplanned releases of radioactivity.'
Proceedings of IIASA Workshop on Procedural and Organisational Measures for Accident Management, Vienna. 1980.
7. Health and Safety Executive. *Health and Safety: Manufacturing and Service Industries,* 1977. HMSO, London. 1978.
8. Taylor, F. E. and Webb, G. A. M. *Radiation Exposure of the UK Population.* National Radiological Protection Board Report R-77. HMSO, London. 1978.
9. Statement from the 1978 Stockholm Meeting of the ICRP. *Annals of the ICRP,* 2 (1). Pergamon Press, Oxford. ICRP. 1978.

High Risk Safety Technology
Edited by A. E. Green
© 1982 John Wiley & Sons Ltd

Chapter 3.5

Biological Hazards

D. J. Hurrell

Introduction

The distinguishing feature of biological, as opposed to chemical or physical, hazards is the living nature of the causative agent. The majority of work-related biological hazards are caused by microbial pathogens rather than macroscopic fauna such as large carnivores. The dangers in the latter case are of course more readily apparent except where the danger arises from the role of such macroscopic fauna as a vector in the transmission of microbial infection.

One aspect of living (biological) hazards which has considerable relevance to their management is the lack of a well defined dose-response curve such as is normally found with chemical (toxic) or physical (irradiation) hazards.

With a living organism, the escape of a single unit, which may then multiply in the environment, can lead to the same end as the release of a much larger number. Furthermore the number of organisms required to produce an infection may vary considerably from individual to individual and be dependent on many factors.

The enormous increase in activity in microbiology and associated laboratories, which has occurred since the late 1940s, has been accompanied by a similar increase in the number of laboratory-acquired infections. Accident rates as high as $11.8/10^6$ man-hours (United States Department of Labour, 1958) and infection rates as high as $50/10^6$ man-hours[1] have been reported. In Reference 1 is reported the occurrence of a number of laboratory epidemics, including a *Brucellosis* epidemic caused by leakage of infected material from a centrifuge. Of 94 persons infected, many were not laboratory staff and included a secretary, a plumber, and a visiting salesman. One of the victims died.

Even a brief search of the literature leaves the impression that every infectious agent which has been studied has also at one time or another caused a laboratory-acquired infection. There are instances where the only known source of a disease is by laboratory-acquired infection. The reported death rate from laboratory acquired infection[2] exceeds by tenfold the death rate in other industrial activity.

Recognition of the special hazards resulting from work with pathogens very dangerous to humans has led in the UK to the formation of a specialist advisory group. The Dangerous Pathogens Advisory Group (DPAG), which publishes guidelines for the safe conduct of such work and gives detailed advice in individual cases. A similar body, The Genetic Manipulation Advisory Group

(GMAG), is concerned with the safety of experimental work involving genetic manipulation.

Laboratory Acquired Infections

There are certain preconditions for there to be a risk of infection:

1. *A causative agent* which may be viral, bacterial, or fungal. The hazard presented by any particular aetiological agent will depend on such factors as the type, infectivity, virulence, pathogenicity, and physical form of the organism; also its tolerance of the laboratory environment and its resistance to antibiotics and disinfectants which may be used in an attempt to control it.

2. *A reservoir of the causative agent* which may be an *in vitro* culture or an *in vivo* culture in a deliberately infected laboratory animal, plant, or human volunteer created as part of the work programme and therefore recognized as a potential hazard and subject to careful control.

   The reservoir may also be present inadvertently and outside normal experimental control. With a few notable exceptions the reservoir of diseases that attack man is of human or animal origin. People may provide reservoirs of infection during the incubation period of a disease before the onset of clinical symptoms, during convalescence as either a temporary or chronic carrier, as a contact carrier who will at no stage show clinical manifestation of the disease, or during a period of overt disease, which may be acute or chronic, when the causative organism may be released through the faeces, urine, mouth, nose, ears, eyes, or lesions which may be present. Animals may act as reservoirs in exactly the same way and are frequently contact carriers of human disease as in the case of *Macaca mulatta* and *M. philippensis* acting as carriers of simian B virus. Inanimate reservoirs also exist in water, soil, dust, air, sewage, food, etc. and may include infective species of *Pseudomonas, Proteus, Clostridium* and so on.

3. *Escape from the reservoir* which may be more or less likely depending on the size and nature of the reservoir. Where the reservoir is an animal, infective organisms may escape in many ways and the possible modes of escape may have a definite effect on the method of control required. Mechanical escape by a biting insect withdrawing and transferring infected blood would require particular attention to laboratory infestation to control potential vectors.

   There is a great variety of laboratory procedures which seem to be designed to facilitate the escape of infective organisms from the material under study. Aerosol generation is very common and frequently particles of appropriate size for human infection are generated which potentially contain infective agents. Equipment giving rise to aerosols includes: washing machines, centrifuges, homogenizers, shakers, blenders, tissue grinders, ultrasonic generators, etc. Aerosols are also generated by opening bottles[34] and ampoules, clearing the last drop from a pipette or syringe, sterilizing an inoculating loop in a naked flame instead of a properly guarded device[35], and even by the discharge from the drain of badly equipped or wrongly operated steam sterilizers.

The behaviour of aerosols and the use of containment and ventilation to control them are an important part of prevention of laboratory acquired infection and one aspect of the containment of microorganisms under precise control which is the essence of safe microbiological techniques.

Historically the primary reason for such control was to protect the integrity of the system under investigation from contamination which could affect the validity of the experiment. Once the aetiology of microbial diseases was understood, similar measures were used to provide protection for laboratory workers against infection from cultures or infected animals, although the efficacy of such measures has frequently been over estimated. Even today the importance of control proportionate to the extant level of risk is not always recognised.

4. *Transfer to the potential host* which may be, in part, dependent on the method of escape from the reservoir. Transfer may be direct, as exampled by ingestion as a result of mouth pipetting, or inhalation of an aerosol, or inoculation by broken glassware, or by misuse of a hypodermic needle and syringe, or may be indirect through an animal or insect vector.

5. *Entry into the potential host* which may be by inhalation; by ingestion directly or indirectly by way of food, drink or cigarettes which for this reason should be banned from the laboratory; by inoculation through simple contact with, say, a spillage, or contact with an open wound, or by injection either from a bite from an insect or animal or accidentally as suggested in 4 above. Inspiration of infected aerosols is undoubtedly the most common route of infection but, as with the other routes of entry, can be controlled by proper use of physical containment.

6. *Host susceptibility* which will allow the disease to develop when the infective agent has been transferred to the host. The level of risk will depend on the disabling effect of the disease, the ease of diagnosis, the adequacy of available therapeutic measures and the degree of immunity of the host, whether naturally occurring or by deliberate immunization. The converse approach of debilitating the potentially infective organism, such that there is no naturally occurring host in which it can survive, is the method known as biological containment, frequently employed in organisms subjected to genetic manipulation.

The above outline provides the factors which should be considered in all aspects of prevention of laboratory-acquired infection. The risk of infection arises from any procedure which releases infective organisms into the environment, or affords access for such organisms to the human body. Many of the risks are obvious as, for example, when accidentally self-inflicted wounds occur during a post-mortem examination procedure on an infected animal. The risks from other sources are less obvious. The transplacental route, by which cytomegalo virus infects the developing embryo during the first trimester of pregnancy, requires 'in-depth' knowledge of the aetiology of the disease. Within the limits of practicability, the worst should always be assumed of any potentially infective agent.

Control of Airborne Contamination

The most widespread and least obvious mechanism for laboratory-acquired infection is by airborne dispersal of the infective agent. Adequate control of airborne contamination requires some knowledge of the properties of aerosols, and a reasoned application of that knowledge.

Properties of Aerosols

An aerosol may be simply defined as particles suspended in a body of air; such particles normally occur in sizes up to $20\,\mu m$ diameter. Many studies have been made on the physical properties of airborne particulates and are the subject of specialist texts[3]-[5].

Particles suspended in air are subject to a number of forces influencing their behaviour. A particle under gravitational acceleration will start to fall. The rate of fall will increase until the opposing frictional force is in equilibrium and maximum, or terminal, velocity is attained. Maximum velocity is given by Stokes's Law, but since changes in viscosity of air in normally encountered temperature and pressure ranges are negligible, this may be simplified to

$$V_{max} = 3.2 \times 10^5 \rho d^2 \, cm\, s^{-1} \qquad (3.5\text{-}1)$$

where V_{max} = terminal velocity, ρ = particle density $(g\,cm^{-3})$, d = particle diameter (cm). However, for particle sizes below $10\,\mu m$, the particle diameter is approaching the value of the mean free path of the gaseous molecules in the atmosphere $(6.5 \times 10^{-6} \, cm)$, with a consequent reduction in frictional resistance and thus a deviation from Stokes's Law occurs. This deviation is allowed for by applying Cunningham's correction, i.e. adding $0.08\,\mu m$ to the diameter in order to calculate V_{max} (or subtracting when calculating the size from the observed mean rate of fall).

The observed density range of airborne microorganisms is $0.9-1.3\,g\,cm^{-3}$ and the mean value of $1.1\,g\,cm^{-3}$ is sufficiently good for calculations. Figure 3.5-1 shows the relationship between V_{max} and particle size.

Airborne particles are also in constant collision with gaseous molecules in the atmosphere and are thus subject to Brownian motion. Although the motion of a single particle cannot be predicted, the average behaviour of a number of particles can be considered. Einstein derived an equation for the root mean square value of

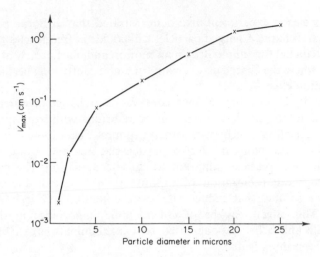

Figure 3.5-1 Terminal velocity (V_{max}) of particles of density
$1.1\,\mathrm{g\,cm^{-3}}$ in still air at 20°C

the displacement of particles from the edge of a cloud of particles ($\overline{X}$), which may be simplified to:

$$\overline{X} = 5.0 \times 10^{-6} \sqrt{(tr^{-1})}\,\mathrm{cm\,s^{-1}} \qquad (3.5\text{-}2)$$

where t = time (s) r = radius (cm)

from which it may be shown that diffusion by Brownian motion is at least an order of magnitude less than the rate of fall due to gravity, and for particles of 10–20 μm diameter may be four orders of magnitude less. Einstein's equation may also be used to show that particles less than 0.1 μm, or greater than 1 μm, diameter are held in the alveoli of the human lung to a much greater extent than particles in the size range 0.1–1.0 μm. The calculated results were experimentally confirmed by Beekmans[6].

Particles that come into close contact may aggregate due to the forces of molecular cohesion. Collisions due to Brownian motion are a function of the square of concentration such that

$$\frac{dn}{dt} = -Kn^2 \text{ number s}^{-1} \qquad (3.5\text{-}3)$$

where K = coagulation constant

This expression is only valid if the particle size remains constant, which it must be apparent is not the case. Coagulation is a second order process and it may be shown that the process is inconsequential for particulate concentrations below 10^6 cm^{-3}, and as a continuous process influencing physical loss of particles can be neglected in most practical situations.

Millikan's classic experiment also demonstrated that airborne particles can have an electric charge. Charged particles can collide at frequencies greater than would be predicted for simple Brownian motion and, particularly at humidities below 30 % where the charge may remain for significantly long periods, increase the coagulation effect.

The humidity effect is caused by a continuous exchange of water molecules adsorbed on to the particle surface in equilibrium with atmospheric water vapour. Similar effects occur with other vapours. The vapour surrounding a particle may cause a change in effective particle size, and may promote or inhibit contact with other particles sufficient to cause coagulation. The particle may contain a higher concentration of vapours than expected, due to selective dissolving of solutes that reduce the surface tension (Gibbs's Adsorption Equation). This effect may be of considerable importance in the case of formaldehyde and other vapours used to decontaminate areas, but has not apparently been investigated.

Movement of particles is also influenced by radiation and temperature. Where a temperature gradient exists particles tend to migrate towards the colder region, but even for a thermal gradient of 20 °C such migration would be neglible for particles larger than 1 μm diameter. Radiation effects are less noticeable and act by producing differential thermal effects.

Consideration has so far only been given to aerosols in still air, whereas in reality almost all air is in motion. The suspended particles forming the aerosol will gain and lose kinetic energy as a result of air movement. Furthermore their inertial properties will differ markedly from the disperse medium so that a change in direction of the air movement will not be followed by the particles. The particles will move across the air stream. This is not only important for design considerations of safety cabinets and air handling plants, but is also of considerable relevance in the design of samplers used to estimate particulate contamination and the construction of filters. In this context the human upper respiratory tract may be considered as a filter, which depends for its efficiency not only on the fine hairs and moist conditions, but also on the impaction effect due to the inertial characteristics of suspended particles as inspired air is forced through rapid changes in direction.

Persistence of Airborne Contamination

The physical characteristics of aerosols outlined above obviously have considerable influence on the rate at which infective agents discharged into the laboratory atmosphere, as aerosols will settle out by coagulation, gravity, and impaction. Bacterial and fungal organisms in the aerosol condition are in a dynamic state and are liable to diminution of the population in response to the stress imposed. The ability of a microorganism to resist such stresses varies from species to species, with the culture conditions preceding release as an aerosol, and with environmental conditions prevailing within the aerosol. Bacterial spores are generally found to survive better than vegetative cells, but there are exceptions such as *Micrococcus radiodurans*, which maintains viability as well as *Bacillus subtilis* var. *niger* spores. The apparent survival is influenced by the adequacy of sampling techniques and the difficulties experienced in recovering microbial aerosols.

Assay of Living Microbial Aerosols

There is no absolute method for determining the number of organisms in a given sample of air. Comparative evaluations using different sampling methods and estimation of the fractional recovery of deliberately generated microbial aerosols can give some measure of efficiency. In general, the recovery level is not sufficiently good to allow monitoring of the atmosphere as a reliable guide of whether or not an infection hazard exists.

Various physical counting methods[32] have been used to estimate the particulate contamination of the atmosphere, but such methods give no indication of the presence of viable organisms.

Sampling methods should be isokinetic, but this represents an unachievable ideal and thus all samples are, to a greater or lesser extent, biased. Further variation is caused by the choice of media employed to recover organisms, which have been collected either by impaction on to an agar plate, or by filtration with a 0.2 μm filter.

Direct impaction on solid medium constitutes a major sudden environmental change and this, together with the stresses of the preceding airborne condition, gives rise to the need for the media to contain factors which maximize the survival and multiplication of injured cells. Two aspects are of importance: first, the media may require additional growth factors to aid recovery of the injured cells; and second, and often overlooked, media which afford optimal growth for non-traumatized cells may yet be inhibitory to the recovery of damaged cells. Great care is thus needed in selecting an appropriate recovery method which will be specific for one or at most a limited range of species.

No method exists which can provide a total viable count of all the microorganisms present in a given atmosphere. For some organisms there is no known *in vitro* culture method, e.g. *Treponema pallidum* or *Mycobacterium leprae*, and the use of *in vivo* culture in laboratory animals is necessary.

'Sentinel' animals have been used as a very sensitive sampling method. Riley[7] directed the airflow from hospital wards through a chamber containing guinea pigs and found one infective particle causing tuberculosis in 340 cubic metres of air from the ward. Considered against a background of 1–350 viable colony-forming units per cubic metre of air normally found throughout the hospital, the sensitivity of the guinea pig as a detector of *Mycobacterium tuberculosis* among

10^4 viable particles is many orders of magnitude better than any available mechnical sampler[36].

Isolation and Ventilation

In the absence of any reliable method of obtaining an instant, accurate indication of the presence and/or extent of environmental microbial contamination, the major effort has been directed towards providing physical barriers to the dissemination of contamination which is assumed to be ubiquitous. The performance of physical barriers can be adequately, and rapidly if not continuously, monitored by surveillance of critical physical parameters.

Before deciding on the method, or level, of containment which would be appropriate, it is necessary to identify the nature and degree of hazard presented by microorganisms under study or potentially present within the experimental system. In the UK microorganisms are classified according to the hazards they offer[8] and a similar system exists in the United States (US Public Health Service[9]). The World Health Organization (WHO)[10] has also published guidelines which differentiate risk categories on the basis of hazard to the individual worker and hazard to the community at large.

The level of containment needs to be appropriate to the potential hazard involved if it is to be neither inadequate, and thus enhancing the risk by providing a feeling of false security, nor a serious impediment to essential work (e.g. in a diagnostic laboratory), nor uneconomic.

There is a large measure of agreement concerning the containment facilities required within each of the three categories adopted in the UK. Category C work in which there is no potential hazard, provided that high standards of microbiological technique and safety are observed, may be undertaken in the basic laboratory. Good microbiological practice can only be achieved if all personnel receive adequate instruction, which must include aseptic handling techniques and the biology of the experimental organism to give an appreciation of the potential hazard. Such training should be repeatedly reinforced as part of a continuing programme, rather than as a 'once and for all' event at the start of employment. A well defined emergency procedure to deal with accidental contamination of personnel and the environment should be established. Where work is with a known pathogen for which there is an effective vaccine, workers should be vaccinated and, in appropriate circumstances, serological monitoring of workers should be undertaken routinely. Detailed guidance on procedures to be adopted in laboratory work to minimize contamination risks have been published by a number of sources[9][11]-[13].

For work involving genetic manipulation, a different set of guidelines exist. The Ashby working party reporting in December 1974 said that a Code of Practice was required, and to formulate such a code the Williams working party was established. The Williams report[14] included proposals for setting up a genetic manipulation advisory group (GMAG), with terms of reference including 'a continuing assessment of risks and precautions (and in particular of any new methods of physical or biological containment) and of any newly developed techniques for genetic manipulation and to advise on appropriate action'.

The Williams report also outlined the requirements for four levels of containment laboratory, related to the level of risk involved with particular procedures, to augment the biological containment, which must be the primary objective in each case. Workers intending to undertake genetic manipulation work need to submit proposals to GMAG for categorization and for work in containment categories III and IV, an initial site visit by members of GMAG, together with representatives of the Health and Safety Executive, to assess the facilities will be made. Although the group may appear to have the role of a policing authority, many of the members are eminent and experienced workers in this field of enquiry and are an excellent source of advice.

This general scheme of containment methods is paralleled by the scheme used for dangerous pathogens and by other classification systems.

Laboratories for moderate physical containment (DPAG category B1; GMAG category III; US Public Health Service class III; and WHO Risk Group III) are basically similar, whatever the particular application. In all cases the building must be of a good standard of construction, separated from public areas and not adjacent to a known fire hazard or sited in an area liable to flooding. The laboratory must be animal and insect proof, sealable to permit fumigation and with all surfaces (walls, floors, ceilings, benches) smooth, durable, and easily cleanable. Entry to the laboratory should be through a changing room, with exit and entrance doors interlocked so that only one can be open at any given time, divided into clean (on the 'street' side) and restricted (on the laboratory side) zones separated by a boot barrier. The restricted zone of the changing room should be equipped to allow proper handwashing and for removing and discarding protective apparel for personnel leaving the laboratory.

The laboratory and changing room, which functions as an airlock, are ventilated by a plenum and exhaust air system. The exhaust air should be ducted from the laboratory to the outside of the building and discharged to atmosphere through a HEPA filter. The laboratory and the exhaust ducting as far as the distal side of the filter must be maintained under negative pressure (at least 8.0 mm of water) to ensure that no egress of potentially contaminated air can occur. Great care, skill, and experience are needed in the design of ventilation systems if the flow direction is not to be compromised by, for example, eddy currents created in the reverse direction when a door is opened. Ventilation of the laboratory in this manner protects the external environment by ensuring that the flow of air into the laboratory is at a greater rate than the possible outward diffusion of airborne particles, and by filtering the discharged air to remove particulate contamination. Despite the remarkable efficiency of HEPA filters, it is still possible for some airborne particulates to escape. In most circumstances this does not present a hazard, because the low number of particles emitted in a very large volume of air provides a dilution factor so that the likelihood of any person inhaling a minimum infective population or more of the infective agent is so remote as to be negligible. (Where particularly hazardous agents are involved, heat treatment of the discharged air may be a necessary decontamination procedure.)

Category A or Class IV containment facilities require similar standards of ventilation but the air lock/changing room must incorporate a shower facility between clean and restricted zones.

In both cases, the performance of the air conditioning system should be carefully established by detailed commissioning tests and then monitored routinely to confirm continuing compliance with the design performance.

The ventilation of the laboratory in this way provides little protection for staff working in the unit, other than by the high dilution factor achieved by the through flow of air and the minimum space requirements (24 m^3/worker) which are recommended, and additional protection is needed.

This may be provided by means of a microbiological safety cabinet, used to isolate hazardous procedures and equipment. British Standard 5726[15] specifies three classes of cabinet to meet differing requirements and levels of risk. Such cabinets should not be confused with laminar flow work stations built to BS 5295[16]. Whichever class of microbiological safety cabinet is chosen as appropriate, it is important that it is correctly installed in a suitable position, where there will be no disturbance of the airflow to compromise the working of the cabinet (Class I and II) and where the filtered exhaust air can be conveniently ducted to a suitable point outside the building for discharge. After installation and before use the cabinet, and any associated duct work, should be subjected to commissioning tests to establish that the required flow rates and operator protection factors are achieved and by challenging with an aerosol of sodium chloride, or dioctyl phthalate smoke, that the filter assembly and duct work are patent. The HEPA filters should be tested in accordance with British Standard 3928[17] and demonstrate 99.997% arrestance. Full commissioning tests should be repeated whenever any major maintenance work is done, e.g. replacement of filter, fan motor, etc. As with other essential equipment, a log book should be kept to record, results of weekly tests. The tests should include monitoring of flow rates, face velocities, and differential pressures as appropriate. The equipment should be subject to a 'Permit to Work' scheme, under the control of the safety officer.

The choice of cabinet (Figure 3.5-2) will be influenced by a number of considerations. The Class III cabinet affords the highest level of protection to the operator and also provides the highest protection to the integrity of the system under study. Where this level of protection is not essential Class III cabinets suffer the disadvantage of being less convenient to use. The Class II cabinet is a partially recirculating type which is designed to afford protection to both the operator and the product (Figure 3.5-2(d)). Considerable controversy exists about the

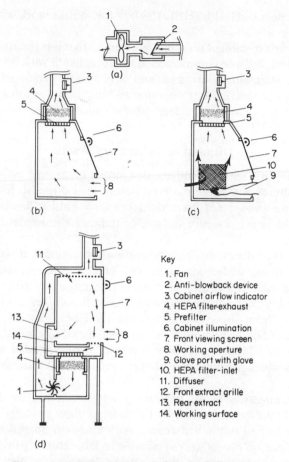

Key

1. Fan
2. Anti-blowback device
3. Cabinet airflow indicator
4. HEPA filter-exhaust
5. Prefilter
6. Cabinet illumination
7. Front viewing screen
8. Working aperture
9. Glove port with glove
10. HEPA filter-inlet
11. Diffuser
12. Front extract grille
13. Rear extract
14. Working surface

Figure 3.5-2 Three classes of microbiological safety cabinet with general arrangement of airflow for each class. (a) Arrangement for discharge to atmosphere. (b) Class I (exhaust protective cabinet). (c) Class III (glove box). (d) Class II (freestanding model)

adequacy of the protection afforded by this system, even when type tests and commissioning tests have demonstrated a particular cabinet as providing the 10^5 protection factor demanded by BS 5726. (That is the operator should be exposed to 100 000 times less airborne contamination when using the cabinet than he would have been while working at the open bench with the same level of aerosol generation.)

Class I cabinets are the standard exhaust protective cabinets with which most microbiologists will have had a long acquaintance and have been recommended by the Department of Health (HEL 1980)[18] for routine work with Category B organisms.

Whichever class of cabinet is used, it is essential that it is the right type for the job, correctly installed and commissioned for compliance with BS 5726, properly monitored and maintained during use and only used by personnel trained to use such equipment properly. Poor operator technique can seriously compromise even the finest microbiological safety cabinet's protection.

Disposal of Laboratory Waste

At the end of work with potentially dangerous organisms, it is necessary to sterilize all cultures and waste material before final disposal. Even potentially infected material should not leave the laboratory until it has been effectively decontaminated, or is in a secure and safe container for transport to an autoclave or incinerator.

In high risk procedures requiring maximum containment, it is mandatory for there to be a double ended autoclave with entry in the restricted area of the laboratory and exist in a clean area outside the laboratory, so that all material leaving the facility can be decontaminated; this includes all effluent. Even effluent including water from showers and sinks in the laboratory suite must not be released to the public sewer until it has been subjected to a suitable sterilization process. For material that cannot be autoclaved, a double ended 'dunk-tank' has to be provided, which permits removal of material through a suitable disinfectant solution.

The decontamination of all laboratory waste, whether by disinfection or sterilization, should be supervised by the Safety Officer or a designated deputy and should be carried out in accordance with a written operational procedure. Detailed guidance on procedures applicable to laboratories providing less than total containment are given in the Code of Practice for the Prevention of Infection in clinical laboratories[13]. The advice is equally applicable to research and industrial laboratories.

Disinfection and Sterilization

Sterilization may be defined as freeing from all living micro-organisms. Sterility is then, necessarily, an absolute term and comparison of the effectiveness of various sterlizing processes is therefore limited to consideration of the degree of certainty with which it may be known that the product is sterile[18]. Major differences in activities and effects of the available microbiocidal processes are recognized in respect of bacterial spores when compared with vegetative forms. Processes which are only effective against the latter are regarded as disinfection. Disinfection, although etymologically suggestive of freedom from infection, can only achieve sterilization if there is a fortuitous absence of resistant organisms.

The choice of process will be influenced by such factors as the type, extent, and frequency of microbial contamination; the likely risks associated with such contamination; the nature of the contaminated material and whether it is needed for re-use or can be destroyed.

Disinfection

In choosing a disinfection process, firm preference should be given to physical processes such as hot water, or low temperature steam. Chemical disinfectants should be regarded as a last resort. If disinfectants are to be effective they require great skill and care in their use[19]. The efficacy of chemical disinfectants is dependent on a number of factors:

1. Nature of contaminating organism — see Table 3.5-1.
2. Inactivation of disinfectant by other materials present – see Table 3.5-1.
3. Bacterial resistance (which may be induced by misuse of disinfectants as a ubiquitous 'cure all' cleaning agent).
4. Direct contact between the disinfectant and the contaminating organism – and hence the importance of cleaning, wherever possible, prior to the use of disinfectant to prevent shielding by soiling.
5. Activity of the disinfectant, which will vary with contact time, concentration, temperature, volume, and pH.

The choice of disinfectant needs to be made very carefully and the activity should be independently tested against the organisms of concern by, for example, the Kelsey–Sykes method[20], noting particularly that many non-phenolic disinfectants are bacteriostatic and can give encouraging results until a suitable inactivator is found

Most of the factors affecting the efficacy of chemical disinfectants cannot conveniently be monitored during use. Possibly the most useful monitoring system is that described by Kelsey and Maurer[21] who suggest in-use testing to police the use of disinfectant, an approach also supported by Ayliffe[22]. Experience in hospitals suggests that nearly all varieties of disinfectant can become contaminated with resistant organisms and have been incriminated in outbreaks of infection[23][24].

Fumigation

It is frequently found necessary to decontaminate microbiological safety cabinets, working areas, and entire laboratory suites, either as a precautionary measure following a minor incident, or prior to allowing access to the laboratory for maintenance staff, etc. The method almost universally adopted uses formaldehyde gas generated from a 36–40 % aqueous solution (formalin). Detailed instructions for the use of formaldehyde are given in a publication of the Public

Table 3.5-1 Properties of commonly used disinfectants

Type	Example	Trade name	Inactivated by	Spores	G+ve	Acid fast	G−ve
Alcohol	Methanol	I.M.S.	h.o.p.	0	5	5	5
Aldehydes	Formaldehyde	Formalin	h.o.p.	3–4	5	5	5
	Glutaraldehyde	Cidex	o.p.	4	5	5	5
Halogens	Hypochlorite	Milton	h.O.p.	5	5	5	5
	Iodophore	Betadine	h.O.p.	3	5	5	5
Phenolic	Clear Soluble	Stericol	h.o.p.	1	5	3	5
QACs	Cetrimide	Cetavlon	H.O.P.	0	5	0	4
QAC+ diguanide	Cetrimide + Chlorhexidine	Savlon	H.O.P.	0	5	0	4
Diguanide	Chlorhexidine	Hibitane	H.O.P.	0	5	0	4
Pine fluid			H.O.?	0	1	0	1
Heat	Low temperature steam at 73°C	----	-------	5	5	5	5

Key

Inactivated				Antibacterial activity
Seriously	Slightly	by		under optimum conditions
H	h	Hard water		0 – none
0	o	Organic material		1
P	P	Synthetic materials		2 – slight
				3
				4 – fair
				5 – good

No disinfectant can be depended upon to kill all bacteria present. (A 4 decade reduction would be regarded as good.)

Health Laboratory Service[25]. Users of the method should be aware of current Threshold Limit Values published by the Health and Safety Executive and take whatever measures are necessary to ensure that these can be met. The limitation of the method should also be known and the shortcomings of formaldehyde disinfection were comprehensively researched by Nordgren[26] whose work is still a valuable reference.

Sterilization

The established method of choice is sterilization by steam. In the UK the recommended combinations of time and temperature were published by a Medical Research Council Working Party[27] – the lowest temperature recommended was 121°C for 15 minutes.

Efficacious sterilization by steam requires the direct contact of pure dry saturated steam (i.e. at phase boundary conditions) at the required temperature for the required time in the absence of air. Problems arising in the use of steam sterilizers can be divided into two categories, each related to the nature of the load to be sterilized, i.e. dry loads and fluids.

Sterlization of dry loads

Dry loads including linen (such as laboratory clothing), wrapped goods, discard culture plates, swabs, petri dishes, empty used glassware, etc. present the problem of air removal. Unless all the air is removed from the load, steam penetration will be inhibited and sterilization conditions will not be achieved. The method of packaging used for the load is of prime importance, as is the type of machine.

Sterilizers complying with British Standard 3970 Part I 1969 (porous load machines)[27] have an operating cycle which incorporates a vacuum assisted air removal stage prior to steam admission for sterilization, and incorporate an air detector to monitor the adequacy of the air removal process. Most laboratory autoclaves do not have an automatic air removal stage, but rely on gravitational displacement of air by the incoming steam through a thermodynamic trap with a near-to-steam element. The shortcomings of the latter system are reviewed by Perkins[28] in his comprehensive text on sterilization methods. Some of these difficulties can be resolved by careful attention to packaging. Deep buckets with solid bottoms, necessary to prevent spillage in transit, filled with small specimen bottles may need to be autoclaved for many hours in a laboratory sterilizer of the downward displacement pattern before steam penetration and total air removal is achieved.

A further problem arises in that the discharge from the chamber of both types of machine is greatest at the beginning of the cycle, when sterilizing conditions have not been achieved within the chamber. There may then be considerable danger of escape and widespread dispersal of infective particles from the load in the chamber.

The best solution currently available seems to be a machine incorporating a sealed drain to retain condensate in the chamber until the end of the sterilizing period, with forced air removal, either by vacuum pump or steam injector, through a bacteria retentive filter situated in the top of the chamber.

Sterilization of fluids

Fluids in sealed, and unsealed, bottles usually present no problems of air removal from the load. They do, however, present a problem related to their relatively large thermal mass. The temperature attained in a fluid filled bottle will be lower than the temperature of the chamber for some time, until an equilibrium is attained. For a 1 litre bottle this may take between 20 and 60 minutes. Unless, during commissioning, careful measurement is made, using thermocouples and a potentiometric recorder, of the lag time for the various shapes and sizes of container to be processed it is possible to process material through a 'sterilizing' cycle without that material ever achieving a temperature of even 100 °C. A similar problem exists at the end of the cycle. When the chamber is vented to atmosphere, steam pressure is lost rapidly and the temperature indicated by the chamber temperature gauge quickly drops to 50 °C or 60 °C. It then appears that it may be safe to open the sterilizer. However the contents of a load of 1 litre bottles in the chamber, that had just been processed at 121 °C, would still be considerably above 100 °C and could in a medium size sterilizer take some six hours to achieve a safe temperature (i.e. below 80 °C). Ideally the sterilizer should be equipped with a bottle simulator, i.e. a device with thermal characteristics similar to the largest bottle to be processed, which can provide a reference point for an interlock, preventing the door being opened until a safe temperature is achieved. In all cases, it is essential that the sterilizer is properly commissioned for each type of load for which it may be used.

Comprehensive advice on the choice, installation, commissioning, and routine testing and maintenance of steam sterilizers has been published by the Department of Health[28]; the standards suggested are minima and yet many current installations fall far below these standards.

Other methods of sterilization

Where steam sterlization is not appropriate, other physical methods are the obvious choice. Dry heat is well established, well documented, and trouble free, provided sufficient time is allowed for the load to attain sterilizing temperature.

Other methods such as gamma and electron beam irradiation are unlikely to find application in the laboratory.

The efficacy, and thus activity, of irradiation with ultraviolet light is controversial. Although there is an undoubted microbiocidal effect, the practical difficulties and uncertainties associated with the method are, in the author's view, sufficient to discount the process except as a vehicle of delusion.

The remaining methods are all chemical and suffer many of the disadvantages of disinfectants. The two methods most commonly used are ethylene oxide gas[29] and low temperature steam with formaldehyde.[30]

Incineration

Where total destruction of the material to be decontaminated is acceptable, incineration may be the method of choice. It is certainly the most nearly absolute of all the sterilization methods, if carried out properly. The importance of a correctly designed, installed, and operated incinerator cannot be overemphasized. An incinerator with a poorly designed after burner can, for example, distribute even moderately resistant organisms from the incinerator load into the atmosphere.

Overall Control

Proper control can only be achieved through carefully planned and considered decisions at every stage of the operation. Many of the problems of containment can be minimized by correct design of the facility[31] and by informed choice of equipment, not only for its suitability for its intended purpose, but also for its ease of cleaning and decontamination, both routinely and prior to maintenance. All equipment and facilities should be properly commissioned after installation and regularly monitored throughout their life for compliance with the performance parameters specified[33].

All critical equipment should have a log book conscientiously maintained with details of results of routine tests, maintenance works, defects, etc. Similarly, every critical procedure should be the subject of a written protocol.

Quite reasonably the overall responsibility for safety must rest with the head of the particular facility and by delegation to the Biological Safety Officer, who should be responsible for the implementation, coordination, and supervision of safety management in the broadest sense. However, safety, in the final analysis, can only be the responsibility of the individual, and this responsibility is best fostered by training. It is difficult to overemphasize the importance of continuing training at all levels as the best safeguard against an accident.

The views expressed in this chapter are the personal opinions of the author and do not commit the Department of Health and Social Security in any way.

Questions 3.5

1. What are the essential differences between a biological hazard due to living microorganisms and that due to a chemical (toxic) or physical (irradiation) cause?
2. Discuss the risk of infection arising from infections which may be acquired in laboratories. Trace possible routes involving a causative agent and a reservoir for the agent and the transfer to a potential host. How does the death rate from laboratory-acquired infection compare with that in other industrial activity?
3. Select from the considerations in Question 2 the part played by airborne contamination and state how this may be controlled.
4. Consider the persistence of airborne contamination and the extent to which isolation and ventilation enter into the problem. What type of laboratory arrangements should be provided for moderate physical containment?
5. Discuss the different classes of microbiological safety cabinets. What consider-

ations would enter into choosing the appropriate safety cabinet to afford protection to both the operator and the product?

6. What are the different methods with their advantages and disadvantages for the disposal of laboratory waste involving potentially dangerous organisms?

7. Discuss the basic methods of sterilization. Consider the problems involved with dry loads and fluids and comment on the properties of commonly used disinfectants.

8. What are the basic principles which should be applied from a biological safety point of view to achieve proper overall control in a laboratory or similar area where potentially dangerous organisms are involved?

References 3.5

1. Phillips, G. B. 'Safety in the chemical laboratory. XIII Microbiological hazards in the laboratory'. *Journal of Chemical Education*, **42**(1), A43–A44 and A46–A48, **42**(2), A117–A120, A122, A124, A126, A128, A130. 1965.
2. Sulkin, S. E. and Pike, R. M. 'Survey of laboratory acquired infections.' *American Journal of Public Health*, **41**, 769–781. 1951.
3. Davies, C. N. (Ed.) *Aerosol Science*. Academic Press, London and New York. 1966.
4. Fuchs, N. A. *The Mechanics of Aerosols*. Pergammon Press, Oxford. 1964.
5. White, P. A. F. and Smith, S. E. (Eds) *High Efficiency Air Filtration*. Butterworth and Co., London. 1964.
6. Beekmans, J. M. 'Deposition of aerosols in the respiratory tract.' *Canadian Journal of Physiology and Pharmacology*, **43**, 157–172. 1965.
7. Riley, R. L., Mills, C. C., O'Grady, E., Sultan, L. U., Wittstadt, F., and Shivpuri, D. N. 'Infectiousness of air from a tuberculous ward.' *American Review Respiratory Diseases*, **85**, 511–525. 1962.
8. DHSS, *Control of Laboratory Use of Pathogens very Dangerous to Humans*. HMSO, London. 1976.
9. US Department of Health, Education and Welfare. 1974(a) September *Laboratory Safety at the Centre for Disease Control*. Publication No. CDC 75-8118. 1974(b) October *Safety Standards for Research Involving Oncogenic Viruses*. Publication No. NIH 75-790.
10. World Health Organization. *Weekly Epidemiological Record*, **54**, 337–342. 1979.
11. Collins, C. H., Hartley, E. G., and Pilsworth, R. Public Health Laboratory Service Monograph Series No. 6. *The Prevention of Laboratory Acquired Infection*. HMSO, London. (ISBN 0-11-880206-2). 1974.
12. WHO. WHO Regional Publication European Series No. 4. *Hospital-Acquired Infections: guidelines to laboratory methods*. World Health Organization Regional Office for Europe, Copenhagen. (ISBN 92-9020-104-5). 1978.
13. HMSO *Code of Practice for the Prevention of Infection in Clinical Laboratories and Post Mortem Rooms*. HMSO, London. 1978.
14. HMSO. Cmnd 6600. *Report of the Working Party on the Practice of Genetic Manipulation*. 1976.
15. British Standard 5726. *Specification for Microbiological Safety Cabinets*. British Standards Institution, 2 Park Street, London W1H 2BS. 1979.
16 British Standard 5295. *Environmental Cleanliness in Enclosed Spaces* British Standards Institution, 2 Park St, London W1H 2BS. 1976.

17. British Standard 3928. *Method for Sodium Flame Test for Air Filters.* British Standards Institution, 2 Park St, London W1H 2BS. 1969.
18. Kelsey, J. C. 'The myth of surgical sterility'. *Lancet*, 16 Dec 1972. 1301–1303. 1972.
19. Maurer, Isobel M. *Hospital Hygiene.* Edward Arnold Ltd, London. (ISBN 0-7131-4228-6). 1974.
20. Kelsey, J. C. and Sykes, G. 'A new test for the assessment of disinfectants with particular reference to their use in hospitals.' *Pharmaceutical Journal*, **202,** 606. 1969.
21. Kelsey, J. C. and Maurer, Isobel M. An in-use test for hospital disinfectants.' *Monthly bulletin of the Ministry of Health and the Public Health Laboratory Service*, **25,** 180. 1966.
22. Ayliffe, G. A. J. and Prince Jeab. 'In-use testing of disinfectants in hospitals.' *Journal of Clinical Pathology*, **25,** 586. 1972.
23. Bassett, D. C. J. 'Causes and prevention of sepsis due to gram negative bacteria: common source outbreaks.' *Proceedings of the Royal Society of Medicine*, **64,** 980. 1971.
24. Burden, D. W. and Whitby, J. L. 'Contamination of hospital disinfectants with *Pseudomonas* species.' *British Medical Journal*, **2,** 153. 1967.
25. Public Health Laboratory Service. (Committee on Formaldehyde Disinfection) 'The practical aspects of formaldehyde fumigation.' *Monthly Bulletin of the Ministry of Health and the Public Health Laboratory Service.* 1958.
26. Nordgren, G. 'Investigations on the sterilisation efficiency of gaseous formaldehyde.' *Acta Path. Microbiol., Scand., Suppl.* **40.** 1939.
27. Medical Research Council, 'Report' by working party on pressure steam sterilisers.' *Lancet*, **1,** 425. 1959.
28. Perkins, J. J. *Principles and Methods of Sterilisation in Health Sciences.* Charles C Thomas, Illinois. (ISBN 0-398-01478-7). 1973.
29. Kelsey, J. 'Use of gaseous antimicrobial agents with special reference to ethylene oxide.' *Journal of Applied Bacteriology*, **30**(1), 92–100. 1967.
30. Hurrell, D. J. 'Low temperature steam disinfection and low temperature steam and formaldehyde sterilisation.' *Sterile World*, **2**(4), 13–18. 1980.
31. Runkle, R. S. and Phillips, G. B. *Microbial contamination control facilities.* Van Nostrand Reinhold Co., New York. 1969.
32. Green, H. L. and Lane, R. W. *Particulate Clouds: Dusts, Smokes and Mists.* (2nd edition) E. and F. N. Spon Ltd, London. 1964.
33. Reitman, M. and Wedum, A. G. Microbiological Safety. *Public Health Reports*, **71,** 659–665. 1956.
34. Tomlinson, A. J. H. 'Infected air-borne particles liberated on opening screw capped bottles.' *British Medical Journal*, **II** (5035), 15–17, 1957.
35. Johansson, K. R. and Ferris, D. H. 'Photography of airborne particles during bacteriological plating operations.' *Journal of Infectious Diseases*, **78,** 238–252. 1946.
36. May, K. R. 'Physical aspects of sampling airborne microbes.' *Airborne Microbes – 17th Symposium of the Society for General Microbiology.* Cambridge University Press, London. 1967.

PART 4

FUTURE AIMS: PHILOSOPHY, LEGISLATION, STANDARDS

High Risk Safety Technology
Edited by A. E. Green
© 1982 John Wiley & Sons Ltd

Chapter 4.1

The General Scene in the UK

F. R. Farmer

Overall Review

Consideration of reliability in complex technology gained considerable momentum in the UK from the 1960s associated with the development of various reactor systems as support for a UK nuclear power programme.

It was recognized somewhat earlier that safety was a major factor in this programme but the link between safety and reliability was perhaps slow to blossom although linked in 1955 at the 1st Geneva Atoms for Peace Conference by such authorities as McCullough, Mills, and Teller.

> In any new field of technology, it is important to investigate, quantitatively if possible, as many features in the field as seem pertinent for human welfare – with all the inherent safeguards that can be put into a reactor, there is still no foolproof system. Any system can be defeated by a great enough fool. The real danger occurs when a false sense of security causes a relaxation of caution. Problems of reliability, adequate control, supervision must be included.

Hence although the need for reliability as an essential component of safety in complex systems was identified in the early development of atomic energy. Little effort was made to follow that route both for lack of data and technical back-up, but also through a reluctance to accept the concept of reliability which also carries with it the possibility of risk or failure. This is not easily accepted when penalty of failure is severe in terms of cost or hurt to people or the environment.

For two decades the growth of complex technology in nuclear plant –and at the same time in the chemical industry, and other potentially hazardous operations – took the form of 'best possible practice' essentially based on past experience even when past experience was limited as noted by McCulloch[1].

> One of the current difficulties in evaluating reactor hazards is this lack of experience with reactor accidents. So far, there have been no reactor accidents leading to serious consequences. For this reason statistical information about reactor accidents, although all favourable, does not suffice to give useful statistical information of the type needed by insurance companies, for example, in evaluating the nature of hazards.

This was again emphasized by Hinton[2].

> All other engineering technologies have advanced not on the basis of their successes but on the basis of their failures. The bridges that have collapsed under load have added more to our knowledge of bridge design than bridges which have been successful; the boilers that have blown up have added more to our knowledge of boiler design than those which have been free from accident; and the turbo alternator rotors which have failed have taught us more of turbo alternator design than those which continued in satisfactory operation. Atomic energy, however, must forego this advantage of progressing on the basis of knowledge gained by failures.

During the decade centring on 1960, the move to reliable and safe equipment and practices relied mainly on historical development – that is the learning process to which Hinton referred – and the continued updating of codes of practice and procedures. Most codes of practice bring together a common experience of good practice – as the name implies – but evolve too slowly to assist newly developing technologies, as nuclear or deep sea drilling. In both of these activities although past experience should be taken into account where relevant, it may not be sufficient. A failure rate may be tolerable when failure does not lead to severe, harmful consequences but not acceptable if the consequences are severe or unusual, as from radiation. There has been little experience of deep sea exploration under conditions in the North Sea so that codes of practice in design and welding of structures as used elsewhere are not immediately applicable.

In the absence of relevant experience, two of the several routes which have been followed are:

1. To choose a process and a design to achieve inherent safety. If this could be done then reliability would be relevant mainly to save money as it is otherwise intended that any unreliability in the equipment would lead to a safe shutdown.

   Hardly any activity can be inherently safe – in an absolute sense; most phenomena that lead to a reversal of a potentially dangerous condition can also exacerbate some other condition.
2. The provision of containment to prevent the release to the atmosphere of any harmful material which might otherwise result from equipment failure or energetic reactions.

This route also had problems, first in regard to the reliability of the containment – which might have many services passing through the boundary, and second in regard to the capability of the containment to be adequate for all plant failure modes.

Hence the difficulties in following routes 1 or 2 led to more careful consideration of the need to achieve reliability in the plant itself. This still leaves open the question of degree of reliability, whether progressive improvement is good enough and will it take account of a changing pattern of industry and society? Manufacturing processes get larger (as does transport) and potentially more dangerous while at the same time society is less willing to accept the risks and standards of the last decade.

The turning point came in atomic energy through the difficulty in transferring design and operational standards and practices from one system to another essentially different one. If a process plant – chemical, nuclear – has appeared and progressively developed for 10 years or more and been shown to be relatively free from fault, it may continue and be updated without being seriously questioned as to whether all the units which make up the whole are adequately reliable. However, the introduction of a new species may lead to the question as to whether all supporting systems are adequate for their function.

This happened when gas cooled reactors were becoming established while alternative systems were being introduced, e.g. steam generating heavy water and fast breeder reactors.

A design organization would propose schemes for feedwater, auxiliary power, control systems, etc., and seek approval. The designs were based on 'best practice' but how could the designer, operator, approval agency decide that they were good enough for a new application. There was little or inadequate data on performance success rate or failure, hence the systems could not be assessed quantitatively notwithstanding McCullough's[1] recommendation to assess quantitatively if possible. Furthermore the designer did not have a defined target –other than an attempt to achieve zero failure rate or as low as reasonably possible or achievable.

Although data collection (as in earlier chapters) for the use in the design of nuclear reactors was set up initially within the Atomic Energy community, it was clear that the relatively short history of a few installations would not give an adequate basis – or would be very limited in application. But 80% or more of a nuclear reactor is external to the reactor and has many components equal to or of the same family (valves, instruments, motors, etc.) as those extensively used in other industries – particularly the chemical industry. It was found that the growth in plant size, in throughput, and in materials in process could lead to major accidents in the chemical industry – as at Flixborough. The industry began to establish its own data base and analytical tools to assess plant reliability.

There has been a tendency to rate highly the data obtained through accidents which might have been caused by equipment failures, but this is changing and the early historic basis is being widened by bringing in data not only from accidents but for plant or equipment failures which have not led to accidents.

The Institution of Chemical Engineers in the UK has for a number of years sponsored an Information Exchange Scheme dealing with accidents, hazards, incidents, and near-miss situations in the process industries. The use of fault trees or failure mode and effects analysis is becoming increasingly common and some tentative steps have been taken to establish target objectives. However, in the UK as in other countries there is considerable hesitation in opening, presenting, and discussing assessments or targets through the fear that these may become sanctified, applied more extensively and vigorously than either might be practical or effective.

Part of the current reluctance to quantification is due to a mistaken belief that all analyses have to be done in full to establish and support a numerical conclusion, i.e. that the chance of an uncontained reaction is 10^{-7} per year. This is not so, much of the value received from performing a reliability analysis is derived from the act of doing the analysis rather than from the results themselves.

A further hesitation arises from fear of publication when reliability analysis is applied to a *risk* situation. Identifying a potential risk often demands a speculative approach from which the first output may be a suspicion based on as yet unproven and tenuous hypotheses. The more imaginative the probing the more likely is this situation to arise. Unfortunately early publication of as yet weakly founded suspicion can cause unnecessary alarm especially if seized upon the sensationalized. It should be recognized that failure to deal rationally with what may be as yet highly speculative suspicion could be a real deterrent to the undertaking of investigations that go beyond what might reasonably be expected.

This is important particularly in view of the trend of the last decade in sharpening the interest in hazard assessment arising from the probing of Royal Commissions, Select Committees, of public inquiries – of organized sections of the public and the greater pressures of institutional organizations of industry and government.

The report of the Robens Committee[3] recommended *inter alia*, the setting up of a National Authority for Safety and Health at work; the bringing together of the many inspectorates. The Health and Safety Commission through its Executive is substantially meeting this and other recommendations and is implementing proposals made by its Advisory Committee on Major Hazards[4]. In its first report it recommended that 'occupiers of certain types of installations, should be required to send to the Health and Safety Executive specified details of their activities'. In relation to Hazard Surveys:

> Much of our thinking has been influenced by the spirit of the Robens report and by the legislative form which has been given to it in the Health and Safety at Work etc. Act. We believe that it is the duty of the company operating a notifiable installation to survey the hazards to which its undertaking gives rise, to identify its own problems and to set up appropriate machinery and procedures for solving them.

Part of this procedure will often be the application of reliability analysis.

The Health and Safety Commission published as a Consultative Document 'Hazardous Installations (Notification and Survey) Regulations 1978' and giving a schedule for notification and outlining proposals for the hazard survey[5].

Today and in the Future

The importance of reliability in complex systems is widely recognized as essential to achieve success and avoid failures to meet targets in performance and safety.

Analytical techniques are available for most applications and they will be improved, simplified, and extended to meet the range of requirements. The database will improve. There is an acute shortage of skilled assessors but of greater concern is still the lack of reliability engineering in most of our educational and training establishments. High reliability will be achieved through the foresight and training of engineers and operators using analytical techniques as naturally as they now use design codes. I quote from Kelly[6]: 'As the level of detail required by the reliability analyst increases so do his demands on the designer's time and experience. At some point it becomes more effective to train the designer in reliability techniques than to train the reliability analyst in design techniques.'

A feature of recent years has been an increase in open government, open debate, request for more publication. It is essential then to achieve a common understanding of any analysis which purports to examine success and failure of processes and equipment — both to continue to upgrade the quality of information and of analytical techniques and to reduce the confusion which inevitably arises when results are presented in such a general form (unlikely, incredible) that the interpretation of optimists and pessimists may differ by many orders of magnitude.

A competent technical analysis plays an important part in open communication.

Questions 4.1

1. In the field of nuclear technology what has been the pattern which has emerged in considering the evaluation of reactor hazards? Can the experience gained be used to indicate methods of approach which may be used for safety and reliability technology in non-nuclear applications?
2. What reactions have been indicated to the use of the quantification of risk and reliability in safety studies? Indicate how the Health and Safety Commission has emerged and its activities in connection with major hazards.
3. Consider an industry of your own choice where there may be a risk of severe consequences. What basic techniques and principles would you expect to apply for hazard analysis and how would you expect these to develop in the future?

References 4.1

1. McCullough, C. R. *et al.* 'The safety of nuclear reactors.' *Proc. Int. Conf. on the Peaceful Uses of Atomic Energy.* Geneva 13, p. 79. 1955.
2. Hinton, C. 'The future for nuclear power.' Axel AX: Son Johnson Lecture, Stockholm 15 March 1957.

3. Robens. *Safety and Health at Work*. Report of the Committee 1970–72. HMSO 10 150 340 7.
4. Advisory Committee on Major Hazards.
   First Report 1976. HMSO ISBN 011 880884 2.
   Second Report 1979. HMSO ISBN 011 883299 9.
5. Hazardous Installations (Notification and Survey) Regulations 1978. HMSO ISBN 011 883205 0.
6. Kelly *et al*. *The Role of Probabilistic Analysis in the GCFR Safety Programme*. GAA 15463. 1979.

High Risk Safety Technology
Edited by A. E. Green
© 1982 John Wiley & Sons Ltd

Chapter 4.2

Europe

W. Vinck

Historical Survey of Safety Philosophy Development in Various Countries

Siting and safety

There was a time when nuclear installations, specifically nuclear power plant's (NPPs) safety was examined mainly as a function of siting. Concern was concentrated on the problem of knowing whether the NPP could be sited in sparsely populated regions or close to towns, or even in the towns themselves. In Europe, sparsely populated sites that are not too far from energy consuming centres and that possess adequate means of cooling are not superabundant.

This implies that the safety of a nuclear installation is determined essentially by the choice of the overall concept and its development, by the quality of the design, of the construction, and of operation.

The result is that in the development of NPP concepts or in the introduction of new concepts, the question – from the point of view of nuclear safety – is not so much 'Where may the NPP be sited?' but rather 'Is the NPP safe enough for the available sites?'

It has become common practice, when discussing NPP safety arguments, to describe how abnormal conditions are provided for and to demonstrate that the various systems of the plant can react satisfactorily to the numerous disturbances – of internal or external origin – that can occur.

Basic principles of nuclear safety

Potential dangers

The dangers from radiation in a normally operating NPP are reduced to permissible levels by protective shields. The design of these shields is influenced by health, economic and operational parameters and technology in this respect are very highly developed. However, in accident conditions, fission products can leave the fuel and the primary circuit and, in this case, additional protection is necessary.

Containing the danger

Experience has shown that no safety problem arises as long as the fission products are retained within the fuel elements.

In general, there are four barriers that constitute the containment:

1. The fuel which retains the fission products, depending on the temperature distribution and the diffusion characteristics.
2. The fuel cladding, most generally formed by an airtight tube of an especially appropriate material.
3. The reactor vessel and the pressurized primary circuit, i.e. the primary boundary.
4. The variable degrees of secondary reactor containment and the associated handling operations (for example, the irradiated fuel handling operation).

The barriers are not necessarily independent of each other, since the failure of one of them can progressively cause the failure of the others. For example, the break of the primary circuit can cause the fuel to overheat and the cladding to rupture, thus releasing the fission products and, in certain cases, can cause the break of the secondary containment (for example, by missile effects or pressure waves resulting from metal/water interactions, steam explosions or gas explosions).

Defence-in-depth

The general range of methods for safety analysis practised to varying degrees and utilizing techniques that are sometimes diversified is currently called 'defence-in-depth'.

This defence-in-depth has three levels:

First level   To ensure that design includes the maximum of safety in normal operation and the maximum of precautions in respect of the failure of equipment (for example, by the application of the 'fail-safe' principle). It is thus important for failures encountered during the construction, the testing, and the operation of the equipment to be recorded, analysed, and communicated to the constructors and operators of other power stations, so that the same type of failure can be corrected or avoided to the extent possible in the future.

Second level   To consider that accident conditions will take place despite the quality of the design, the construction, and the operation; to implement safety systems with the aim of protecting the workers and the public; and to avoid or decrease detrimental effects if these accidents take place.

Third level   To apply additional safety systems based on an evaluation of the effects of postulated accidents, in the event of certain protective systems being supposed to have failed or become unavailable simultaneously with the accident against which they were supposed to provide protection.

This third level constitutes an additional safety margin, confronting the concept with serious accident conditions.

The analysis of these credible accidental events leads to sequences of accidents which will be selected as a basis for the concept and which are generally called design basis accidents (DBA's).

As an example with regard to light-water reactors, accidents involving the loss of primary coolant (LOCA) have received the most attention in this respect, but other accidents can also be mentioned, for example, accidents of external origin such as seismic events, missile effects, explosions, floods; or of internal origin such as failure of the primary pumps or of the steam generators, overpressure and failure of the reactor shutdown system, over-power transients.

The current practice in respect of light-water reactors has been to consider (see page 597) among the design basis accidents, that (or those) which is (are) determining for the effectiveness analysis of the physical containment also as a reference accident involved in the siting analysis and in the intervention plan (emergency plan).

The same principle of 'defence-in-depth' is also used – and has also developed with time – for the analysis of each type of potential accident condition. The hypotheses considered are generally 'realistic', 'conservative' or 'pessimistic'. Calculation models are represented by 'best estimate models' on the one hand, and by 'evaluation models' on the other. The analyses for each accident are being refined to an ever increasing extent.

Development of safety concepts (review, prospects) Deterministic/probabilistic approach

The methods called 'deterministic' have sometimes been opposed to the 'probabilistic' methods.

It is not correct to oppose the one to the other, since, in reality, all that is being done is to define progressively, with more quantitative accuracy, the probability factors in the product.

Risks resulting from accidents

$$= \text{Event probabilities (frequency)} \times \text{Consequences probabilities}$$

The safety philosophies and methods have been developed since the industrial advent of nuclear energy in ways that differed from one country to another. These variations were influenced by various parameters such as the practices in conventional industry, inventive imagination in study and development, socio-economic conditions, the peculiarities of the character of the people, the demographic aspects of power plant sites.

It could thus be said, for example, that the concept of defence-in-depth has, in reality, always been applied in one form or another and, of course, with increasing refinement. In Great Britain, and doubtless also in the USSR, nuclear energy for peaceful uses, in general, and nuclear power stations, in particular, were developed from military applications and by extrapolating conventional industrial techniques in a relatively simple manner. This led to an approach which at first devoted much more attention to level 1 and – to a certain extent – to level 2 of the defence-in-depth concept than to level 3. On the other hand, in the United States, attention was at first more especially devoted to level 1 and level 3 and not so much to level 2.

An example in this respect is the greater attention paid at first to the secondary containment provided by the physical containment around the reactor rather than to the performance, i.e. reliability and effectiveness, of the emergency cooling systems of the reactor core in the event of a primary coolant loss accident (LOCA). This is a situation which was reversed towards 1970 and then reinstated afterwards partly because of the consideration of possible impacts arising from outside.

In addition, the factor that distinguishes nuclear energy safety from that of conventional activities that are reckoned to be hazardous (for example chemical and petrochemical industry, handling of explosives, toxic gases) is that, in respect of the latter, a halt is most often made at the first level, or perhaps the second level of the defence-in-depth concept.

The notion of 'maximum credible accident' (MCA), used in the 1960s, is often interpreted as if only one accident condition was being analysed. In reality, this was only another name for, and a less refined approach to, what we might now call the most serious design basis accident or the reference accident resulting from level 3 of the defence-in-depth concept.

Accident scenarios having still more serious consequences belong to the category called 'unforeseen accidents' and can have 'catastrophic' consequences.

In the safety analyses that form the basis of the siting, construction and operation licences and in the emergency planning, accidents of this type are currently ruled out in practice (not necessarily in theory; see page 597), taking into account the following:

1. An adequate design and sufficient margins of safety in design, construction, and operation.
2. A high degree of guarantees and quality control (maintenance, inspections, testing, monitoring).
3. The experience acquired with certain vital equipment, either in the conventional field and extrapolated towards the nuclear field or in the nuclear field itself.
4. Results of research programmes.

Although previously the initiating mode of an accident, the sequence of the events (development) and finally the consequences were assessed in an empirical or semi-quantitative manner. To an increasing extent – for about the last 10 years – probabilistic techniques have been developed. These techniques use available information concerning the frequency of equipment failures and continue quantitatively by means of fault-tree analyses or event-tree analyses.

The opposite of the 'maximum credible accident' (MCA) is an 'incredible' accident. The opposite of the 'design basis accident' or 'reference accident' is an 'unforeseen accident' or rather an 'unscheduled accident' that could have more serious consequences. In the probabilistic approach, these notions have the advantage of being calculated in a more or less accurate manner. The figures can concern the initiating phenomenon of the accident (for example, the bursting of a pipeline or of the vessel), the phenomena occurring in the accident scenarios under consideration (for example, the non-availability of the emergency cooling systems) and the consequences (for example, radioactive material dispersion, the probabilities of somatic and genetic effects).

The point from which an accident is 'unscheduled' (once called 'incredible') is essentially defined in a 'deterministic' manner on the basis of a value judgement that becomes increasingly quantified.

Table 4.2-1, based on the data from the 1975 studies of the 'Commissie Reactor Veiligheid' in the Netherlands, shows how a deterministic approach can be placed within the probabilistic analyses (including the risks of radioactive product release, i.e. part of the sequence of the accidents).

Table 4.2-1   Activities released in the event of variable probability accidents (LWR)

Activity released to the atmosphere (expressed in thousands of curies)	Probability in 10^{-6} per reactor-year				
	Core meltdown			No core meltdown	
	$1 \times$	$15 \times$	$60 \times$	$100 \times$	$400 \times$
Rare gases	250 000	120 000	900	110	30
Iodine	250 000	15 000	5	0.15	0.012
Caesium, rubidium	5 500	400	0.1	—	—
Tellurium	60 000	5 600	2	—	—
Strontium, barium	19 000	1 300	0.3	—	—
Ruthenium	10 000	1 300	0.3	—	—
Others	5 000	700	0.3	—	—
			DBA pessimistic case		DBA realistic case

Note
The following uncertainties are estimated in the figures given in Table 4.2-1: a factor of 3 in respect of the probabilities, a factor of 2 in respect of the figures for radioisotopes other than the rare gases and iodine.

The Development and Application of Legal Acts, Rules, and Regulations; the Development and Application of Safety Standards (Criteria, Guidelines, Code-Requirements, etc.)

Definitions

With regard to definitions, it is hardly possible to make a clear-cut distinction between a number of terms used in nuclear regulation and industrial standardization because the national structures, practices, and languages have marked the terminology. Even within the single English language there are discrepant designations.

At the onset of the CEC activities in harmonization, an effort was made to give clear definitions (in various languages) of terms used in nuclear practice such as (in English) regulation, criterion, guide, standard, specification; (code). The only thing which might be said is that generally the mandatory (obligatory) nature diminishes going along those terms, with a special mention for the term 'code' which could be qualified as a hybrid term covering both a regulation and a standard.

The boundary between regulation and pure standardization (so-called voluntary standardization) results in communication between these two areas by way of reference to standards.

This method of reference to standards is used in particular by the European Communities who themselves establish by 'Directives' (in the conventional area) the constraints which the governments of member countries are required to impose, but leave the methods of measurement and preferred rules of the art which will be accepted everywhere without the need for justification, etc. to standards – provided these are harmonized throughout the Community countries. The question of directives is further discussed in Chapter 4.2, page 577. National standards in fact represent hindrances to trade whenever they depart from standards in the rest of the world; they represent less serious hindrances than standards established by government regulations.

Harmonization of regulations is much more difficult than harmonization of industrial standards because governments are much less flexible in harmonizing their points of view, which is understandable because the texts concerned involve much greater constraints than those involved in the industrial standards.

The national scene: licensing, regulation, standardization, and their connections

General background

The licensing procedures in the various countries have developed – and are continuing to develop – along different lines, depending on:
— the political and administrative structures and the laws applicable in each country;

— the organizational characteristics of each country and its regions.

These procedures are the vehicle for the development and/or application (again with variations) of the technical practices and methods concerning nuclear safety as well as the relevant safety requirements mostly spelled out in regulations, criteria, guides, standards, and codes for site, plant structures, systems, and components which one may call in common 'rules'.

Technical methods and 'rules' (regulations)

Against a background of evolution over the last 20 years in a rather wide range of safety analysis methods (e.g. MCA concepts, successive barriers concepts, defence-in-depth concepts, deterministic approaches, quantified probabilistic approaches), more specifically since 1965 and initially in certain of the most advanced countries in the nuclear field, systematic efforts to develop 'rules' have been made.

How can the rôle to be played by these rules be defined? The rules represent a systematic and disciplined codification of the engineering and, simultaneously, a review of the lessons provided by experience.

If these rules are applied by the industrial architect who designs and by those who construct and operate the nuclear power station, they can contribute to the confidence that the operator can have in the reliability, the availability, and the safety of the plant. In addition, they contribute to decreasing the cost of design and of manufacture and to facilitating the 'planning' of manufacturing and construction.

This codified good practice (rules) provides the licensing authorities and the safety and control organizations with a high degree of confidence that the design basis and the performance requirements of the equipment on which they based their safety analysis and their approval will effectively be achieved.

Rules can be evolved by a regulatory authority and/or associated safety and control organizations, in which case they find expression in mandatory (obligatory) requirements or indicative (quasi-mandatory) requirements. As an example one could quote 10 CFR 50 as mandatory and the USNRC regulatory guides as indicative rules.

On the other hand, they may evolve on a voluntary basis, i.e. by a joint effort on the part of experts of the electricity producers, industrial architects and builders, and the licensing and regulatory authorities together with the associated safety and control bodies (example ANS, ANSI, ASTM, ASME codes and standards). They are then usually called industrial 'standards'.

One can speak about a hierarchy in nuclear 'rules' or 'standards' illustrated in the following pyramid:

These rules (standards) varying in 'status' may in substance cover the design of the plant as a whole and its siting conditions, the structures (e.g. steel pressure vessels and concrete structures such as safety containments), systems or sub-

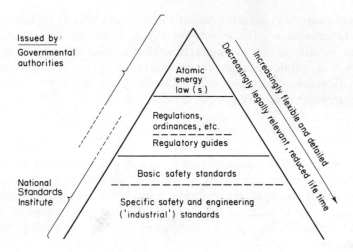

Figure 4.2-1 Hierarchy of nuclear rules/standards

assemblies (e.g. primary boundary, cooling systems, containment spray systems, effluent gas treatment systems, reactor protection systems, power supply systems) or components (tubing, pumps, valves, electromechanical components, electronic components, etc.).

They may refer, at the same time or separately, to design, fabrication and construction, assembly, testing and inspection, periodic inspection and monitoring, normal operating conditions, transient behaviour, malfunctioning or accident conditions.

Finally, the rules have to be adapted to take account of advances in technological knowledge (e.g. the results of safety research) and the evolution of safety methods (e.g. the use of probabilistic methods of analysis). This means that they have to be periodically revised.

Review of developments in industrialized countries, with brief examples

The following review is of necessity incomplete and will be largely concerned with the reactor type which in Western countries is the mostly developed industrially, i.e. the light-water reactors (LWR).

To begin with, it may be pointed out that there are some countries which, although they have themselves significantly developed nuclear energy, have less felt the need to formulate many 'rules'. These countries are the United Kingdom and Canada. This can no doubt be explained by an inherent pragmatism and by the evolution in reactor types selected in those countries, namely the GCR, the AGR, the SGHWR, and the PTR of the CANDU type.

On the other hand, the two countries most advanced in the development of rules are undoubtedly the USA (the homeland of the LWR) and the Federal Republic of Germany.

The main body which coordinates the preparation of indicative (quasi-mandatory) rules of a technological nature in the Federal Republic of Germany is the Kerntechnischer Ausschuss (KTA). Up to a point these rules are comparable with the Regulatory Guides drawn up by the USNRC. However, it should be noted that these rules are the outcome of a voluntary agreement between licensing and regulatory authorities and the associated safety bodies, operators, and vendors; in this respect the role of the KTA is more akin to that of ANSI in the United States.

Though their efforts are relatively modest compared with the USA and the German Federal Republic, more systematic attempts are being made to draw up rules by other countries such as France, Italy, the Netherlands, and the Scandinavian countries; here, however, they are mostly of the mandatory requirement or indicative rule type, i.e. they are chiefly drawn up by the regulatory authorities and safety bodies, usually in consultation with industry. In some cases these rules are evolved in line with the characteristics of the sites in the country in question (e.g. geological and seismic conditions, external, man-made or natural impact conditions, population densities). Often, however, these rules are revised versions of rules of American origin which in time are still evolving.

To a great extent however, these countries (and others such as Belgium and Spain) still adopt the method of applying the rules of the country of origin for projects and plant of installations of imported design, whether imported directly or under licence. It would be wrong, however, to think that this practice is necessarily easy of application. It is necessary to:
— keep constantly up to date with amendments to the rules;
— understand and interpret them correctly (language difficulties);
— correlate them with any 'indigenous' rules that may exist;
— build in any requirements arising from special site characteristics;
— make a true assessment of any technical differences in relation to 'reference plants'.

The international scene: development of rules (regulatory and industrial standards); harmonization of practices and rules; associated difficulties and limitations

General

If one recognizes the need to develop nuclear energy and the industries associated with it, it seems essential to evolve harmonized cross-frontier approaches and techniques in nuclear health and safety matters. It is true to say that:

1. There is no reason to allow the development of different overall levels of safety to which individuals of different nationalities are exposed.

2. In view of the growing international exchange of projects and technical equipment (structures, systems, sub-assemblies and individual components), disparities in the rules, i.e. in the regulatory requirements and in industrial standards, constitute a barrier to industrial development and energy production.
3. The development of standardized methods to deal with safety problems and their environmental implications for the design, fabrication, and assembly of equipment can shorten lead times for the setting up of nuclear power stations, and speed up their being put into service.

The nature of the rules, and the way in which they have been developed at national level (see Chapter 4.2 page 577) show that, whatever merits the systematic approaches applied may have, the outcome is a very complex situation which is rendered all the more difficult in view of harmonization by the magnitude of the problem and by the national efforts dealing with it. Any international attempt at harmonization which is not sufficiently selective and judicious in the choice of options and priorities would fail to make any impact.

The three most noteworthy approaches – in the technological area (leaving aside the radiation protection area, widely covered by the ICRP and also within the EC) – to the problem at international, worldwide and European levels are those undertaken by the International Atomic Energy Agency (IAEA), the International Standards Organization (ISO) (together with its sister organization the International Electrotechnical Commission (IEC), and the European Communities (EC).

The International Atomic Energy Agency – IAEA (Nuclear Safety Standards programme; NUSS programme)

This work was started in 1974 and is geared to the needs of countries in the process of industrial development. Its purpose is to develop a set of safety recommendations (non-mandatory) based on national and international practices, rules, and standards under the form of so-called 'Safety codes' and 'Safety guides'. These recommendations are to set a standard frame of reference to which especially developing countries can refer to in taking main decisions on nuclear safety of nuclear power plants with thermal neutron reactors. They can represent a substantial assistance in identifying problems and in establishing minimum requirements for safety and suggesting acceptable methods to achieve them.

At the time of writing, five safety codes (in the areas of Governmental Organization, Siting, Design, Quality Assurance, Operations) have been published and about 18 safety guides in these respective areas have been published or have been approved.

The end product (recommendations of member states) represents a common denominator of minimal requirements of limited practical use in industrially developed countries in the nuclear area but it provides an excellent synthesis of commonly agreed basic practice, from which one can further construct in a more detailed manner.

Although the final approval of the end product is at the regulatory authorities level, in the process of elaboration of the documents, industry experts (i.e. from the utilities and vendors) are also closely associated in the work together with experts from the regulatory bodies: this provides a good opportunity of a joint discussion in a systematic way on specific subjects classified in a precise order and therefore there is general interest in the undertaking. Furthermore this safeguards sufficiently the interest of the industries with respect to the international market (especially towards developing countries).

In principle, it was agreed at the onset that the IAEA–NUSS programme would concern itself with rules covering the regulatory (mandatory oriented) aspects, while the ISO Technical Committee 85 programme (see below) would give its attention to rules covering the voluntary aspects (i.e. industrial standards). In practice though and because of the inherent safety-oriented requirements of the nuclear technology it has proven difficult, if not impossible, to keep a clear-cut distinction between the technical subjects calling for mandatory rules and those that should be covered by rules (standards) of the voluntary type. So interphase between the IAEA and ISO/IEC programmes is inevitable. This can be partly solved by applying the system of reference to the standards (see Chapter 4.2, page 577).

The International Standardization Organization-ISO (and the International Electro – technical Commission-IEC) programme

The programme of ISO, optimized at the development of internationally accepted industrial (voluntary type) standards is, in the nuclear area, concentrated in its Technical Committee 85 with its five sub-committees and several working groups. The main sub-committees are those working in the areas of 'Power Reactor Technology' (SC 3) and 'Nuclear Fuel Technology' (SC 5). Links exist with the work of the IEC, in particular with TC 45 (Nuclear Instrumentation).

After an interruption of several years work was more actively resumed at the end of 1974, but is progressing at a relatively slow pace especially on certain subjects.

The European Communities (EC)

In the area of radiation protection, the EC accomplishes its obligations in accordance with Chapter III of the Euratom treaty. This implies essentially:

1. The setting up and periodic revision of the 'Basic standards' for the protection against ionizing radiation (article 31); i.e. an EC Directive largely inspired by

the ICRP recommendations (to be) implemented by the EC Member States in their legislation within a given period of time.
2. The application of article 37: examination of whether a nuclear project could give rise to a normal or accidental effluents release that would unduly contaminate a neighbouring country.

A systematic effort to achieve gradually harmonization of technological safety practices and requirements for nuclear power plants, more specifically those using light water reactors, was launched in 1973. A permanent working group (WG No. 1) was set up for this purpose, on which the licensing (and regulatory) authorities and associated safety and control organizations are represented on one hand and the utilities and vendors on the other.

This permanent working group has set itself an order of priority for dealing with the various subjects and has in the course of 1977 worked out a revised working scheme for its second phase of work.

The basic support, at the political level, for these harmonization efforts is provided by the European Community Council of Ministers Resolution on nuclear safety technology of 22 July 1975 which among others:
— calls for a strengthening of effort at harmonization within the European Community;
— requests the Member States to notify the Commission of any draft laws, regulations or provisions of similar scope concerning the safety of nuclear installations in order to enable the appropriate consultations to be held at Community level at the initiative of the Commission;
— provides that, at the appropriate time, Community recommendations shall be published (article 124 of the Euratom-Treaty) on those subjects which lend themselves best to a harmonized approach.

A schematic survey is given of:
(a) the tasks undertaken (Table 4.2-2);
(b) the subjects (to be) dealt with (Table 4.2-3)
For the subject of external (natural or man-made) hazards, most of the studies and findings in this area are not only applicable for LWRs, but also apply to other reactor types and to a certain degree even to nuclear fuel cycle facilities.

Although initially slow progress was made, the first stage of work has allowed the various parties involved to get better acquainted with the detailed 'how and why' of the different national practices and rule-making processes and contents and to define systematically and also in detail – for various topics already – the extent, and in some cases the reasons, of existing convergency and divergency.

It is to be hoped that in the second stage of work it will be possible to lay down more precisely what specific safety requirements can be commonly agreed to in general and in sufficient detail and to mitigate the divergent issues of importance which may subsist.

Table 4.2-2   Overall tasks undertaken by the CEC Working Group no. 1 (Water reactor safety; methodologies, criteria, standards)

1. Exchange of information on applied safety rules, criteria, codes, and standards; working towards a certain degree of harmonization by mutual information.
2. Identification of the general safety criteria, codes and standards and the approaches to specific LWR safety problems applicable and/or under development in the various member countries.
3. Classification and statement of the areas of divergence and those of common requirements.
4. Identification of the reason of differences (administrative, industrial, geographic, etc.); suggestion of formulations to overcome the litigious points.
5. Evaluation of priority – issues to be considered.
6. Consultation on drafts of national rules, criteria, codes, etc. in EC member countries.
7. Consultation on drafts of internationally established codes and guides, specifically IAEA–NUSS.
8. Examination of the necessity of uniform rules, criteria, codes, standards, etc. and the respective range of applicability.
9. Examination of codes, guides, and standards, etc. elaborated by international organizations with regard to their applicability in EC member countries including subsequent necessary amendments and/or additions.
10. Elaboration of recommendations (art. 124 EURATOM Treaty) on items of common interest and considered mature for that purpose.

Table 4.2-3   Subjects dealt with in the frame of Working Group no. 1 (CEC)

External Hazards
Protection of NPPs against aircraft crashes
Protection of NPPs against external explosions including flammable vapour clouds
Protection of NPPs against earthquakes
Protection of NPPs against floods

Internal Hazards
LOCA–ECCS
— Mechanical and thermal hydraulic effects
— Reference accident assumptions and radiological consequences
Spectrum of steamline breaks inside and outside containment
— PWR steamline break outside containment
— PWR steamline break inside containment
— BWR steamline break outside containment
Turbine missiles
Coolant pump flywheel integrity
Fuel handling accident
Anticipated transients without scram (ATWS)

Methodology
Quantitative methods applied in preparation and assessment of Safety Reports (frequency of event or risk concept)
Comparison of practices (safety reports) with regard to fault and accident conditions that determine the design of NPPs

Safety Design and Operational Provisions
Classification of safety function systems and components of PWR
General design criteria
Quality assurance (comparison of QA requirements)
Reactor coolant pressure boundary
(correlation study on criteria, codes, etc. for primary boundary)
Overpressure protection of primary circuit

Table 4.2-3 (*Continued*)

In-service inspection
Reactor vessel internals
Reactor protection systems and control (feasibility study)
Containment structure and engineering safeguards
(methods and procedures applied in containment leak testing)
Protection of NPPs against loss of electric power supply
(emergency power supply)
Fuel storage and handling systems
Ventilation system
Control room
Reactor core concepts and physics
Radioactive waste treatment systems
Service water systems and ultimate heat sink
Fire protection
Technical safety aspects of decommissioning

Siting and Emergency Planning
Site suitability criteria; technical bases of emergency planning; accident hypothesis

Incidents Reporting Requirements and Systems
Consultation on criteria, standards, codes, etc.
— Consultation of national criteria, standards, etc.
— Information on national programmes for development of criteria, codes, etc.
Consultation on IAEA codes and guides
Consultation on documents transmitted by the Member countries to the CEC in the frame of the
Council Resolution of 22 July 1975 on the technological problems of nuclear safety – Article 6
Periodically updated 'Nuclear Standards – Catalogue and Classification'.
Survey of licensing procedures for the construction and operation of nuclear plants in the various
Community countries
Survey of licensing procedures in certain non-member countries

With regard to the worldwide IAEA–NUSS programme and to a certain extent also the ISO/TC 85 work, the CEC activities have the necessary connections with the 'code', 'guide', and 'standards' writing-efforts there.

As the CEC 'in depth' comparative examinations often go further into technical detail, it is likely that the IAEA–NUSS end products can in many instances be used as a 'substratum' which can be built upon. On certain occasions also the reference to the standard method may be applied (see page 582).

Transfer of Nuclear Safety Technology and Concepts Related to Export/Import of Designs and Equipment

The existence and development of national safety practices, regulatory requirements, and rules and industrial standards evidently pose problems in the transfer of nuclear technology and concepts related to export/import of nuclear installations designs and equipments. These problems have been mitigated so far mostly in a bilateral manner. International efforts on a wider and systematic scale (see page 585) so far have contributed inherently to a better understanding and perhaps converging trend of the national positions of the industrially developed countries but still have to demonstrate their explicit effectiveness with regard to easing the transfer of nuclear safety technology among those developed countries themselves and with the developing countries.

However, the problem is not so much that of technology transfer but how to assure that the degree of protection of the professionally exposed and of the public at large is at a sufficiently equivalent level in neighbouring countries and on a wider scale. This is related to the acceptability and actual acceptance of nuclear power production and its associated fuel cycle.

Correlations (and Lack of Correlations) between Regulatory Rules and Regulations, Safety Standards and Practices, Possibilities and Limitations in Harmonization; some Worked Examples for NPP's

A survey of the EC actions in harmonization of technological safety matters is given in Chapter 4.2, page 577. From the various subjects outlined in Table 4.2-3 only two have been selected here to provide worked examples. They refer respectively to an example of protection against a natural external effect and to an example of protection against internally originated potential accident conditions. For both of these, noticeable progress was made in understanding the various positions taken, in delineating as clearly as possible the convergencies and divergencies in national approaches, and in defining the extent to which harmonization can reasonably be accomplished in each case.

EXAMPLE 4.2-1 Protection of Nuclear Power Plants against Seismic Effects

The effects of seismicity, as they are observed in European countries may, under certain circumstances, lead to significant nuclear hazards, so that every effort must be undertaken to protect nuclear power plants against such abnormal natural hazards. This is necessary not only in countries with high seismicity, but also in regions with medium or even low seismicity.

Consequently, it is necessary to identify those structures, systems, and components of a nuclear power plant which must be designed to maintain their safety-related function under the maximum potential seismic load, and to define the methodology which must be followed to ensure an acceptable earthquake-resistant design.

Most Member States of the European Community (EC) base their anti-seismic safety and design concept and even specific design parameters on the current concept of the USA which has been developed with respect to the specific characteristics of the seismicity of North America.

There are, however, important discrepancies between the two continents in the seismological and geological conditions as well as in the general design demands.

This is why many European countries realize now that they have to develop their own specific safety concepts adapted to the typical seismological and geological conditions in the respective countries.

The work performed is based on an inventory of the applied national practices, the existing specifications, regulations, and guidelines applied in the design, construction, and safety assessment of structures, systems, and components to withstand potential earthquake effects. These specifications and guidelines are compared with those for the USA and Japan and due notice is taken of the work performed within other international organizations such as ISO and IAEA. Points of agreement were identified and divergencies discussed in particular with reference to the US practice.

As a basis for design two different reference earthquakes are defined; the first more severe earthquake (the safe shutdown earthquake, SSE) according to US definitions is associated with the safe shutdown of nuclear power plant, while the second (the operating basis earthquake, OBE) according to US definitions is associated with its reliable operation. The procedures applicable and actually in use for establishing the reference earthquakes for a given site are the deterministic approach as well as the probabilistic approach. Both approaches rely on subjective judgments based on seismological and geological experience.

The SSE is based upon an evaluation of the maximum earthquake potential considering the regional and local sub-surface material. It is the earthquake which produces the maximum vibratory ground motion for which those structures, systems, and components are designed to remain functional and to assure the following:

1. The integrity of the reactor coolant pressure boundary.
2. The capability of shutting down the reactor and maintaining it in a safe shutdown condition.
3. The capability of preventing or mitigating the consequences of accidents which could result in potential off-site exposures comparable to the USNRC guideline exposures of 10 CFR 100.

Since the SSE is related to the maximum earthquake potential at the site, it has a low probability of occurrence during the lifetime of a nuclear power plant. Nevertheless, in most European countries the OBE is determined in dependence of the SSE assuming that the maximum vibratory ground acceleration of the OBE shall be at least one-half of the maximum vibratory ground acceleration of the SSE.

Only according to French practice the larger reference earthquake which is called 'Séisme Majoré de Sécurité SMS' is determined by adding a certain safety margin to the intensity of the smaller reference earthquake which is named "Séisme Maximal Historiquement Vraisemblable SMHV' (Maximum Plausible Historical Earthquake). The safety margin applied is usually one degree of the macroseismic intensity measured with the MSK-scale.

At present, preference is given to the deterministic approach in the evaluation of both the SSE and OBE, although the assessment of the OBE with statistical and probabilistic methods on the basis of available data in Europe appears feasible.

Recent discussions under experts indicate that for the definition of the lower level earthquake no agreement seems to be possible on a deterministic basis; there may, however, be a chance to agree on a probabilistic basis.

EXAMPLE 4.2-2 *Protection of Nuclear Power Plants against fuel handling accidents*

A comparison of the different methods and procedures applied by European countries, in the analysis of the fuel handling accident was performed in order to investigate convergencies and divergencies in safety assessments. It can be demonstrated that existing differences in the assumptions of the basic conditions (as are decay time, or mode of operation) are only of minor importance. The assessment of a fuel drop is in all European countries based on the damage of all fuel rods (of one fuel element). However, there appears to be a tendency at least in some European countries to assume that only the fuel rods of an outer array are damaged. This was also assumed for the evaluation of the radiological consequences of a German nuclear power plant. Also the basic assumptions concerning the quantity of fission products released from the fission gas plenum to the pool water differ from country to country, with the most conservative assumption in the US Regulatory Guide 1.25. According to RG 1.25 the assumed releases are as follows:

noble gases   10%
krypton-85   30%
Iodine   10%

More realistic assumptions used in France or Germany are based on experimental results of iodine release (organic and inorganic) less than 10% down to 1%.

Minor differences are observed also with regard to the decontamination factors for storage pool water

$$\text{(Decontamination factor} = \frac{\text{Activity release to pool water}}{\text{Activity release to building atmosphere.}}$$

There again the assumptions published in the US Regulatory Guide 1.25 are extremely conservative.

Off-site radiological consequences of the accident are limited by the construction of the storage building and the filter equipment of this building. In the USA these filters obviously are not mandatory. They are, however, provided in all European countries, also in those which apply the US rules and guides in general.

The design of the fuel handling devices have in all nuclear power plants the same objective: to avoid the fuel handling accident. The fuel handling transfer machines are designed to preclude mishandling, the designed safety features being as follows:

1. Lifting of the fuel assembly is not possible until the grips are additionally locked mechanically. This locking mechanism must be redundant and diversified.
2. Lifting and travelling motions must be preclusive, the relevant locking mechanism being either positive or redundant.
3. Travelling motion must be impossible until a prescribed minimum lifting height has been achieved.

An important difference exists in some European countries (especially Germany, France, and Belgium) where the fuel handling machine must be protected against external events (e.g. earthquakes, aeroplane crashes, explosion pressure waves). The effects of such external events are analysed locally and the relevant structural components are so designed as to preclude fuel handling accident due to external events.

Other important divergencies exist with respect to a potential fuel cask drop inside the storage building. These divergencies result from the different location of the fuel storage pool either within the containment as in German plants, or outside the containment in a separate fuel storage building as in US, French, or Belgian plants. In older German plants the fuel cask is lowered directly into the spent fuel pool. In US, French, and Belgian plants there are three different pools, the storage pool, the decontamination pool, and the loading pool (without fuel elements). This applies primarily to existing power stations, whereas the differences appear to be eliminated in German power stations under construction or planned where a separate fuel cask pool is foreseen within the containment next to the fuel storage pool.

Recent and Foreseeable Developments for some Important Issues that will Reflect in Evolution of Health and Safety Rules, Regulations, Standards, and Practices

The application of deterministic methods versus quantified reliability and risk analysis in the nuclear and other areas

Man-made installations behave in essence in a deterministic way. Human intervention in the operational processes of the installations is largely non-deterministic. Nevertheless for both the behaviour of installations and human behaviour a probabilistic approach is conceivable in order to try structurizing uncertainty. In both cases the probability approach is induced by lack of precise knowledge.

It is with much prudence that reliability analysis of structures and systems and especially the quantified risk-concept gropes actually its way in nuclear safety practices and rules, especially from the licensing and regulatory standpoint.

The care with which such concepts need to be handled can be illustrated with some actual tendencies in the area of potential major hazards (low frequency–high consequence) situations. Although the intermediate (with regard to frequency and consequences) hazard cases are probably more relevant, the extreme cases are the ones that are the most difficult to handle from all standpoints: in the technical assessment itself, in risk appraisal and acceptance, and in policy and decision making.

In most countries there is a tendency to adopt the position that the probability of an accident with serious consequences, caused by an external influence (natural such as seismic effects and floods, or man-made such as aircraft crashes and gas-cloud explosions) has to be small compared with the probability of serious accidents with internal causes.

The derived guideline is then that the design protection of the plant must be such that external influence with a frequency of the order of 10^{-7}–10^{-6} per year or more which could lead to extreme radioactive releases can be resisted. One can speak about target or limit-values for the acceptability of risk or about design basis probability values (DBPV).

However such values are applied and interpreted in a rather confused manner for a variety of reasons:

1. The error-band due to limitations in statistical data available and the uncertainties in predictive analysis.
2. Such values are applied in two different meanings:
   (a) as a limit for the probability of events, which could cause a radioactive release above some predetermined limit;
   (b) as a limit for the composed probability consisting of the probability of an external event multiplied by the conditional probability that this event causes (rather undefined) unacceptable consequences.
3. The relatively short duration of the data-base which is more or less accurately registered (e.g. floods, seismic phenomena, aircraft crashes of different types).
4. The fact that over the past and future years there are environmental changes that should reflect in changes in the assumed data-base (e.g. natural and man-made changes in the flood-basins, in air-traffic conditions).

Furthermore such target limit-values are insufficiently assessed with regard to the cost–benefit implications of possible supplementary protective devices to reduce further the risk.

Much remains to be done before such approaches will be fully satisfactory. Meanwhile the deterministic approach is still the main tool in nuclear safety management. The strength of a structure and the efficiency of a system are based on a maximum 'expected' load. The determination of the maximum expected load is done considering (qualitively, possibly quantitatively) the probability of the influences from external and internal events causing such a load: by qualified judgement a maximum credible event is introduced. If more than one event is of importance for the design, the notion design basis events is applied.

Safety margins compensate for lack of knowledge regarding the effects of
— load uncertainties,
— uncertainty in material properties and/or systems behaviour,
— calculational and constructional deviations,
— expected severity of the consequences of structures or systems-failures.

Quantification of safety margins for industrially developed reactors such as LWRs is considered important and even now this quantification is an important part of confirmatory nuclear regulatory research programmes.

One can also see that inherently the deterministic approach has also probabilistic aspects, though mostly intuitively instead of systematically.

Likewise in the non-nuclear licensing and regulation, quantified risk analysis is increasingly applied over the range going from the high frequency–low consequence situation to the low frequency–high consequence situation.

Air-traffic and aerospace applications are probably the activities where risk analysis has been furthest developed. In fact in these instances target values for the acceptability of risk have been applied (e.g. automatic landing systems to respect the criteria of failure of less than 10^{-7} per landing).

For a number of activities where reasonably good statistics exist, e.g. road-traffic, fossil fuelled power plants, conventional hazardous factories, safety provisions, authorizations and decision making could be increasingly established with the support of risk analysis. This tendency is indeed gradually developing. Examples are as follows:

1. To evaluate risks in the medical field, e.g. risks in the use of pharmaceuticals and vaccinations (e.g. smallpox).
2. To evaluate risks of storage and handling of toxic gases (e.g. ammonia, chlorine, acrylonitrile, ethylene oxide, LNG, LPG).
3. To evaluate the risks of insecticide, pesticide, fertilizer and other aspects of the chemical industry.

Even for problems, such as those related to natural phenomena where with regard to the statistical data-base the situation is not much better than in the nuclear area, the quantified risk concept has been applied in regulation and in decision making. Typical examples of this are:
— Determination of height and strength of dikes to protect against exceptionally high floods (e.g. Dutch DELTA dikes, the Thames flood defences).
— Codes of practice for buildings and dams to protect against 100 or a 1000 year exceptional wind or wave force.

Finally, it is worth while to point out that for other major hazards industries, severe consequences (say potentially hundreds of dead) are associated with event-frequencies of the order of 10^{-3}–10^{-4} per installation-year (to be compared with the 10^{-6}–10^{-7} 'targets' for nuclear). This is one of the reasons why extreme consequence-modelling distorts the picture, unless it is clearly specified in which contexts such analyses have to be seen. This is further dealt with in the following text.

Accident hypotheses for site selection and emergency planning

The practice applied so far has been roughly outlined earlier on page 577 However since the increased development of quantified risk assessments for NPPs (1975) and the March 1979 TMI-2 events, accident hypotheses beyond Design-Basis accidents (DBAs), i.e. 'unforeseen' accidents including extremely severe environmental consequences, are being given much more attention in site – selection and in the planning basis for emergency measures. This can be considered a rather unexpected by-product of quantified risk assessments and is highly debatable with regard to the extrapolation made from the TMI-2 sequence of events.

As a corollary of this tendency, supplementary structural protection devices to mitigate such very low probability of occurrence of large-consequence accidents (frequency: order of 10^{-6}–10^{-7}/reactor-year) are also being seriously considered (examples: increased volume capacity containment, vented filter containment, blast/missile protected containment, molten core retention devices, supplementary hydrogen control); they render their installations increasingly complex, imply additional cost and do not necessarily contribute to the overall safety of the plant.

At the same time there are increasing calls for the gradual development of quantitative safety goals (how-safe-is-safe-enough?), an exercise in which the nuclear activities would be put into perspective with other public risks (e.g. other energy sources) and in which the cost–benefit (risk-reduction) aspects of supplementary protection and accident consequence mitigating devices would be taken into account. Such an approach tends to de-emphasize the importance given to nuclear 'unforeseen' accidents, as in non-nuclear major hazards activities equivalent severe-consequence accidents are in fact 'foreseen' on the basis of actual statistical data of occurrence (of the order of 10^{-2} to 10^{-4} per installation-year).

At the present point in time it is unclear how the situation in these regards will develop.

Application of ICRP publication 26 and the ALARA (as low as reasonably achievable) principle for occupational personnel

The application of the ALARA principle for the control of normal effluents releases from nuclear installations has lead to differences in the limits of exposure to the public. However as these limits are very low indeed, the resulting discrepancies can be considered as marginal.

In view of the more significant risk to which occupational personnel (operators and transient workers) are exposed in nuclear installations, the application of the ALARA principle in divergent ways may in this case lead to more significant discrepancies which are significant because:

— on one hand they may lead to different equipment design requirements which could develop into a hindrance to trade exchanges for the vendors;
— on the other hand they may lead to different operational procedures and requirements (e.g. degrees of decontamination before repair and maintenance jobs) that could hinder exchange of specialized transient personnel internationally.

Harmonization in the EC Framework of Rules/Standards in the Non-Nuclear Area

The by now 13-year-old programme for the removal of technical barriers to trade within the EC is product oriented. It includes the concern of protecting consumers health and of promoting user safety. By mid-1979, 180 directives (to be) implemented in national regulations have been adopted by the Council, while about 60 directives are still under discussion: 130 directives deal with industrial products, 50 with foodstuffs.

More relevant however to the type of problems tackled in the preceding text are the following two directives

1. *A proposed Council Directive on the major accidents hazards of certain industrial activities*
   This proposal was submitted to the EC Council by the Commission on 19 July 1979 and is still under discussion.
   The aim of the directive is to prevent major accidents which might result from certain industrial activities (essentially chemical process industries and storage of dangerous substances but excluding mining operations, explosives factories, and military installations). In general terms, the directive requires the Member States to take in their regulatory dispositions the necessary measures to protect the health and safety of the professionally exposed and the public and to protect the environment: safety reports, training and drills, incidents reporting, emergency planning provisions, etc.
   It includes also reporting requirements to the EC Commission.

2. *A proposed Council Directive concerning the assessment of the environmental effects of certain projects*
   This has been the subject of preparatory discussions with the relevant national authorities over two years (1978–1979) and would also cover certain nuclear power plants and industrial installations of the nuclear fuel cycle besides conventional activities. It refers to the practice started in the USA since the 1971 Calverts' Cliff decision of establishing environmental impact statements

(EIS) by an applicant and of the regulatory review of such statements. In the nuclear area this is for many countries already covered by the usual licensing procedure. This proposed directive is related to (1) above and to a proposed 'Regulation' concerning the introduction of a Community consultative procedure in respect of power stations (nuclear and non-nuclear) likely to affect the territory of another Member State (COM(79)269 final of 17 May 1979) which is still under discussion at the EC Council level.

Generally speaking such a directive would tend to promote for non-nuclear major hazards industries in-depth safety and environmental protection practices and regulating procedures largely equivalent to those already in force for nuclear activities.

Questions 4.2

1. Discuss the development of the defence-in-depth principle as applied to a nuclear reactor plant and illustrate the same principle applied to a specific hazardous type of non-nuclear plant.
2. What is meant by the 'maximum credible accident' concept? How has deterministic safety analysis been extended to include probabilistic techniques in the European Economic Community (EEC) for the assessment of plant?
3. Discuss the 'harmonization' programme to clarify definitions in the European Community. What are the problems involved in defining standards and regulations and how have these been tackled?
4. What safety benefits accrue in the European Community by having common rules for the design of plant such as nuclear power stations?
5. Select a hazardous industrial plant then illustrate and discuss the hierarchy of 'rules' and standards which could apply to cover its design and siting. What approaches and philosophies need to be introduced so as to make these 'rules' and standards suitable for use by the European Community?
6. Discuss the problems involved at the international level in developing rules involving regulatory and industrial standards. Taking the nuclear reactor field as an example, illustrate the type of developments which have taken place to solve such problems.
7. What are the advantages and disadvantages of 'Safety Codes' and 'Safety Guides' based on national and international practices? If such codes and guides are made mandatory would they contribute to the achievement of safety or stifle the development of new plant designs?
8. Outline the obligations of the European Community taking as an example the area of radiation protection. Illustrate the type of systematic effort employed to achieve the harmonization of technological safety practices and requirements.

9. Discuss the needs and means for the transfer of safety technology at the international level. How has this been basically achieved in the European Community?

10. To what extent is it seen that a harmonized safety technology will emerge in the European Community which will be of a generic nature? What steps would be considered necessary to ensure that such a technology would have people to operate it so that adequate safety standards can be maiantained?

Further Reading 4.2

1. Vinck, W. 'Evolution in safety methodologies.' Lectures given at University of Brussels in a cycle on Nuclear Safety 1976.
2. Vinck, W. *et al.* 'Regulating practices and standards: the international scene and trends.' Paper presented at the International Conference on Regulating Nuclear Energy (AIF–FONUBEL) Brussels, May 1978.
3. Becker, K. Nuclear Standards – Catalogue and Classification, 1981.
4. Vinck, W. *et al.* 'Technical bases for regulating review; comparison of practices, standards and guides.' Paper presented at Specialist Meeting on Regulatory Review (NEA–CSNS); Madrid, November 1979.
5. CEC–JRC Seminar on Risk and Safety assessment in industrial activities. June 1979, Ispra.
6. Vinck, W. 'Practices and rules for nuclear power stations; the role of the risk concept in assessing acceptability.' CEC/CERD Seminar on Technological Risk; Berlin 1–3 April 1979.
7. Journées d'étude sur les positions prises à la conception et en exploitation pour réduire la radioexposition professionnelle dans les centrales nucléaires à eau ordinaire. SFRP, Paris, Décembre 1979.
8. Removal of technical barriers to trade
    a) Communication to the European Parliament; COM (80) 30 finae
    b) European filei 12/79 June 1979.
9. Proposal for a Council Directive on the major accident hazards of certain industrial activities; Official Journal of the EC No. C 212/4; 24/8/79.

High Risk Safety Technology
Edited by A. E. Green
© 1982 John Wiley & Sons Ltd

Chapter 4.3

Safety Regulations in the USA

D. Okrent and R. Wilson

Background

Between 1957 and 1977 the US Congress passed 7909 laws of which 179 were laws
to regulate hazards. There are in consequence 92 government agencies, of which
26 were created in 1969–1976 alone. These agencies issued over 6000 regulations
in this period.

The principal agencies are the Nuclear Regulatory Commission, the Federal
Energy Agency, Department of Transportation, Federal Aviation Agency,
Environmental Protection Agency, Consumer Product Safety Commission,
Food and Drug Administration, Occupational Safety and Health Admin-
istration. The laws of each agency are different and there are often different laws
for different activities of the same agency.

In this short chapter, therefore, we have no space even for a *listing* of all these
laws and regulations. We therefore pick some of the regulations and judicial
opinions supporting them which seem to us to be most important.

Strategies with Examples and Effects

There are in general three strategies for controlling a risk: (1) ban the activity or
substance; (2) use a best technology fix; and (3) use a risk/cost/benefit analysis.

Strategy (1) has the advantage of simplicity and has worked when the only risks
calculated were large ones. In primitive societies it was a taboo. The obvious
modern example is for carcinogenic substances. A ban will work if the only
carcinogenic substances are not widely used.

There are two contrary trends in the development of present US policy. The
first is to demand that safety be the first priority regardless of cost – and almost of
feasibility – strategy (2). The second is to demand a risk–benefit analysis – strategy
(3). Contrary to a commonly held view, the actual statutes do not prescribe any
one of these views. The choice between them is usually left to regulatory
discretion, subject to review by the courts.

We first pick out an array of laws governing carcinogenic substances, shown in
Table 4.3-1, taken from an OTA report[1]. There are three reasons for
concentration upon carcinogens. First, cancer is a disease which is particularly
dreaded by a large section of the public. Second, it is increasingly important as a

Table 4.3-1   Federal regulation of carcinogenic substances

	(a) Administered By	(b) Type of Substances Regulated	(c) Specific Procedures for Regulating Carcinogens?	(d) If 'c' Does Not Apply. How are Carcinogens Regulated?	(e) Benefit Risk Analysis of Consideration of Factors Other Than Safety	(f) Discretion in Regulating	(g) Relationship to Other Federal Statutes
1(a) Federal Food Drug and cosmetic Act – food provisions	Food and Drug Administration. DHEW	Foods, food additives, other substances of food residues in food	Yes, in several sections (food additives colour additives, residues of animal drugs)	For other sections, general safety is the criterion	Risks dominate, no such analysis permitted of colour or food additives or residues from animal drugs are carcinogenic. If a naturally occurring substance in food is carcinogenic, technological feasibility of removing it may be weighed against the health risk	Carcinogenic food and colour additives, and foods with carcinogenic residues of animal drugs,* must be banned otherwise discretion is not prohibited	The Act takes precedence in areas of foods and related substances; for residues from pesticides there is an inter agency memorandum of agreement between FDA and EPA
1(b) Federal Food, Drug, and Cosmetic Act drug provisions	Food and Drug Administration – DHEW	Drugs and substances in drugs	No	Carcinogenicity is considered as a risk of the drug; used in weighing safety against usefulness	Explicitly required, the benefits and the risks (safety) of a drug must be considered in regulating	Yes, FDA may permit carcinogenic drugs or substances in drugs to be marketed if the benefits outweigh the risks	Takes precedence in the area of foods
1(c) Federal Food, Drug, and Cosmetic Act cosmetic provisions	Food and Drug Administration – DHEW	Cosmetics and substances in cosmetics	No	Action is taken on the basis of adulteration (unsafe or injurious)	No benefits to health are presumed; risks predominate in analysis, those 'cosmetics' claiming positive health benefits are treated as drugs	Banning takes place based on the discretion allowed by the adulteration sections of the Act; public health is only criterion	Takes precedence in the area of cosmetics
2. Toxic Substances Control Act (TOSCA)	Environmental Protection Agency	Substances such as foods, drugs, cosmetics, tobacco are not covered; all non excluded substances are covered but if	Carcinogenic and certain other substances are to receive priority	Toxicity; cancer regarded as a priority class of toxicity	Explicitly required by the Act	All regulatory actions are at the discretion of EPA	See Column 'b'

	Act	Agency	Scope		Definition / Criterion	Risk/benefit analysis	Regulatory action	Precedence
				attention; a ruling must be made on carcinogens within a specified time, but regulatory action is based on toxicity				other Acts cover such substances those Acts take precedence
3.6	Clean Air Act; Water Pollution Control Act Safe Drinking Water Act. Federal Insecticide Fungicide and Rodenticide Act (FIFRA)	Environmental Protection Agency	Pollutants in the respective areas of the environmental	No	As environmental pollutants posing danger to public health; toxicity	Permitted	All regulatory actions are at the discretion of the Commission	At the discretion of the EPA, these Acts take precedence over the Toxic Substances Control Act
7.	Consumer Product Safety Act	Consumer Product Safety Commission	Substances used by consumers (at home, in recreation, etc.)	No	As hazardous products, imminent hazards	Explicitly required by the Act	All regulatory actions are at the discretion of the Commission	Not applicable to substances covered by Food and Drug Act; close relationship to Hazardous Substances Act
8.	Federal Hazardous Substances Act	Consumer Product Safety commission	Hazardous substances (in effect, if primarily covers household products)	No	As hazardous substances; toxicity is criterion	Not explicitly mentioned; has been interpreted as allowing if, and the Commission uses such analyses	Banning is at the discretion of the Commission; certain labeling requirements are non discretionary	Not applicable in substances covered by Food and Drug Act
9.	Occupational Safety and Health Act	Occupational Safety and Health Admin. Dept. of Labor	Hazardous substances in the workplace	No	As toxic substances; these are proposed implementing regulations dealing specifically with carcinogens	Permitted by the Act, required by the implementing regulations	Yes	Tables action when other Federal agencies have not, for workplace hazards

* There is some judicial opinion that for animal drug residues, if regulated under general safety some risk/benefit analysis must be made, even if carcinogenicity is indicated

cause of death, since other causes, such as epidemics, have been removed; it is now responsible for 15 % of all deaths. Third, scientists tend to believe that there is no safe threshold below which a carcinogen is harmless.

All the laws in Table 4.3-1 apply to toxic substances as well as carcinogens. For toxic substances, a 'no effect level' is usually defined where there was no identifiable toxic effect in either human studies or animal experiments. Then a dose level 100 times below the no effect level is taken as a safe level[2]. Although operationally this could be done for carcinogens also, the (theoretical) extrapolation to low dose usually does not assume a 'no effect' threshold.

The Food, Drug and Cosmetic Act[3] is the oldest of those laws involving toxic substances. For example, 21 CFR 201(s) defines a food additive as any substance 'the intended use of which results or may reasonably be expected to result, directly or indirectly in its becoming a component or otherwise affecting the characteristics of any food'. Any food additive is deemed *unsafe* unless there is 'a regulation issued under this section' (21 CFR 409(a)(2)) 'prescribing the conditions under which such additive may be safely used'.

However the next clause, first introduced by Representative Delaney and called the Delaney clause, says 'that no additive shall be deemed to be safe if it is found to induce cancer when ingested by man or animal' (21 CFR 409(c)(3)(A). This sounds like a complete ban; but it is hedged by the statement that 'this proviso shall not apply to the use of a substance as an ingredient of feed for animals which are raised for food production, if the secretary finds that no residue of the additive will be found (by methods of examination approved by the secretary . . .) in any edible portion of such animal after slaughter'. This is commonly called the DES proviso, because it was enacted to allow DES to be used in controlled quantities as an animal drug. The Commissioner of FDA has proposed a set of criteria for these methods of examination[4]. The method will be adequate if it could detect a carcinogen which poses a *lifetime* risk of one in a million to the most sensitive member of the public. This is very stringent. By the method of calculations suggested by FDA for additives such as DES, aflatoxin B1 in corn, milk and nuts gives over 1000 times the risk that would be acceptable for DES even though there is some FDA regulation of these foods.

A question hence arises about other 'accidental' food additives – aflatoxin in corn, milk, and nuts, or migration of vinyl chloride monomer from PVC plastic The former is regulated, but not banned, because it is 'unavoidable'. A recent Appeal Court case discusses the latter. The FDA had banned a plastic Coca Cola bottle on the basis that acrylonitrile monomer *can* reach into the Coca Cola. The court[5] rejected the FDA calculation that acrylonitrile migrates from the plastic to the liquid as too tenuous, and also argued that FDA has discretion to allow a minute risk. Although these are interesting clarifications of the law, the plastic bottle has not been reapproved.

Before a food additive is marketed the petitioner has to prove that the proposed food additive is safe. Once it is on the market FDA has the responsibility for presenting evidence that would lead to reconsideration of safety. Under the 'Delaney clause' FDA's responsibility has to be satisfied as soon as it finds that a food additive is carcinogenic.

There have been no recent examples of laws enacted with the stark simplicity of the Delaney clause, but the thinking involved in this clause has pervaded much of the regulatory apparatus. The clause makes most sense when applied to *deliberate* food additives and particularly those added to food so that it may be seen, tasted or smelled, rather than to preserve it. Simple chemistry suggests that to do any of these things, the additive must be present at least in an amount of 1 molecule in 10 000. A common food can be 10% of the diet, leading to a pollutant concentration in the diet of 1 in 100 000. To test for carcinogenicity, we typically use 100 rats; if we feed the rats a high dose of 10% in the diet, we find a barely significant fraction of 5% cancers in the rat, and using proportionality, we find that 1 in 200 000 rats (and by analogy people) could be getting cancer in their lifetime without detection. This is 14 per year and cannot be ignored.

For accidental additives, the Delaney clause can produce incentives contrary to common sense unless the DES proviso is widely used (which it is not). Industry has incentives to do bad and insensitive experiments to avoid detecting weak carcinogens, and has no incentive to develop sensitive detection systems for these pollutants.

A clear example of the second (best available technology) approach is provided by the safe drinking water acts[6] which have been, in one form or another, present for nearly a century. A century ago many communities in the US had bad water, as many communities in the world still have today. This led to disease and epidemics. The danger was so real and so large that the public demanded pure water at any cost, and the law specified the best available technology. Yet even here, the *scientifically* best available technology and the purest water were not in fact demanded or achieved. We have always known how, if we wish, to distil our water, and that gives us cleaner and purer water than comes out of any US tap. Yet we have contented ourselves with chlorination which is much cheaper and achieves one originally desired effect – elimination of epidemics – although it may not follow the rigid letter of the law. A recent court case emphasizes this flexibility. The court approved balancing of risk and benefit 'the task of the agency . . . is largely one of line drawing. Agency expertise and judgement must be applied in determining the optimal balance between promotion of the public welfare and avoidance of unnecessary expense'[7].

Of course, more recently chlorination has become suspect as a source of carcinogens in water. And there appears to be a significant quantity of carcinogens and other unhealthful pollutants in drinking water, enough to show statistically significant changes in cancer rate among communities having different water supplies.

The third approach, a risk/benefit analysis, seems to be mandated by several recent acts. The Toxic Substances Control Act (TOSCA)[8], the Federal

Insecticide, Fungicide, and Rodenticide Act (FIFRA)[9], the Consumer Product Safety Act[10], and an older one, the Flood Control Act of 1936[11].

For example, under FIFRA the pesticide in question 'must perform its intended function without unreasonable adverse effects on the environment' and this is is further defined to mean 'taking account of the economic, social and environmental costs and benefits of the use of any pesticide'. Under CPSC safety standards must be 'reasonably necessary to prevent an unreasonable risk of injury' and CPSC must consider 'the probable effect of (any) rule upon the utility, cost, or availability of such products to meet the need'.

However none of these acts specifies any procedure for comparing risks and costs, nor do they specify that procedure must be quantitative.

If we can indeed work out the risks of the various chemicals and the costs of all alternative control strategies, this third approach of comparing risks and benefits seems most logical. But it is clear from the legislative and legal discussions of the various acts, that the first two approaches are often used just because in practice we *cannot* see all alternatives.

The 'best available technology' is also very appropriate for situations where a single control method is generic to several hazards. If we wish to purify a water supply, using a catalyst or charcoal absorber removes many impurities at once. However the only case where this appears to have occurred is the water quality standard referred to previously.

A major problem with specification of 'best available technology' is that, from a cost/benefit point of view, it is hard to be consistent. This is illustrated by contrasting EPA actions under the same law. The Clean Air Act[12] does not allow economic costs to be considered in establishing primary (to protect public health) and secondary (to protect the environment) air quality standards. It has been confirmed by the US Supreme Court that economic and technological infeasibility need not be considered. The regulations under the Act demand the Best Available Control Technology (BACT) for new stationary sources of sulphur pollution even in the absence of a sulphate standard. Present estimates of controlling sulphur emissions from electricity generating plants are over $10 000 million per year. However if we believe the large numbers given by some retrospective statistical studies of air pollution[13] – that 50 000 people die early in the US because of air pollution, and allow a sum of $1 million to save a life, an expenditure of $50 000 million would be justified, and the use of a risk/benefit analysis would lead to a similar result (with more work).

But the use of the clean air acts to control relatively small quantities of industrial chemicals is less clear. There is a proposal to regulate carcinogens under section 112(b)(1)(A) of this Act, using 'best available technology'[14]. Detailed comparison of cost and benefits is explicitly rejected. As proposed for the regulation of benzene from a particular type of chemical plant[15] this would mean a specification by EPA of the control technology for old plants, and a switch of process for new plants, even though EPA estimates less than one leukaemia per year for the total exposed population of a million (see Table 4.3-2). The procedure is more draconian than is being suggested for the risk of sulphate pollution which is 100 000 times larger for society as a whole. It will be interesting to watch the development of this case.

Table 4.3.-2 Comparison of EPA actions under clean air act

	SO_x	Benzene from MA plant	Benzene from gas stations
Proof of hazard at x10 exposure	Yes	No	No
Proof of hazard at exposure	No	No	No
Direct *evidence* of hazard at exposure	Yes	No	No
Calculated number of persons dying per year	50 000	< 1	~ 10
Number of persons exposed	150 000 000	500 000	5 000 000
Annual risk to heavily exposed group	10^{-3}	5×10^{-7}	3×10^{-6}
Best Available Technology (BAT) proposed for:			
New source	10-fold reduction	Change technology	Nothing proposed
Old source	Only tall stacks	30-fold reduction	Nothing proposed

The Occupational Safety and Health Act[16] (US Code 651 *et seq.*) seems to allow no leeway in that it requires promulgation of 'the standard which assures the greatest protection of the safety or health of the affected "employees" but the act also insisted in section 3§8 that "the standard be reasonably necessary" '.

A federal Appeal Court[17] has indeed construed the Act as requiring that any regulation be based on quantified costs and benefits. In this case OSHA had proposed to reduce the allowable level of benzene in the workplace from 10 ppm to 1 ppm. Neither in the hearing before promulgating the standard nor before the court was evidence presented to show that not even one person is being injured by exposure to the presently allowed level. OSHA was content to argue that since benzene gives leukaemia at high exposures (100 ppm) the exposure must be reduced to the 'lowest feasible level'. and refused to perform any risk analysis. Industry had presented a calculation showing on a pessimistic proportional relationship, that less than one leukaemia victim would be saved each year for (OSHA's) estimated cost of $150 million per year. The Appeal Court insisted that OSHA must show a 'reasonable relationship' between cost and benefit.

The Supreme Court ruled against OSHA[17] on the ground that OSHA failed to establish that the standard 'is reasonably necessary or appropriate to provide safe or healthful employment and places of employment' (section 3§8 of the Act). The court believed that the risk would have to be significant to show this. Since such a threshold determination was not made, the Supreme Court did not have 'occasion to determine whether costs must be weighed against benefits'. Chief Justice Burger in a concurring opinion went on to say 'the Secretary is well admonished to remember that a heavy responsibility burdens his authority. Inherent in this statutory scheme is authority to refrain from regulation of insignificant or *de minimis* risks'.

Another interesting court case was the criminal suit against Ford Motor Company[18] for allowing the Pinto car to be built with a less secure gasoline tank than was technically feasible. The Court decided in favour of Ford because the company had made prompt attempts to recall the car after safety hazards had been brought to its attention.

In the newspaper discussion of this case, great attention was paid to the fact that a document of the National Highway Transportation Administration[19] suggested that safety improvements be considered if they cost less than $180\,000 per life saved. Although this was not considered by the court, many people believe the case would never have come to court if an engineer had not considered safety improvements and Ford had not filled in appropriate NHTSA forms. It is unfortunate that this counter incentive to truth and honesty exists.

In the principle 'as low as reasonably achievable,' the Nuclear Regulatory Commission has applied category 3 for regulating the routine release of radiation for nuclear power plants. In a rule-making hearing[20] RM-30-2, it specifically states (and there has been no objection) that reduction of radiation dose should be considered if it can be done for a cost of $1000/man-rem – roughly equal to $5 million per life saved calculated on a linear hypothesis. Yet the legislative mandate makes no mention of costs, and Congress merely exhorts NRC to put safety above all other considerations.

The decision is important because it is the first explicit recognition of a procedure. The procedure is to evaluate:

Value = Benefit (or cost depending on the question) − α risk

and to proceed if this value is positive and stop if the value is negative. In this case $\alpha = \$5$ million.

There are few decisions affecting future safety that are made in any logical way at all. Method 3 might be to calculate the economists net present value.

$$\text{NPV} = \sum_t \frac{B(t) - \alpha R(t)}{(1+d)^t} \tag{4.3-1}$$

where d is the discount rate. If d is the usual $5\% - 10\%$, then we argue that we would be willing to spend less now to save a life in the future than we are at present, and would, for example, prefer to use the money to find a cure for cancer. This procedure is often implicitly used in decisions and ignores anyone beyond two generations. For example, we talk of storing chemical wastes for only 100 years!

Yet for nuclear wastes, EPA has come out with a set of proposed regulations that ignore this completely. They propose, first, that no risk be imposed on a future generation that we are unwilling to bear ourselves, and second, a procedure is proposed that is related solely to technical feasibility and not to a calculated risk level. It seems likely that NRC will follow this approach. Thus although in RM-50-2 NRC follows a cost–benefit approach (Method 3), for waste disposal it tends to follow BAT (Method 2).

The NRC regulates nuclear power under two major statutes, the Atomic Energy Act of 1954 as amended[21] and the Energy Reorganization Act of 1974[22]. These do not require any risk analysis but the NRC, like the AEC before it, has interpreted its mandate as a requirement to balance to some extent nuclear power benefits against health and safety.

But neither the NRC commissioners nor the AEC commissioners before them have defined any quantitative risk acceptance criteria for nuclear power reactor safety. In 1965 a kind of quantitative benchmark was provided by the AEC commissioners in connection with their review of the application for a construction permit for a large pressurized water reactor at Malibu, California. The application was contested by a vigorous, well-funded intervenor group that hired technical consultants who argued that permanent surface displacement might occur on a small fault under the reactor containment building during a large earthquake on a nearby large fault. The AEC regulatory staff and the applicant for the construction permit, the Los Angeles Department of Water and Power, disagreed. However, the Atomic Safety and Licensing Board, on hearing the case, agreed with the intervenor. The AEC staff appealed to the Atomic Energy Commission. In their ruling upholding the Atomic Safety and Licensing Board, the Commissioners said that the fact that geologic evidence indicated that there had been no displacement along the fault under the containment building in at least 10 000 years was not sufficient to rule out a need to design for displacement along the fault. Thus the ruling provided implicit guidance that reactors should be designed for events having a recurrence interval shorter than 10 000 years.

The next suggestion of a quantitative safety criterion for power reactors came in 1973 from the AEC regulatory staff in connection with their report on anticipated transients without scram[23]. In this report they defined a safety goal that, assuming a population of 1000 LWRs in the early twenty-first century, the probability of serious reactor accidents should be less than 1 in 1000 per year from this reactor population (or one in a million per reactor year). The regulatory staff defined a serious accident as one whose off-site doses exceeded those in the regulations[24], namely 25 rem whole body or 200 rem to the thyroid. Mr L. Manning Muntzing, the Director of Regulation of the AEC, presented this position to the US Congress. There was no comment by the Congress or the AEC commissioners.

In 1976, the Advisory Committee on Reactor Safeguards (ACRS), in responding to a question from a committee of the US Congress, proposed the use, as an interim criterion, of 10^{-6} per reactor year for a serious accident (which the ACRS defined as equivalent to a fatal crash of a loaded commercial aircraft).

The NRC commissioners have stated that they have never provided a quantitative definition of acceptable risk. In 1979 the ACRS recommended to the NRC that the NRC develop a quantitative risk acceptance criterion, and substantial efforts are underway on the matter within and outside of the NRC.

In addition to these specific laws and agencies there is now a general law – the National Environmental Policy Act (NEPA) of 1969[25]. Section 102(c) of this Act declares that 'all agencies of the Federal Government shall include in every recommendation or report on proposals for legislation and other major Federal actions significantly affecting the quality of the environment, a detailed statement by the responsible official on (i) the environmental impact of the proposed action, (ii) any adverse environmental effects which cannot be avoided should the proposal be implemented, (iii) alternatives to the proposed action . . . '. The courts have interpreted 'environment' very widely and this act has been used to alter many proposals for energy installations. It is therefore correctly called the most important *energy* legislation of the decade.

It has been interpreted by the courts, particularly, in the Calvert Cliffs case[26], as requiring the agency to perform a 'case by case balancing judgement' in which 'the particular economic and technical benefits of the planned action must be assessed and then weighed against the environmental costs'. However, the Act does not describe, nor have the courts demanded, any particular way, quantitative or otherwise, of doing so. This law can be, and is used, to control safety in cases where no other exists. One of the uses has been to restrict the locations for Liquefied Natural Gas facilities. The first importation terminal in the USA was in Everett, near Boston. The original environmental impact statement of the Federal Power Commission was 12 pages long, mostly about economics and only one paragraph on safety[27]. Now proposed new terminals involve environmental impact statements of 400 pages, with detailed calculations of risk of catastrophe. As a result, terminals as close to population centres as Everett are no longer proposed. For southern California, sites in Los Angeles Harbour and Oxnard have been rejected and a facility is likely to be built at Point Conception which is unpopulated. There has been no one stated reason; the *probability* of catastrophe was calculated to be small *if sabotage was excluded.*

Dams have not been regulated to any appreciable extent. This is in spite of a historical record of more than 10^{-4} per dam-year for catastrophic failure[28]. Dam construction is an art almost as much as a science, but both the art and science have improved since 1930 so that its practitioners argue that 'a dam properly built will not fail'[29]. Even if we were to agree that this statement were true, it is of limited practical utility since there is no written objective criterion for 'properly built'.

However the siting of dams is being regulated under NEPA and to the extent that dam siting contributes to safety, this is included. For example, a proposed dam at Auburn, California on the Sacramento River was halted, at least temporarily, in part because if it had failed, its failure would have strained Folsom dam downstream and caused it to fail – and then flooded Sacramento.

Thus the uses of NEPA, for LNG and perhaps for dam safety, have been to choose sites where the *consequences* of an accident can be minimized; but usually there has been no appreciable regulation of the design to ensure that the probability is small.

In 1972 the National Dam Safety Act was passed, which mandates Federal Inspections of all dams by the Army Corps of Engineers. Like many safety laws, this one was passed after a major dam failure (Buffalo Creek) which cost 125 lives; but the funding for the inspections was not provided until another dam failed (Loccoa, Georgia.) where 39 lives were lost in 1977. So far there is no regulation on how to deal with the 30% of dams that are found to be substandard.

There are the beginning of efforts to develop quantitative safety design criteria for dams within the Army Corps of Engineers and there is a Federal inter-agency group beginning to try to develop some kind of quantitative approach to design safety.

Storage of chemicals and fuels generally is not well regulated. The insurance industry set up in 1907 the National Fire Prevention Association and its guides became standard industry practice. For example, its guide NFPA 59A, on storage of liquefied natural gas (LNG) compares LNG safety with petroleum fuels. However, largely because of public protest, many state governments have set up more restrictive guidelines.

For other chemicals, explosive and toxic standards are few. When it comes to long-term disposal of hazardous chemical waste, they are almost non-existent. The Office of Solid Waste of the EPA was set up under the Resource Conservation and Recovery Act[30]. Unlike radioactive wastes, which have definite half-lives, no one is sure whether chemicals break down in time or not; they have either been put in chemical dumps (benzene, toluene, etc.) where the problem may merely be postponed, or scattered (arsenic) where small doses may plague large numbers of people. There seems at present no scientific consensus on what the problems are, let alone the solutions, and this leaves the regulators and politicians completely in the air.

One of the issues in all these cases is the objectivity of the scientist or expert workers who make the assessment of risk. Judge Bazelon states a jurist's view of this.[31] He and so many others have stressed the need for separation of the scientific and value decisions. This is not completely possible. For example, the risk analyst's choice of a proportional model for relating cancers to dose is usually more a desire to be cautious than a scientifically based belief in its truth. Objectivity can be maintained, however, if the assumptions are clearly stated.

Summing Up

We summarize by noting tht there is a state of flux. While it is generally realized that there is no such policy as zero risk, and that we want to get the most life saving activity for our safety dollar, there is still a lot of legislative and regulatory feeling, and some precedent, to feel that we have not reached the limit on the supply of safety dollars, and that both mortality and practical considerations demand an absolute approach. The extent to which US society decides in favour of risk analysis for decision making will probably depend on the ability of analysts to present honest, believable risk calculations. In their absence, regulatory cost alone is likely to be the criterion.

Questions 4.3

1. What are, in general, the basic strategies which have emerged in the USA for controlling risk? Discuss some examples of the US Federal regulation of carcinogenic substances and the relationship with the basic strategies for controlling risk.
2. Discuss the 'Delaney Clause' with reference to food additives in connection with the US Food, Drug, and Cosmetic Act. What types of criteria have emerged for carcinogenicity in people when accepting a proposed food additive is safe?
3. How did the bringing into being of Safe Drinking Water Acts lead to acceptably pure water?
4. What part does risk/benefit analysis play in controlling risk? Give examples of how this analysis has been implemented. How does the strategy of specifying the 'best available technology' conflict with that of risk/benefit analysis and can legislation deal with this matter?
5. Consider the actions taken by the US Environmental Protection Agency (EPA) in dealing with the control of relatively small quantities of chemicals in the air to produce 'clean air'. Discuss the comparison of these actions by the EPA under the·Clean Air Act for benzene.
6. In applying the principle 'as low as reasonably possible' how has the Nuclear Regulatory Commission applied risk/cost/benefit analysis for regulating the routine release of radiation for nuclear power plants?
7. List a number of activities in the USA which could be potentially hazardous to people such as transportation, the construction of dams, the operation of nuclear power plant, and comment on the legislation and regulations which have been imposed.
8. From the considerations given in Question 7 what types of criteria have emerged which recognize that there is no such policy as 'zero risk'? To what extent is it visualized that quantitative risk criteria will develop in the future in the USA as part of safety technology so that the saving of human life can be maximized?

References 4.3

1. Cancer Testing Technology and Saccharins, Office of Technology Assessment, US Congress, October 1977, Table I.
2. *Code of Federal Regulations*, Volume 21, Section 121.5; usually abbreviated 21CFR 121.5.
3. 21CFR 201; 21CFR 409.
4. *Federal Register*, Volume 44, page 17070, No. 55, Tuesday 20 March 1979 usually abbreviated 44 Fed. Reg. 17070.
5. *Monsanto Corp. and others* v. *Kennedy* (FDA) 613 F 2d 947 (DC Circuit) 1979.
6. *US Code (of laws)*, Title 42, paragraphs 300; usually abbreviated 42 USC §300 f et seq.
7. *Environmental Defense Fund* v. *Costle* 578 F 2d 337 (D.C. Circuit 1978).
8. 15 USC §2601 *et seq.*
9. 7 USC §136 *et seq.*
10. 15 USC §§2051 *et seq.*
11. 33 USC §701a *et seq.*
12. 42 USC §1857 *et seq.*
13. Lave, L. and Seskin, E. *Air Pollution and Human Health*. Van Nostrand. 1977.
14. Proposed policy and procedures for identifying, assessing and regulating air borne substances posing a risk of cancer 44 Fed. Reg. 58642, 10 October 1979.
15. National Emission Standard for Hazardous Air Pollutants, benzene emissions from maleic anhydride plants (proposed rule and amendment to 40 CFR61) 45 Fed. Reg. 25688, 18 April 1980.
16. Title US Code §651 *et seq.*
17. *American Petroleum Institute* v. *OSHA et al.*, 581 F 2d (DC Circuit) 1978. 100 S. Ct. 2844 (1980)
18. *State of Indiana* v. *Ford Motor Co.* Pulaski County Court, Winamac, Indiana, 13 March 1980.
19. 1975 Societal cost of motor vehicle accidents. National Highway & Traffic Safety Administration HS-802119 Dec 1976.
20. Final Ruling by the commissioners in 'As low as practicable hearing'. RM-30-2.
21. 42 USC §2011 *et seq.*
22. 42 USC §5801 *et seq.*
23. *Anticipated Transients without Scram (ATWS)*. Atomic Energy Commission Report WASH-1270.
24. 10 CFR 100.
25. National Environmental Policy Act of 1969, Public Law No. 91–190. (Often merely abbreviated PL 91–190.)
26. *Calvert Cliffs Coordinating Committee* v. *AEC*, 449 F 2d 1109 (DC Circuit 1971).
27. Wilson, R. 'National gas is a beautiful thing.' Richard Wilson, *Bulletin Atomic Scientists*, **29**, 35, 1973.
28. *Energy in Transition 1985–2010*. p. 459. Committee on Nuclear and Alternative Energy Systems (CONAES). National Academy of Sciences, National Research Council (1980).
29. Wilson, R. Private communication from Prof. Arthur Casagrande, Harvard University.
30. PL 92–500.
31. Bazelon, D. L. 'Risk and responsibility.' *Science*, **213**, 277, 1979.

Answers to Questions

Chapter 1.1

4.   1 in 10^5

Chapter 1.3, Section 1.3.2

4.

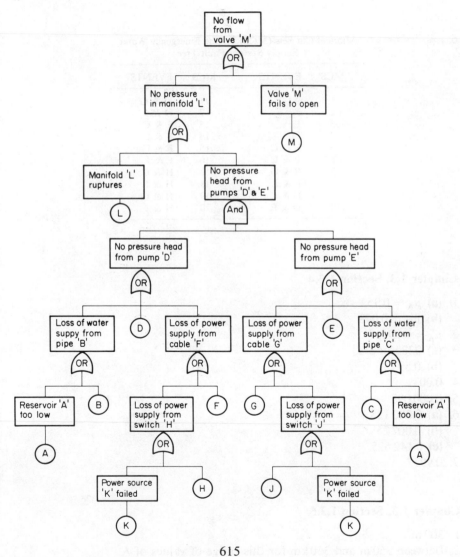

5. Note that no protection has been specified to prevent a pump starting when it has no water supply to it. If a demand occurs when the reservoir is low, both of the pumps could be damaged, and the repair of the system might well be dependent upon the repair of the pumps, rather than upon the repair of the reservoir level. There is a similar problem with ruptures of the inlet pipes and perhaps one should also consider the effects of pump overspeed if a demand occurs when the manifold is ruptured.

6.

Minimal Cut Sets (MCS) for the Emergency Water
Supply System Fault Tree

MCS	EVENTS	MCS	EVENTS
1	A	11	D & G
2	K	12	D & J
3	L	13	F & C
4	M	14	F & E
5	B & C	15	F & G
6	B & E	16	F & J
7	B & G	17	H & C
8	B & J	18	H & E
9	D & C	19	H & G
10	D & E	20	H & J

Chapter 1.3, Section 1.3.4

1. (a) $\mu_A = 0.952$
   (b) $\mu_D = 0.048$
2. $\mu_A = 0.994$
3. (a) 2222 hours
   (b) 0.593
4. 0.003
5. 0.000 098
6. (a) 0.000 125
   (b) 0.007 25
   (c) 0.142 625
7. 0.9977

Chapter 1.3, Section 1.3.5

1. 300 m
2. Between 350 m and 350 km for this range of values of Λ

Chapter 1.6, Section 1.6.1

6. (a) Average number of failures during one calibration act approx. 100×10^{-2}
= 1 per test, but they will in general be corrected immediately because they affect the following steps.
   (b) Approx. 10^{-2} per test
   (c) Approx. 2×10^{-2} per test
   (d) Approx. 2×10^{-2}
   (e) (a), (b), and (c) will not be affected. In (d) the estimate will be modified by factor 10^{-2}, but now the result in the range 10^{-4} per test should call for a more careful analysis of the task, since omission of step 86 and 87 will no longer mask more subtle errors; for instance, the quality of test equipment maintenance may influence the result.
   (f) The risk of introducing a fault leaving the channel inoperable is most significant. A probability of overlooking an equipment failure during a test of 2×10^{-2} is insignificant, since even a probability of 0.5 would only double the fractional dead-time of the channel. This is due to the fact that the fractional dead-time to a first approximation is proportional to the length of the test interval.

Chapter 1.6, Section 1.6.3

3. (a)

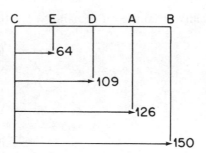

   (b) $P_B = 0.83 \times 10^{-1}$
   $P_A = 0.2 \times 10^{-1}$
   $P_D = 7.3 \times 10^{-3}$
   $P_E = 5 \times 10^{-4}$
   $P_C = 1.1 \times 10^{-5}$
   (c) B is error associated with impractical, stressful situation.
   A is error of omission.
   D is error of commission.
   E is error in regularly performed, simple task.
   C is extraordinary error.

Chapter 1.9, Section 1.9.2

3. (a) $p(\theta_1/D) = 0.308\,01$; $p(\theta_2/D) = 0.691\,99$
   (b) $p(\theta_1/D) = 0.982\,01$; $p(\theta_2/D) = 0.017\,99$
   (c) $p(\theta_1/D) = 0.916\,10$; $p(\theta_2/D) = 0.083\,90$
4. (b) 2.3×10^3 hours
5. (a) $\beta = 1000$; $\alpha = 3$
   (b) $E(\theta_0) = 0.003$; $\sigma^2(\theta_0) = 3 \times 10^{-6}$
   (c) $0.761\,90$ (0.76 by Poisson Graph)
   (d) $E(\theta_0/D) = 0.002$; $\sigma^2(\theta_0/D) = 10^{-6}$
   (e) $0.957\,62$ (0.96 by Poisson Graph)
6. (a) 1.32×10^{-4}/year
   (b) 0.8413
   (c) Prior for failure rate would be log-normal

Chapter 2.1

7. (a) 0.025/year
   (b) Approaches 0.5/year, i.e. the hazard rate is almost the same as the failure rate due to the high demand rate.

Chapter 2.3

1. The overall probability of failure on demand $= 5 \times 10^{-14}$.

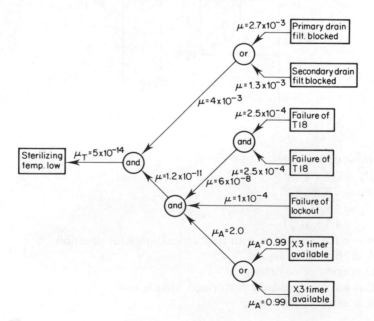

2. Probability of failure on demand $= 0.007$. If repair fault rate is 10% then probability of failure $= 0.008$

Chapter 3.1

5. Chance of exceedance of $25\,\text{m}$ wave $= 16.6\%$.
   For 10 years chance $= 11.1\%$

Glossary of Terms

Absolute Filtration

Filtration through a membrane filter which has pores so small (about 0.2 μ m) that all cells and spores are held back.

Accuracy

The measure of exactness of a component's, equipment's or system's performance compared with some true or standard performance. This measure includes those deviations in performance which may be due to both systematic and non-systematic effects.

Agranulocytosis

An acute disease characterized by a marked reduction in the number of white blood cells associated with ulcerative inflammation of the throat and a grave general condition.

Alkylation Plant

This involves the production of aliphatic fluorine compounds by treating chlorine compounds with hydrogen fluoride in the presence of catalysts usually under pressure.

Anaphylactic Shock

Unusual or exaggerated reaction of the organism to foreign protein or other substances.

Antihistamine

A drug which counteracts the effects of histamine.

Asepsis

Prevention of the access of micro-organisms.

Assessor

A person from whom subjective judgements are solicited.

Automatic Total Flooding System

A fire extinguishing system which operates by filling the fire hazard area with gas, such as carbon dioxide or halon, to deprive the area of oxygen or to interrupt the combustion process.

Availability

The proportion of the total time that a component, equipment, or system is performing in the desired manner.

Blowout

An escape of oil or gas (usually accidental) from a well.

621

Boundary Value

The limiting value of some performance characteristic which may be used for assessment purposes.

Capability

The fundamental ability of a component, equipment, or system to perform in the desired manner.

Carcinogen

Any substance which causes living tissue to become cancerous.

Catastrophic Failure

The failure of a component, equipment, or system in which its particular performance characteristic moves completely to one or the other of the extreme limits outside the normal specification range.

Christmas Tree

An assembly of valves and fittings located at the head of an oil well to control the flow of oil or gas.

Circular Tread

Associated with the method of paired comparisons, a circular tread is produced when the results of three paired trials, namely (i) A with B, (ii) B with C, and (iii) C with A, yield the inconsistency that, for example, A is judged to be more than B, B is judged to be more than C but then C is judged to be more than A; to be consistent, of course, this third judgement should have been A more than C.

Co-carcinogen

Anything which furthers the action of a carcinogen in producing a malignant tumour, such as pre-existing chronic inflammation or the irritating effects of a chemical not itself a carcinogen.

Common Mode/Cause

A descriptive term which represents the statistical dependency between basic component failures.

Communal Mind

A notional term; the population of judgements about a given issue, offered by a group of individuals, is described collectively as the judgement of the 'communal mind' of those individuals.

Confidence

In the statistical sense this expresses the degree of belief that the parameter (mean, standard deviation, etc.) of a given statistical distribution lies between stated values; it is itself a probabilistic quantity ranging from 0 to 1 (or, more popularly, from 0% to 100%) and typically is a function of the distribution type and the sample size of the data forming the distribution.

Confidence Interval

The estimated range or interval for a distribution parameter which is given at a certain confidence level.

Confidence Level

The probability that the assertion made about the value of the distribution parameter, from the estimates of that parameter, is true.

Confidence Limits

The end points of the confidence interval.

Consecutive Testing

The testing of a series of components or equipments one after the other. This may or may not involve a time interval between the completion of the test on one device and the starting of the test on the next device.

Correlation Coefficient

When appearing in a statistical expression which comprises, say, two statistical distributions, the correlation coefficient represents the degree of interaction between the two variables forming the distributions; if the variables are totally independent of each other the corresponding correlation coefficient will be zero.

Cortical Function

Activity of the adrenal cortex which is concerned with the production of numerous steroid hormones, i.e. glucocorticoids concerned with carbohydrate metabolism; mineralocorticoids concerned with electrolyte and water metabolism; and steroids identical to those secreted by the gonads.

Cumulative Probability

The probability that a random variable, x, lies between some lower limit, which might be 0 or $-\infty$, and some upper limit determined by a particular value of x. For continuous random variables, the cumulative probability is given by the integration of the density function between the limits required, e.g.

$$p(x) = \int_{-\infty}^{x} f(x)\,\mathrm{d}x$$

for discrete variables, the cumulative probability is given by the sum of the appropriate probability function over the range of values required, e.g.

$$p(r) = \sum_{j=0}^{r} f(j)$$

Curie

Unit of radioactivity which is defined as 3.7×10^{10} disintegrations per second. One disintegration per second is called a Bicquerel.

Cut Set

A cut set is a group of component faults that lead to the Top Event of the fault tree.

Data Bank

This is essentially a data-base with two additional features; (i) a formalized system for accepting and storing additional data, (ii) a formalized system of data-base interrogation, enabling the user to locate efficiently and read (if present) the stored information relevant to his enquiry.

Data-Base

A reference collection of performance information, assembled as either or both (i) descriptive text (qualitative form), (ii) listed events with associated probabilities or frequencies (quantitative form).

Data Density

A qualitative concept embodying the ideas of (i) the designed degree of separation between adjacent data cells in a given data-bank structure, and (ii) the proportion of data cells actually filled.

Dead Time

(*see* Fractional Dead Time)

Delphi Technique

A method of collecting data in which a group of expert assessors are presented with a series of questionnaires that propose a structure of data and ask for estimates of data values. All responses are combined and the results fed back to each assessor who has the opportunity to revise his original response.

Density Function (Probability Density Function)

If x is a continuous random variable, then the density function of x is defined as the first derivative of the cumulative probability function with respect to x, i.e.

$$f(x) = \frac{d}{dx} p(x)$$

Discriminal Difference

The nominal scale difference along the psychological continuum (for a given single performance dimension between two discriminal processes).

Discriminal Distribution

Distribution of discriminal processes pertaining to one stimulus and one performance dimension; this could apply to the variations produced by a single judge repeatedly addressing the same judgement task, or at the other extreme, to the variations produced by many judges each addressing the judgement task once only.

Discriminal Dispersion

The standard deviation of a discriminal distribution.

Discriminal Process

The cognitive activity of registering a strength of feeling for a given stimulus along an identified performance dimension (psychological continuum) in the mind; for example, if the performance dimension is 'likelihood of winning' and the given stimulus is 'contestant N' the discriminal process is that act of judgement which places B along the judge's scale of likelihood.

Disinfection

Is the process of freeing from infection.

Distributed Network

A network of mini-computers or terminals with limited processing power, geographically distributed over a large physical area. All are connected to a central computer by data communication lines.

Diuretic

A substance which increases the volume of the urine.

Diversity

The performance of the same overall function by a number. of independent and different means.

Early Failure

The failures of components, equipments, and systems which occur during the initial life phase of such devices and which are generally caused by initial production, assembly, test, installation, or commissioning errors.

Engineered Safety Feature

An engineered safety feature (ESF) is any system which has been specifically designed to mitigate the effects of an imagined initiating event.

Enthalpy
The total heat of a substance.

Entrainment
The process by which a plume or 'puff' is diluted by the addition of air.

Error
The deviation that can exist between the actual performance characteristic of a component, equipment, or system and the true or required value of such performance.

Error of Commission
A positive but erroneous human action which directly results in a situation not identifiable as that required, when required; very often an error of commission produces an associated error of omission.

Error of Omission
A human action *not* performed by the time that it is required.

Error Recovery
The action of negating a standing human error before the condition has interacted with the associated system.

Event Tree
A logic method for identifying the various possible outcomes of a given event which is called the initiating event.

Exceedance Probability
The probability that a particular parameter will exceed a set value within a stated time period.

Exchangeable Disc
A magnetic disc pack, used to store data or programs, which can be physically removed from the disc drive. A *fixed* disc is permanently connected to the disc drive and cannot be removed from the disc drive.

Extrapolation Process
In data bank terms, the means by which existing data points may be used to estimate the likely value of an absent data point.

Extreme Value Theory
Predictive technique using extreme values of particular parameters.

Failed State
The condition of a component, equipment, or system during the time when it is subject to a failure.

Failed Time
The time for which a component, equipment, or system remains in the failed state.

Failure
The condition of a component, equipment, or system whereby a particular performance characteristic, or number of performance characteristics, of such a device has moved outside the assessed specification range for that characteristic in such a way that the component, equipment, or system can no longer perform adequately in the desired manner.

High Risk Safety Technology

Fault Tree

A fault tree is a graphical display to show how the basic component failures in a system can lead to a pre-defined system failure state.

Field Experience

The collection of data and information about the performance characteristics of components, equipments, and systems from their actual use in appropriate practical installations.

Flame Acceleration

The rate of change of the flame speed in a burning gas cloud.

Flammable Limit (Lower)

The concentration of flammable vapour in air below which it cannot be ignited.

Flammable Limit (Upper)

The concentration of flammable vapour in air above which it cannot be ignited.

Fractional Dead Time

The proportion of the total relevant time that a component, equipment, or system is in the failed state.

Fréchet Distribution

Technique for the prediction of an extreme value of a particular phenomenon defined by the following

$$F(x) = \exp - \left(\frac{v}{x}\right)^k \quad 0 < x < \infty$$

Gumbel Distribution:

Technique for the prediction of an extreme value of a particular phenomenon, defined by the following

$$F(x) = \exp(-e^{-\alpha(x-u)})$$
$$-\infty < x < \infty$$

Haematology

The study of blood and blood-forming tissues.

Human Error

The performance or non-performance of an act which directly results in a situation not identifiable as that required, when required.

Implicant Set

An implicant set is a group of component states, success and failure, that lead to the top event of the fault tree.

Infant Mortality

A term sometimes used to describe 'Early Failure' (*q.v.*).

Initiating Event

Any failure of the operating system which potentially might result in a hazardous situation.

Maintenance

The art of ensuring that the performance of a component, equipment, or system is kept within a set of predetermined limits.

Majority-vote System

A system made up of *n* identical and independent elements which are so connected that any *m* or more ($m \leqslant n$) of the elements are required to give a correct output before a correct system output is obtained. The system may be considered as being partly redundant. In the particular cases where $m = 1$ the majority-vote system becomes completely redundant (*see* Redundant Elements) and where $m = n$ the system becomes completely non-redundant (*see* Non-redundant Elements).

Mean

The mean, μ, is defined as the sum of all values of a variate, *x*, where each value is weighted by its associated probability of occurrence. For discrete variables:

$$\mu = \sum_{r=0}^{\infty} rf(r)$$

and for continuous variables:

$$\mu = \int_{-\infty}^{\infty} xf(x)\,dx$$

Mean Time Between Failures (m.t.b.f.)

The total measured operating time of a population of components, equipments, or systems divided by the total number of failures. The term is generally used in connection with the exponential distribution of failures where the m.t.b.f. is a time-independent constant.

Mean Time to Failure (m.t.t.f.)

The mean of the distribution of the times to the first failure.

Method of Paired Comparisons

A formalized process for extracting subjective experience in the comparisons domain where the human judgement capability is naturally strong. If it is required to use subjective experience to rank a set of stimuli with respect to some specified dimension of performance, the method requires that the stimuli be offered in all combinations of pairs to the assessor(s); at each offering he (they) then decide which of the pair has the greater performance; from the set of decisions produced it is possible to deduce the required ranking.

Microcode

A set of program instructions which is hardwired to the computer memory to speed up processing.

Middleware

A set of program instructions provided for computer users to do various tasks. Normally, a name is allocated to each task. The user wanting to use any of such tasks provided will call on the name of the task in his program without having to know the detailed programming steps for that task.

Minimal Cut Set

A cut set is minimal if it contains no other cut sets

Mutagenicity

The property of being able to induce genetic mutation.

Monte Carlo Simulation

In the content of reliability analysis, this is the process of generating a chronological record of failure events for a system, using random members operating on the statistical time distributions for the components of the system; these chronological records or 'failure histories' can be produced for any specified time period; when produced they can be subjected to any subsequent analyses that the analyst may choose; normally the simulation is conducted on a computer and therefore it is possible to represent hundreds of system-years of operation in only a few real-time seconds of computation.

Non-redundant Elements

A configuration of elements which can only produce a correct output when each element in the configuration is functioning correctly.

Non-systematic Errors (or Random Errors)

Errors which do not represent a fixed deviation from the true value and have the implication of chance. It is not possible to evaluate a particular error of this sort but such errors in the mass can be understood and calculated using the laws of probability.

Off-site Storage

Storage of essential items away from the computer installation. Such items could range from documentation, programs, and data to pre-printed computer lines.

Oncogenicity

The quality or property of being able to cause tumour formation.

Organogenesis

The development or growth of organs.

Paired Comparisons

A method for collecting data in which the expert is presented with all possible combinations of pairs of systems and is asked to judge which system best meets some comparison requirement.

Pharmacodynamics

The study of the action of drugs on living organisms.

Pharmacokinetics

The study of the absorption, distribution, metabolism, and excretion of drugs.

Phased Mission

This is a task to be performed by a system where the demand on the system changes with time, or where the system configuration changes with time. A separate fault tree is therefore required for each phase of the mission.

Plume Rise

The rise into the atmosphere of a buoyant gas.

Population

The total collection of any items or events related to any particular common consideration. In statistics, the terms 'population', 'universe' or 'parent population' are generally synonymous and are used to describe an exceedingly large or infinite number of items or events from which samples may be taken for statistical analysis.

Prime Implicant Set

An implicant set is prime if it contains no other implicant sets.

Probability

If the assertion is made that the probability of an event occurring as an outcome of a series of random and independent experiments is given by some value P, then the value of the probability, P, is defined as the limit of the ratio of the number of occurrences of the event to the total number of experiments (i.e. the relative frequency of occurrence) as the total number of experiments tends to infinity.

Probability Density Function (p.d.f.)

(*see* Density Function)

Producer's Risk

The probability of the manufacturer of a component, equipment or system rejecting such devices when their failure-rate, or other performance characteristic, is better than some agreed or stated value.

Proof Testing

A method of ensuring that a component, equipment or system possesses all the required performance characteristics and is capable of responding to input conditions in the manner desired.

Psychological Continuum

The notional axis along which strengths of feeling, in respect of given stimuli and a particular performance dimension, are registered in the mind. Every performance dimension, for example frequency of occurrence, in being comprehended by the mind thereby establishes its own psychological continuum.

Psychological Scaling

In the context of subjective judgements, this is a rationalized quantitative representation of the strengths of conviction stimulated in the mind by various judgement situations.

Pyrogen

Toxic substance formed by micro-organisms which cause a rise in temperature when injected into animals. Distilled water pyrogens are filterable thermostable substances produced by water-borne bacteria which may be present in distilled water unless special precautions are taken.

Quantification

The allocation of probabilities or frequencies to defined changes of state.

Random

A term used to describe the unpredictable occurrence of events in space, in time or in both space and time.

Ranking

The ordering of events in relation to their individual ratings with respect to some defined dimension of performance.

Redundancy

The performance of the same overall function by a number of independent but identical means.

Redundant Elements

A configuration of a number of identical and independent elements connected in such a way that the correct functioning of any one element, irrespective of the state of the others, produces a correct output.

Refrigerant-12

Defined by the chemical formula $C \cdot Cl_2 F_2$.

Reliability

That characteristic of an item expressed by the probability that it will perform its required function in the desired manner under all the relevant conditions and on the occasions or during the time intervals when it is required so to perform.

Reliability Spectrum

A numeric and often pictorial presentation of the individual reliabilities of the components of a system so that weak system areas may be readily identified and compared.

Reliability Technology

The scientific study of the trustworthy nature of devices or systems in practical and industrial situations.

Return Period

The average time between particular statistically independent events.

Revealed Failure

A failure of a component, equipment, or system which is automatically brought to light on its occurrence.

Righting Reflex

The ability to assume the optimal position when there has been a departure from it, e.g. small animals rapidly revert to the 'standing' position when placed on their backs.

Risk

The product of the probability of an event occurring and the consequences of that event.

Risk Aversion

The reaction of an individual person or group of people which implies their adverse reaction to risk is not a function of risk alone.

Risk (Individual) of Death

The probability per year that the process being considered will cause the death of an identifiable individual.

Risk (Societal) of Death

fN where f is the frequency with which the process under consideration is predicted to cause the deaths of N or more people.

Rubella

German measles.

Scaled Ranking

The spacing of ranked events along a scale so that the scale separations numerically represent the rotative differences between the events.

Simultaneous Testing

The testing of a number of components, equipments, or systems at the same time.

Staggered Testing

A special case of consecutive testing (*q.v.*) where there is normally a time interval between the completion of a test on one device and the starting of a test on the next device.

Sterility

The degree of uncertainty with which it may be known that a product is sterile.

Sterilization

The process of freeing from all living micro-organisms.

Stimulus

An event, a continuing existence, a change of state capable of resolution by the human mind.

Stochastic Process

A time-dependent random process.

Structure (Data Bank)

The multidimensional attributes system within which data are classified and stored; for example, a simple two-dimensional system could be pictured as a table of quantities (the data-base) with adjacent vertical (y) and horizontal (x), discrete-value, attributes scales; the cells of the table are arranged so that the data corresponding to a given pair of x, y attributes values are located at the intersection of the horizontal and vertical lines drawn through these values.

Subjective Data

Normally refers to quantitative data (probabilities, frequencies) which are produced by the exercise of subjective judgement.

Subjective Judgement

A constructive thought process within the human mind which involves the recall, matching, and weighing of remembered knowledge and experience in order to resolve some expressible prediction about a future outcome.

Synthesis

The combinations of probabilities or frequencies in accordance with the logical structure of the system model.

Systematic Errors

An error which represents a predictable deviation from the true or standard value.

System Modelling (Safety)

The process of identifying and representing the various sequences of events which take a system from a normal operational state into some defined unacceptable condition.

Technological System

That engineered complex which converts a given quantity of a certain range of facets such as wealth, materials, resources, facilities, and information with specified characteristics into a given quantity of a certain range of information, power, products, services or wealth with specified characteristics under a certain set of space and time conditions.

Teratology

The study of abnormal development and congenital malformation.

Top Event

The pre-defined system failure state for which the fault tree was constructed.

Toxicology

The study of toxic effects of drugs

Transparent Monitoring

The technique of observing human behaviour by a means which, although non-secret, does not excite awareness to the extent of significantly altering this behaviour.

Tumorogenic

Causing or producing tumours.

Unit Normal Deviate

The scale value along the horizontal axis of the cumulative normal distribution which corresponds with a given probability value (0 to 1) on the vertical axis; since, by virtue of its symmetry, the cumulative normal distribution is centred on a probability value of 0.5, corresponding to a unit normal deviate value of $x = 0$, the total range of the unit normal deviate is $-\infty \leqslant x \leqslant +\infty$.

Unrevealed Failure

A failure of a component, equipment, or system which remains hidden until revealed by some thorough proof-testing procedure.

Useful Life Phase

That part of a component's, equipment's, or system's life which lies between the phase of early failure and the phase of wear-out failure. In some instances the useful life phase is characterized by a constant average failure rate.

Vasomotor

Regulating the tension of vessels, particularly blood vessels.

Wear-out Failure

The failures of components, equipments, and systems which occur after the end of their useful life.

Index